T0190186

High Performance Computing in Science
and Engineering '19

Wolfgang E. Nagel · Dietmar H. Kröner ·
Michael M. Resch
Editors

High Performance Computing in Science and Engineering '19

Transactions of the High Performance
Computing Center, Stuttgart (HLRS) 2019

 Springer

Editors
Wolfgang E. Nagel
Zentrum für Informationsdienste und
Hochleistungsrechnen (ZIH)
Technische Universität Dresden
Dresden, Germany

Dietmar H. Kröner
Abteilung für Angewandte Mathematik
Universität Freiburg
Freiburg, Germany

Michael M. Resch
Höchstleistungsrechenzentrum Stuttgart
(HLRS)
Universität Stuttgart
Stuttgart, Germany

Front cover figure: Auto-ignition in an impulsively started jet by iso-surfaces of the dissipation rate of mixture fraction (blue) and ignition temperature (red). The data is taken from a simulation with 675 million cells, which costs 45 million compute-hours. Details can be found in "Fully Resolved Auto-Igniting Transient Jet Flame Simulation" by Eray Inanc and Andreas M. Kempf, Chair of Fluid Dynamics, Institute for Combustion and Gasdynamics (IVG), University of Duisburg-Essen, Germany, on page 249 ff.

ISBN 978-3-030-66794-8   ISBN 978-3-030-66792-4   (eBook)
https://doi.org/10.1007/978-3-030-66792-4

Mathematics Subject Classification: 65Cxx, 65C99, 68U20

This Springer imprint is published by the registered company Springer Nature Switzerland AG
The registered company address is: Gewerbestrasse 11, 6330 Cham, Switzerland

Contents

Molecules, Interfaces, and Solids

Materials Science

Reactive Flows

Computational Fluid Dynamics

Contents

Physics

Prof. Dr. Peter Nielaba

In this section, seven physics projects are presented, which achieved important scientific results in 2018/19 by using Hazel Hen at the HLRS and ForHLR II of the Steinbuch Center.[1]

Fascinating new results are being presented in the following pages on astrophysical systems (simulations of galaxy formation, analysis of the first image of a black hole and binary merger evolutions, and of photoionization and atomic, molecular and optical collision processes), soft matter systems (interaction between proteins and hydrophilic polymers), many body quantum systems (simulations of ultracold quantum systems), and high energy physics systems (hadronic contributions to the muon magnetic moment).

The studies of the astrophysical systems have focused on galaxy formation, analysis of the first image of a black hole and binary merger evolutions, and on photoionization and atomic, molecular, and optical collision processes.

A. Pillepich, D. Nelson, V. Springel, R. Pakmor, L. Hernquist, M. Vogelsberger, R. Weinberger, S. Genel, F. Marinacci, P. Torrey, J. Naiman from Heidelberg (A.P.), Garching (D.N., V.S., R.P.), Cambridge USA (L.H., R.W., J.N., M.V.), New York (S.G.), Bologna (F.M.), and Gainesville (P.T.) present in their project *GCS-DWAR* recent results from the TNG50 run, the third and final volume of the cosmological, magnetohydrodynamical simulations in the IllustrisTNG project (AREPO code), with over 20 billion resolution elements and a spatial resolution of 100 parsecs. The authors report on the results of the recently completed TNG50 simulations, reaching the red shift z = 0, the present day. TNG50 is bridging the gap between large cosmological volumes and better resolved zoom galaxies. Results include the abundances of dwarf galaxies in the first few billion years of the history of the Universe and the kinematic properties of the Hα and molecular gas within and around galaxies at intermediate redshift, as well as predictions for our improved understanding of galaxy evolution. The results will be made publicly available to the astronomy community.

R. Gold and L. Rezzolla from the University of Frankfurt contributed in their project *BBHDISKS* to the analysis of the first image of a black hole by the Event

---

[1]Fachbereich Physik, Universität Konstanz, 78457 Konstanz, Germany, e-mail: peter.nielaba@unikonstanz.de

Horizon Telescope (EHT) collaboration and developed a code further for the study of two black holes in orbit around each other. In their studies, they use a black hole accretion code (BHAC) for solving the general relativistic magntohydrodynamic (GRMHD) equations, which is based on the MPI-AMRVAC toolkit, using a fully adaptive block-based (oct-) tree mesh refinement with a fixed refinement factor of two between successive levels, and which is interfaced with many analysis tools and external codes. The project topics include the general behavior of magnetized, turbulent gas moving around or falling onto a black hole system, the observational appearance of such systems, the predictions for systems with two black holes in magnetized gas, and the advancement of computational tools to tackle such systems in the contexts of both ideal and resistive magnetohydrodynamics on non-homogenous and dynamical grids. A GRMHD image library has been build for the analysis of the EHT results, and the computation of initial data for the space–time metric was an important step for the planned investigations of accreting binary black holes.

B. M. McLaughlin, C. P. Ballance, R. T. Smyth, M. S. Pindzola, P. C. Stancil, J. F. Babb and A. Müller from the Universities of Belfast (B.M.M., C.P.B., R.T.S.), Auburn (M.S.P.), Georgia (P.C.S.), Cambridge USA (J.F.B.), and Giessen (A.M.) investigated in their project *GCS-PMP2* atomic, molecular, and optical collisions on petaflop machines, relevant for astrophysical applications, for magnetically confined fusion and plasma modeling, and as theoretical support for experiments and satellite observations. The Schrödinger and Dirac equations have been solved with the R-matrix or R-matrix with pseudo-states approach, and the time-dependent close-coupling method has been used for the studies of ion-atom excitations and electron-impact ionizations. Various systems and phenomena have been investigated in detail such as the photoionization of Rb^+ and $Rb^{2}+$ ions, the electron impact excitation of Fe II, the proton impact on sodium atoms, the $CS-H^2$ diatomic collision, and the dicarbon formation by radiative association.

Studies of the soft matter systems have focused on the interaction of proteins with hydrophilic polymers.

T. Schäfer, C. Muhl, M. Barz, F. Schmid and G. Settanni from the University of Mainz investigated in their project *Flexadfg* the thermodynamics and kinetics of the interactions between proteins and hydrophilic polymers by Molecular Dynamics simulations, using the program NAMD and CHARMM force fields with CMAP correction, a cell-list algorithm, and a smooth particle mesh Ewald (PME) method, and a modified TIP3P model for water molecules. In particular, the adsorption of poly-(ethylene-glycol) (PEG), poly-sarcosine (PSar), and poly-alanine (PAla) on the surface of proteins was studied in order to determine their usefulness as coating agents for nanoparticles. The authors find that the pattern of interactions with the surface amino acids does not change much for the different polymers, and that the amount of adsorbed polymers and the residence time on the protein surface varies significantly, depending on the conformational properties.

In the last granting period, quantum mechanical properties of high energy physics systems have been investigated as well as the quantum many body dynamics of trapped bosonic systems.

A. U. J. Lode, O. E. Alon, L. S. Cederbaum, B. Chakrabarti, B. Chatterjee, R. Chitra, A. Gammal, S. K. Haldar, M. L. Lekala, C. Leveque, R. Lin, P. Molignini, L. Papariello, and M. C. Tsatsos from the Universities of Wien (A.U.J.L, C.L.), Haifa (O.E.A., S.K.H.), Heidelberg (L.S.C.), Kolkata (B.C.), Kanpur (B.C.), Zürich (R.C., R.L., P.M., L.P.), Sao Paulo (A.G., M.C.T.), Pretoria (M.L.L.) studied in their project *MCTDHB* trapped ultracold atomic systems by their method termed multiconfigurational time-dependent Hartree for bosons (MCTDHB and MCTDH-X software packages). The principal investigators investigated the fermionization and crystallization of ultracold strongly interacting bosons, correlations of bosons with strong dipolar interactions, cavity-induced phase transitions of ultracold bosons in highfinesse cavities, the dimensionality of the variance of Bose Einstein condensates (BECs) in annular confinements, the dynamics of BECs with long-ranged interactions in a bosonic Josephson junction, the dynamics of BECs in an asymmetric bosonic Josephson junction, and the variance of BECs in anharmonic traps.

Two projects (GCS-HQCD and GCS-MUMA) investigated hadronic contributions to the muon magnetic moment with very interesting results. The principal investigators and project contributors of the project GCS-HQCD were M. Ce, A. Gerardin, G. von Hippel, B. Horz, H. B. Meyer, D. Mohler, K. Ottnad, S. Schaefer, J. Wilhelm, and H. Wittig from the Universitiy of Mainz (M.C., A.G., G.v.H.,B.H., H.B.M., D.M., K.O., J.W., H.W.), from DESY (A.G., S.S.) and from the Lawrence Berkeley National Laboratory (B.H.), the principal investigators and project contributors of project GCS-MUMA were K. Szabo, Cs. Torok, K. Miura, L. Parato, and L. Varnhorst. Since only the project GCS-HQCD was participating in the 22nd Results and Review Workshop of the HLRS (2019), only the report of this project can be presented in the present volume. The interest in the anomalous magnetic moment of the muon is motivated by the consequences for a theory beyond the standard model. In project GCS-HQCD, leading systematic uncertainties in standard model predictions from hadronic contributions are computed by Lattice Quantum Chromodynamics simulations, using the Hybrid Monte Carlo (HMC) algorithm in the openQCD code. Calculations of the hadronic vacuum polarization contribution to the anomalous magnetic moment based on Coordinated Lattice Simulations consortium gauge ensembles with 2+1 dynamical flavors with improved Wilson quarks have been done and reveal an effect on the anomalous magnetic moment in the order of about $7 \cdot 10^{-8}$.

The TNG50 Simulation: Highly-Resolved Galaxies in a Large Cosmological Volume to the Present Day

Annalisa Pillepich, Dylan Nelson, Volker Springel, Rüdiger Pakmor, Lars Hernquist, Mark Vogelsberger, Rainer Weinberger, Shy Genel, Federico Marinacci, Paul Torrey, and Jill Naiman

---

A. Pillepich (✉)
Max-Planck-Institut für Astronomie, Königstuhl 17, 69117 Heidelberg, Germany
e-mail: pillepich@mpia-hd.mpg.de; pillepich@mpia.de

D. Nelson · V. Springel · R. Pakmor
Max-Planck-Institut für Astrophysik, Karl-Schwarzschild-Str. 1, 85741 Garching, Germany
e-mail: dnelson@mpa-garching.mpg.de

V. Springel
e-mail: vspringel@mpa-garching.mpg.de

R. Pakmor
e-mail: rpakmor@mpa-garching.mpg.de

L. Hernquist · R. Weinberger · J. Naiman
Harvard-Smithsonian Center for Astrophysics, 60 Garden Street, Cambridge, MA 02138, USA
e-mail: lars@cfa.harvard.edu

R. Weinberger
e-mail: rainer.weinberger@cfa.harvard.edu

J. Naiman
e-mail: jill.naiman@cfa.harvard.edu

M. Vogelsberger
MIT Kavli Institute for Astrophysics and Space Research, Department of Physics, Massachusetts Institute of Technology, Cambridge, MA 02139, USA
e-mail: mvogelsb@mit.edu

S. Genel
Center for Computational Astrophysics, Flatiron, 162 Fifth Avenue, New York, NY 10010, USA
e-mail: shygenelastro@gmail.com

F. Marinacci
Department of Physics and Astronomy, University of Bologna, via Gobetti 93/2, 40129 Bologna, Italy
e-mail: federico.marinacci2@unibo.it

P. Torrey
Department of Physics, University of Florida, 2001 Museum Rd., Gainesville, FL 32611, USA
e-mail: paul.torrey@ufl.edu

© Springer Nature Switzerland AG 2021
W. E. Nagel et al. (eds.), *High Performance Computing in Science and Engineering '19*,
https://doi.org/10.1007/978-3-030-66792-4_1

Abstract Large-volume cosmological hydrodynamical simulations of galaxy formation enable us to theoretically follow the co-evolution of thousands of galaxies while directly outputting the observable signatures that result from the complex and highly non-linear process of cosmic structure formation. Here we present the first results from the TNG50 run, an unprecedented 'next generation' cosmological, magnetohydrodynamical simulation that we have recently brought to completion on the Hazel Hen supercomputer. TNG50 is the third and final volume of the IllustrisTNG project. With over 20 billion resolution elements it resolves spatial scales down to ~ 100 parsecs, following the co-evolution of dark matter, gas, stars, supermassive black holes and magnetic fields across the history of the Universe.

1 Introduction

The evolution and physical properties of galaxies depend on a rich set of physical ingredients: the laws of gravity; the nature of dark matter; the details of the growth of cosmological structures on the largest spatial scales; the interaction between radiation and cosmic gas and hence gas cooling and heating; the chemical and thermodynamical properties of the gas which forms stars and feeds the growth of super massive black holes; the evolution and death of stellar populations; and the non-linear effects and coupling of energy, momentum, and radiative feedback from stars and black holes. The diversity and complexity of the relevant physical processes can be followed in full generality only through cosmological hydrodynamical computer simulations. However, the enormous range of spatial and time scales, as well as the complexity of the physical processes involved, makes this a remarkable computational challenge. Significant progress has been made in this direction, as embodied in large-volume hydrodynamical projects such as Illustris [1–4], EAGLE [5, 6], and Horizon-AGN [7]. These have begun to generate plausible and diverse galaxy populations by combining ab-initio calculations with sub grid prescriptions of small-scale phenomena: run at 'kilo-parsec' spatial resolutions, these numerical experiments have reproduced a number of fundamental scaling relations and properties of observed galaxy populations. This zeroth order agreement has buttressed many theoretical investigations and predictions. At the same time, however, it has revealed many shortcomings in the current generation of models.

IllustrisTNG is a 'next generation' series of large, cosmological, gravo-magneto-hydrodynamical simulations incorporating a comprehensive model for galaxy formation physics [8, 9]. It has been conducted over the past three years on the Hazel Hen machine at the High Performance Computing Center Stuttgart (HLRS) and supported by two Gauss Centre for Supercomputing allocations (GCS-ILLU in 2014, and GCS-DWAR in 2016). IllustrisTNG includes three flagship runs: TNG50, TNG100, and TNG300. The latter two have been completed within the first allocation and presented last year [10–14]. TNG50 is the most computational demanding of the three simulations by far: with a total of about 135M core hours, TNG50 has been recently completed, reaching $z = 0$ (the current epoch, or present day) in April 2019. Here we showcase the first results from this project.

2 IllustrisTNG: Physical and Numerical Developments

The IllustrisTNG simulation project[1] extends the original Illustris simulation in two key ways. First, it alleviates most of the previous model deficiencies [15], i.e. tensions with respect to available observational data. Second, it expands upon the scope in all directions by executing simulations with higher resolution, of larger volumes, and with new physics.

2.1   Galaxy Formation in TNG: The Numerical Code

The IllustrisTNG simulation suite uses the AREPO code [16], which solves for the coupled evolution of self-gravity and magnetohydrodynamics [MHD; [17, 18]]. The former is computed with the spatially split Tree-PM approach, while the latter is based on a finite-volume method whose spatial discretization is a dynamic, unstructured, Voronoi tessellation. The scheme is quasi-Lagrangian (ALE) and second order in both space and time. It achieves high dynamic range through an individual particle time-stepping approach. In contrast to past cosmological simulations, IllustrisTNG now fundamentally includes a treatment of magnetic fields under the assumptions of ideal MHD [17, 18].

The AREPO code has been architected to execute large parallel astrophysical simulations. For instance, the TNG50 simulation reviewed here has been run on 16320 cores. At this scale the highly coupled, high dynamic range of the galaxy formation problem is particularly challenging: TNG50 captures a spatial dynamic range of $\sim 10^7$, while the time hierarchy necessitates evolution on timescales which differ by $\sim 10^4$. For numerical optimization reasons, over the past several years and in preparation for the TNG simulations, (i) the previous MUSCL-Hancock time integration scheme has been replaced with an approach following Heun's method [19], (ii) the method for obtaining gradients of primitive fluid quantities has been replaced with an iterative least-squares method, (iii) the long-range gravity FFT calculation now uses a new, column-based MPI-parallel FFT which improves scaling at high core numbers, and (iv) the gravity solver incorporates a new, recursive splitting of the N-body Hamiltonian into short- and long- timescale particle systems.

---

[1] http://www.tng-project.org.

2.2   Galaxy Formation in TNG: Physical Model

Cosmological hydrodynamical simulations such as TNG acknowledge that physics below a given spatial scale, of order a hundred to a few hundred parsecs, cannot be resolved and must be treated by approximate, sub-resolution models. This includes, most importantly, the process of star formation, the detailed action of individual supernova events, the formation and growth of supermassive blackholes, and the near-field coupling of blackhole feedback energy to the surroundings. Together, these components make up the updated TNG model for galaxy formation, which is described in [9] and [8]. We employ it unchanged (and invariant with numerical resolution) in all TNG simulations, including TNG50.

The physical framework includes models of the most important physical processes for the formation and evolution of galaxies; (i) gas radiative microphysics, including primordial (H/He) and metal-line cooling and heating with an evolving ultraviolet/x-ray background field, (ii) star formation in dense interstellar medium (ISM) gas, (iii) the evolution of stellar populations and chemical enrichment, tracking supernovae Ia, II, and AGB stars, and individual species: H, He, C, N, O, Ne, Mg, Si, and Fe, (iv) galactic-scale outflows launched by stellar feedback, (v) the formation, binary mergers, and gas accretion by supermassive blackholes, (vi) blackhole feedback, operating in a thermal mode at high accretion rates and a kinetic 'wind' mode at low accretion rates. Aspects (iv) and (vi) have been substantially revised in TNG, and we described the key changes in our previous report [20]. In short, galactic-scale outflows generated by stellar feedback are modeled using a kinetic wind approach [21] based on the energy available from Type II (core-collapse) supernovae. In TNG the directionality, thermal content, energy budget scaling with metallicity, and minimum launch velocity were all redesigned in order to better reflect available data constraints [full details in [8]]. Additionally, supermassive black holes (SMBHs) form in massive halos and subsequently inject large amounts of energy, as allowed by their instantaneous mass accretion rates as derived from their immediate neighborhoods. In TNG we introduced a new low-state kinetic wind feedback model, in the form of a time-pulsed, oriented, high-velocity 'wind', suggested to be a possibly crucial mechanism by recent theoretical as well as observational work [full details in [9]].

2.3   Early Results from TNG, Model Confirmations, and Predictions

Over the past year and a half, the TNG model has been shown to produce results that are consistent with a wide range of observational constraints, including regimes beyond those adopted for the model development. With respect to galactic structural and stellar population properties these include: the shape of the red sequence and blue cloud of SDSS galaxies [12]; the spatial clustering of active and passive galaxies at the 1-halo and 2-halo term scales [11]; galaxy stellar mass functions up to $z \sim 4$ [10]; stellar sizes out to $z \sim 2$ split by star-forming vs. quiescent populations [22]; the scatter of Europium abundance in metal-poor stars in Milky Way-like systems [23]; the quenched galaxy population at both low [24] and high [25] redshift; the gas-phase mass-metallicity relation [26]; the dark matter fractions within massive galaxies at $z = 0$ in comparison to SLUGGS results [27]; and the visible light morphologies of galaxies versus Pan-STARRS data [28].

The IllustrisTNG model also produces a range of less common galaxies, i.e. it samples tails of the galaxy population. These include low surface brightness (LSB) galaxies [29] and ram-pressure stripped 'jellyfish' systems [30]. With respect to massive galaxy cluster, intra-cluster and circumgalactic medium properties we find broad agreement in: the relationship between total radio power and X-ray luminosity, total mass, and Sunyaev-Zel'dovich signal [13]; the distribution of metals versus radius in the intra-cluster medium [ICM; [31]]; the observed fraction of cool core vs. non-cool core clusters [32]; and the OVI content of the circumgalactic media around galaxies versus the COS-Halos and eCGM surveys [33].

IllustrisTNG is also returning novel insights on the formation and evolution of galaxies: for instance, on the universality of stellar mass profiles [10]; how star-forming and quenched galaxies take distinct evolutionary pathways across the galaxy size-mass plane [22]; that galaxies oscillate around the star formation main sequence and the mass-metallicity relations in an anti-correlated, time synchronized fashion [26]; that jellyfish galaxies are signaled by large-scale bow shocks in the surrounding intra-cluster medium [30]; how the metal enrichment of cluster galaxies is higher than field counterparts at fixed mass, even prior to infall [34]; the way in which baryonic processes modulate the matter power spectrum [11] and steepen the inner total density profiles of early-type galaxies [35]; and the properties of OVII, OVIII [33] and NeIX [36] absorption systems as detectable by future X-ray telescopes like ATHENA.

We have also generated mock 21-cm maps [37] and estimates of the molecular hydrogen (H2) abundance [38], also as a function of environment, in the local [39] and high-redshift Universe accessible with ALMA [40]. Finally, TNG provides a test bed to explore future observational applications of techniques such as machine learning: for example, using Deep Neural Networks to estimate galaxy cluster masses from Chandra X-ray mock images [41] or CNN-based optical morphological classification versus SDSS [42].

2.4 TNG50 Project Scope

TNG50 is the third and final volume of the IllustrisTNG project. This simulation occupies a unique combination of large volume and high resolution—its details and numerical properties are given in Table 1 in comparison to the TNG series as a whole, while Fig. 1 gives a visual comparison of the volumes. Our 50 Mpc box is sampled by 2160^3 dark matter particles (with masses of $4 \times 10^5 \, M_\odot$) and 2160^3

Table 1 Details of the TNG50 simulation in comparison to its two larger volume counterparts

Run Name	Volume	N_{GAS}	N_{DM}	N_{TR}	m_{baryon}	m_{DM}	$\epsilon_{gas,min}$	$\epsilon_{DM,stars}$
	[Mpc^3]	–	–	–	[M$_\odot$]	[M$_\odot$]	[pc]	[pc]
TNG50	51.7^3	2160^3	2160^3	2160^3	8.5×10^4	4.5×10^5	74	**288**
TNG100	110.7^3	1820^3	1820^3	2 × 1820^3	1.4×10^6	7.5×10^6	185	740
TNG300	302.6^3	2500^3	2500^3	2500^3	1.1×10^7	5.9×10^7	370	1480

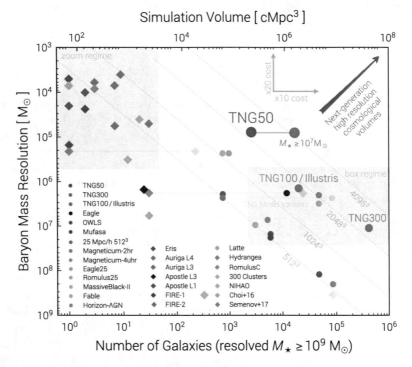

Fig. 1 The TNG50 simulation is a unique cosmological hydrodynamical simulation: it includes 2 × 2160^3 resolution elements, implying a baryonic mass resolution of 8.5 × 10^{4}M$_\odot$ with adaptive gas softening down to 74 comoving parsecs. This approaches, or exceeds, the resolution of modern 'zoom' simulations of individual galaxies, while maintaining the statistical power and unbiased sampling of the full ~50 cMpc cosmological volume. Here we show TNG50 (dark blue) in comparison to other cosmological volumes (circles) and zoom simulation suites (diamonds) at the current cosmological epoch (i.e. $z \sim 0$), based on the total number of resolved galaxies (a proxy for volume and statistics). Pushing towards the upper right corner represents the frontier of galaxy formation simulations, as well as extreme computational difficulty

initial gas cells (with masses of 8 × 10^4 M$_\odot$). The total number of resolution elements is therefore slightly over 20 billion. The *average* spatial resolution of star-forming interstellar medium (ISM) gas is ~90 (~140) parsecs at $z = 1$ ($z = 6$). TNG50 has 2.5 times better spatial resolution, and 15 times better mass resolution, than TNG100 (or equivalently, original Illustris). This resolution approaches or exceeds that of modern 'zoom' simulations of individual galaxies [43, 44], while the volume contains $\sim$ 20,000 resolved galaxies with $M_\star > 10^7\,M_\odot$ (at $z = 1$).

At the time of writing, the TNG50 simulation has been evolved from the initial conditions of the Universe all the way to the current epoch, $z = 0$ (13.8 billion years after), and it is hence completed.

TNG50 contains roughly 200 Milky Way and Andromeda analogs, enabling detailed comparisons to our own galaxy at $z = 0$. It also hosts two massive galaxy clusters with a total mass $\sim 10^{14}\,M_\odot$, i.e. Virgo-like analogs, and dozens of group sized halos at $\sim 10^{13}\,M_\odot$. All of these massive objects are simulated at higher numerical resolution than in any previously published study, enabling studies not only of the gaseous halos and central galaxies, but also of the large populations of their satellite galaxies.

3   The TNG50 Simulation: Current Results and Outlook

The TNG50 simulation has been presented in the scientific literature with two introductory papers, focusing, respectively, on the internal structural and kinematical properties of star-forming galaxies across time [45] and on the gaseous outflows resulting from stellar and black hole feedback [46]. Another study based on the combination of TNG50 with the first two runs of the IllustrisTNG project (TNG100 and TNG300) has too been submitted for peer-review with the goal of quantifying the evolution of the stellar mass and luminosity functions of galaxies in anticipation of the observations with the James Webb Space Telescope (JWST) [47]. In the next Sections, we summarize selected highlights from these early TNG50 analyses and showcase ongoing and future investigations tailored at maximally exploit the TNG50 run and explored in a number of papers currently in preparation.

We can now analyze a fully representative, simulated galaxy population spanning $10^7 < M_\star/M_\odot < 10^{11.5}$ across time, $0 < z < 10$. The high resolution of TNG50 is specifically exploited to investigate scales, regimes, and scientific questions not addressable using other cosmological simulations. This coverage in redshift range and galaxy stellar mass enables us to make quantitative predictions for signatures observable with JWST, now anticipated to launch in 2021, as well as recent ground-based IFU instruments such as MUSE and SINFONI, in addition to capturing the dynamics of gas, dark matter, and magnetic fields within and between galaxies. The key science drivers of TNG50 focus not only on the present day ($z = 0$), but also at earlier epochs, from cosmic noon ($z \sim 2$) through reionization ($z \sim 6$).

3.1 Uncovering Galactic Structure and Galaxy Kinematics

Studies of the stellar kinematics of star-forming galaxies are now common in the local Universe e.g. with integral field spectroscopy (IFS) data from SAMI, MANGA, and CALIFA. However, this is not yet viable at earlier cosmic epochs. The progenitors of present-day galaxies are observed at $z \sim 1 - 2$ and characterized in their kinematics from ground based telescopes using adaptive-optics techniques: these enable kiloparsec scale resolution but are based on tracing bright emission lines from hydrogen (such as Hα) or metals (i.e. OIII). In other terms, such high-redshift observations trace the kinematics of galactic gas instead of stars, as in [48]. In practice, at intermediate and early cosmic time (redshift $z > 0.3$), the observational analysis of galaxies encounters a difficulty: galaxy morphologies are typically obtained through multiwavelength imaging surveys that trace the stellar light, while galaxy kinematics are commonly obtained through Hα spectroscopy. Using TNG50, we are able to provide model predictions for projected radial profiles and resolved 2D maps of stellar and gas density, star formation rate (Hα), stellar and gas line-of-sight velocity and velocity dispersions [45]. We uncover outcomes of TNG50 for which the model has not been in any way calibrated and is thus predictive. Furthermore, we can contrast structural versus kinematical features, as well as the properties of the stellar versus gaseous components of galaxies, by focusing on a redshift regime where such comparisons are currently prohibitive in observations, though soon to emerge, therefore maximizing the predictive return of the TNG50 calculation.

In Fig. 2 we show the matter distribution and velocity fields of a randomly selected galaxy at $z = 2$ from TNG50, one of thousands (see www.tng-project.org/explore/gallery/ for a more comprehensive set of examples). From top to bottom, the panels show the stellar component of the galaxy (in edge-on and face-on projections) and its star-forming and gaseous component (also in edge-on and face-on projections). The line-of-sight velocity of the galaxy in the edge-on projections (mid column, first and third rows) is a proxy for the rotation of the galaxy and the corresponding rotation curve is shown in the rightmost columns (first and third rows). The line-of-sight velocity dispersion of the galaxy in the face-on projections (mid column, second and bottom rows), on the other hand, represent the contribution of random (i.e. disordered, non rotational, or even turbulent) local motions of stars and gas. Small scale structures at sub-kiloparsec scales are easily resolved by TNG50, revealing rich morphological and kinematical features. We can see, for example, how outflows generated from the nuclear regions of disks leave signatures in the gas left behind, evidenced in the central depressions because of the expulsion of gas through black hole feedabck. In fact, despite their strong ordered rotation, galactic disks at intermediate redshifts $z \sim 1 - 3$ are highly turbulent gaseous reservoirs, characterized by velocity dispersion fields that are remarkably less coherent in space than their stellar analogs (bottom vs. second rows).

Galactic disks, however, settle with time. This is shown in Fig. 3 in terms of the balance between ordered and disordered motions in TNG50 star-forming galaxies as a function of redshifts. Orange curves and markers denote gas-based kinematics,

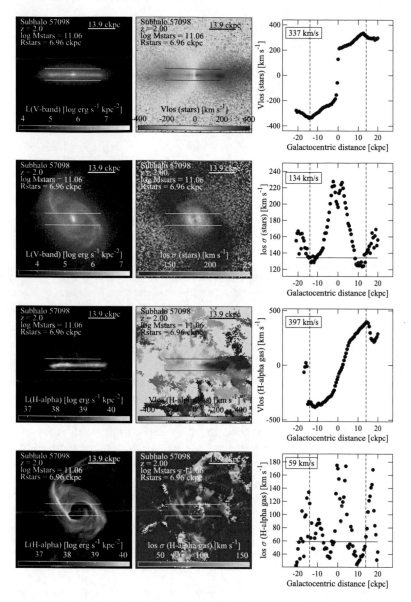

Fig. 2 One example of a massive galaxy at redshift 2 from the TNG50 simulation, randomly chosen among thousands [45]. The panels show V-band light maps, velocity maps and velocity profiles for the stellar component of the galaxy in edge-on (top row) and face-on (second row from the top) projections, together with the analog Hα light maps, velocity maps and velocity profiles for its star-forming and gaseous component (bottom two rwos). A more comprehensive set of examples is available on the IllustrisTNG website: www.tng-project.org/explore/gallery/

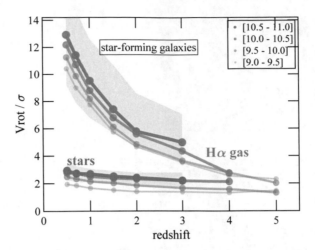

Fig. 3 Degree of ordered vs. turbulent motion (Vrot/σ) in TNG50 galaxies as a function of red-shift in bins of galaxy stellar mass [45]. Solid curves and markers denote medians of the TNG50 Vrot/σ, for the Hα-emitting gas (orange) and the stellar component (blue), separately. Overall, the balance between ordered and disordered motions of the gaseous bodies increases substantially as the Universe evolves, and more so than the stellar counterparts

while blue curves indicate stellar-based kinematics, in bins of galaxy stellar mass. TNG50 star-forming galaxies host strongly rotating gaseous disks, more rotationally-supported the older the Universe: these trends are qualitatively consistent with current observations and constitute a non-trivial confirmation of the underlying physical model. Additionally, for the first time, this plot demonstrates that, at all times and masses, the dense gas component of star-forming galaxies is characterized by larger circular motions than the stellar material, with differences as large as a factor of several at low redshift and high mass. Such contrast, as a function of galaxy mass and cosmic time, can be directly tested against upcoming observational programs.

3.2   Gas-Dynamical Processes and Feedback-Driven Outflows

The kinematical and thermodynamical state of the gas within and around galaxies is a sensitive probe of galaxy formation physics, plasma physics, and even cosmology. Within very massive haloes, observations at X-ray and radio wavelengths reveal a rich level of detail and sub-substructure [49]. On galactic scales, energetic outflows driven out of galaxies reveal astrophysical feedback processes in action. These are associated with supernova explosions and to the activity of super massive blackholes residing at the center of galaxies—both are considered fundamental for the regulation of star formation in galaxies. The resolution of TNG50 enables a detailed study of

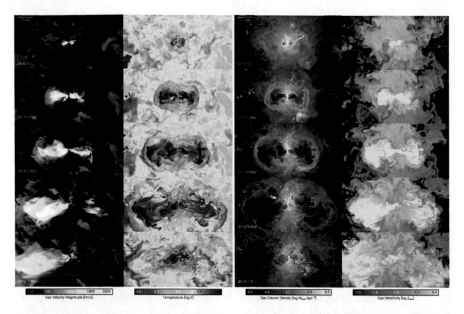

Fig. 4 We show a black hole feedback event driving a large-scale, galactic outflow in TNG50 [46]. From top to bottom shows the time evolution, with five snapshots $\sim$ 100 Myr apart, starting from $z \simeq 2$. Depicted is a single massive galaxy with a stellar mass of $10^{11.4}$ $M_\odot$, which is currently 'quenching' its star formation. From left to right: gas velocity, temperature, density, and metallicity, all on the halo scale (virial radius as white circles). The galaxy itself is oriented vertically, edge-on, and is visible as the small, cold disk at the center of each image. The central black hole has a mass of $10^{8.7}$ $M_\odot$ and is driving a large-scale collimated outflow in the kinetic 'wind' feedback mode

the properties of gas motions within and around galaxies across an unprecedented range of galaxy types, masses, environments, and cosmic epochs.

In [46] we use TNG50 to quantify the properties of galactic outflows in the cosmological setting and with respect to the galaxies from which they arise, by focusing on the way in which outflows shape the galaxy population as a whole, modulate galaxy evolution, and generate associated observational signatures. Figure 4 visualizes, for example, the time evolution of a strong outflow driven by a massive black hole in TNG50 originating from a massive galaxy at $z \sim 2$. Maps show gas velocity, gas temperature, gas density and gas metallicity (from left to right), with time progressing downwards, each row roughly 100 Myr apart. In our model, energy injection from the black hole produces a high-velocity, large-scale, and highly collimated (directional) outflow, which reaches speeds exceeding 2500 km/s even as it crosses the halo virial radius, qualitatively similar to some observed in the real Universe. By showing the highly resolved structure of an individual galaxy, we emphasize here that TNG50 allows us to connect small-scale (i.e. few hundred pc) feedback and large- scale (i.e. few hundred kpc) outflows.

Our TNG50 calculation predicts that gaseous outflow velocities increase with a galaxy's stellar mass and that outflows are faster at higher redshift. The phase structure of galactic winds is also complex, and we demonstrate that the TNG model can produce high velocity, multi-phase outflows which include cool, dense components. Importantly, we show how the relative simplicity of model inputs (and scalings) at the injection scale produces complex behavior at galactic and halo scales. For example, despite isotropic wind launching, outflows exhibit natural collimation and an emergent bipolarity [46].

3.3 Demographics of Galaxies in the First Few Billion Years

Galaxies form and evolve not in isolation but within the large-scale structure that emerges according to the theory of hierarchical assembly and collapse [50]. Energetic feedback processes (see Section above) result in radiation from their constituent stars and black holes, enabling galaxies to alter the ionization state of the surrounding gas, driving the cosmic Reionization of the intergalactic medium that is believed to have occurred within the first billion year of cosmic evolution. However, to quantitatively pin down the details of this final *phase transition* in the history of the Universe, we must first confront theoretical predictions with observations of the most basic quantity that defines the galaxy population: its abundance, i.e. the galaxy luminosity function, measuring the number density of galaxies as a function of their luminosities at different wavelengths and at different redshifts. This in turn can be used to quantify the escape fraction of ionizing radiation from galaxies that could have re-ionized the Universe.

The upcoming JWST promises to open a new window into the high redshift Universe to study faint and distant galaxies during the epoch of Reionization and later. Particularly, JWST will quantify the galaxy population and galaxy luminosity functions at higher redshifts than ever before: it will also decidedly increase the statistical sample sizes of high redshift galaxies. As TNG50 evolves all of the relevant baryonic and non-baryonic components self-consistently together and up to spatial scales of tens of mega-parsec scales, we can provide expectations for the high redshift galaxy population and Universe in general.

In [47], we exploit the large dynamic range of the whole IllustrisTNG simulation suite, TNG50, TNG100, and TNG300, to derive multi-band galaxy luminosity functions from $z = 1$ to $z = 10$. We put particular emphasis on the exploration of different dust attenuation models to determine galaxy luminosity functions for the rest-frame ultraviolet (UV), and apparent wide NIRCam bands. In Fig. 5, we visualize the outcome of our most detailed dust model, based on continuum Monte Carlo radiative transfer calculations employing observationally calibrated dust properties. The top row depicts light maps a random sample of TNG50 galaxies at $z \sim 1$, based on a combination of apparent F070W, F090W, F115W filter fluxes. The high numerical resolutions of TNG50 reveals sub-kpc details as well as lanes of dust that absorb light from the central bright regions. The bottom left panel shows the intrinsic and dust attenuated spectral energy distributions of two random TNG50 galaxies at $z = 2$, together with the filters of the NIRCam instrument on board JWTS. The bottom right

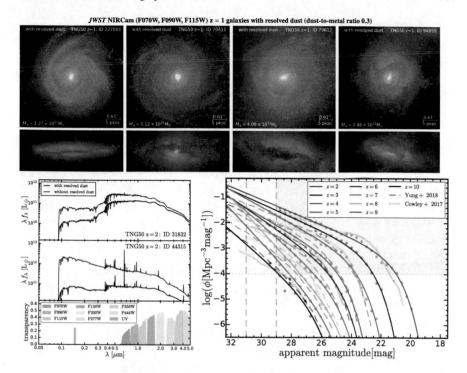

Fig. 5 Top: Face-on and edge-on images of a few randomly-chosen TNG50 galaxies at $z = 2$, as would be seen by JWST NIRCam [47]. These images include the effects of resolved dust attenuation calculated via a post-processing, Monte Carlo dust radiative transfer calculation employing 10^7 photon packets per wavelength on a wavelength grid spanning $0.05 \mu m$ to $5 \mu m$. Dust radiative transfer effects are critical: in fact, the light from the central bright regions of the various galaxies is strongly absorbed and scattered due to dust. Bottom left: Spectral energy distribution for two example galaxies from TNG50 at $z = 2$, including the intrinsic spectral energy distribution (blue) and the dust attenuated spectrum (red). The bottom inset shows the relevant transmission functions of the different bands, including the ultraviolet and eight wide JWST NIRCam filters. Bottom right: Apparent luminosity functions at different redshifts predicted by the IllustrisTNG calculations, as will be observed in the JWST NIRCam F200W band. Markers show simulation data, and lines functional fits, including Schechter fits from currently-available observations for comparison

panel provides the F200W band luminosity functions and best-fit Schechter function parameters for the predicted NIRCam wide filter apparent luminosity functions as predicted by the IllustrisTNG model. For the F200W NIRCam band, we predict that JWST will detect about 80 ($\sim$200) galaxies with a signal-to-noise ratio of 10 (5) within the NIRCam field of view for a total exposure time of 10^5s in the redshift range $z = 8 \pm 0.5$. These numbers will drop to about 10 (40) for a shorter exposure time, e.g. with 10^4s [47].

3.4 Further Ongoing TNG50 Explorations

The synthetic Universe of TNG50 encompasses thousands of realistic and well-resolved galaxies, spanning a variety of masses, environments, evolutionary and interaction stages, and fully described in terms of their gaseous, stellar, dark matter, and black hole content. The aforementioned scientific highlights represent a sample of the diverse applications that the TNG50 calculation enables. With a physical resolution which is unprecedented for such a cosmological volume, TNG50 bridges the gap towards 'zoom' simulations of individual galaxies and resolves systems as small as the 'dwarf' satellites of galaxies like our own Milky Way. Simultaneously, the large volume of the simulation box enables statistically significant and unbiased analyses of the galaxy population and of the large scale structure. With TNG50 brought to completion in April 2019 (i.e. to the current epoch, $z = 0$), we can now tackle all the originally identified fundamental science drivers of the TNG50 calculation: (1) The formation, evolution, and properties of dwarf galaxies; (2) Milky Way-like galaxies and their satellites systems; (3) Galaxy evolution in massive cluster environments and the intra-cluster medium; and (4) Low-density circumgalactic and intergalactic gas, from halo to cosmological scales (Fig. 6).

Numerous projects based on the TNG50 output are currently in progress, and extend in topic far beyond the originally identified science goals. These include the quantification of the turbulence in the diffuse gas of the universe and within the intra-halo medium of groups and clusters of galaxies; the amplification of magnetic fields

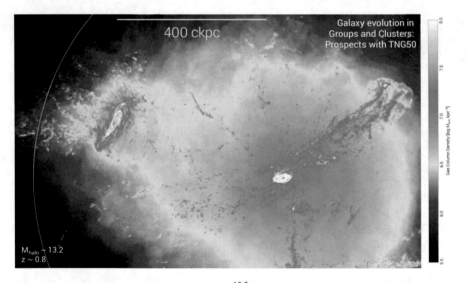

Fig. 6 The gas density in projection within a $10^{13.2}$ M$_\odot$ halo at $z \sim 0.8$ from the TNG50 cosmological volume. The virial radius of the halo extends to hundreds kilo parsecs distances from its central massive galaxy. Other galaxies are visibly undergoing 'stripping' of their gas reservoirs into long tails due to the interaction with the central object and the diffuse intra-cluster medium

in high-redshift galaxies and the magnetic field topology within galaxy clusters; the properties, emergence and comparison to observations of stellar haloes surrounding Milky Way-like galaxies; the onset of bars in disk-like galaxies; the X-ray scaling relations of the gas within massive elliptical galaxies; and the stellar metallicity gradients in low-surface brightness outskirts. We anticipate that the TNG50 dataset will have a long-lasting legacy value and constitute a platform for future programs which make detailed comparisons to astronomical observations as well as advance key aspects of galaxy formation theory. After a proprietary period of roughly one year, it will be made publicly available within the framework already developed for the first runs of the IllustrisTNG project [51].

4 Conclusions

Despite the tremendous theoretical and numerical achievements of recent large volume cosmological simulations such as Illustris, Eagle, or even TNG100 and TNG300, their limited mass and spatial resolution complicates the study of the structural details of galaxies less massive than a few times $10^9\,M_\odot$. In contrast, projects focused on higher-resolution galaxy 'zoom' simulations have been less useful in broadly testing the outcome of their underlying physical models against population-wide morphological observed estimators because of their small sample sizes. For the most massive galaxy clusters, simulations with sufficient resolution to simultaneously model the co-evolving population of satellite galaxies have been prohibited by the large computational requirements as well as the complexity of the physical mechanisms which shape the circumgalactic and intracluster gas.

The TNG50 calculation that we have recently completed on the Hazel Hen machine is redefining the state-of-the-art of cosmological hydrodynamical simulations of galaxy formation by bridging the gap between large cosmological volumes and better resolved zoom galaxies. We have already showcased a few scientific applications of the TNG50 simulation and demonstrated that TNG50 is already proving to be an instrumental theoretical tool for the comparison—via mock observations of the simulated data—with existing and upcoming observational datasets. These include, for example, the abundances of dwarf galaxies in the first few billion years of the history of the Universe (with HST, JWST) and the kinematic properties of the $H\alpha$ and molecular gas within and around galaxies at intermediate redshift (with e.g. integral field spectroscopy and other surveys, SINS/zC- SINF, PHIBBS, KMOS3D, ASPECS). Our analyses of TNG50 have also uncovered novel predictions, shedding new light on our understanding of galaxy evolution and providing a foundation for theoretical interpretation. Even beyond our currently-ongoing immediate scientific investigations, the TNG50 simulation will be a unique platform to pursue as of yet unimagined future projects, as we are now able to treat cosmological simulations as almost open ended laboratories for studying galaxy formation physics. The synthetic Universe of TNG50 will be a long term resource for the analysis, exploration and interpretation of observations, one that we will make publicly available to the whole astronomy community.

Acknowledgements The authors acknowledge the Gauss Centre for Supercomputing (GCS) for providing computing time for the GCS Large-Scale Projects GCS-ILLU (2014) and GCS-DWAR (2016) on the GCS share of the supercomputer Hazel Hen at the High Performance Computing Center Stuttgart (HLRS). AP and DN acknowledge additional simulations and analysis carried out on supercomputers at the Max Planck Computing and Data Facility (MPCDF, formerly RZG).

References

1. M. Vogelsberger, S. Genel, V. Springel, P. Torrey, D. Sijacki, D. Xu, G. Snyder, D. Nelson, L. Hernquist, MNRAS **444**, 1518 (2014). https://doi.org/10.1093/mnras/stu1536
2. M. Vogelsberger, S. Genel, V. Springel, P. Torrey, D. Sijacki, D. Xu, G. Snyder, S. Bird, D. Nelson, L. Hernquist, Nature **509**, 177 (2014). https://doi.org/10.1038/nature13316
3. S. Genel, M. Vogelsberger, V. Springel, D. Sijacki, D. Nelson, G. Snyder, V. Rodriguez-Gomez, P. Torrey, L. Hernquist, MNRAS **445**, 175 (2014). https://doi.org/10.1093/mnras/stu1654
4. D. Sijacki, M. Vogelsberger, S. Genel, V. Springel, P. Torrey, G.F. Snyder, D. Nelson, L. Hernquist, MNRAS **452**, 575 (2015). https://doi.org/10.1093/mnras/stv1340
5. J. Schaye, R.A. Crain, R.G. Bower, M. Furlong, M. Schaller, T. Theuns, C. Dalla Vecchia, C.S. Frenk, I.G. McCarthy, J.C. Helly, A. Jenkins, Y.M. Rosas-Guevara, S.D.M. White, M. Baes, C.M. Booth, P. Camps, J.F. Navarro, Y. Qu, A. Rahmati, T. Sawala, P.A. Thomas, J. Trayford, MNRAS **446**, 521 (2015). https://doi.org/10.1093/mnras/stu2058
6. R.A. Crain, J. Schaye, R.G. Bower, M. Furlong, M. Schaller, T. Theuns, C. Dalla Vecchia, C.S. Frenk, I.G. McCarthy, J.C. Helly, A. Jenkins, Y.M. Rosas-Guevara, S.D.M. White, J.W. Trayford, MNRAS **450**, 1937 (2015). https://doi.org/10.1093/mnras/stv725
7. Y. Dubois, S. Peirani, C. Pichon, J. Devriendt, R. Gavazzi, C. Welker, M. Volonteri, MNRAS **463**, 3948 (2016). https://doi.org/10.1093/mnras/stw2265
8. A. Pillepich, V. Springel, D. Nelson, S. Genel, J. Naiman, R. Pakmor, L. Hernquist, P. Torrey, M. Vogelsberger, R. Weinberger, F. Marinacci, MNRAS **473**, 4077 (2018). https://doi.org/10.1093/mnras/stx2656
9. R. Weinberger, V. Springel, L. Hernquist, A. Pillepich, F. Marinacci, R. Pakmor, D. Nelson, S. Genel, M. Vogelsberger, J. Naiman, P. Torrey, MNRAS **465**, 3291 (2017). https://doi.org/10.1093/mnras/stw2944
10. A. Pillepich, D. Nelson, L. Hernquist, V. Springel, R. Pakmor, P. Torrey, R. Weinberger, S. Genel, J.P. Naiman, F. Marinacci, M. Vogelsberger, MNRAS **475**, 648 (2018). https://doi.org/10.1093/mnras/stx3112
11. V. Springel, R. Pakmor, A. Pillepich, R. Weinberger, D. Nelson, L. Hernquist, M. Vogelsberger, S. Genel, P. Torrey, F. Marinacci, J. Naiman, MNRAS **475**, 676 (2018). https://doi.org/10.1093/mnras/stx3304
12. D. Nelson, A. Pillepich, V. Springel, R. Weinberger, L. Hernquist, R. Pakmor, S. Genel, P. Torrey, M. Vogelsberger, G. Kauffmann, F. Marinacci, J. Naiman, MNRAS **475**, 624 (2018). https://doi.org/10.1093/mnras/stx3040
13. F. Marinacci, M. Vogelsberger, R. Pakmor, P. Torrey, V. Springel, L. Hernquist, D. Nelson, R. Weinberger, A. Pillepich, J. Naiman, S. Genel, ArXiv e-prints (2017)
14. J.P. Naiman, A. Pillepich, V. Springel, E. Ramirez-Ruiz, P. Torrey, M. Vogelsberger, R. Pakmor, D. Nelson, F. Marinacci, L. Hernquist, R. Weinberger, S. Genel, ArXiv e-prints (2017)
15. D. Nelson, A. Pillepich, S. Genel, M. Vogelsberger, V. Springel, P. Torrey, V. Rodriguez-Gomez, D. Sijacki, G.F. Snyder, B. Griffen, F. Marinacci, L. Blecha, L. Sales, D. Xu, L. Hernquist, Astronomy Comput. **13**, 12 (2015). https://doi.org/10.1016/j.ascom.2015.09.003

16. V. Springel, MNRAS **401**, 791 (2010). https://doi.org/10.1111/j.1365-2966.2009.15715.x
17. R. Pakmor, A. Bauer, V. Springel, MNRAS **418**, 1392 (2011). https://doi.org/10.1111/j.1365-2966.2011.19591.x
18. R. Pakmor, V. Springel, MNRAS **432**, 176 (2013). https://doi.org/10.1093/mnras/stt428
19. R. Pakmor, V. Springel, A. Bauer, P. Mocz, D.J. Munoz, S.T. Ohlmann, K. Schaal, C. Zhu, MNRAS **455**, 1134 (2016). https://doi.org/10.1093/mnras/stv2380
20. D. Nelson, A. Pillepich, V. Springel, R. Weinberger, L. Hernquist, R. Pakmor, S. Genel, P. Torrey, M. Vogelsberger, G. Kauffmann, F. Marinacci, J. Naiman,
21. V. Springel, L. Hernquist, MNRAS **339**, 289 (2003). https://doi.org/10.1046/j.1365-8711.2003.06206.x
22. S. Genel, D. Nelson, A. Pillepich, V. Springel, R. Pakmor, R. Weinberger, L. Hernquist, J. Naiman, M. Vogelsberger, F. Marinacci, P. Torrey, MNRAS **474**, 3976 (2018). https://doi.org/10.1093/mnras/stx3078
23. J.P. Naiman, A. Pillepich, V. Springel, E. Ramirez-Ruiz, P. Torrey, M. Vogelsberger, R. Pakmor, D. Nelson, F. Marinacci, L. Hernquist, R. Weinberger, S. Genel, MNRAS **477**, 1206 (2018). https://doi.org/10.1093/mnras/sty618
24. R. Weinberger, V. Springel, R. Pakmor, D. Nelson, S. Genel, A. Pillepich, M. Vogelsberger, F. Marinacci, J. Naiman, P. Torrey, L. Hernquist, ArXiv e-prints (2018)
25. M. Habouzit, S. Genel, R.S. Somerville, D. Kocevski, M. Hirschmann, A. Dekel, E. Choi, D. Nelson, A. Pillepich, P. Torrey, L. Hernquist, M. Vogelsberger, R. Weinberger, V. Springel, ArXiv e-prints (2018)
26. P. Torrey, M. Vogelsberger, F. Marinacci, R. Pakmor, V. Springel, D. Nelson, J. Naiman, A. Pillepich, S. Genel, R. Weinberger, L. Hernquist, ArXiv e-prints (2017)
27. M.R. Lovell, A. Pillepich, S. Genel, D. Nelson, V. Springel, R. Pakmor, F. Marinacci, R. Weinberger, P. Torrey, M. Vogelsberger, L. Hernquist, ArXiv e-prints (2018)
28. V. Rodriguez-Gomez, G.F. Snyder, J.M. Lotz, D. Nelson, A. Pillepich, V. Springel, S. Genel, R. Weinberger, S. Tacchella, R. Pakmor, P. Torrey, F. Marinacci, M. Vogelsberger, L. Hernquist, D.A. Thilker, MNRAS **483**, 4140 (2019). https://doi.org/10.1093/mnras/sty3345
29. Q. Zhu, D. Xu, M. Gaspari, V. Rodriguez-Gomez, D. Nelson, M. Vogelsberger, P. Torrey, A. Pillepich, J. Zjupa, R. Weinberger, F. Marinacci, R. Pakmor, S. Genel, Y. Li, V. Springel, L. Hernquist, MNRAS **480**, L18 (2018). https://doi.org/10.1093/mnrasl/sly111
30. K. Yun, A. Pillepich, E. Zinger, D. Nelson, M. Donnari, G. Joshi, V. Rodriguez-Gomez, S. Genel, R. Weinberger, M. Vogelsberger, L. Hernquist, MNRAS (2018). https://doi.org/10.1093/mnras/sty3156
31. M. Vogelsberger, F. Marinacci, P. Torrey, S. Genel, V. Springel, R. Weinberger, R. Pakmor, L. Hernquist, J. Naiman, A. Pillepich, D. Nelson, MNRAS **474**, 2073 (2018). https://doi.org/10.1093/mnras/stx2955
32. D.J. Barnes, M. Vogelsberger, R. Kannan, F. Marinacci, R. Weinberger, V. Springel, P. Torrey, A. Pillepich, D. Nelson, R. Pakmor, J. Naiman, L. Hernquist, M. McDonald, MNRAS **481**, 1809 (2018). https://doi.org/10.1093/mnras/sty2078
33. D. Nelson, G. Kauffmann, A. Pillepich, S. Genel, V. Springel, R. Pakmor, L. Hernquist, R. Weinberger, P. Torrey, M. Vogelsberger, F. Marinacci, MNRAS (2018). https://doi.org/10.1093/mnras/sty656
34. A. Gupta, T. Yuan, P. Torrey, M. Vogelsberger, D. Martizzi, K.V.H. Tran, L.J. Kewley, F. Marinacci, D. Nelson, A. Pillepich, L. Hernquist, S. Genel, V. Springel, MNRAS **477**, L35 (2018). https://doi.org/10.1093/mnrasl/sly037
35. Y. Wang, M. Vogelsberger, D. Xu, S. Mao, V. Springel, H. Li, D. Barnes, L. Hernquist, A. Pillepich, F. Marinacci, R. Pakmor, R. Weinberger, P. Torrey, ArXiv e-prints (2018)
36. D. Martizzi, M. Vogelsberger, M.C. Artale, M. Haider, P. Torrey, F. Marinacci, D. Nelson, A. Pillepich, R. Weinberger, L. Hernquist, J. Naiman, V. Springel, ArXiv e-prints (2018)
37. F. Villaescusa-Navarro, S. Genel, E. Castorina, A. Obuljen, D.N. Spergel, L. Hernquist, D. Nelson, I.P. Carucci, A. Pillepich, F. Marinacci, B. Diemer, M. Vogelsberger, R. Weinberger, R. Pakmor, ApJ **866**, 135 (2018). https://doi.org/10.3847/1538-4357/aadba0

38. B. Diemer, A.R.H. Stevens, J.C. Forbes, F. Marinacci, L. Hernquist, C.d.P. Lagos, A. Sternberg, A. Pillepich, D. Nelson, G. Popping, F. Villaescusa-Navarro, P. Torrey, M. Vogelsberger, ApJS **238**, 33 (2018). https://doi.org/10.3847/1538-4365/aae387

39. A.R.H. Stevens, B. Diemer, C.d.P. Lagos, D. Nelson, A. Pillepich, T. Brown, B. Catinella, L. Hernquist, R. Weinberger, M. Vogelsberger, F. Marinacci, ArXiv e-prints (2018)

40. G. Popping, A. Pillepich, R.S. Somerville, R. Decarli, F. Walter, M. Aravena, C. Carilli, P. Cox, D. Nelson, D. Riechers, A. Weiss, L. Boogaard, R. Bouwens, T. Contini, P.C. Cortes, E. da Cunha, E. Daddi, T. Díaz-Santos, B. Diemer, J. González-López, L. Hernquist, R. Ivison, O. Le Fevre, F. Marinacci, H.W. Rix, M. Swinbank, M. Vogelsberger, P. van der Werf, J. Wagg, L.Y.A. Yung, arXiv e-prints (2019)

41. M. Ntampaka, J. ZuHone, D. Eisenstein, D. Nagai, A. Vikhlinin, L. Hernquist, F. Marinacci, D. Nelson, R. Pakmor, A. Pillepich, P. Torrey, M. Vogelsberger, ArXiv e-prints (2018)

42. M. Huertas-Company, V. Rodriguez-Gomez, D. Nelson, A. Pillepich, M. Bernardi, H. Domínguez-Sánchez, S. Genel, R. Pakmor, G.F. Snyder, M. Vogelsberger, arXiv e-prints (2019)

43. R.J.J. Grand, F.A. Gómez, F. Marinacci, R. Pakmor, V. Springel, D.J.R. Campbell, C.S. Frenk, A. Jenkins, S.D.M. White, ArXiv e-prints (2016)

44. P.F. Hopkins, A. Wetzel, D. Keres, C.A. Faucher-Giguere, E. Quataert, M. Boylan-Kolchin, N. Murray, C.C. Hayward, S. Garrison-Kimmel, C. Hummels, R. Feldmann, P. Torrey, X. Ma, D. Angles-Alcazar, K.Y. Su, M. Orr, D. Schmitz, I. Escala, R. Sanderson, M.Y. Grudic, Z. Hafen, J.H. Kim, A. Fitts, J.S. Bullock, C. Wheeler, T.K. Chan, O.D. Elbert, D. Narananan, ArXiv e-prints (2017)

45. A. Pillepich, D. Nelson, V. Springel, R. Pakmor, P. Torrey, R. Weinberger, M. Vogelsberger, F. Marinacci, S. Genel, A. van der Wel, L. Hernquist, arXiv e-prints arXiv:1902.05553 (2019)

46. D. Nelson, A. Pillepich, V. Springel, R. Pakmor, R. Weinberger, S. Genel, P. Torrey, M. Vogelsberger, F. Marinacci, L. Hernquist, arXiv e-prints arXiv:1902.05554 (2019)

47. M. Vogelsberger, D. Nelson, A. Pillepich, X. Shen, F. Marinacci, V. Springel, R. Pakmor, S. Tacchella, R. Weinberger, P. Torrey, L. Hernquist, arXiv e-prints arXiv:1904.07238 (2019)

48. N.M. Förster Schreiber, A. Renzini, C. Mancini, R. Genzel, N. Bouché, G. Cresci, E.K.S. Hicks, S.J. Lilly, Y. Peng, A. Burkert, C.M. Carollo, A. Cimatti, E. Daddi, R.I. Davies, S. Genel, J.D. Kurk, P. Lang, D. Lutz, V. Mainieri, H.J. McCracken, M. Mignoli, T. Naab, P. Oesch, L. Pozzetti, M. Scodeggio, K. Shapiro Griffin, A.E. Shapley, A. Sternberg, S. Tacchella, L.J. Tacconi, S. Wuyts, G. Zamorani, ArXiv e-prints (2018)

49. H.R. Russell, J.S. Sanders, A.C. Fabian, S.A. Baum, M. Donahue, A.C. Edge, B.R. McNamara, C.P. O'Dea, MNRAS **406**, 1721 (2010). https://doi.org/10.1111/j.1365-2966.2010.16822.x

50. H.J. Mo, S. Mao, S.D.M. White, MNRAS **295**, 319 (1998). https://doi.org/10.1046/j.1365-8711.1998.01227.x

51. D. Nelson, V. Springel, A. Pillepich, V. Rodriguez-Gomez, P. Torrey, S. Genel, M. Vogelsberger, R. Pakmor, F. Marinacci, R. Weinberger, L. Kelley, M. Lovell, B. Diemer, L. Hernquist, arXiv e-prints arXiv:1812.05609 (2018)

Long-term Simulations of Magnetized Disks and Jets Around Supermassive Black-hole Binaries in General Relativity

Roman Gold, Luciano Rezzolla, Ludwig Jens Papenfort, Samuel Tootle, Hector Olivares, Elias Most, Ziri Younsi, Yosuke Mizuno, Oliver Porth, and Christian Fromm

Abstract We describe our utilization of HLRS resources in 2019. Chief among this has been important modeling work closely related to the theoretical interpretation of the first-ever image of a black hole. A large library of theoretical model images was built from computationally expensive magnetohydrodynamical simulations of a magnetized fluid around a black hole. This was an international effort and the HLRS contributed to it. Additionally we developed and tested our algorithms further with direct benefits for the funded project. Lastly we describe key development milestones that were achieved that will enable such matter simulations around two orbiting black holes within the same efficient computational framework.

GOETHE
UNIVERSITÄT
FRANKFURT AM MAIN

Allocation:
44149 BBHDISKS

Research field:
General relativity/Astrophysics

R. Gold (✉) · L. Rezzolla · L. J. Papenfort · S. Tootle · H. Olivares · E. Most · Z. Younsi · Y. Mizuno · O. Porth · C. Fromm
Institute for Theoretical Physics (ITP), Goethe University, Max-von-Laue-Str. 1, 60438 Frankfurt am Main, Germany
e-mail: gold@itp.uni-frankfurt.de

© Springer Nature Switzerland AG 2021
W. E. Nagel et al. (eds.), *High Performance Computing in Science and Engineering '19*,
https://doi.org/10.1007/978-3-030-66792-4_2

1 Introduction

The allocation "44149 BBHDISKS" on the system CRAY XC40 (HAZEL HEN) at HLRS has been awarded to our research group in August 2018 for one year. In Oct 2018 all users and software was set up. The scientific research carried out in the group is broad, covering a number of active topics in the field of general relativistic astrophysics and compact objects. Specifically these topics include:

- the general behaviour of magnetized, turbulent gas moving around or falling onto a black hole systems
- the observational appearance of said systems
- the predictions from alternative theories of gravity
- the predictions for systems with two black holes in magnetized gas
- advancing our computational tools to tackle the above systems in the contexts of both ideal and resistive magnetohydrodynamics on non-homogenous and dynamical grids.

Our studies exclusively probe systems governed by coupled, non-linear partial differential equations of Magnetohydrodynamics in strong-field gravity both GR and non-GR. In our research we therefore heavily rely on numerical methods and large scale computational facilities such as the ones provided by HLRS, which are a very important asset to us.

From November 2018 until April 10th 2019 our group was working on the limit in the context of the first 6 papers published by the Event Horizon Telescope collaboration including the first-ever image of a black hole [1–6]. During this very active time several opportunities were also taken to improve and test our algorithms against independent codes [7] and to further the capabilities of our numerical set up in the BHAC code. Specifically, the implementation of staggered magnetic fields and detailed investigations of Cartesian vs spherical grids proved to be very fruitful. As a result the long term stability and accuracy is now under better control, which are key ingredients also for our accreting binary black hole project. We have made use of HLRS resources for these production phase runs, which meant the long queue waiting times did not hamper us that much. For the continued development of the binary black hole elements of the code much shorter turnaround times are required to test new implementations and get quicker feedback. These tests were carried out on our local machines. Once we enter production phase for our binary simulations, we will need the HLRS resources but can cope with longer queue waiting times.

2 Numerical Methods

The simulations were performed using the recently developed code BHAC [8], which is based on the public MPI-AMRVAC Toolkit [9]. BHAC was built to solve the general-relativistic MHD equations in the ideal limit (i.e., for plasmas with infinite conductivity) on spacetimes that are fixed but arbitrary and can therefore model BH spacetimes in any theory of gravity [10, 11]. The code is currently used within the Event Horizon Telescope collaboration, and in particular, the ERC synergy grant "BlackHoleCam" [12] to model mm-VLBI observations of the supermassive BH M87* [1–6] and the candidate at the galactic center Sgr A* to simulate plasma flows onto compact objects [8, 13]. BHAC offers a variety of numerical methods, fully adaptive block based (oct-) tree mesh refinement with a fixed refinement factor of two between successive levels, and is already interfaced with many analysis tools or external codes used by our group. In all the above studies so far the spacetime metric is analytically given and manifestly stationary.

In the current project BBHDISKS we are taking the next steps to spacetime metrics describing two black holes in orbit around each other. The necessary substantial code development is under way. The first non-trivial step is concerned with having valid (i.e. solving Einstein's field equations) initial data. Binary BH initial data are being computed within the conformal thin-sandwich approximation (CTS), which assumes that the binary spacetime has a helical Killing vector (i.e., the binary is stationary in a co-rotating frame) and is conformally flat [14]. To this scope, we are using, improving, and extending the general-purpose spectral solver Kadath [15] to generate highly accurate solutions of the time independent CTS spacetime metric. Furthermore, by implementing an interface to the library functions of the latter directly into BHAC, and exploiting the use of the spectral coefficients for the reconstruction of the metric and extrinsic curvature, we will have a spectrally accurate, fully general-relativistic spacetime at each moment in time with the only approximation being that GW losses are ignored. For the predecoupling models we plan to use 8 AMR levels such that the coarsest resolution is roughly one gravitational radius (i.e., M, where M is mass of the supermassive BH) and on the finest level the coordinate diameter of the black hole is covered by roughly 100 grid cells. During code development we have identified an additional opportunity to increase both efficiency and accuracy of our calculations: We are developing a setup to compute the solution in the rotating frame of the binary. In this frame the metric is explicitly stationary and we do not need to track the binary motion with our grid. Note that BHAC has the ability to do so, but at an unnecessary computational cost. We feel that the additional development time and delayed production phase will ultimately pay off greatly.

When considering the feasibility of the simulations we note that BHAC's parallelism is already very efficient, scaling very well up to 8,000 cores in uniform-grid setups and up to 6,000 cores when an AMR strategy is employed.

We have participated three times in the HLRS code optimisation workshop (recently in November) to achieve optimal performance on the Cray XC40 system and plan to do so again in the course of this allocation when transitioning to the new Hawk system.

3 Job Statistics and Resources Usage

The typical job size for our projects is largely dependent on the type of physical system we are modelling, the microphysical processes involved and of course the level of accuracy needed, which sets constraints on numerical parameters such as the resolution, domain size and order of the method. Nonetheless we can provide some estimates. A typical job can require the use of 50 nodes (1200 cores) for a period of 24 h at low resolution and twice as much (100 nodes/2400 cores) for several days at mid-high resolution. In the hybrid MPI/OpenMP parallelization scheme that we employ, this results in a total of 50 to 100 MPI processes (2 per node), each of them spawning 12 threads each. We normally do not exploit hyperthreading and attempt to bind each thread to a separate physical core.

A grand total of 34865152 core hours were awarded to us, and we have used 112131 node hours (7.72%) at this point into our allocation as of 13/05/2019. The long queue waiting times have prevented us from using more of our allocation at this point, because much shorter turn around times were critical when components of our software were tested. Instead many test simulations have been run on local resources where faster iterations between runs were possible. This will change substantially during the upcoming production phase. At present 7 researchers within our group have access to computing resources at HLRS.

4 Completed Projects and Publications

4.1 Building the Library of GRMHD Images for the EHT

Three-dimensional general relativistic magnetohydrodynamic simulations of accretion onto supermassive black holes were performed at the HazelHen supercomputer using the code BHAC, as part of the efforts to build a library of simulations and images to aid in the interpretation of the first horizon-scale radio images of a supermassive black hole candidate [1–6]. Figure 1 shows a snapshot of one such simulations on the meridional plane. The left panel shows the spatial distribution of particle number density, while the right panel shows the quality factor of the magneto rotational instability $Q_{\mathrm{MRI}}^{(\theta)}$, a measure of how well magnetic turbulence is resolved. It is important that a quality factor larger than ~ 6 is achieved, in order to obtain a realistic angular momentum transport that sustains accretion [16].

The experience gained from these detailed comparisons make us confident that we have a competitive numerical package to take on the next set of challenges.

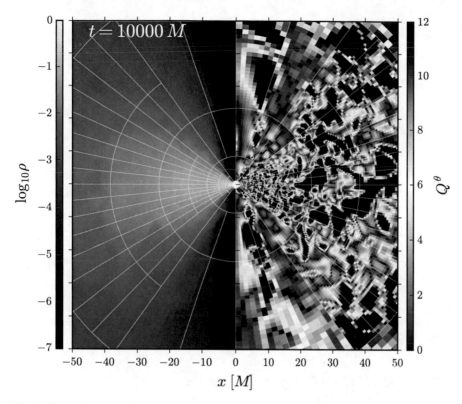

Fig. 1 Central rest-mass density (left panel) of the MHD torus evolved with BHAC and MRI quality factors (right panel) indicating a resolution metric

4.2 Resolution Study and Staggered B-Field Test

In order to investigate the reliability of results obtained, several simulations of the same system were performed employing different algorithms, resolutions and coordinate systems. For instance, Fig. 2 shows a comparison between the mass accretion rate and magnetic flux through the event horizon for simulations of the same accretion problem in spherical polar coordinates (MKS192-fCT and MKS192-UCT) and in Cartesian coordinates (CKS8-UCT), employing two different methods to evolve the magnetic field, namely flux-interpolated constrained transport [17] and upwind constrained transport (UCT) [18, 19], showing consistent results for all of the three simulations.

The improvements in the evolution of the induction equation will serve us well in the future also for the accreting black hole binary project.

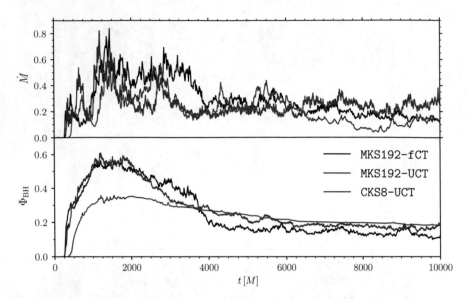

Fig. 2 Rest-mass flux through the horizon, i.e. accretion rates (upper panel) and magnetic flux on the horizon (lower panel) as a function of coordinate time for different implementations of the induction equation and different coordinate systems

4.3 GRMHD Code Comparison Project

In [7] we have published the results of an exhaustive comparison of various indepenent GRMHD codes in an effort to quantify the degree of agrrement among different codes when evolving the same initial conditions. Such comparisons are very challenging for a number of reasons and a significant computational effort. The main difficulty is that the underlying dynamics features turbulent motion that is exponentially sensitive to the initial data and therefore cannot give formally convergent results as each code and each method will give a slightly different turbulent realization that can at best be compared in an ensemble-averaged sense. Despite such complexities a quantitative comparison was achieved and models generated from codes within this comparison show predictive power for actual radio VLBI observations [6]. This is in itself a remarkable achievement by the collaboration. In particular the strong evidence for the Blandford Znajek process was a main conclusion in the theory paper [6] and is derived from the comparison of observational data and these models.

4.4 Initial Data for the Binary Project

A key requisite to evolve accreting black hole binary systems is the generation of valid initial data, i.e. initial conditions for the spacetime metric featuring two black

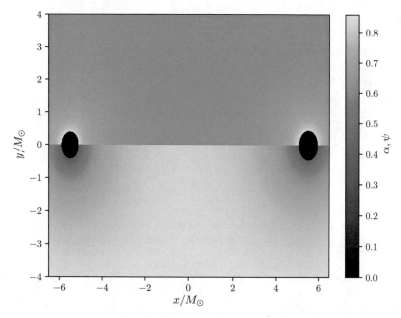

Fig. 3 Lapse metric function (lower panel) and conformal factor (upper panel) (as a rough proxy for spacetime curvature) in the equatorial plane of the irrotational binary black hole system

holes orbiting around each other that satisfies Einstein's field equations of General Relativity. Such initial conditions cannot be found by simply superimposing two known black hole solutions as would be possible in Newtonian gravity. Instead, rather complex non-linear elliptic equations have to be solved at very high accuracy.

We have achieved to compute such initial conditions, see Fig. 3 with a general framework for solving such elliptical problems, called `Kadath` [20]. The initial data can in this form be read-in to BHAC to set the correct spacetime metric featuring the black hole binary system we want to study.

Once our implementation of the comoving frame and the implementation of our matter initial data is finished we will be in the position to evolve the system as intended with BHAC. These constitute the last steps before the final testing and full production phase can be carried out on the HLRS Hazelhen system.

5 Conclusion

The relativistic astrophysics group in Frankfurt has a number of active projects benefiting from the 44149 BBHDISKS allocation on the HazelHen cluster. Despite a number of difficulties and setbacks mainly due to unforeseeable developments within the EHT both in the scientific part of said projects and their computational implementation, much progress and major achievements have been made. The GRMHD image library that was a central component to the seminal EHT results, in particular [3], was partially build by our group using HLRS resources. Furthermore, our lessons learned from the GRMHD code comparison are published [7]. These findings are putting us in an excellent position for the continued development of our code towards accreting binary black holes, where key milestones are being reached in the form of having valid initial data for the spacetime metric. We will proceed with the last steps of code development and enter final testing and production phase on the HLRS HazelHen cluster soon.

References

1. The Event Horizon Telescope Collaboration, First M87 event horizon telescope results. I. The shadow of the supermassive black hole. Astrophys. J. **875**(1):L1 (2019)
2. The Event Horizon Telescope Collaboration, First M87 event horizon telescope results. II. Array and instrumentation. Astrophys. J. **875**(1), L2 (2019)
3. The Event Horizon Telescope Collaboration, First M87 event horizon telescope results. III. Data processing and calibration. Astrophys. J. **875**(1):L3 (2019)
4. The Event Horizon Telescope Collaboration, First M87 event horizon telescope results. IV. Imaging the central supermassive black hole. Astrophys. J. **875**(1):L4 (2019)
5. The Event Horizon Telescope Collaboration, First M87 event horizon telescope results. V. Physical origin of the asymmetric ring. Astrophys. J. **875**(1):L5 (2019)
6. The Event Horizon Telescope Collaboration, First M87 event horizon telescope results. VI. The shadow and mass of the central black hole. Astrophys. J. **875**(1):L6 (2019)
7. O. Porth, K. Chatterjee, R. Narayan, C.F. Gammie, Y. Mizuno, P. Anninos, J.G. Baker, M. Bugli, C.-k. Chan, J. Davelaar, L. Del Zanna, Z.B. Etienne, P.C. Fragile, B.J. Kelly, M. Liska, S. Markoff, J.C. McKinney, B. Mishra, S.C. Noble, H. Olivares, B. Prather, L. Rezzolla, B.R. Ryan, J.M. Stone, N. Tomei, C.J. White, Z. Younsi, The event horizon telescope collaboration. The event horizon general relativistic magnetohydrodynamic code comparison project. Astrophys. J. Supplem. Seri. **243**(2):26 (2019)
8. O. Porth, H. Olivares, Y. Mizuno, Z. Younsi, L. Rezzolla, M. Moscibrodzka, H. Falcke, M. Kramer, The black hole accretion code. Comput. Astrophys. Cosmol. **4**, 1 (2017)
9. O. Porth, C. Xia, T. Hendrix, S. P. Moschou, R. Keppens, MPI-AMRVAC for solar and astrophysics. Astrophys. J., Supp. **214**:4 (2014)
10. R. Konoplya, L. Rezzolla, A. Zhidenko, General parametrization of axisymmetric black holes in metric theories of gravity. Phys. Rev. D **93**(6), 064015 (2016)
11. L. Rezzolla, A. Zhidenko, New parametrization for spherically symmetric black holes in metric theories of gravity. Phys. Rev. D **90**(8), 084009 (2014)
12. C. Goddi, H. Falcke, M. Kramer, L. Rezzolla et al., BlackHoleCam: fundamental physics of the galactic center. Int. J. Modern Phy. D **26**, 1730001–239 (2017)
13. Y. Mizuno, Z. Younsi, C. M. Fromm, O. Porth, M. De Laurentis, H. Olivares, H. Falcke, M. Kramer, L. Rezzolla, The current ability to test theories of gravity with black hole shadows. Nat. Astron. (2018)
14. Harald P. Pfeiffer, James W. York, Extrinsic curvature and the Einstein constraints. Phys. Rev. D **67**, 044022 (2003)
15. Philippe Grandclement, Kadath: a spectral solver for theoretical physics. J. Comput. Phys. **229**, 3334–3357 (2010)
16. T. Sano, S.-I. Inutsuka, N.J. Turner, J.M. Stone, Angular momentum transport by magnetohydrodynamic turbulence in accretion disks: gas pressure dependence of the saturation level of the magnetorotational instability. Astrophys. J. **605**, 321–339 (2004)

17. G. Toth, The div b=0 constraint in shock-capturing magnetohydrodynamics codes. J. Comput. Phys. **161**, 605–652 (2000)
18. L. Del Zanna, O. Zanotti, N. Bucciantini, P. Londrillo, ECHO: a Eulerian conservative high-order scheme for general relativistic magnetohydrodynamics and magnetodynamics. Astron. Astrophys. **473**, 11–30 (2007)
19. P. Londrillo, L. Del Zanna. On the divergence-free condition in godunov-type schemes for ideal magnetohydrodynamics: the upwind constrained transport method. J. Comput. Phy. *195*(1), 17–48 (2004)
20. P. Grandclément, KADATH: a spectral solver for theoretical physics. J. Comput. Phy. **229**, 3334–3357 (2010)

PAMOP2: State-of-the-Art Computations for Atomic, Molecular and Optical Processes

B. M. McLaughlin, C. P. Ballance, R. T. Smyth, M. S. Pindzola, P. C. Stancil, J. F. Babb, and A. Müller

Abstract Our prime computational effort is to support current and future measurements of atomic photoionization cross-sections at various synchrotron radiation facilities, and ion-atom collision experiments, together with plasma, fusion and astrophysical applications. We solve the Schrödinger or Dirac equation using the R-matrix method from first principles. A time-dependent close coupling approach is used to investigate ion-atom excitation and electron-impact ionization. Finally, we present cross-sections and rates for molecule-molecule interactions and radiative association reactions between atoms of interest in astrophysics.

B. M. McLaughlin (✉) · C. P. Ballance · R. T. Smyth
Centre for Theoretical Atomic, Molecular and Optical Physics (CTAMOP), School of
Mathematics & Physics, Queen's University Belfast, Belfast BT7 1NN, UK
e-mail: bmclaughlin899@btinternet.com; bmcl104@gmail.com

C. P. Ballance
e-mail: c.ballance@qub.ac.uk

R. T. Smyth
e-mail: rsmyth41@qub.ac.uk

M. S. Pindzola
Department of Physics, Auburn University, 206 Allison Laboratory, Auburn, AL 36849, USA
e-mail: pindzola@physics.auburn.edu

P. C. Stancil
Department of Physics and Astronomy and the Center for Simulational Physics, University of
Georgia, Athens, GA 30602-2451, USA
e-mail: stancil@physast.uga.edu

J. F. Babb
Institute for Theoretical Atomic, Molecular and Optical Physics, Center for Astrophysics |
Harvard & Smithsonian, 60 Garden Street, Cambridge, MA 02138, USA
e-mail: jbabb@cfa.harvard.edu

A. Müller
Institut für Atom- und Molekülphysik, Justus-Liebig-Universität Giessen, 35392 Giessen,
Germany
e-mail: Alfred.Mueller@iamp.physik.uni-giessen.de

© Springer Nature Switzerland AG 2021   33
W. E. Nagel et al. (eds.), *High Performance Computing in Science and Engineering '19*,
https://doi.org/10.1007/978-3-030-66792-4_3

1 Introduction

Our research efforts continue to focus on the development of computational methods to solve the Schrödinger and Dirac equations for atomic and molecular collision processes. Access to leadership-class computers such as the Cray XC40 at HLRS, for example, enable us to benchmark theoretical solutions against experiments at synchrotron radiation facilities such as the Advanced Light Source (ALS), ASTRID II, BESSY II, SOLEIL and PETRA III and to provide atomic and molecular data for ongoing research in ion-atom collisions, plasma and fusion science. High end computer architectures are trending towards exaflop speeds. For example, the hybrid CPU/GPU Summit architecture currently installed at Oak Ridge National Laboratoty (ORNL), has a computing speed of 0.2 exaflops. The new Frontier CPU/GPU hybrid computer, currently being built, and to be installed at ORNL in 2021 will have a top speed of 1.5 exaflops. Our highly efficient codes are positioned to take account of this trend towards exascale computing.

In order to have direct comparisons with experiment, and to achieve spectroscopic accuracy, semi-relativistic, fully relativistic R-matrix, or R-matrix with pseudo-states (RMPS) computations, an exponentially increasing number of target-coupled states is required and such computations could not be even attempted without access to high performance computing (HPC) resources such as those available at computational centers in Europe (HLRS) and the USA (NERSC, NICS or ORNL).

The motivation for our work is multi-fold; (a) Astrophysical applications [1–4], (b) Magnetically-confined fusion and plasma modeling, (c) Fundamental studies and (d) Support of experiment and satellite observations. For photoabsorption by heavy atomic systems [5–7], little atomic data exists and our work provides results for new frontiers on the applications of the R-matrix semi-relativistic (BREIT- PAULI) or fully relativistic Dirac Atomic R-matrix (DARC) parallel suite of codes. Our highly efficient R-matrix codes have evolved over the past decade and have matured to a stage now that large-scale collision calculations can be carried out in a timely manner for electron or photon impact of heavy systems where inclusion of relativistic effects is essential. These codes are widely applicable for the theoretical interpretation of current experiments being performed at leading synchrotron radiation facilities. Examples of our results are presented below in order to illustrate the predictive nature of the methods employed compared to experiment.

2 Valence Shell Studies

For comparison with the measurements made at the leading light sources such as, ALS, ASTRID, BESSY and PETRAIII, state-of-the-art theoretical methods using highly correlated wavefunctions were applied that include relativistic effects. An efficient parallel version of the DARC [8–10] suite of codes continues to be developed and applied to address electron and photon interactions with atomic systems, providing for hundreds of levels and thousands of scattering channels. These codes are presently running on a variety of parallel high performance computing architectures world wide [11–16]. The input wavefunctions for the DARC codes are determined by the GRASP0 code [17–19].

GRASP0 [17, 18] is used to construct a bound orbital basis set for the residual ion of the system of interest, with the Dirac—Coulomb Hamiltonian H_D, via the relation,

$$H_D = \sum_i -ic\alpha \nabla_i + (\beta - 1)c^2 - \frac{Z}{r_i} + \sum_{i<j} \frac{1}{|r_j - r_i|}, \tag{1}$$

where the electrons are labelled by i and j and the summation is taken over all electrons of the system. The matrices α and β are directly related to the Pauli spin matrices, c is the speed of light and the atomic number is Z. The relativistic orbitals are described with a large component, $\mathcal{P}_{n\ell}$ and small component $\mathcal{Q}_{n\ell}$. The residual ion target wavefunctions are appropriately defined on a radial grid for input into the relativistic R-matrix -codes (DARC) [10].

The total cross-section σ for photoionization (in Megabarns, 1 Mb=10^{-18} cm^2) by unpolarized light is obtained by integrating over all electron-ejection angles $\hat{k}$ and averaging over photon polarization [19] to give

$$\sigma = \frac{8\pi^2 \alpha a_0^2 \omega}{3(2J_i + 1)} \sum_{l,j,J} |\langle \Psi_f^- |M_1|\Psi_i\rangle|^2, \tag{2}$$

where M_1 represents the dipole length operator and the equations are simplified by use of the Wigner-Eckart theorem [19]. The cross-section σ can be cast in both length (L) and velocity gauges (V) through the dipole moment operator M_1, where for the velocity gauge $\omega = 2\pi\nu$, with ν the frequency of the photon is replaced by ω^{-1}, and a_0 is the Bohr radius, α is the fine structure constant, and $g_i = (2J_i + 1)$ is the statistical weighting of the initial state, with Ψ_i, Ψ_f^- being the initial and final state scattering wavefunctions.

For the case of electron-impact excitation (EIE), the collision strength between an initial state i and a final state j can be obtained from the EIE cross-section $\sigma_{j\rightarrow i}^e$,

$$\Omega_{i\rightarrow j} = \frac{g_i k_i^2}{\pi a_0^2} \sigma_{j\rightarrow i}^e. \tag{3}$$

The effective collision strength is defined as a function of the electron temperature in Kelvin (K) as,

$$\Upsilon_{i\rightarrow j} = \int_0^\infty \Omega_{i\rightarrow j} \, \exp(-\epsilon_f/kT) \, d\left(\frac{\epsilon_f}{kT}\right) \tag{4}$$

where ϵ_f is the energy of the electron, where $1{,}000 \leq \mathrm{T} \leq 100{,}000$ in K, and $k=$ 8.617$\times 10^{-5}$ eV/K is Boltzmann's constant.

2.1 Photoionization of Atomic Rubidium Ions: Rb^+ and Rb^{2+}

Rb I optical pumping lines have been observed in AGB stars, and both isotopes of Rb I ([85]Rb I and [87]Rb I) are present based on the presence of s-process elements such as Zr [20]. A major source of discrepancy is the quality of the atomic data used in the modelling [21–25].

The motivation for the current study of this *trans*-Fe element, Rb II, is to provide benchmark PI cross section data for applications in astrophysics. High-resolution measurements of the photoionization cross section of Rb^+ were recently performed at the ALS synchrotron radiation facility in Berkeley, California [26], over the photon energy range 22–46 eV at a resolution of 18 meV FWHM. Many excited Rydberg states have been identified in the energy (wavelength) range 22 eV (564 Å) to 46 eV (270 Å). Large-scale DARC PI cross section calculations when compared with previous synchrotron radiation (SR)and dual laser plasma (DLP) experimental studies [27, 28] indicate excellent agreement.

The GRASP0 code generated the target wave functions employed in the present work. All orbitals were physical up to $n=3$, $4s$, $4p$ and $4d$. We initially used an extended averaged level (EAL) calculation for the $n = 3$ orbitals. The EAL calculations were performed on the lowest 13 fine-structure levels of the residual Rb III ion. In our work we retained all the 456—levels originating from one, two and three–electron promotions from the $n=4$ levels into the orbital space of this ion. All 456 levels arising from the six configurations were included in the DARC close-coupling calculation, namely: $3s^23p^63d^{10}4s^24p^5$, $3s^23p^63d^{10}4s4p^6$, $3s^23p^63d^{10}4s^24p^44d$, $3s^23p^63d^{10}4s4p^54d$, $3s^23p^63d^{10}4s^24p^34d^2$, and $3s^23p^63d^{10}4s^24p^24d^3$.

The scattering calculations were performed for photoionization cross sections using the DARC codes with large-scale configuration interaction (CI) target wave-functions as input to the parallel DARC suite of R-matrix codes. The latest examples of the DARC R-matrix method, implemented in our parallel suite of codes to predict accurate photoionization cross sections are the recent experimental and theoretical studies on the Zn II trans—Fe ion [29] and the Ca II ion [30].

Single photoionization cross sections for Kr-like Rb^+ ions are reported in the energy (wavelength) range 22 eV (564 Å) to 46 eV (270 Å). Theoretical cross section calculations for this *trans*-Fe element are compared with measurements from the ASTRID radiation facility in Aarhus, Denmark and the dual laser plasma (DLP) technique, at respectively 40 meV and 35 meV FWHM energy resolution. In the photon energy region 22–32 eV the spectrum is dominated by excitation autoionizing resonance states. Above 32 eV the cross section exhibit classic Fano window resonances features, which are analysed and discussed. Large-scale theoretical photoionization cross-section calculations, performed using a Dirac Coulomb R-matrix approximation are bench marked against these high resolution experimental results. Comparison of the theoretical work with the experimental studies allowed the identification of resonance features and their parameters in the spectra in addition to contributions from excited metastable states of the Rb^+ ions.

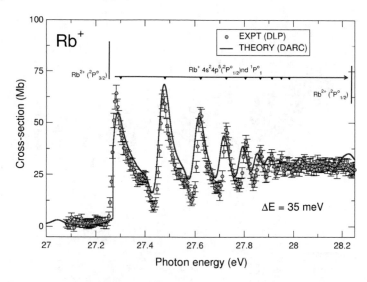

Fig. 1 (Colour online) Single photoionization cross section of Rb$^+$ in the photon energy region 27–28.25 eV. Experimental measurements (solid cyan circles) were obtained using the dual laser plasma technique (DLP) taken at a photon energy resolution of 35 meV FWHM compared with results obtained from the 456-level DARC calculations. The Rb$^+$ $4s^2 4p^5 (^2P^o_{1/2})nd$ $^1P^o_1$ (inverted solid black triangles) are identified in the spectra. The DARC photoionization cross sections (solid red line) have been convoluted with a Gaussian distribution having a 35 meV FWHM profile and an appropriate admixture used (see text for details) for the ground state and the metastable states

Theoretical results from large-scale DARC photoionization cross section calculations were used to interpret the experimental data from the DLP and ASTRID facilities. From our DARC results, a resonance analysis of both the Feshbach and the Fano window resonances illustrate excellent agreement with the available experimental data.

Prior dual laser plasma (DLP) measurements at Dublin City University (DCU) made at 35 meV and synchrotron measurements performed on ASTRID at 40 meV [27] are compared with our DARC calculations. These comparisons are illustrated in Fig. 1 and Fig. 2 respectively. Figure 1 shows the comparison of our present DARC calculations with the DLP measurements taken at 35 meV FWHM in the photon energy region 27–28.6 where excellent agreement between theory and experiment is seen.

In Fig. 2, a comparison with our present DARC PI calculations and the measurements from the ASTRID radiation facility in Aarhus (taken at a photon resolution of 40 meV) is made in the photon region where the prominent Fano window resonances are located. Here again excellent agreement between theory and experiment is observed. The good agreement with the available experimental measurements provides further confidence in our theoretical cross section data for astrophysical applications. Further details can be found in the recent publication by McLaughlin and Babb [7].

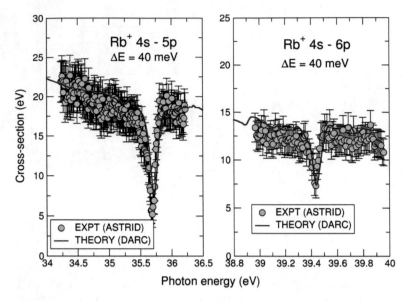

Fig. 2 (Colour online) Single photoionization cross section of Rb^+ in the photon energy regions of the two major Fano window resonances, respectively, 34–36.5 eV and 38.8–40 eV. Experimental measurements (solid cyan circles) were obtained from the ASTRID radiation facility, in Aarhus, Denmark, at a photon energy resolution of 40 meV FWHM, compared with results obtained from the 456-level (DARC) approximation. The DARC photoionization cross sections (solid red line) have been convoluted with a Gaussian distribution having a 40 meV FWHM profile and an appropriate admixture used (see text for details) for the ground and the metastable states

In the case of the Rb^{2+} ion, absolute photoionization cross sections for the Rb^{2+} ion were recently measured at high resolution over the energy range 37.31–44.08 eV, with autoionizing Rydberg resonance series identified, using the photon-ion merged-beam setup at the Advanced Light Source (ALS) in Berkeley. Detailed photon-energy scans taken at 13.5 ± 2.5 meV bandwidth illustrated multiple Rydberg resonance series associated with the ground and metastable states. Here we present theoretical cross section results obtained using the Dirac-Coulomb R-matrix approximation with a detailed analysis of the resonances. The calculations were performed for the $3d^{10}4s^2 4p^5$ $^2P^o_J$, $J = \frac{3}{2}$ ground state and the corresponding $3d^{10}4s^2 4p^5$ $^2P^o_J$, $J = \frac{1}{2}$ metastable level. Results from the large-scale calculations are benchmarked against the ALS high-resolution measurements and reproduce the dominant resonance features in the spectra, providing confidence in the theoretical work for astrophysical applications.

The GRASP0 code was used to generate the residual Rb^{3+} target wave functions employed in our collision work. All orbitals were physical up to $n = 3$, and in addition the $4s$, $4p$, $4d$, $5s$, $5p$ and $5d$ orbitals were included. We began by performing an extended averaged level (EAL) calculation for the $n = 4$ orbitals and extended these calculations with the addition of the $n = 5$ orbitals. All EAL calculations were performed on the lowest 18 fine-structure levels of the residual Rb^{3+} ion in order to generate target wave functions for our photoionization studies. In our work we retained all the 687—levels originating from one, and two electron promotions from the $n = 4$ levels into the orbital space of this ion. All 687 levels arising from the sixteen configurations were included in the DARC close-coupling calculation, namely, the one-electron promotions, $3d^{10}4s^24p^4$, $3d^{10}4s4p^5$, $3d^{10}4s4p^44d$, $3d^{10}4s4p^45s$, $3d^{10}4s4p^45p$, $3d^{10}4s4p^45d$, $3d^{10}4s^24p^34d$, $3d^{10}4s^24p^35s$, $3d^{10}4s^24p^35p$ and $3d^{10}4s^24p^35d$. In addition we include the two-electron promotions, $3d^{10}4s^24p^24d^2$, $3d^{10}4s^24p^25s^2$ $3d^{10}4s^24p^25p^2$, $3d^{10}4s^24p^25d^2$, $3d^{10}4p^44d^2$, and $3d^{10}4p^45s^2$.

Photoionization cross section calculations were performed for the Rb^{2+} ion in the $3d^{10}4s^24p^5\ ^2P^o_{3/2}$ ground state and in the $3d^{10}4s^24p^5\ ^2P^o_{1/2}$ metastable level using the DARC codes with the above Rb^{3+} residual ion target wave functions.

We carried out large-scale PI cross section calculations using the parallel version of the DARC codes. Our statistically averaged cross sections for the ground and metastable states show excellent agreement with the recent measurements from the ALS [31] radiation facility from thresholds to about 41.4 eV as seen from Fig. 3. The present theoretical cross sections in the photon region 41.4–44 eV are on average about 20% higher than experiment, but they are consistent with previous Breit-Pauli calculations performed on this complex [31]. An analysis of the Auger Rydberg resonances series using the eigenphase sum derivative approach show excellent agreement with previous results within the experimental resolution [31, 32]. Comparison between theory and experiment for resonance energies and quantum defects in provide further confidence in our theoretical data for applications in astrophysics. The present theoretical work may be incorporated into astrophysical modelling codes like Cloudy [33, 34], Xstar [35] and AtomDB [36] used to numerically simulate the thermal and ionization structure of ionized astrophysical nebulae. Further details can be found in the recent publication by McLaughlin & Babb [37].

3 Electron Impact Exciation of Fe II

Absorption and emission lines of the iron-peak species Fe II are prominent in the infrared, optical and ultraviolet spectra of a myriad of astrophysical sources, requiring extensive and highly reliable sets of atomic structure and collisional data for an accurate quantitative analysis. However, comparisons among existing calculations reveal large discrepancies in the effective collision strengths, often up to factors of three, highlighting the need for further steps towards new converged calculations. Here we report a new 20 configuration, 6069 level atomic structure model, calculated using the multi-configurational Dirac-Fock method. Collision strengths and effective collision strengths are presented, for a wide range of temperatures of astrophysical relevance, from substantial 262 level and 716 level Dirac R-matrix calculations, plus a 716 level Breit-Pauli R-matrix calculation. Convergence of the scattering calculations is discussed, and results are critically compared with existing data in the literature, providing us with error estimates for our data. As a consequence, we assign an uncertainty of $\pm15\%$ to relevant forbidden and allowed transitions encompassed

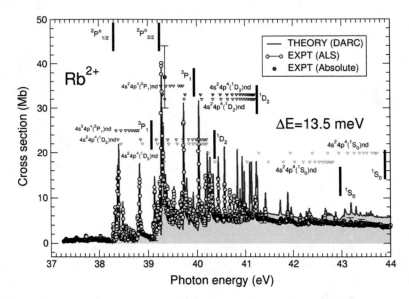

Fig. 3 (Colour online) Single photoionization cross section of Rb^{2+} in the energy region 37–44 eV. Experimental measurements (solid yellow circles) on an absolute scale with the absolute cross section measurements (solid magenta circles) made at selected energies used to normalized the photo-ion yield spectra. See the work of Macaluso and co-workers for further details [31, 32]. Cross section results from the merged beam measurements at the ALS radiation facility, obtained at a photon energy resolution of 13.5 ± 2.5 meV FWHM, are compared with those from an 687-level Dirac R-matrix (DARC) approximation. The DARC photoionization cross sections (solid blue line) were convolved with a Gaussian distribution having a 13.5 meV FWHM profile and statistically averaged for the ground state $3d^{10}4s^2 4p^5 \, {}^2P^o_{3/2}$ and $3d^{10}4s^2 4p^5 \, {}^2P^o_{1/2}$ metastable state (see text for details). The inverted triangles are the energy positions of the various $4s^2 4p^4 nd$ Rydberg resonance series found in the Rb^{2+} photoionization spectrum, originating from the ${}^2P^o_{1/2}$ and ${}^2P^o_{3/2}$ initial states, for the energy region investigated. The solid black vertical lines indicate the photon energies required to ionize the Rb^{2+} ${}^2P^o_{1/2}$ and ${}^2P^o_{3/2}$ states and the limits for the each of the $4s^2 4p^4 nd$ Rydberg series converging to ${}^3P_1$, ${}^1D_2$, or ${}^1S_0$ states of the residual Rb^{3+} ion

within a 50 level subset of the 716 level Dirac R-matrix dataset. To illustrate the implications of our new datasets for the analysis of astronomical observations of Fe II, they are incorporated into the Cloudy modelling code, sample Fe II spectra are generated and compared.

Our initial MCDF calculation included the $3d^7$; $3d^6[\,4s, 4p, 4d, 5s, 5p\,]$; $3d^5[\,4s^2, 4p^2, 4d^2, 4s4p, 4s4d, 5s^2, 5p^2\,]$; $3p^4[\,3d^9, 4d^9\,]$; $3p^6 3d^1 4d^6$; $3p^6 3d^2 4d^5$; and $3p^6 3d^3 4d^4$ configurations, and all orbitals were variationally determined by employing the extended average level (EAL) method, once for Co III and then once for Fe II. We note that the initial Co III calculation was necessary to determine a suitable starting point for the Dirac-Hartree-Fock iterative process for the more complex Fe II case. One final EAL calculation with the $3d^5 4s4d$; $3p^4 4d^9$; $3p^6 3d^1 4d^6$; $3p^6 3d^2 4d^5$; and $3p^6 3d^3 4d^4$ configurations removed, the $3p^5 3d^7[\,4s, 4p, 5s\,]$; $3p^5 3d^6[4s^2, 4s5s\,]$; $3p^4 3d^8 4s$; and $3p^4 3d^7 4s^2$ configurations included, and also holding the $3d$ and

$4d$ orbitals fixed, yielded our final Fe II atomic structure. The full 20 configuration, 6069 level model was taken through to the Dirac R-matrix calculations. However, we chose to retain only the first 262 and 716 levels of our target structure in the close-coupling expansions of the collision wavefunctions. When compared with the NIST [40] tabulated values our 262 level target model has an overall average error of 9.3%. Even parity levels have an average error of 10.9%, ranging from 0.8% to 21.8% with four individual outliers ranging from 41.0% to 46.2% due to the $3d^6(^5D)4s$ ^4D_J ($J = \frac{7}{2}, \frac{5}{2}, \frac{3}{2}, \frac{1}{2}$) levels. The odd parity levels are well represented, with an average error of 7.9%, ranging from 0.8% to 18.7%. Similarly, our 716 level target model has an overall average error of 11.3%, with averages of 13.8% and 10.2% for the even parity and odd parity levels, respectively. Main sources of error are due to a small number of some highly excited levels in the target description.

We note that our current 20 configuration GRASP0 target model considerably improves upon the GRASP0 model of Bautista et al. [41] which bears an average error of 20%, this compared to the average errors of 9.3% and 11.3% for our 262 and 716 level models, respectively. In particular, we note that these existing GRASP0 calculations fail to include a fully variationally determined 4d orbital, which is in contrast to our present 20 configuration calculation. The implications of such an omission has been highlighted in Bautista et al.[41], showing that without a properly calculated 4d orbital to account for the relaxation effects of the 3d orbital, poor agreement with observed spectra will be seen. Additionally, statistically averaging our fine-structure resolved atomic structure model reveals (when the first 16 terms are considered) an average error of 15%. Comparing with the calculated term energies of Bautista et al. [41] we see good agreement with the results of their 16 configuration "NewTFDAc" model, showing only a 6% overall difference with the present statistically averaged GRASP0 results. Good agreement is also seen with the 26 configuration LS-coupled CIV3 model of Ramsbottom et al. [42], exhibiting an overall difference of 9%.

Collision strengths for the $3d^6(^5D)4s$ $^6D_{5/2}$ - $3d^7$ $^4F_{9/2}$ forbidden transition from the ground state complex are presented in Fig. 4. Overall we see very good agreement among the present BP716, DARC262 and DARC716 results. However, across portions of the shown energy range the dense resonances structures from BP716 are smaller than those from both DARC262 and DARC716. There are also little discrepancies with two out of the three collision strengths of Nussbaumer & Storey [39], while BP716, DARC262 and DARC716 show very good overall agreement with the results of Ramsbottom et al. [38]. However, since the target levels of Ramsbottom et al. [38] were not shifted to experimental values, the resonances near the excitation threshold lie slightly further up the energy range. Furthermore, given the high level of agreement between the 262 and 716 level calculations, as seen from Fig. 4, we can deduce that our close-coupling expansion has certainly converged for low-lying forbidden transitions. It is clear that we have also reached convergence in terms of the employed energy mesh sizes with no discernible differences in resolution between a mesh with 16000 energy points and one with 22300.

The corresponding effective collision strengths are presented in Fig. 5. There is very good agreement between the current DARC262 and DARC716 values and those of Ramsbottom et al. [38]. However, at lower temperatures BP716 is, at most,

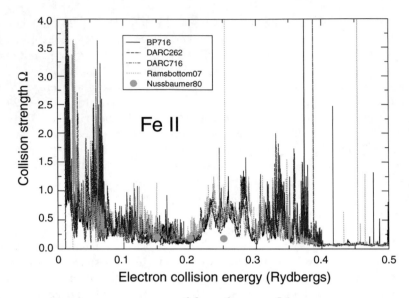

Fig. 4 Fe II collision strengths for the $3d^6(^5D)4s\ ^6D_{5/2} - 3d^7\ ^4F_{9/2}$ forbidden transition. The solid black line is the result from BP716, the dashed red line is from DARC262, the blue dash-dot line is from DARC716, the dotted green line is from Ramsbottom et al. [38], and the orange circles are from Nussbamer & Storey [39]

a factor of 1.3 times smaller than these three sets of effective collision strengths, with agreement becoming much better as the electron temperature increases. This discrepancy at the lower temperatures is most likely due to the differences in the heights of the dense resonance structures near the excitation threshold as mentioned previously. At the highest temperatures there is excellent agreement. Across the shown temperature range we have an average difference of 5% between DARC262 and DARC716, and an average of 12% between DARC716 and BP716. Similar to before, comparisons of BP716, DARC262 and DARC716 with the results of Zhang & Pradhan [44] show that there is good agreement in terms of magnitude and shape from 5000 K onwards. At the lowest temperatures there is very little agreement with the results of Tayal & Zatsarinny [43], owing to the sensitivity of the effective collision strengths on the complex resonance structures within the corresponding collision strengths. However, agreement in shape and magnitude improves at the temperature increases, with their results clearly converging to the same point as the present DARC262, DARC716 and BP716. Again, the results of Keenan et al. [45], Berrington et al. [46] and Bautista et al. [41] are much too low. As discussed above, this may be due to the omission of odd parity target states in the close-coupling expansions.

In Fig. 6 we present synthetic Fe II spectra across a portion of the UV and visible spectral regions from 100 to 500 nm. Reasonable agreement is found among all three datasets at wavelengths greater than about 200 nm, with some line peaks absent in

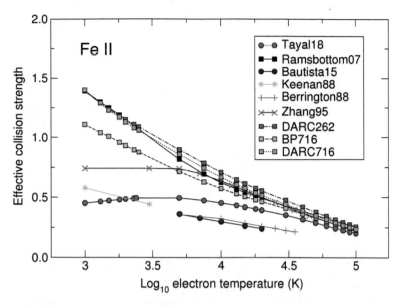

Fig. 5 FeII effective collision strengths for the $3d^6(^5D)4s\ ^6D_{5/2} - 3d^7\ ^4F_{9/2}$ forbidden transition. Cyan squares with the dashed line are the current BP716 results, green squares with the dash-dot line are from DARC262 and orange squares with the dotted line are from DARC716. Patterned circles are results of Tayal & Zatsarinny [43]; solid red circles are from various models of Bautista et al. [41]; black squares with the solid line are from Ramsbottom et al. [38]; green crosses are from Zhang & Pradhan [44]; orange stars are from Keenan et al. [45]; and the pink plus signs are from Berrington et al. [46]

the DARC262 plot due to missing transitions. Comparisons between DARC716 and C371 show better agreement, but with discrepancies in peaks and DARC716 providing additional lines around 300 nm. It is evident that the synthetic spectrum obtained from using the DARC262 dataset is very sparse across the lower UV region from 100 to 150 nm. This wavelength range only contains 1018 lines, with approximately 6% of these having radiative rates larger than 10^5 s^{-1}. Such few lines from DARC262 (and from existing models of comparable size) will obviously significantly limit the modelling of objects which show prominent Fe II emission features in the UV spectral region. Similarly, employing the existing C371 dataset, we have a total of 4167 lines across the same 100 to 150 nm range. However, with DARC716 we have a total of 23791 lines across the same 50 nm interval, with more than 20% of these having radiative transition rates larger than 10^5 s^{-1}. Of these, 308 are particularly strong, having radiative transition rates larger than 10^8 s^{-1}. In the future, we plan to undertake more detailed Cloudy simulations of Fe II spectra using our new DARC716 dataset, and assess and quantify its implication for the analysis of astronomical sources.

The present work investigated the electron-impact excitation of the iron-peak species Fe II. Discrepancies among existing effective collision strengths from earlier calculations in the literature have been highlighted, and addressed using three large-

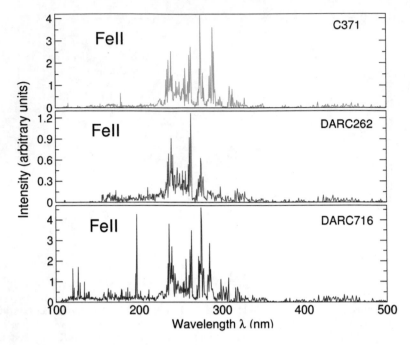

Fig. 6 Synthetic Fe II spectra, calculated with the Cloudy code [33, 34], appropriate to AGN conditions. Top panel is the model calculated using the existing 371 level dataset of Verner et al. [47]; middle panel is that using the DARC262 dataset; and the bottom panel is that using the DARC716 dataset

scale R-matrix scattering models, the target descriptions of which are sufficiently extensive enough to reach energies required for modelling astrophysical sources displaying strong UV emission and/or absorption lines of Fe II. The calculations presented here provide the most extensive set of high quality atomic data currently available for Fe II, and will be useful for future astrophysical modelling applications. Further details can be found in the recent publication of Smyth and co-workers [48].

4   Time Dependent Close-Coupling

A time-dependent close-coupling method is developed and used to calculate proton impact excitation cross sections for atoms. Cross sections for $3s \rightarrow n\ell$ excitations of the Na atom are calculated for proton impact energies ranging from 1.0 keV to 100.0 keV. The time-dependent close-coupling cross sections are found to be in agreement for some energies with recommended cross sections from overall fits to theoretical and experimental atomic data.

$$i \frac{\partial P_{\ell m}(r, t)}{\partial t} = T_\ell(r) P_{\ell m}(r, t)$$

$$+ \sum_{\ell' m'} W_{\ell m, \ell' m'}(r, R(t)) P_{\ell' m'}(r, t) . \tag{5}$$

The kinetic, nuclear, and atomic core operator is given by:

$$T_\ell(r) = -\frac{1}{2} \frac{\partial^2}{\partial r^2} + \frac{\ell(\ell+1)}{2r^2} - \frac{Z_t}{r} + V_\ell^{HX}(r) , \tag{6}$$

where Z_t is the target nuclear charge and $V_\ell^{HX}(r)$ is a Hartree-local exchange potential for the atomic core. $W_{\ell m, \ell' m'}(r, R(t))$ is the electron-projectile coupling operator and $P_{\ell m}(r, t)$ the associated Legendre function.

4.1 Proton Impact on Sodium Atoms

For excitation of the $3s$ subshell, the Na$^+$ ($1s^2 2s^2 2p^6$) ground configuration is calculated in the Hartree-Fock approximation [51]. Using a 360 point radial coordinate mesh with $\Delta r = 0.20$, $T_\ell(r)$ of Eq.(6) is diagonalized. For each ℓ a local exchange potential parameter is adjusted so that the single particle energies for the active outer subshells match the experimental binding energies from the NIST tabulated values [40]. For the $3s$ orbital with a binding energy of -5.13 eV and the $3p$ orbital with a binding energy of -3.04 eV, pseudo-orbitals are then generated by smoothly removing the inner nodes of the wavefunctions. Using these 2 orbitals, the radial Schrodinger equation is then inverted to obtain ℓ dependent pseudo-potentials for $\ell = 0, 1$. The use of pseudo-potentials prevents unphysical excitation of the $1s$, $2s$, and $2p$ filled subshells during propagation of Eq. (5).

The TDCC calculations were carried out for $3s \rightarrow 3p$, $3d$ and $3s \rightarrow 4s, 4p, 4d, 4f$ excitations of the Na atom over an incident energy range from 1–100 keV. The TDCC cross sections were found to be in agreement for some energies with the recommended cross sections obtained by overall fits of Igenbergs et al. [52] to theoretical and experimental atomic data.

The time-dependent close-coupling (TDCC) cross sections for the $3s \rightarrow 3p$ excitation are compared with the atomic-orbital close-coupling (AOCC) calculations of Shingal and Bransden [49] and experimental cross sections of Allen et al. [50] in Fig. 7. Except at 2.0 keV and 3.0 keV, the TDCC and AOCC cross sections are in good agreement, while the experimental cross sections are slightly above the theoretical cross sections over most of the energy range. The experimental cross sections of Jitschin et al. [53] are higher than the experimental cross sections of Allen et al.[50], but are not shown since their uncertainty is unknown.

The $3s \rightarrow 4s$, $3s \rightarrow 4p$, $3s \rightarrow 4d$, and $3s \rightarrow 4f$ excitation cross sections are presented in Fig. 8. For comparison purposes, we include the recommended cross

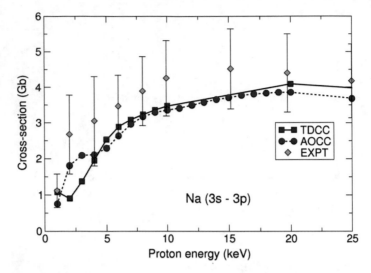

Fig. 7 Proton-impact excitation of Na ($3s \to 3p$) over the energy range 1–25 keV. Solid line with squares (red): TDCC cross sections, dashed lines with circles (blue): AOCC cross sections of Shingal and Bransden [49], diamonds (green): experimental cross sections of Allen et al. [50] (1.0 Gb = 1.0×10^{-15} cm^2)

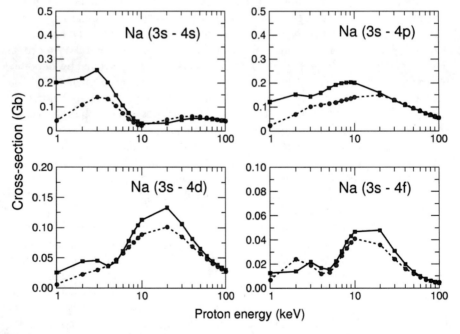

Fig. 8 Proton-impact excitation of Na over the energy range 1–100 keV Solid lines with squares (red): TDCC cross sections, dashed lines with circles (blue): recommended cross sections [52]. Top-left graph: $3s \to 4s$, top-right graph: $3s \to 4p$, bottom-left graph: $3s \to 4d$, bottom-right graph: $3s \to 4f$ (1.0 Gb = 1.0×10^{-15} cm^2)

sections of Igenbergs et al. [52]. The TDCC and recommended cross sections are in agreement for some energies for all the excitations over a wide energy range.

In the future we plan to calculate proton-impact excitation cross-sections for many other atoms and low-charged atomic ions using the TDCC method. The cross-sections will be used to help model a variety of plasmas found in solar flares and the interstellar medium. Further details can be found in the recent publication by Pindzola and Loch [54].

5 Diatom—Diatom Collisions: CS-H$_2$

We report a six-dimensional (6D) potential energy surface (PES) for the CS-H$_2$ system computed using high-level electronic structure theory and fitted using a hybrid invariant polynomial method. Full-dimensional quantum close-coupling scattering calculations have been carried out using this potential for rotational and, for the first time, vibrational quenching transitions of CS induced by H$_2$. State-to-state cross sections and rate coefficients for rotational transitions in CS from rotational levels j_1=0-5 in the ground vibrational state are compared with previous theoretical results obtained using a rigid-rotor approximation. For vibrational quenching, state-to-state and total cross sections and rate coefficients were calculated for the vibrational transitions in CS($v_1 = 1$, j_1)+H$_2$($v_2 = 0$, j_2) → CS($v_1' = 0$, j_1')+H$_2$($v_2' = 0$, j_2') collisions, for j_1=0-5. Cross sections for collision energies in the range 1 to 3000 cm^{-1} and rate coefficients in the temperature range of 5 to 600 K are obtained for both para-H$_2$ (j_2=0) and ortho-H$_2$ (j_2=1) collision partners. Application of the computed results in astrophysics is also discussed.

We constructed a 6-dimensional (Fig. 9) potential energy surface for CS-H$_2$ based on high-level electronic structure calculations. State-to-state cross sections for rotational and rovibrational transitions in CS in collisions with H$_2$ are reported using

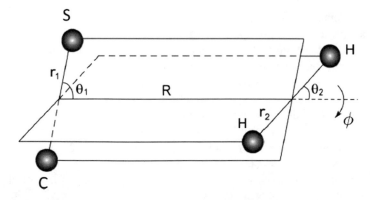

Fig. 9 The six-dimensional Jacobi coordinates for the CS-H$_2$ system

full-dimensional quantum scattering calculations and an analytical fit of the potential surface. Cross sections and rate coefficients for pure rotational transitions from the present study are found to be in good agreement with the rigid-rotor approximation calculations of Denis-Alpizar et al. [55, 57] using a 4-dimensional potential surface. The vibrational quenching cross sections and rate coefficients have been reported for the first time. In future work, we plan to extend the current calculations to higher rotational and vibrational states of CS and include the effect of vibrational excitation of the H_2 molecule.

The ab initio electronic structure computations of potential energies were performed using the explicitly correlated coupled-cluster (CCSD(T)-F12b) method [58, 59]. All the calculations employed aug-cc-pV5Z (for H and C atoms) [60] and aug-cc-pV(5+d)Z (for S atom) orbital basis set [61], and the corresponding MP2FIT (for H and C atoms) and aug-cc-pwCV5Z (for S atom) auxiliary bases [62, 63] for density fitting. The aug-cc-pV6Z-RI auxiliary bases (without k functions) [64] were used for the resolutions of the identify and density-fitted Fock matrices for all orbital bases. No scaled triples correction was used in the CCSD(T)-F12 calculation. The interaction PES was corrected for basis set superposition error (BSSE) [65] using the counter-poise (CP) [66] method. Benchmark calculations at this CCSD(T)-F12 level were carried out on selected molecular configurations and results were compared with those from the conventional CCSD(T) method using aug-cc-pV5Z. The CP corrected interaction energy agrees closely with those derived from CCSD(T)/aug-cc-pV5Z.

The full-dimensional CS-H_2 interaction potential, referred to as VCSH2 (see Fig. 10), is a hybrid one combining a fit to the full ab initio data set (denoted V_I and a fit to the long-range data (denoted V_{II}) .

The full-dimensional quantum close-coupling scattering calculations were performed using the TwoBC code [67]. In the full-dimensional rovibrational scattering calculations with the TwoBC code [67] the log-derivative matrix propagation method of Johnson [68] was employed to propagate the close-coupling equations. Sufficient number of partial waves were included to ensure the convergence of the cross sections.

In Fig. 11 the state-to-state quenching cross sections of CS from initial state j_1=5 from our calculations are compared with the results of Denis-Alpizar et al. [55]. Results are presented for both para-H_2 (j_2=0) and ortho-H_2 ($j_2 = 1$) for final states j_1'=0, 1, 2, 3, and 4. Figure 11a shows that for para-H_2, for collision energies below ~ 30 cm^{-1}, all state-to-state cross sections display resonances due to the decay of quasibound states supported by the van der Waals well of the PES. At collision energies below 200 cm^{-1}, the cross sections generally increase with increasing j_1' with the smallest cross section corresponding to the $j_1 = 5 \rightarrow 0$ transition, i.e., the largest $|\Delta j_1 = j_1' - j_1|$. While for collision energies above 200 cm^{-1} the state-to-state quenching cross sections are dominated by $\Delta j_1 = -2$ transitions. Above ~ 1000 cm^{-1}, the cross sections for $j_1 = 5 \rightarrow 1$ becomes larger than $j_1 = 5 \rightarrow 2$. The agreement between our results and the results of Denis-Alpizar et al. [55] is reasonable, though differences are seen in the low collision energy region due to the presence of scattering resonances. These differences tend to disappear as the collision energy is increased.

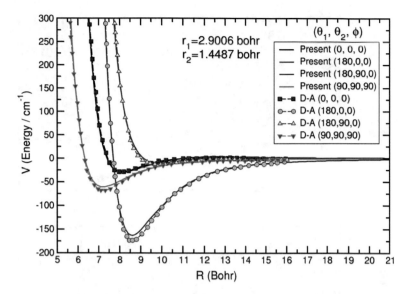

Fig. 10 The R dependence of the CS-H$_2$ interaction potential, VCSH2 for $(\theta_1, \theta_2, \phi)$= (0°, 0°, 0°), (180°, 90°, 0°), (90°, 90°, 90°), and (180°, 0°, 0°). The bond lengths of CS (r_1) and H$_2$ (r_2) are fixed at equilibrium value and vibrationally averaged value in the rovibrational ground state, respectively. Symbols are for the PES of Denis-Alpizar et al. [55] (D-A). Further details can be found in the recent work of Yang et al. [56]

In Fig. 12 we present total rate coefficients for the vibrational quenching of CS compared to the same transitions for CO [69] and SiO [56]. Previously we have provided a comparison of CS-H$_2$ rate coefficients with that of CO-H$_2$ [69]) and for SiO-H$_2$ [56]. It was shown that for the same transitions, the total quenching rate coefficients of CO-H$_2$ are typically $\sim$2-3 orders of magnitude smaller than that of CS-H$_2$ and SiO-H$_2$. The large magnitude of SiO-H$_2$ rate coefficients are likely due to the high anisotropy of the SiO-H$_2$ PES. Interestingly, at high temperatures the CS-H$_2$ rate coefficients merge with those of SiO-H$_2$. Furthermore, generally the rate coefficients increase with increasing well depth of the interaction PES, the SiO-H$_2$ rate coefficients are largest with a well depth of 293.2 cm^{-1} and CO-H$_2$ rate coefficients are smallest with a well depth of 93.1 cm^{-1}.

Vibrationally excited molecules, which are excited by collision or infrared radiation, can be used to probe extreme physical conditions with high gas densities and temperatures. The first observation of vibrationally excited CS in the circumstellar shell of IRC + 10216 was reported by Turner [70] through transitions $j_1 = 2 \rightarrow 1$ and $5 \rightarrow 4$ in $v_1 = 1$. In 2000, Highberger et al. [71] reobserved these lines and also detected new transitions of $j_1 = 3 - 2, 6 - 5$, and $7 - 6$ of vibrationally excited CS ($v = 1$) toward IRC+10216. Using Submillimeter Array, Patel et al. [72] detected the CS $v_1 = 2$, $j_1 = 7 - 6$ transition from the inner envelope of IRC+10216. Finally, vibrational absorption lines for CS have been detected for its fundamental band near 8

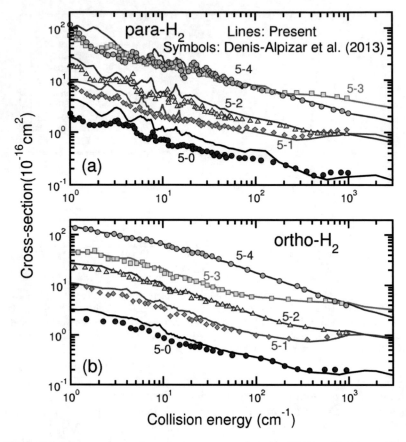

Fig. 11 Rotational state-to-state de-excitation cross sections for $CS(j_1) + H_2(j_2) \rightarrow CS(j_1') + H_2(j_2')$, $j_1 = 5$, $j_1' < j_1$. Lines are for the present results and symbols are for the results of Denis-Alpizar et al. [55]. **a** CS in collision with para-H_2, $j_2 = j_2' = 0$; **b** CS in collision with ortho-H_2, $j_2 = j_2' = 1$

μm in IRC +10216 [73]. These observations used the Texas Echelon-cross-Eschelle Spectrograph (TEXES) on the 3 m Infrared Telescope facility. Further details are available in the recent publication Yang and co-workers [56].

6   Dicarbon Formation by Radiative Association

Radiative association cross sections and rates have been computed, using a quantum approach, for the formation of C_2 molecules during the collision of ground state $C(^3P)$ atoms. The results are compared and constrasted with previous results obtained from a semi-classical approximation. New *ab initio* potential curves and

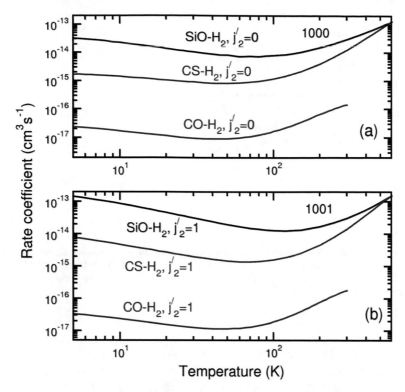

Fig. 12 Total rate coefficients for the vibrational quenching of CS compared to the same transitions for CO [69] and SiO [56]; **a** from (1000) to $v'_1=0$ + para-H$_2$($v'_2=0$, $j'_2=0$). **b** from (1001) to $v'_1=0$ + ortho-H$_2$($v'_2=0$, $j'_2=1$). Further details can be found in the recent work of Yang et al. [56]

transition dipole moment functions have been obtained for the present work using the multi-reference configuration interaction approach with the Davidson correction (MRCI+Q) and aug-cc-pCV5Z basis sets. Applications of the current computations to various astrophysical environments are briefly discussed focusing on these rates.

The radiative association (RA) rate was originally estimated to have a rate coefficient $k_{C_2} \approx 1 \times 10^{-17}$ cm^3/s [75, 76] for theoretical models of interstellar clouds and subsequent semi-classical calculations [77] found comparable values with a weak temperature dependence increasing from 3.07×10^{-18} cm^3/s at 300 K to 1.65×10^{-17} cm^3/s at 14, 700 K. In recent studies using a quantum approach on systems such as SiP [78], SiO [79, 80] and CS [81, 82], it was found that the semi-classical calculations [77, 83] underestimated the cross sections and rates, particularly at low temperatures.

Figure 13 illustrates a sample of the experimentally observed bands [84–87] connecting eleven singlet, triplet, and quintet states of the C$_2$ molecule that contribute to the overall radiative association rate coefficient for this molecule. In the ejecta of SN1987A and other core-collapse supernovae, CO and SiO were detected, see [88,

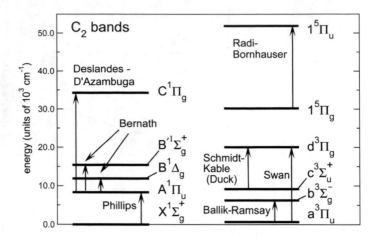

Fig. 13 Experimentally observed C_2 band systems connecting eleven singlet, triplet, and quintet electronic states dissociating to ground state carbon atoms; schematic illustration of electronic state term energies T_e in cm^{-1} calculated in the present work. Further details may be found in the recent publication of Babb and co-workers [74]

89], through fundamental ($\Delta\nu = 1$) bands. Moreover, recent three-dimensional mapping of CO and SiO in the SN 1987A core ejecta with the Atacama Large Millimeter/submillimeter Array (ALMA) shows a clumpy mixed structure calling for improvements beyond one-dimension in hydrodynamical and chemical modeling of molecular formation [90]; a reliable description of dicarbide formation might improve such future calculations. Finally, understanding the origins of cosmic dust and the roles played by supernovae in contributing to extragalactic dust depends on progress in modeling dust formation [89, 91].

In the present study we obtain results from a quantum approach to estimate the cross sections and rate coefficients for C_2 formation by radiative association using new highly accurate ab initio molecular data for the potential energy curves (PEC's) and transition dipole moments (TDM's) coupling the states of interest. Results from our present quantum approach are compared with the previous semi-classical results of [77] and conclusions are drawn.

The potential energy curves (PECs) and transition dipole moments (TDMs) for the eighteen single, triplet and quintet electronic states are calculated within an MRCI+Q approximation for the approach of ground state carbon atoms. We use a state-averaged-multi-configuration-self-consistent-field (SA-MCSCF) approach, followed by multi-reference configuration interaction (MRCI) calculations together with the Davidson correction (MRCI+Q) [92]. Potential energy curves and transition dipole moments as a function of internuclear distance R are calculated starting from a bond separation of $R = 1.5$ Bohr extending out to $R = 20$ Bohr. The basis sets used in the present work are the augmented correlation consistent polarized core valence quintuplet [aug-cc-pcV5Z (ACV5Z)] Gaussian basis sets. The use of such large basis sets is well known to recover 98% of the electron correlation effects in molecular structure calculations [92]. All the PEC and TDM calculations for the C_2 molecule were performed with the quantum chemistry program package MOLPRO 2015.1 [93], running on parallel architectures.

The potential curves for C_2 singlet, triplet and quintet states are shown in Fig. 14. It should be noted that several of the quintet states show barriers, which may be expected to lead to dramatic effects in the radiative association cross sections. At bond separations beyond $R = 14$ Bohr, the PECs are smoothly fitted to functions of the form

$$V(R) = \frac{C_5}{R^5} - \frac{C_6}{R^6},\qquad(7)$$

where for the particular electronic state, C_5 is the quadrupole-quadrupole electrostatic interaction [94, 95] and C_6 is the dipole-dipole dispersion (van der Waals) coefficient (we use atomic units unless otherwise specified). For $R < 1.5$ Bohr, short-range interaction potentials of the form $V(R) = A \exp(-BR) + C$ are fitted to the *ab initio* potential curves. Estimates of the values of the quadrupole-quadrupole coefficients C_5 were given by [94, 96].

The TDMs for the C_2 molecule (Fig. 15) are similarly extended to long- and short-range internuclear distances. For $R > 14$ a functional fit of the form $D(R) = a \exp(-bR) + c$ is applied, while in the short range $R < 1.5$ a quadratic fit of the form $D(R) = a'R^2 + b'R + c'$ is adopted.

The quantum mechanical cross section for the radiative association process $\sigma_{i \to f}^{QM}(E)$, where the initial i and final f electronic states are labeled by their molecular states (e.g. $d^3 \Pi_g$) can be calculated using perturbation theory (see, for example, [100–102]). The result is

$$\sigma_{i \to f}^{QM}(E) = P_i \sum_{v'J'} \sum_{J} \frac{64}{3} \frac{\pi^5}{137.036^3} \frac{v^3}{2\mu E} S_{JJ'} |M_{iEJ,fv'J'}|^2,\qquad(8)$$

where the sum is over initial partial waves with angular momenta J and final vibrational v' and rotational J' quantum numbers, $S_{J,J'}$ are the appropriate line strengths [103, 104] or Hönl-London factors [105], 137.036 is the speed of light in atomic units, μ is the reduced mass of the collision system, and $M_{iEJ,fv'J'}$ is given by the integral

$$M_{iEJ,fv'J'} = \int_0^\infty F_{iEJ}(R)D_{if}(R)\Phi_{fv'J'}(R)dR.\qquad(9)$$

Due to presence of identical nuclei and the absence of nuclear spin in $^{12}C_2$, the rotational quantum numbers of the $^1\Sigma_g^+$ states are even and those of the $^1\Sigma_g^-$ states are odd and for any given value of $\Lambda = 1$ or 2, only one lambda-doubling level is populated [106]. Thus, the statistical weight factor P_i is given by

$$P_i = (2S_i + 1)/81,\qquad(10)$$

Fig. 14 Potential energy
curves (eV), as a function of
internuclear distance (Bohr)
for C$_2$ molecular states
dissociating to ground state
carbon atoms, **a** singlet, **b**
triplet, and **c** quintet states.
Results were obtained using
the quantum chemistry
package MOLPRO and
aug-cc-pCV5Z basis for each
atom

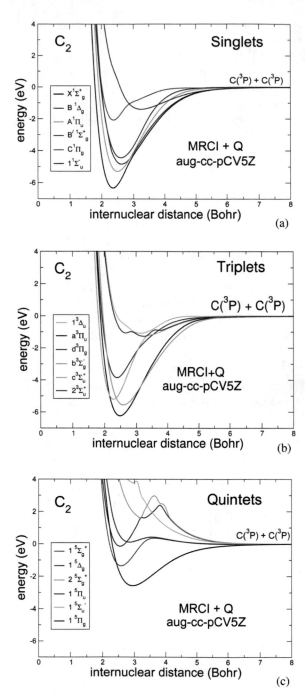

Fig. 15 Transition dipole moments (TDMs) for singlet transitions in C_2 in atomic units, **a** Phillips band ($A^1\Pi_g$-$X^1\Sigma_g^+$), **b** Bernath bands ($B^1\Delta_g$-$A^1\Pi_u$ and $B'\Sigma_g^+$-$A^1\Pi_u$) and **c** Deslandes-D'Azambuga band ($C^1\Pi_g$-$A^1\Pi_u$). For the Phillips band we compare the present MRCI+Q work with results from the prior studies [97–99] Further details may be found in the recent publication of Babb and co-workers [74]

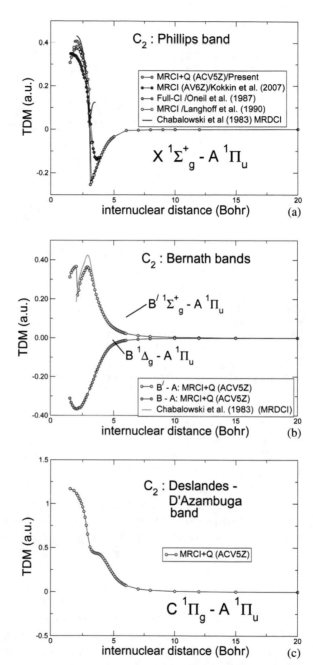

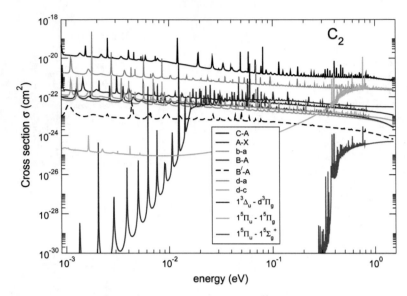

Fig. 16 Radiative association cross sections (units of cm^2) as a function of the collision energy E (eV) for the collision of two ^{12}C(^3P) atoms. Results are shown for the singlet, triplet, and quintet transitions listed in Table 1. To gauge the cross sections, all the data are plotted with the statistical factor $P_i = 1$. We see that the Deslandes-D'Azambuga (C$^1\Pi_g$-A$^1\Pi_u$), Swan (d$^3\Pi_g$-a$^3\Pi_u$) and Radi-Bornhauser ($1^5\Pi_u$-$1^5\Pi_g$) bands have the largest cross sections Further details may be found in the recent publication of Babb and co-workers [74]

where S_i is the total spin of the initial molecular electronic state (here 1, 3, or 5), and there are for two C(^3P) atoms $3^4 = 81$ molecular states labeled by Λ and S. Thus, for the ^{12}C$_2$ molecule considered here, $P_i = \frac{1}{81}, \frac{3}{81}$ or $\frac{5}{81}$, respectively, for $i = A^1\Pi_u$, $b^3\Sigma_g^+$, or $1^5\Pi_g$.

In Fig. 16, results are shown for the radiative association cross section as a function of energy for several singlet, triplet, and quintet transitions. We plot the cross sections with the statistical factor P_i set equal to unity for all states. Numerous shape resonances are visible. For the $1^3\Delta_u$-d$^3\Pi_g$ and $1^5\Pi_u$-$1^5\Pi_g$ cross sections, resonance tunneling features are visible, corresponding to potential barriers (local maxima) in the entrance channels.

In Fig. 17 we compare our Maxwellian averaged quantal rates with those determined from the previous semi-classical approximation by [77] over the temperature range 10–17,500 K. The quantal rates have the appropriate statistical population included so a comparison could be made directly with the previous semiclassical results of [77]. We see that our present quantal rates are larger than those from the previous semiclassical results of [77] at all temperatures, particularly as the temperature is lowered. We attribute this enhancement of the rate due to the presence of resonances in the quantal cross sections.

Accurate cross sections and rates for the radiative association process in the C$_2$ molecule have been computed for transitions from several excited electronic states

Table 1 C_2 transitions studied in this work. The transitions are listed in order of decreasing contribution to the total radiative association cross section

Initial molecular state	Final molecular state	C_2 Band Name
$C^1\Pi_g$	$A^1\Pi_u$	Deslandres-d'Azambuja
$d^3\Pi_g$	$a^3\Pi_u$	Swan
$1^5\Pi_u$	$1^5\Pi_g$	Radi-Bornhauser
$2^3\Sigma_u^+$	$d^3\Pi_g$	–
$1^3\Delta_u$	$d^3\Pi_g$	–
$b^3\Sigma_g^-$	$a^3\Pi_u$	Ballik-Ramsey
$d^3\Pi_g$	$c^3\Sigma_u^+$	Schmidt-Kable
$A^1\Pi_u$	$X^1\Sigma_g^+$	Phillips
$B'^1\Sigma_g^+$	$A^1\Pi_u$	Bernath B'
$B^1\Delta_g$	$A^1\Pi_u$	Bernath B
$2^5\Sigma_g^+$	$1^5\Pi_u$	–

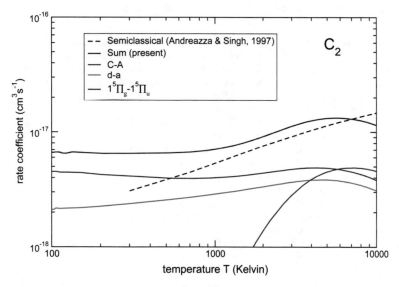

Fig. 17 Maxwellian averaged radiative association rates (cm³/s) as a function of temperature (Kelvin) for the C_2 molecule. Results are shown for several singlet and triplet transitions with their appropriate statistical factor included. The total quantal rate (dashed violet line) is seen to lie significantly above the previous total semiclassical rate [77] (dashed black line) at all temperatures, particularly as the temperature is lowered. The enhancement of the rate in the quantal calculations is due to the presence of strong resonance features in the various cross sections, which are not present in the semiclassical results

using new *ab initio* potentials and transition dipole moment functions. The computed cross sections and rates for C_2 are suitable for applicability in a variety of interstellar environments including diffuse and translucent clouds, circumstellar disks, and protoplanetary disks. Further details may be found in the recent publication of Babb and co-workers [74].

7 Summary

Having access to leadership computer architectures is essential to our research work (like the Cray-XC40 at HLRS, the Cray-XC40 at NERSC, the Cray-XK7 or Summit architectures at ORNL). We are grateful to the various computational centres for access to such high end computing resources enabling our theoretical investigations on the atomic, molecular and optical collision processes outlined above. The calculations could not be addressed without access to these state-of the-art computational facilities.

Acknowledgements A Müller acknowledges support by Deutsche Forschungsgemeinschaft under project number Mu-1068/20. B M McLaughlin acknowledges support from the US National Science Foundation through a grant to ITAMP at the Harvard-Smithsonian Center for Astrophysics, the University of Georgia at Athens for the award of an adjunct professorship, and Queen's University Belfast for a visiting research fellowship (VRF). ITAMP is supported in part by a grant from the NSF to the Smithsonian Astrophysical Observatory and Harvard University. M S Pindzola acknowledges support by NSF and NASA grants through Auburn University. P C Stancil acknowledges support from NASA grants through University of Georgia at Athens. The authors gratefully acknowledge the Gauss Centre for Supercomputing e.V. (www.gauss-centre.eu) for funding this project by providing computing time on the GCS Supercomputer HAZEL HEN at Höchstleistungsrechenzentrum Stuttgart (www.hlrs.de). This research also used computational resources at the National Energy Research Scientific Computing Center (NERSC) in Berkeley, CA, USA. The Oak Ridge Leadership Computing Facility at the Oak Ridge National Laboratory, provided additional computational resources, which is supported by the Office of Science of the U.S. Department of Energy under Contract No. DE-AC05-00OR22725. The Advanced Light Source is supported by the Director, Office of Science, Office of Basic Energy Sciences, of the US Department of Energy under Contract No. DE-AC02-05CH11231.

References

1. B.M. McLaughlin, J.M. Bizau, D. Cubaynes, S. Guilbaud, S. Douix M.M. and Al Shorman, M.O.M. El Ghazaly, I. Sahko, M.F. Gharaibeh, K-shell photoionization of O^{4+} and O^{5+} ions: experiment and theory. Mon. Not. Roy. Astro. Soc. **465**, 3690 (2017)
2. B.M. McLaughlin, Inner-shell photoionization, fluorescence and auger yields, in *Spectroscopic Challenges of Photoionized Plasma*, ed. by G.J. Ferland, D.W. Savin, vol. 247, pp. 87, San Francisco, CA, USA, 2001. Astronomical Society of the Pacific, ASP Con f. Series
3. T.R. Kallman, Challenges of plasma modelling: Current status and future plans. Space Sci. Rev. **157**, 177 (2010)

4. B.M. McLaughlin, C.P. Ballance, Photoionization, fluorescence and inner-shell processes, in *Yearbook of Science and Technology*, ed. by McGraw Hill, p. 281. New York, NY, USA, 2013. McGraw Hill
5. B.M. McLaughlin, C.P. Ballance, Photoionization cross section calculations for the halogen-like ions Kr^+, Xe^+. J. Phys. B: At. Mol. Opt. Phys. **45**, 085701 (2012)
6. B.M. McLaughlin, C.P. Ballance, Photoionization cross sections for the trans-iron element Se^+ from 18 to 31 eV. J. Phys. B: At. Mol. Opt. Phys. **45**, 095202 (2012)
7. B.M. McLaughlin, J.F. Babb, Single photoionization of the Kr-like Rb II ion in the photon energy range 22—46.0eV. Mon. Not. Roy. Astro. Soc. **486**, 245 (2019)
8. C.P. Ballance, D.C. Griffin, Relativistic radiatively damped R-matrix calculation of the electron-impact excitation of W^{46+}. J. Phys. B: At. Mol. Opt. Phys. **39**, 3617 (2006)
9. P.H. Norrington, I.P. Grant, Low-energy electron scattering by Fe XXIII and Fe VII using the Dirac R-matrix method. J. Phys. B: At. Mol. Phys. **20**, 4869 (1987)
10. C.P. Ballance, Darc codes website url. http://connorb.freeshell.org 2018
11. B.M. McLaughlin, C.P. Ballance, Petascale computations for large-scale atomic and molecular collisions, in *Sustained Simulated Performance 2014*, ed. by M.M. Resch, Y. Kovalenko, E. Fotch, W. Bez, H. Kobaysah, pp. 173–190 (Springer, Berlin, Germany, 2015)
12. B.M. McLaughlin, C.P. Ballance, M.S. Pindzola, A. Müller, PAMOP: petascale atomic, molecular and optical collisions, in *High Performance Computing in Science and Engineering'14*, ed. by W.E. Nagel, D.H. Kröner, M.M. Resch, pp 23–40 (Springer, Berlin, Germany, 2015)
13. B.M. McLaughlin, C.P. Ballance, M.S. Pindzola, A. Müller, PAMOP project : Petaflop computations in support of experiments, *High Performance Computing in Science and Engineering'15*, ed. by W.E. Nagel, D.H. Kröner, M.M. Resch, pp. 51–74 (Springer, Berlin, Germany, 2016)
14. B.M. McLaughlin, C.P. Ballance, M.S. Pindzola, P.C. Stancil, S. Schippers, A. Müller, PAMOP project: Petaflop computations in support of experiments, in *High Performance Computing in Science and Engineering'16*, ed. by W.E. Nagel, D.H. Kröner, M.M. Resch (Springer, Berlin, Germany, 2017), pp. 33–48
15. B.M. McLaughlin, C.P. Ballance, M.S. Pindzola, P.C. Stancil, J.F. Babb, S. Schippers, A. Müller, PAMOP: Large-scale calculations supporting experiments and astrophysical applications, in *High Performance Computing in Science and Engineering'17*, ed. by W.E. Nagel, D.H. Kröner, M.M. Resch, pp. 37–59(Springer, Berlin, Germany, 2018)
16. B.M. McLaughlin, C.P. Ballance, M.S. Pindzola, P.C. Stancil, J.F. Babb, S. Schippers, A. Müller, PAMOP2: Towards Exascale Computations Supporting Experiments and Astrophysics, *High Performance Computing in Science and Engineering'18*. pp. 37–59. Springer, Berlin, Germany, 2019
17. K.G. Dyall, C.T. Johnson, I.P. Grant, F. Parpia, E.P. Plummer, GRASP: A general-purpose relativistic atomic structure program. Comput. Phys. Commun. **55**, 425 (1989)
18. I.P. Grant, *Relativistic Quantum Theory of Atoms and Molecules: Theory and Computation* (Springer, New York, USA, 2007)
19. P.G. Burke, *R-Matrix Theory of Atomic Collisions: Application to Atomic* (Molecular and Optical Processes. Springer, New York, USA, 2011)
20. J. Darling, The 87 Rubidium atomic clock maser in giant stars. Research Notes of the AAS **2**(1), 15 (2018)
21. C. Sneden, R.G. Gratton, D.A. Crocker, Trends in copper and zinc abundances for disk and halo stars. Astron. & Astrophys. **246**, 354 (1991)
22. T.V. Mishenina, V.V. Kovtyukh, C. Soubiran, C. Travaglio, M. Busso, Abundances of Cu and Zn in metal-poor stars: Clues for galaxy evolution. A&A **396**, 189 (2002)
23. I.U. Roederer, A.F. Marino, C. Sneden, Characterizing the heavy elements in globular cluster M22 and an empirical s-process abundance distribution derived from the two stellar groups. Astrophys. J. **742**, 37 (2011)
24. I.U. Roederer, C. Sneden, I.B. Thompson, G.W. Preston, S.A. Shectman, Characterizing the chemistry of the milky way stellar halo: Detailed chemical analysis of a metal-poor stellar stream. Astrophys. J. **711**, 573 (2010)

25. A. Frebel, J.D. Simon, E.N. Kirby, Segue 1: An unevolved fossil galaxy from the early universe. Astrophys. J. **786**, 74 (2014)
26. A. Mueller, D. Macaluso, N. Sterling, A. Juarez, I. Dumitriu, R. Bilodeau, E. Red, D. Hardy, A. Aguilar, Absolute photoionization of Rb^+ and Br^{2+} ions for the determination of elemental abundances in astrophysical nebulae, in *APS Division of Atomic, Molecular and Optical Physics Meeting Abstracts* Bull. Am. Phys. Soc. **58**, 178 (2013)
27. D. Kilbane, F. Folkman, J.M. Bizau, C. Banahan, S. Scully, H. Kjeldsen, P. van Kampen, M.W.D. Mansfiled, J.T. Costello, J.B. West, Absolute photoionization cross-section measurements of the Kr I isoelectronic sequence. Phys. Rev. A **75**, 032711 (2007)
28. A. Neogi, E.T. Kennedy , J.P. Mosnier, P. van Kampen, G. Costello J T O'Sullivan, M.W.D. Mansfiled, V. Demekhin Ph, B.M. Lagutin, V.L. Sukhorukov, Trends in autoionization of rydberg states converging to the 4s threshold in the Kr - Rb^+ - Sr^{2+} isoelectonic sequence: theory and experiment. *Phys. Rev. A* **67**, 042707 (2003)
29. G. Hinojosa, V.T. Davis, A.M. Covington, J.S. Thompson, A.L.D. Kilcoyne, A. Antillón, E.M. Hernández, D. Calabrese, A. Morales-Mori, A.M. Juárez, O. Windelius, B.M. McLaughlin, Single photoionization of the Zn II ion in the photon energy range 17.5–90.0 eV: experiment and theory. *Mon. Not. Roy. Astro. Soc.* **470**, 4048 (2017)
30. A. Müller, S. Schippers, R.P. Phaneuf, A.M. Covington, A. Aguilar, G. Hinojosa, J. Bozeck, M.M. Sant'Anna, A.S. Schlachter, C. Cisneros, B.M. McLaughlin, Photoionisation of Ca^+ ions in the valence energy region 20–56 eV: experiment and theory. J. Phys. B: At. Mol. Opt. Phys. **50**, 205001 (2017)
31. D.A. Macaluso, K. Bogolub, A. Johnson, A. Aguilar, A.L.D. Kilcoyne, R.C. Bilodeau, M. Bautista, A.B. Kerlin, N.C. Sterling, Absolute single photoionization cross-section measurements for Rb^{2+} ions : theory and experiment. J. Phys. B: At. Mol. Opt. Phys. **49**, 235002 (2016)
32. D.A. Macaluso, K. Bogolub, A. Johnson, A. Aguilar, A.L.D. Kilcoyne, R.C. Bilodeau, M. Bautista, A.B. Kerlin, N.C. Sterling, Corrigendum: Absolute single photoionization cross-section measurements for Rb^{2+} ions : theory and experiment. J. Phys. B: At. Mol. Opt. Phys. **50**, 119501 (2017)
33. G.J. Ferland, K.T. Korista, D.A. Verner, J.W. Ferguson, J.B. Kingdon, E.M. Verner, CLOUDY 90 : numerical simulations of plasmas and their spectra. *Pub. Astron. Soc. Pac. (PASP)* **110**, 761 (1998)
34. G.J. Ferland, Quantative spectroscopy of photoionized clouds. Ann. Rev. of Astron. Astrophys. **41**, 517 (2003)
35. T.R. Kallman, Photoionized modeling codes. Astrophys. J. Suppl. Ser. **134**, 139 (2001)
36. A.R. Foster, L. Ji, R.K. Smith, N.S. Brickhouse, Updated atomic data and calculations for X-Ray spectroscopy. Astrophys. J. **756**, 128 (2012)
37. B.M. McLaughlin, J.F. Babb, Photoionization of Rb^{2+} ions in the valence energy region 37 eV - 44 eV. J. Phys. B: At. Mol. Opt. Phys. **52**, 125201 (2019)
38. C.A. Ramsbottom, C.E. Hudson, P.H. Norrington, M.P. Scott, Electron-impact excitation of Fe II: Collision strengths and effective collision strengths for low-lying fine-structure forbidden transitions. Astron. & Astrophys. **475**, 765 (2007)
39. H. Nussbaumer, P.J. Storey, Atomic data for FeII. Astron. & Astrophys. **89**, 308 (1980)
40. A.E. Kramida, Y. Ralchenko, J. Reader, NIST ASD Team, NIST Atomic Spectra Database (version 5.6.2). National Institute of Standards, Technology, Gaithersburg, MD, USA, 2018
41. M.A. Bautista, V. Fivet, C. Ballance, P. Quinet, G.J. Ferland, C. Mendoza, T.R. Kallman, Atomic data and spectral models for Fe II. Astrophys. J. **808**, 174 (2015)
42. C.A. Ramsbottom, C.J. Noble, V.M. Burke, M.P. Scott, P.G. Burke, Configuration interaction effects in low-energy electron collisions with Fe II. J. Phys. B: At. Mol. Opt. Phys. **37**, 3609 (2004)
43. S.S. Tayal, O. Zatsarinny, Electron-impact excitation of forbidden and allowed transitions in Fe II. Phys. Rev. A **98**, 012706 (2018)
44. H.L. Zhang,m A.K. Pradhan, Atomic data from the Iron Project. VI. Collision strengths and rate coefficients for Fe II. *Astron. & Astrophys.* **293**, 953 (1995)

45. F.P. Keenan, A. Hibbert, P.G. Burke, K.A. Berrington, Fine-structure populations for the ^{6}D ground state od Fe II. Astrophys. J. **332**, 539 (1988)

46. K.A. Berrington, P.G. Burke, A. Hibbert, M. Mohan, K.L. Baluja, Electron impact excitation of Fe$^+$ using the R-matrix method incorporating fine-structure effects. J. Phys. B: At. Mol. Phys. **21**, 339 (1988)

47. E.M. Verner, D.A. Verner, K.T. Korista, J.W. Ferguson, F. Hamann, G.J. Ferland, Numerical Simulations of Fe II Emission Spectra. Astrophys. J. Suppl. Ser. **120**, 101 (1999)

48. R.T. Smyth, C.A. Ramsbottom, F.P. Keenan, G.J. Ferland, C.P. Ballance, Towards converged electron-impact excitation calculations of low-lying transitions in Fe II. *Mon. Not. Roy. Astro. Soc.* **483**, 654 (2018)

49. R. Shingal, B.H. Bransden, Charge transfer, target excitation and ionizationin H$^+$ - Na(3s) collisions. J. Phys. B: At. Mol. Phys. **20**, 4815 (1987)

50. J.S. Allen, L.W. Anderson, C.C. Lin, Cross-sections for excitation of sodium by impact of H$^+$, H$_2^+$, H$_3^+$, and H$^-$ ions. Phys. Rev. A **37**, 349 (1988)

51. C.F. Fischer, *The Hatree-Fock Method for Atoms* (Wiley, New York, USA, 1977)

52. K. Igenbergs, J. Schweinzer, I. Bray, D. Bridi, F. Aumayr, Database for inelastic collisions of sodium atoms with electrons, protons, and multiply charged ions. At. Data Nucl. Data Tables **94**, 981 (2008)

53. W. Jitschin, S. Osimitsch, D.W. Mueller, H. Reihl, R.J. Allan, O. Scholler, H.O. Lutz, Excitation of the Na 3p state by proton impact. J. Phys. B: At. Mol. Phys. **19**, 2299 (1986)

54. M.S. Pindzola, S.D. Loch, Proton Impact Excitation of the Na atom. J. Phys. B: At. Mol. Opt. Phys. **52**, 025202 (2019)

55. O. Denis-Alpizar, T. Stoecklin, P. Halvick, M.L. Dubernet, Rotational relaxation of CS by collision with ortho- and para-H$_2$ molecules. J. Chem. Phys. **139**, 204304 (2013)

56. B. Yang, P. Zhang, C. Qu, P.C. Stancil, J.M. Bowman, N. Balakrishnan, R.C. Forrey, Inelastic vibrational dynamics of CS in collision with H$_2$ using a full-dimensional potential energy surface. Phys. Chem. Chem. Phys. **20**, 28425 (2018)

57. O. Denis-Alpizar, T. Stoecklin, S. Guilloteau, A. Dutrey, New rate coefficients of CS in collision with para and ortho-H$_2$ and astrophysical implications. Mon. Not. Roy. Astro. Soc. **478**, 1811 (2018)

58. T.B. Adler, G. Knizia, H.J. Werner, A simple and efficient CCSD(T)-F12 approximation. J. Chem. Phys. **127**, 221106 (2007)

59. H.-J. Werner, T.B. Adler, F.R. Manby, General orbital invariant MP2-F12 theory. *J. Chem. Phys.* **126**, 164102 (2007)

60. T.H. Dunning Jr., Gaussian basis set for use in correlated molecular calculations, I. The atoms boron through neon and hydrogen. *J. Chem. Phys.* **90**, 1007 (1989)

61. T.H. Dunning Jr., K.A. Peterson, A.K. Wilson, Gaussian basis sets for use in correlated molecular calculations. X. The atoms aluminum through argon revisted. *J. Chem. Phys.* **114**, 9244 (2001)

62. F. Weigend, A. Köhn, C. Hättig, Efficient use of the correlation consistent basis sets in resolution of the identity MP2 calculations. J. Chem. Phys. **116**, 3175 (2002)

63. C. Hättig, Optimization of auxiliary basis sets for RI-MP2 and RI-CC2 calculations: Core–valence and quintuple-f basis sets for H to Ar and QZVPP basis sets for Li to Kr. Optimization of auxiliary basis sets for RI-MP2 and RI-CC2 calculations: Core–valence and quintuple - ζ basis sets for H to Ar and QZVPP basis sets for Li to Kr. *Phys. Chem. Chem. Phys.* **7**, 59 (2005)

64. EMSL basit set exchange. https://bse.pnl.gov/bse/portal 2018

65. D. Feller, K.A. Peterson, J.G. Hill, Calibration study of the CCSD(T)-F12a/b methods for C$_2$ and small hydrocarbons. J. Chem. Phys. **133**, 184102 (2010)

66. S.F. Boys, F. Bernardi, The calculation of small molecular interactions by the differences of separate total energies. Some procedures with reduced errors. *Mol. Phys.* **19**, 553 (1970)

67. R.V. Krems, *TwoBC - quantum scattering program* (University of British Columbia, Vancouver, Canada, 2006)

68. B.R. Johnson, Multichannel log-derivative method for scattering calculations. J. Comp. Phys. **13**, 445 (1973)
69. B.H. Yang, P. Zhang, X. Wang, P.C. Stancil, J.M. Bowman, N. Balakrishnan, R.C. Forrey, Inelastic vibrational dynamics of CO in collision with H_2 using a full-dimensional potential energy surface. Nat. Commun. **6**, 6629 (2015)
70. B.E. Turner, Vibrationally excited CS in IRC + 10216. Astron. & Astrophys. **182**, L15 (1987)
71. J.L. Highberger, A.J. Apponi, J.H. Bieging, L.M. Ziyrys, J.G. Mangun, Millimeter observations of vibrationally excited CS toward IRC +10216: a new circumstellar maser? Astrophys. J. **544**, 881 (2000)
72. N.A. Patel et al., Submillimeter narrow emission lines from the inner envelope of IRC + 10216. Astrophys. J. **692**, 1205 (2009)
73. J.P. Fonfría, M. Agúndez, J. Cernicharo, M.J. Richter, J.H. Lacy, Carbon chemistry in IRC+10216: infrared detection of diacetylene. Astrophys. J. **852**, 80 (2018)
74. J.F. Babb, R.T. Smyth, B.M. McLaughlin, Dicarbon formation in collisions between two carbon atoms. Astrophys. J. **876**, 38 (2019)
75. S. Prasad, W.T. Huntress, A model for gas phase chemistry in interstellar clouds. I - The basic model, library of chemical reactions, and chemistry among C, N, and O compounds. *Astrophys. J. Suppl. Ser.* **43**, 1 (1980)
76. T.J. Millar, A. Bennett, J.M.C. Rawlings, P.D. Brown, S.B. Charnley, Gas phase reactions and rate coefficients for use in astrochemistry - The UMIST ratefile. Astrophys. J. Suppl. Ser. **87**, 585 (1991)
77. C.M. Andreazza, P.D. Singh, Formation of Si_2, C_2, C_2^+ and N_2^+ by radiative association. Mon. Not. Roy. Astro. Soc. **287**(2), 287–292 (1997)
78. N.V. Golubev, D.S. Bezrukov, M. Gustafsson, G. Nyman, S.V. Antipov, Formation of the SiP Radical through Radiative Association. J. Phys. Chem. A **117**(34), 8184–8188 (2013)
79. R.C. Forrey, J.F. Babb, P.C. Stancil, B.M. McLaughlin, Formation of silicon monoxide by radiative association: the impact of resonances. J. Phys. B: At. Mol. Opt. Phys. **49**(18), 184002 (2016)
80. M. Cairnie, R.C. Forrey, J.F. Babb, P.C. Stancil, B.M. McLaughlin, Rate constants for the formation of SiO by radiative association. Mon. Not. Roy. Astro. Soc. **471**(2), 2481 (2017)
81. R.J. Pattillo, R. Cieszewski, P.C. Stancil, R.C. Forrey, J.F. Babb, J.F. McCann, B.M. McLaughlin, Photodissociation of CS from excited rovibrational levels. Astrophys. J. **858**(1), 10 (2018)
82. R.C. Forrey, J.F. Babb, P.C. Stancil, B.M. McLaughlin, Rate constants for the formation of CS by radiative association. Mon. Not. Roy. Astro. Soc. **479**(4), 4727 (2018)
83. C.M. Andreazza, E.P. Marinho, P.D. Singh, Radiative association of C and P, and Si and P atoms. Mon. Not. Roy. Astro. Soc. **372**, 1653 (2006)
84. A. Tanabashi, T. Hirao, T. Amano, P.F. Bernath, The swan system of C_2: a global analysis of fourier transform emission spectra. Astrophys. J. Suppl. Ser. **169**, 472–484 (2007)
85. P. Bornhauser, R. Marquardt, C. Gourlaouen, G. Knopp, M. Beck, T. Gerber, J.A. van Bokhoven, P.P. Radi, Perturbation-facilitated detection of the first quintet-quintet band in C_2. J. Chem. Phys. **142**(9), 094313 (2015)
86. R.M. Macrae, Puzzles in bonding and spectroscopy: the case of dicarbon. Sci. Prog. **99**(1), 1–58 (2016)
87. T. Furtenbacher, I. Szabó, A.G. Császár, P.F. Bernath, S.N. Yurchenko, J. Tennyson, Experimental energy levels and partition function of the $^{12}C_2$ molecule. Astrophys. J. Suppl. Ser. **224**(2), 44 (2016)
88. I. Cherchneff, A. Sarangi, Molecules in Supernova Ejecta. In José Cernicharo and Rafael Bachiller, editors, *The Molecular Universe of IAU Symposium*, Vol. 280, pp. 228–236 (2011)
89. A. Sarangi, M. Matsuura, E.R. Micelotta, Dust in supernovae and supernova remnants I: formation scenarios. Space Sci. Rev. **214**(3), 63 (2018)
90. F.J. Abellán et al., Very deep inside the SN 1987a core ejecta: molecular structures seen in 3d. Astrophys. J. **842**(2), L24 (2017)
91. A. Sluder, M. Milosavljević, M.H. Montgomery, Molecular nucleation theory of dust formation in core-collapse supernovae applied to SN 1987A. Mon. Not. Roy. Astro. Soc. **480**, 5580–5624 (2018)

92. T. Helgaker, P. Jørgensen, J. Olsen, *Molecular Electronic-Structure Theory* (Wiley, New York, USA, 2000)
93. H.J. Werner, P.J. Knowles, G. Knizia, F.R. Manby, M. Schütz, et al., MOLPRO, version 2015.1, a package of *ab initio* programs, 2015. See http://www.molpro.net
94. J.K. Knipp, Quadrupole-quadrupole interatomic forces. Phys. Rev. **53**, 734–745 (1938)
95. T.Y. Chang, Moderately long - range interatomic forces. Rev. Mod. Phys. **39**(4), 911–942 (1967)
96. M. Boggio-Pasqua, A.I. Voronin, P. Halvick, C.J. Rayez, Analytical representations of high level ab initio potential energy curves of the C_2 molecule. *J. Molec. Struct.: THEOCHEM* **531**(1-3), 159–167 (2000)
97. S.V. ONeil, P. Rosmus, H-J. Werner, The radiative lifetime of A ^1u C_2. *J. Chem. Phys.* **87**(5), 2847–2853 (1987)
98. S.R. Langhoff, C.W. Bauschlicher, A.P. Rendell, A. Komornicki, Theoretical study of the radiative lifetime of the A $^1\Pi_u$ state of C_2. J. Chem. Phys. **92**(5), 3000–3004 (1990)
99. D.L. Kokkin, G.B. Bacskay, T.W. Schmidt, Oscillator strengths and radiative lifetimes for C_2: Swan, Ballik-Ramsay, Phillips, and $d^2\Pi_g \leftarrow c^3\Sigma_u^+$ systems. J. Chem. Phys. **126**(8), 084302 (2007)
100. J.F. Babb, A. Dalgarno, Radiative association and inverse predissociation of oxygen atoms. Phys. Rev. A **51**, 3021–3026 (1995)
101. F.A. Gianturco, Giorgi P. Gori, Radiative association of LiH ($X^1\Sigma^+$) from electronically excited lithium atoms. Phys. Rev. A **54**(5), 4073–4077 (1996)
102. J.F. Babb, K.P. Kirby, Molecule Formation in Dust-poor Environments, in *Molecular Astrophysics of Stars and Galaxies, page 11*, ed. by T.W. Hartquist, D.A. Williams (Clarendon Press, Oxford, UK, 1998)
103. R.D. Cowan, *The Theory of Atomic Structure and Spectra* (University of California Press, Berkeley, California, USA, 1981)
104. L.J. Curtis, *Atomic Structure and Lifetimes: A Conceptual Approach* (Cambridge University Press, Cambridge, UK, 2003)
105. J.K.G. Watson, Hönl-London factors for multiplet transitions in Hund's case a or b. J. Mol. Spec. **253**, 5 (2008)
106. C. Amiot, Fourier spectroscopy of the $^{12}C_2$, $^{13}C_2$, and $^{12}C^{13}C$ (0 − −0) Swan bands. Astrophys. J. Suppl. Ser. **52**, 329–340 (1983)

Thermodynamics and Kinetics of the Interactions Between Proteins and Hydrophilic Polymers

Timo Schäfer, Christian Muhl, Matthias Barz, Friederike Schmid, and Giovanni Settanni

Abstract Hydrophilic polymers are being investigated as possible coating agents for therapeutic nanoparticles because of their capacity to reduce immune response and increase circulation life time. The mechanism of action of these coatings is not well understood although it is clear that they unspecifically reduce the amount of proteins adsorbing on the nanoparticle surface coming in contact with biological fluids. Here we have investigated, using state-of-the-art atomistic molecular dynamics simulations, the equilibrium and kinetic properties of the interactions forming between human serum albumin, the most abundant protein in the blood stream, and two different and promising polymers poly(ethylene glycol) and poly-sarcosine and we have compared the results with a polymer which is an isomer of poly-sarcosine but has a totally different behavior in terms of adsorption, poly-alanine, because of its well-known aggregation propensity. The results show how the two hydrophilic polymers have a very similar behavior in terms of the amount of polymers adsorbed on the protein surface, pattern of interactions and the kinetics of the adsorption process, with differences emerging due to the different flexibility of the two molecules. In contrast poly-alanine adsorbs significantly more strongly on the protein surface, with a slower kinetics, and a quantitatively, but not qualitatively, different interaction pattern with the surface amino acids with respect to the hydrophilic polymers.

1 Introduction

More and more sophisticate manufacturing techniques have opened the way to nano-materials reaching the world of consumer products. Commonly used items such as clothes and sporting goods, skin care products, batteries, flame retardants etc. can contain nanomaterials to improve their functionality, for example by killing bacteria, increasing mechanical strength, improving skin penetration, electrodes efficiency,

T. Schäfer (✉) · F. Schmid · G. Settanni
Institut für Physik, Johannes Gutenberg University, Mainz, Germany
e-mail: settanni@uni-mainz.de

C. Muhl · M. Barz
Institut für Organische Chemie, Johannes Gutenberg University, Mainz, Germany

© Springer Nature Switzerland AG 2021
W. E. Nagel et al. (eds.), *High Performance Computing in Science and Engineering '19*,
https://doi.org/10.1007/978-3-030-66792-4_4

etc. More specifically, nanomaterials are finding applications as drug delivery systems, because of the envisaged opportunity to design them with the ability to fulfill several different functions like for example load the drug, protect it from degradation, transport it through the host organism and finally release it on the target tissue [1]. The increase in the use on nanomaterials in everyday life and in particular their use in medicine raises the question about their toxicity [2]. What could be the short and long term consequences on our health of the exposure to nanomaterials? What happens when nanomaterials come into contact with our organism? Only recently research has started to investigate in these directions.

In the case of nanoparticles coming in contact with the blood stream, it has been shown that they are immediately covered by a layer of proteins and biological molecules, the so-called protein corona. The size, nature and composition of this corona determines the reaction of the organism to the presence of the nanoparticles and, eventually, its fate in terms of circulation time, cell uptake, inflammatory response etc. [3, 4]. Indeed the adsorption on the nanoparticle surface of opsonins, specific markers for the immune system, may trigger the activation of an inflammatory response, thus favoring disposal and, consequently, reduction of the circulation time.

In order to improve the reaction of the organism to the introduction of nanoparticles a common strategy consists in coating them with specific materials which eventually modify the nanoparticle protein corona. Some of these materials show a so-called stealth effect, that is, they contribute to make the nanoparticle "invisible" to the immune system [5]. The mechanism of action of these materials is not fully understood. They definitely help and reduce the amount of proteins adsorbing on the nanoparticle surface but they probably do it in a specific way which may reduce in particular the amount of opsonins adsorbed. One of the most used materials in this context is poly-(ethylene-glycol) (PEG), a highly soluble and non-toxic polymer. Notwithstanding its positive properties, PEG cannot preclude completely the formation of a protein corona around the coated nanoparticles. In addition it has been shown that, since PEG is not metabolized by the organism, it may lead to dangerous accumulation [6]. For these reasons other materials are being investigated which could replace PEG as stealth coating. Hydrophilic polymer like poly-sarcosine (PSar) [7] or poly(N-(2-hydroxypropyl) methacrylamide) [8] represent promising substitute for PEG.

Molecular dynamics (MD) simulations represent a powerful tool for the investigation of nano-materials-biomolecule interactions in that they allow for the dissection of the interaction steps involved in the observed phenomena and provide them with a molecular level description. This has become only recently possible, both thanks to advances in available computational power and, very importantly, in the molecular models and force fields for the description of interactions between biomolecules and materials [9, 10]. Recently we used MD to describe the interactions between PEG and several blood proteins [11], and we revealed how these interactions can be rationalized in terms of the amino acid composition of the protein surface [12]. In these report we will discuss the extension of these studies in several directions [13]. On one hand we will discuss more in detail the thermodynamic and the kinetics of the

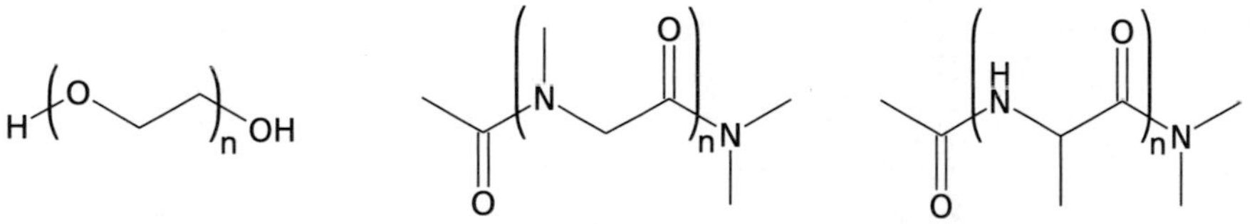

Fig. 1 Chemical formulas of PEG (left), PSar (middle) and PAla (right). Adapted from Ref. [13] licensed under CC BY

adsorption process of polymers on the protein surface, on the other hand, beside PEG, we will consider other polymers and in particular PSar and poly-alanine (PAla). PSar is a polypeptoid, which has a chemical structure similar to a polypeptide (i.e. normal proteins) but with the residue attached to the backbone nitrogen rather than to the C_α atom (Fig. 1). PSar is hydrophilic and very soluble, and has shown characteristics similar to PEG in terms of the capacity to reduce unspecific interactions with the biological milieu when used to coat nanoparticles [14–18]. In addition, sarcosine is naturally found in our organism and, unlike PEG, can be metabolized [19]. PSar, thus, represents a promising replacement for PEG in nanotechnological applications. PAla is the polypeptide isomer of PSar. The only difference between the two molecules is the position of the residue (a methyl group), which, in PAla, is attached to the C_α atom. Notwithstanding the small difference in the chemical structure, PAla shows completely different characteristics with respect to PSar. PAla is known to aggregate and, when added to protein solutions, it favors the aggregation of proteins. Thus analyzing the behavior of those two polymers, beside PEG, provides a useful term of comparison to help and dissect which characteristics of the polymers are important for the stealth effect.

2   Methods

MD simulations were performed using the program NAMD [20]. The program uses MPI parallelization and has been compiled to exploit the hardware and low-latency interconnect available on Hazelhen. The CHARMM force field with CMAP correction [21] was used for the simulations with additional parameters to describe PEG [22] and PSar [23]. The time step for the simulations was set to 1 fs, although a multiple time step scheme [24] was adopted where non-bonded interactions were updated every 2 time steps and long-range electrostatics every 4 time steps. Direct-space non-bonded interactions were truncated at 1.2 nm with a switch function from 1.0 to 1.2 nm. A cell-list like algorithm [25] was used to reduce the computational burden of the neighbor list update. The simulations were carried out with periodic boundary conditions. Long range electrostatic interactions were evaluated using a smooth particle-mesh Ewald (PME) approach [26]. Water molecules were explicitly included in the simulation boxes using a modified TIP3P model [27]. Pressure and temperature were regulated at 1 atm and 300 K, respectively, using a Langevin piston [28, 29].

Table 1 List of the analyzed simulations. Adapted from Ref. [13] licensed under CC BY

System	Box size (Å)	N. Atoms	Polymer length	Polymer molecules	Concentration (g/ml)	Simulation time (ns)
PEG1[a]	98.8	100881	4	214	0.08	4 × 200
PEG2[a]	98.2	99301	4	292	0.11	4 × 200
PEG3[a]	108.6	134134	4	424	0.12	4 × 200
PEG4[a]	109.0	134778	7	88	0.04	5 × 200
PEG5[a]	118.2	172541	4	560	0.12	5 × 100
PSar1[b]	98.5	99894	4	83	0.060	4 × 200
PSar2[b]	98.5	100128	4	103	0.074	5 × 200
PAla1[b]	98.3	99331	4	66	0.047	5 × 200
PAla2[b]	98.3.2	99413	4	76	0.055	5 × 200

[a]Trajectories from Ref. [12]
[b]Trajectories from Ref. [13]

Few water molecules were replaced by sodium and chlorine ions during simulation setup to neutralize the charge of the simulation box and to achieve the physiological ion concentration of [NaCl] = 0.15M. In our previous work we demonstrated that the pattern of interactions of PEG with the surface amino acids does not depend on the specific protein considered [12]. So, in this work we concentrated on the interactions between three different polymers, PEG, PSar and PAla and human serum albumin [30] (pdbid 1A06), the most abundant protein in the blood stream. HSA has 578 amino acids and a total charge of −15e. The systems to be simulated were prepared by immersing the protein in pre-equilibrated simulation boxes containing polymers at different concentrations. Atoms overlapping with the protein were removed. Only short polymers of 4 to 7 monomers were considered to favor fast diffusion. The simulation boxes were cubic with a side length between 10 and 12 nm. The list of the simulated systems is provided in Table 1. The systems were minimized for 10,000 steps using steepest descent with positional constraints on the the heavy atoms of the protein. Then, the systems were equilibrated using MD in the NPT ensemble for 1 ns by gradually releasing the constraints. Finally the production runs were started. Previous analysis [12] showed that the number of polymer molecules adsorbing on the protein surface has a correlation time of up to 10 ns, so the first 10ns of each run were discarded during analysis to allow for proper distribution of the polymers on the protein surface. The size of the simulated systems reached and/or exceeded the 100,000 atoms which is sufficient to observe good scaling on Hazelhen using up to 100 nodes per job. Typical job lengths do not exceed 3 h to avoid the need for writing intermediate restart files, an operation that, in our experience, may lead to job failure due to delayed disk write. This set up allowed for the collection of between 3 to 4 ns of trajectory per hour on 100 Hazelhen nodes. Several independent simulations were collected for each system to increase the statistical significance of the results. The trajectories were analyzed using the program VMD [31] and WORDOM [32].

3   Results and Discussion

All the simulations show an accumulation of polymers on the surface of the protein, where the polymer density is larger than in the bulk solution. This points out to an effective attractive interaction between protein and polymers. The excess density with respect to bulk is, however, strongly dependent on the polymer being considered (Fig. 2). In the case of PEG the excess density is only observed in a shell of 0.5 nm around the protein and it exceeds twice the bulk density only in the case of PEG_7. PSar shows an excess density of slightly less than twice as bulk in the 0.5 nm shell, however the excess density extends beyond 0.7 nm from the protein surface, possibly because, unlike PEG, it is a branched polymer, thus bulkier than PEG. PAla, on the other hand, shows significantly larger excess densities than the other two polymers extending up to 1.1 nm from the protein surface. Considering the radius of gyration of the polymers which in none of them exceeds 0.5 nm, these data suggest that PEG and PSar form a monolayer on the protein surface, while PAla accumulates over multiple layers. In order to quantify the affinity of the polymers for the protein surface, we have fitted a simple Langmuir-like model of adsorption to the simulation data. In this model we assume that the protein surface has a fixed amount of equivalent sites where polymer atoms can bind. The binding reaction can then be written as $A_s + P_s \rightleftharpoons A_P$, where A_s is a generic (heavy) atom of the polymer in solution, P_s is an empty binding site on the protein surface and A_P is the polymer atom adsorbed on the protein surface. This model leads to a Langmuir-like isotherm [33]:

$$[A_P] = [P_s^{max}]\frac{K_a.[A_s]}{K_a + [A_s]} \tag{1}$$

where $[A_s]$ is the concentration of polymer atoms in the solvent (i.e., the fraction of polymer heavy atoms), $[A_P]$ is the concentration of polymer atoms in the vicinity of the protein (again, measured as the fraction of polymer heavy atoms), $[P_s^{max}]$ is the maximum concentration of binding sites on the protein surface, and K_a is the equilibrium constant of the adsorption reaction. The free parameters of the model

Fig. 2 The radial distribution function of the polymer atoms around the protein surface in the simulations as indicated in the legend. Adapted from Ref. [13] licensed under CC BY

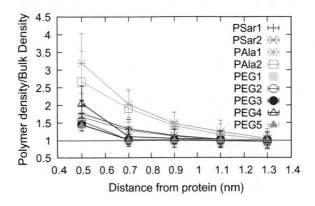

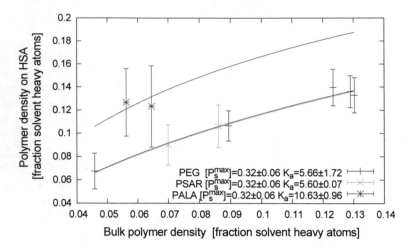

Fig. 3 Density of polymer heavy atoms in a 0.5nm-thick layer of solvent around the protein as a function of overall polymer concentration in the simulation box. PEG, PSar and PAla are represented in red, green and blue, respectively. The error bars represent standard deviations from the simulations. The continuous lines represent fits of the Langmuir-like adsorption model Eq. 1. Adapted from Ref. [13] licensed under CC BY

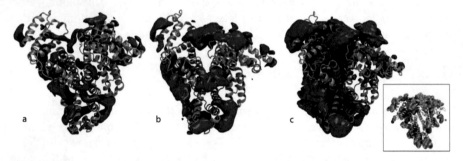

Fig. 4 Cartoon of HSA showing the regions with a polymer density higher than a cutoff, taken to be twice as the overall density of polymer in the box. Data for PEG, PSar and PAla are shown in (**a**), (**b**) and (**c**), respectively. For these data the simulations PEG1, PSar2, PAla2 have been used, where the overall density of polymer in the simulation box is approximately similar. High water density regions, not shown here for clarity, approximately cover the protein surface areas not occupied by the high polymer density regions. Inset: a snapshot of the PAla2 simulation showing the aggregated clumps of PAla (cyan-blue-red) around the protein (yellow). Adapted from Ref. [13] licensed under CC BY

are the equilibrium constant K_a and the maximal concentration of binding sites $[P_s^{max}]$ on the protein surface. We used the data from PEG simulations to fit both parameters, then we assumed that $[P_s^{max}]$ is approximately the same for all the polymers and we fitted the K_a for PSar and PAla. The results (Fig. 3) show that the binding affinity of PEG and PSar is basically the same within the error, while for PAla it is almost twice as large. This further confirms that PAla establishes much stronger interactions with the

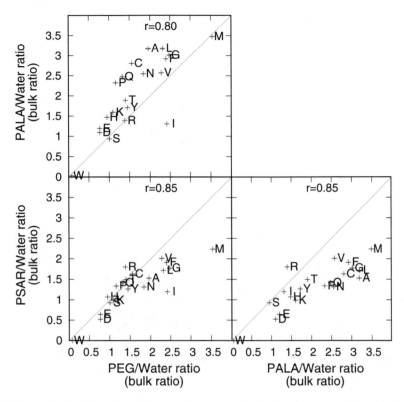

Fig. 5 Comparison of the polymer/water ratios measured around each amino acid type (indicated using the single-letter code) for each pair of polymers considered. Although those values show high correlations (r-values), those from PAla are systematically larger than the others, indicating larger affinity for the protein surface. The lines indicate the identity function. The data come from simulations PEG1, PSar2 and PAla2. Adapted from Ref. [13] licensed under CC BY

protein than the other two polymers. We then analyzed how the polymers distribute on the protein surface (Fig. 4). These data show that the locations of regions with the largest polymer density are similar across the three polymers that we investigated, however, the size of the regions is polymer dependent. PEG shows relatively thin high density regions, PSar shows slightly thicker layers, in agreement with the data from Fig. 2 possibly due to its branched and bulkier structure, while PAla shows much larger high density regions covering large parts of the protein surface. This behavior is determined by the adsorption of entire clusters of PAla on the protein surface (Fig. 4 inset).

We then measured how the various polymers interact with the different amino acids present on the protein surface. This can be done by evaluating the ratio between the number of polymer and water heavy atoms in a shell of 0.5 nm around each amino acid of the same type and comparing it to the same ratio measured in the bulk (Fig. 5 [11, 12]). These data reveal that, qualitatively, the pattern of interactions with the various

Table 2 Kinetics of adsorption. Adapted from Ref. [13] licensed under CC BY

System	Avg. num. ads. mol.	Avg. ads. rate (ns^{-1})	Avg. desorp. rate (ns^{-1})	Avg. Time to desorb. (ns)[a]	Avg. Time to adsorb. (ns)[a]
PEG1	32 ± 5	0.186	1.066	0.3 (+0.9 −0.2)	1.0 (+7.0 −0.9)
PEG2	43 ± 6	0.181	1.074	0.3 (+0.9 −0.2)	1.0 (+7.2 −0.9)
PEG3	42 ± 5	0.123	1.153	0.3 (+0.9 −0.2)	1.2 (+10.1 −1.1)
PEG4	13 ± 3	0.057	0.365	0.8 (+2.7 −0.6)	3.9 (+25.3 −3.4)
PEG5	42 ± 6	0.091	1.257	0.3 (+0.9 −0.2)	1.5 (+13.5 −1.3)
PSar1	16 ± 4	0.062	0.285	1.2 (+3.9 −0.9)	4.7 (+23.7 −3.9)
PSar2	18 ± 4	0.055	0.276	1.3 (+4.0 −1.0)	5.0 (+26.9 −4.2)
PAla1	23 ± 6	0.050	0.102	2.5 (+10.8 −2.0)	5.1 (+27.6 −4.3)
PAla2	22 ± 7	0.039	0.103	2.7 (+12.1 −2.2)	5.5 (+33.9 −4.7)

[a] The geometric mean and the 68% confidence interval are reported

amino acid types is similar in the three polymers, as demonstrated by the relatively high correlation between normalized polymer/water ratios, and as expected from the fact that similar regions of the protein surface are covered by the polymers. However, the data also show that, quantitatively, the polymer/water ratios are systematically larger for PAla than PEG or PSar in particular for the most polymer-attractive amino acids (i.e., mostly hydrophobic amino acids).

To evaluate the kinetics of the adsorption process we first defined as adsorption event of the polymer molecules the time point when a desorbed molecule enters with more than half of its heavy atoms the 0.5 nm shell around the protein surface. A desorption event was defined when all the heavy atoms of a polymer molecule leave the shell. With these definitions we could measure the adsorption and desorption rates of the molecules (Table 2). If we consider only the heaviest polymers (PEG$_7$, PSar and PAla), which diffuse slowly because of their size, the data show that while PSar and PEG show a similar kinetics related to their concentration in the simulations, PAla has a significantly slower desorption rate, indicating its tendency to remain adsorbed to the protein surface. This also explain why the number of PAla molecules adsorbed is in general larger than PSar and PEG$_7$ notwithstanding PAla lower concentrations in the simulations (Table 1). Further analysis reported in our original publication [13] reveals that the number of adsorption/desorption events of PAla is significantly smaller than the other polymers because the diffusion rate of PAla is smaller. This is due to the formation of aggregates of PAla molecules not seen in the other two

Fig. 6 a Cumulative radial distribution function of water oxygen atoms around the heavy atoms of the polymers (data from 3 selected runs). **b** Cartoon of a PAla cluster formed in the simulation of the polymer-water mixture water molecules are omitted for clarity). The black connectors indicate hydrogen bonds. Adapted from Ref. [13] licensed under CC BY

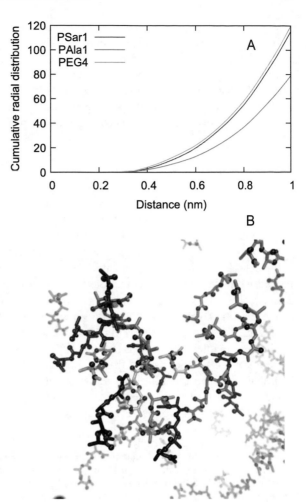

polymers, which, due to their large size, diffuse significantly slower than single molecules. This phenomenon is further discussed below.

In the reported simulations we have three main players: the protein, the polymers and the water molecules (we are neglecting the ions for simplicity). The simulations reveal the presence of effective attractive interactions between the polymers and the protein surface, which are much stronger in the case of PAla than PEG or PSar. The data show, however, that in terms of qualitative patterns of interaction and also direct enthalpic contributions and/or average h-bond formation capabilities with water [13] the three polymer do not differ significantly. The differences emerge in the collective behavior of the polymers in the presence of the water. Indeed, we observed that the amount of water molecules present on average around polymer heavy atoms is much smaller in PAla than in the other polymers (Fig. 6a). This is due to the formation of large clusters of even more than ten PAla molecules(Fig. 6b), where water density is low. The clusters are stabilized by the formation of hydrogen bonded β-bridges between the backbone of pairs of molecules and by hydrophobic contacts between the side chains. These phenomena do not occur in PSar because it lacks the hydrogen bond donor on the backbone and because it has a much larger conformational variability than PAla as documented recently [23, 34]. PEG and PSar on the other hand show a very similar hydrophilic pattern of interactions in agreement with the experimentally evaluated elution properties and contact angles [35].

4 Conclusions

Here, MD simulations, which exploit the highly parallel architecture of Hazelhen, have been shown to provide a detailed description of the adsorption process of polymers on the surface of proteins. The approach has helped identify the features of the polymers which make them amenable for use as coating agents for nanoparticles. Indeed, while the pattern of interactions with the surface amino acids does not change dramatically across the tested polymers, the amount of adsorbed polymers and the residence time on the protein surface varies significantly and is dependent on the aggregation propensities of the polymers, which ultimately, depend on their conformational properties. PSar forms slightly thicker adsorption layers than PEG on the protein surface but, apart from that, it shows similar protein adsorption properties and, thus, may represent a viable replacement as stealth agent. A similar analysis could be performed on a large variety of polymers, opening the way to an *in silico* screening of polymers for use as stealth coating agents.

Acknowledgements TS gratefully acknowledges financial support from the Graduate School Materials Science in Mainz. GS gratefully acknowledges financial support from the Max-Planck Graduate Center with the University of Mainz. We gratefully acknowledge support with computing time from the HPC facility Hazelhen (project Flexadfg) at the High performance computing center Stuttgart and the HPC facility Mogon at the university of Mainz. This work was supported by the German Science Foundation within SFB 1066 project Q1.

References

1. K. Cho, X. Wang, S. Nie, Z.G. Chen, D.M. Shin, Clin. Cancer Res. **14**(5), 1310 (2008). https:// doi.org/10.1158/1078-0432.CCR-07-1441; http://clincancerres.aacrjournals.org/content/14/5/1310.abstract
2. H. Lopez, E.G. Brandt, A. Mirzoev, D. Zhurkin, A. Lyubartsev, V. Lobaskin, (Springer, Cham 2017) 173–206. https://doi.org/10.1007/978-3-319-47754-1_7; http://link.springer.com/10.1007/978-3-319-47754-1_7
3. M.P. Monopoli, C. Aberg, A. Salvati, K.A. Dawson, Nat Nanotechnol **7**(12), 779 (2012). https:// doi.org/10.1038/nnano.2012.207; http://dx.doi.org/10.1038/nnano.2012.207
4. A. Lesniak, F. Fenaroli, M.P. Monopoli, C. Aberg, K.A. Dawson, A. Salvati, ACS Nano **6**(7), 5845 (2012). https://doi.org/10.1021/nn300223w; http://dx.doi.org/10.1021/nn300223w

5. S. Schöttler, G. Becker, S. Winzen, T. Steinbach, K. Mohr, K. Landfester, V. Mailänder, F.R. Wurm, Nat Nanotechnol (2016). https://doi.org/10.1038/nnano.2015.330; http://dx.doi.org/10.1038/nnano.2015.330

6. T. Ishida, M. Ichihara, X. Wang, K. Yamamoto, J. Kimura, E. Majima, H. Kiwada, J. Controll. Rel. 112(1), 15 (2006). https://doi.org/10.1016/J.JCONREL.2006.01.005; https://www.sciencedirect.com/science/article/pii/S016836590600037X?via%3Dihub

7. B. Weber, C. Seidl, D. Schwiertz, M. Scherer, S. Bleher, R. Süss, M. Barz, Polymers 8(12), 427 (2016). https://doi.org/10.3390/polym8120427; http://www.mdpi.com/2073-4360/8/12/427

8. L. Nuhn, M. Barz, R. Zentel, Macromol. Biosci. 14(5), 607 (2014). https://doi.org/10.1002/mabi.201400028; http://doi.wiley.com/10.1002/mabi.201400028

9. M. Ozboyaci, D.B. Kokh, S. Corni, R.C. Wade, Quart. Rev. Biophy. 49, (2016). https://doi.org/10.1017/S0033583515000256; URL https://doi.org/10.1017/S0033583515000256

10. H. Heinz, H. Ramezani-Dakhel, Chem. Soc. Rev. 45(2), 412 (2016). https://doi.org/10.1039/C5CS00890E; http://pubs.rsc.org/en/content/articlelanding/2016/cs/c5cs00890e

11. G. Settanni, J. Zhou, F. Schmid, J. Phys.: Conf. Seri. 921(1), 012002 (2017). https://doi.org/10.1088/1742-6596/921/1/012002; http://stacks.iop.org/1742-6596/921/i=1/a=012002?key=crossref.205805530aa231b2f5ac7c2605e3bfa7

12. G. Settanni, J. Zhou, T. Suo, S. Schöttler, K. Landfester, F. Schmid, V. Mailänder, Nanoscale 9, 6 (2017). https://doi.org/10.1039/c6nr07022a

13. G. Settanni, T. Schäfer, C. Muhl, M. Barz, F. Schmid, Comput. Struct. Biotechnol. J. 16, 543 (2018). https://doi.org/10.1016/J.CSBJ.2018.10.012; https://www.sciencedirect.com/science/article/pii/S2001037018300758

14. A. Birke, J. Ling, M. Barz, Prog. Pol. Sci. 81, 163 (2018). https://doi.org/10.1016/J.PROGPOLYMSCI.2018.01.002; https://www.sciencedirect.com/science/article/pii/S0079670017301442?via%3Dihub

15. B. Weber, A. Birke, K. Fischer, M. Schmidt, M. Barz, Macromolecules 51(7), 2653 (2018). https://doi.org/10.1021/acs.macromol.8b00258; http://pubs.acs.org/doi/10.1021/acs.macromol.8b00258

16. P. Klein, K. Klinker, W. Zhang, S. Kern, E. Kessel, M. Wagner, M. Barz, Polymers 10(6), 689 (2018). https://doi.org/10.3390/polym10060689; http://www.mdpi.com/2073-4360/10/6/689

17. J. Yoo, A. Birke, J. Kim, Y. Jang, S.Y. Song, S. Ryu, B.S. Kim, B.G. Kim, M. Barz, K. Char, Biomacromolecules 19(5), 1602 (2018). 10.1021/acs.biomac.8b00135. URL http://pubs.acs.org/doi/10.1021/acs.biomac.8b00135

18. E. Ostuni, R.G. Chapman, R.E. Holmlin, S. Takayama, G.M. Whitesides, Langmuir 17(18), 5605 (2001). https://doi.org/10.1021/la010384m; https://pubs.acs.org/doi/abs/10.1021/la010384m

19. D. Zhang, S.H. Lahasky, L. Guo, C.U. Lee, M. Lavan, Macromolecules 45(15), 5833 (2012). https://doi.org/10.1021/ma202319g; https://doi.org/10.1021/ma202319g

20. J.C. Phillips, R. Braun, W. Wang, J. Gumbart, E. Villa, C. Chipot, R.D. Skeel, L. Kale, K. Schulten, J. Comput. Chem. 26, 1781 (2005)

21. A.D. Mackerell, M. Feig, C.L. Brooks, J. Comput. Chem. 25(11), 1400 (2004). https://doi.org/10.1002/jcc.20065; http://dx.doi.org/10.1002/jcc.20065

22. H. Lee, R.M. Venable, A.D. Mackerell, R.W. Pastor, Biophys. J. 95(4), 1590 (2008). https://doi.org/10.1529/biophysj.108.133025; http://dx.doi.org/10.1529/biophysj.108.133025

23. D.T. Mirijanian, R.V. Mannige, R.N. Zuckermann, S. Whitelam, J. Comput. Chem. 35(5), 360 (2014). https://doi.org/10.1002/jcc.23478; http://dx.doi.org/10.1002/jcc.23478

24. R.D. Skeel, J.J. Biesiadecki, Ann. Numer. Math. 1, 191 (1994)

25. M.P. Allen, D.J. Tildesley, *Computer Simulation of Liquids* (Clarendon Press., 1987)

26. U. Essmann, L. Perera, M.L. Berkowitz, T. Darden, H. Lee, L.G. Pedersen, J. Chem. Phy. 103(19), 8577 (1995). https://doi.org/10.1063/1.470117; http://scitation.aip.org/content/aip/journal/jcp/103/19/10.1063/1.470117

27. W.L. Jorgensen, J. Chandrasekhar, J.D. Madura, R.W. Impey, M.L. Klein, J. Chem. Phy. 79(2), 926 (1983). https://doi.org/10.1063/1.445869; http://link.aip.org/link/?JCP/79/926/1

28. G.J. Martyna, D.J. Tobias, M.L. Klein, J. Chem. Phys **101**(5), 4177 (1994). https://doi.org/10.1063/1.467468; http://link.aip.org/link/?JCP/101/4177/1

29. S.E. Feller, Y. Zhang, R.W. Pastor, B.R. Brooks, J. Chem. Phys **103**(11), 4613 (1995). https://doi.org/10.1063/1.470648; http://link.aip.org/link/?JCP/103/4613/1

30. S. Sugio, A. Kashima, S. Mochizuki, M. Noda, K. Kobayashi, Prot. Eng. **12**(6), 439 (1999)

31. W. Humphrey, A. Dalke, K. Schulten, J. Mol. Grap. **14**, 33 (1996)

32. M. Seeber, M. Cecchini, F. Rao, G. Settanni, A. Caflisch, Bioinformatics **23**, 19 (2007). https://doi.org/10.1093/bioinformatics/btm378

33. R.I. Masel, *Principles of Adsorption and Reaction on Solid Surfaces* (Wiley, 1996). https://www.wiley.com/en-us/Principles+of+Adsorption+and+Reaction+on+Solid+Surfaces-p-9780471303923

34. A. Prakash, M.D. Baer, C.J. Mundy, J. Pfaendtner, Biomacromolecules **19**(3), 1006 (2018). https://doi.org/10.1021/acs.biomac.7b01813; http://pubs.acs.org/doi/10.1021/acs.biomac.7b01813

35. K.H.A. Lau, C. Ren, T.S. Sileika, S.H. Park, I. Szleifer, P.B. Messersmith, Langmuir **28**(46), 16099 (2012). https://doi.org/10.1021/la302131n; http://pubs.acs.org/doi/10.1021/la302131n

Crystallization, Fermionization, and Cavity-Induced Phase Transitions of Bose-Einstein Condensates

A. U. J. Lode, O. E. Alon, L. S. Cederbaum, B. Chakrabarti, B. Chatterjee,
R. Chitra, A. Gammal, S. K. Haldar, M. L. Lekala, C. Lévêque, R. Lin,
P. Molignini, L. Papariello, and M. C. Tsatsos

Abstract Bose-Einstein condensates (BECs) are one of the cornerstones in the exploration of the quantum many-body physics of interacting indistinguishable particles. Here, we study them using the MultiConfigurational Time-Dependent Hartree

A. U. J. Lode (✉)
Institute of Physics, Hermann-Herder-Str. 3, 79104 Freiburg, Germany
e-mail: auj.lode@gmail.com; axel.lode@tuwien.ac.at

A. U. J. Lode · C. Lévêque
Vienna Center for Quantum Science and Technology, Atominstitut, TU Wien, Stadionallee 2, 1020 Vienna, Austria

Wolfgang Pauli Institute c/o Faculty of Mathematics, University of Vienna, Oskar-Morgenstern Platz 1, 1090 Vienna, Austria

O. E. Alon · S. K. Haldar
Department of Mathematics, University of Haifa, 3498838 Haifa, Israel
e-mail: ofir@research.haifa.ac.il

Haifa Research Center for Theoretical Physics and Astrophysics, University of Haifa, 3498838 Haifa, Israel

L. S. Cederbaum
Theoretische Chemie, Physikalisch-Chemisches Institut, Universität Heidelberg, Im Neuenheimer Feld 229, 69120 Heidelberg, Germany
e-mail: lorenz.cederbaum@pci.uni-heidelberg.de

B. Chakrabarti
Department of Physics, Presidency University, 86/1 College Street, Kolkata 700 073, India

B. Chatterjee
Department of Physics, Indian Institute of Technology-Kanpur, Kanpur 208016, India

R. Chitra · R. Lin · P. Molignini · L. Papariello
Institute for Theoretical Physics, ETH Zürich, 8093 Zürich, Switzerland

A. Gammal
Instituto de Fisica, Universidade de São Paulo, CEP 05315-970, 66318, São Paulo, Brazil

M. L. Lekala
Department of Physics, University of South Africa, P.O. Box-392, Pretoria 0003, South Africa

M. C. Tsatsos
São Carlos Institute of Physics, University of São Paulo, 369, São Carlos, São Paulo 13560-970, Brazil

© Springer Nature Switzerland AG 2021   77
W. E. Nagel et al. (eds.), *High Performance Computing in Science and Engineering '19*,
https://doi.org/10.1007/978-3-030-66792-4_5

for Bosons (MCTDHB) method implemented in the MCTDHB and MCTDH-X software packages using the Cray XC40 system Hazel Hen. In this year we investigated the physics of (i) fermionization and crystallization of ultracold strongly interacting bosons, (ii) correlations of bosons with strong dipolar interactions, (iii) cavity-induced phase transitions of ultracold bosons in high-finesse cavities, (iv) the dimensionality of the variance of BECs in annular confinements, (v) the dynamics of BECs with long-ranged interactions in a bosonic Josephson junction, (vi) the dynamics of BECs in an asymmetric bosonic Josephson junction, (vii) the variance of BECs in anharmonic traps. All these results are novel and intriguing findings that demonstrate the versatility of the MCTDHB method, its implementations in the MCTDHB and MCTDH-X software packages, and how extremely fruitful the computational resources at the HLRS system Hazel Hen were for the MCTDHB project there. For the sake of brevity, we restrict the present report to the results (i)–(iii). We conclude with an outline for possible future avenues in the development revolving around MCTDHB and MCTDH-X.

1 Introduction

Software implementations of the multiconfigurational time-dependent Hartree for bosons (MCTDHB) [1–11] method are available in the MCTDHB, MCTDHB-LAB, and MCTDH-X software packages [12–14], which were used to obtain the results in our present report.

The solutions of the time-dependent many-boson Schrödinger equation computed with MCTDHB on Hazel Hen unveiled exciting insights into the physics of interacting ultracold atoms and were reported in Refs. [15–22].

For the sake of brevity, we restrict ourselves to the presentation of the papers [20–22] about, respectively, the phase transitions of ultracold bosons in a cavity, the reduced density matrices and Glauber correlation functions of strongly interacting dipolar bosons in lattices, and the comparison of the state of crystallized dipolar bosons to the state of fermionized bosons with contact interactions.

We have annually reported our investigations, that were enabled and boosted through the availability of computation time on the HLRS Stuttgart platforms [23–28] and continue our practice with the present report.

The structure of our report is as follows: in Sect. 2, we collect the quantities of interest that we used to investigate the many-boson physics in Refs. [20–22], in Sect. 3 we compare the fermionization and crystallization phenomena, in Sect. 4, we discuss the correlations of strongly interacting dipolar atoms in lattices, in Sect. 5, we discuss the phase transitions of bosons immersed in a high-finesse optical cavity, and in Sect. 6 we conclude with a summary and outlook.

2 Quantities of Interest

The many-body physics that we present in this report use the one- and p-body reduced density matrices (1- and p-RDMs, respectively), the 1-RDM's eigenfunctions and eigenvalues, a so-called crystal order parameter Δ that is a function of the eigenvalues of the 1-RDM, and the p-body Glauber correlation functions. We discuss these quantities in the following.

For an N-body system, the p-RDM is obtained by tracing out $N - p$ particles from the N-particles density operator $[|\Psi\rangle\langle\Psi|]$ ([29]):

$$\rho^{(p)}(\mathbf{r}_1, ..., \mathbf{r}_p, \mathbf{r}'_1, ..., \mathbf{r}'_p; t) = \mathrm{Tr}_{p+1,...,N} [|\Psi\rangle\langle\Psi|] \tag{1}$$

$$= \frac{N!}{(N - p)!} \int d\mathbf{r}_{p+1} \cdots d\mathbf{r}_N \Psi^*(\mathbf{r}'_1, ..., \mathbf{r}'_p, \mathbf{r}_{p+1}, ..., \mathbf{r}_N; t)$$

$$\times \Psi(\mathbf{r}_1, ..., \mathbf{r}_p, \mathbf{r}_{p+1}, ..., \mathbf{r}_N; t).$$

Here, Ψ is the wavefunction of the many-boson system and $\mathbf{r}_k$ the position of the k-th boson.

In the following, we will make use of the diagonal of the p-th order density matrix: $\rho^{(p)}(\mathbf{r}_1, ..., \mathbf{r}_p, \mathbf{r}_1, ..., \mathbf{r}_p; t) \equiv \rho^{(p)}(\mathbf{r}_1, ..., \mathbf{r}_p; t)$ is the probability to find particles $1, ..., p$ at positions $\mathbf{r}_1, ..., \mathbf{r}_p$, respectively. The off-diagonal part of the p-th order reduced density matrix, $\rho^{(p)}(\mathbf{r}'_1 \neq \mathbf{r}_1, ..., \mathbf{r}'_p \neq \mathbf{r}_p, \mathbf{r}_1, ..., \mathbf{r}_p; t)$, determines the p-th order coherence.

To further quantify p-th order coherence, we will use the Glauber p-th order correlation function [30],

$$g^{(p)}(\mathbf{r}_1, ..., \mathbf{r}_p, \mathbf{r}'_1, ..., \mathbf{r}'_p; t) = \frac{\rho^{(p)}(\mathbf{r}_1, ..., \mathbf{r}_p, \mathbf{r}'_1, ..., \mathbf{r}'_p; t)}{\sqrt{\prod_{k=1}^{p} \left[\rho^{(1)}(\mathbf{r}_k, \mathbf{r}_k; t) \rho^{(1)}(\mathbf{r}'_k, \mathbf{r}'_k; t) \right]}}. \tag{2}$$

Essentially, $g^{(p)}$ gives a spatially resolved picture of the representability of the p-th order density matrix by a product of one-body densities.

In order to quantify the degree of condensation [31] of a given many-body state, we resort to the eigenvalues and eigenfunctions of the 1-RDM, ρ_i and φ_k, respectively:

$$\rho^{(1)}(\mathbf{r}, \mathbf{r}'; t) = \sum_i \rho_i(t)\varphi_i^*(\mathbf{r}; t)\varphi_i(\mathbf{r}; t). \tag{3}$$

A state is condensed when a single ρ_i is macroscopic [31]. The state is fragmented, if multiple ρ_i are macroscopic [1, 32–44].

We note that in the following, we will use x instead of $\mathbf{r}$ to emphasize that we study one-dimensional problems. Furthermore, we will also consider the momentum space representations of the p-RDM, $\tilde{\rho}^{(p)}$ and p-th order Glauber correlation functions,

$\tilde{g}^{(p)}$ that are obtained by replacing the state $\Psi(\mathbf{r}_1, ..., \mathbf{r}_N; t)$ by its momentum space representation $\Psi(\mathbf{k}_1, ..., \mathbf{k}_N; t)$ in Eqs. 1 and 2, respectively.

The eigenvalues of the 1-RDM can furthermore be used to characterize different quantum phases in optical lattices. For this purpose, we define the occupation order parameter Δ (see [45]):

$$\Delta = \sum_{i=1}^{M} \left(\frac{\rho_i}{N}\right)^2. \tag{4}$$

For a condensed state, a single eigenvalue ρ_i contributes and Δ is close to unity. For a Mott-insulator, the 1-RDM has as many (quasi-)degenerate eigenvalues as there are sites in the lattice [46]. For an S-site N-particle Mott-insulator $\Delta = N/S$ [47]. In the crystal phase of dipolar bosons, each particle sits in a separate natural orbital and, thus, $\Delta = 1/N$, see [45].

3 Fermionization and Crystallization of Ultracold Strongly Interacting Bosons

Ultracold bosonic atoms, since the realization of Bose-Einstein condensation [48–50], are a versatile tool to study and explore many-body physics. The vessel that holds the atoms as well as their interactions can be controlled almost at will. This control enables investigations into the physically rich, strongly interacting regimes of many-body systems with short- (contact) and long-range (dipolar) interactions.

For the case of strong contact interactions, the bosons' (p-body) density becomes identical to the (p-body) density of non-interacting fermions [35, 46, 51–55]: fermionization takes place.

For the case of strong dipolar interactions, a crystallization process takes place when the dipole-dipole interactions dominate the physics of the many-body system [56–62].

In our work [22], we use MCTDHX [7, 11, 14, 63–71] to compute the crystalline state of dipolar bosons with strong interactions and the fermionized state of bosons with strong contact interactions. See Fig. 1 for a comparison of the 1-RDM.

The 1-RDM for crystalline and fermionized bosons has a correlation hole as a common feature, i.e., $\rho^{(1)}(x, y) \to 0$ for $x \approx y$. Remarkably, however, the behavior of the RDMs as functions of the interaction strength is different for dipolar and contact interactions. For contact interactions (Fig. 1a) N peaks are formed and the RDM remains localized and converges as the interaction strength tends to infinity. In sharp contrast, for the case of dipole-dipole interactions (Fig. 1b) the RDM does not remain localized and diverges as the interaction strength tends to infinity. This divergent property of the (p-body) density of dipolar systems represents one straightforward way for the detection of their crystalline state and is the main result of our work [22].

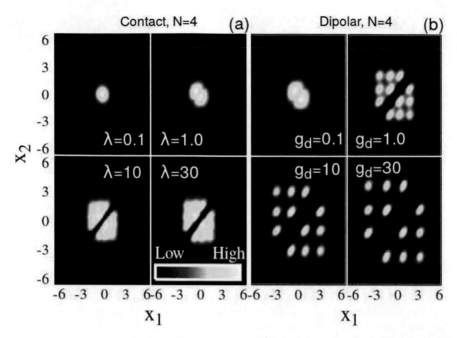

Fig. 1 *Reduced density matrices* $\rho^{(1)}(x = x_1, x' = x_2)$ of $N = 4$ bosons with contact (**a**) and dipolar (**b**) interactions. As the contact interactions become stronger, i.e., $\lambda \to \infty$ in (**a**), the 1-RDM forms a correlation hole $[\rho^{(1)}(x_1, x_2 = x_1) \to 0]$, but the extent of it stays confined and converges as a function of λ—its peaks are clumped together within $(x_1, x_2) \in [-3, 3] \otimes [-3, 3]$, compare the plots for $\lambda = 10$ and $\lambda = 30$ in (**a**). As the dipolar interactions become stronger, the 1-RDM also forms a correlation hole; in contrast to the contact interaction case in (**a**), however, the correlation hole and the 1-RDM continue to expand as a function of g_d: completely separated peaks form and the 1-RDM does not converge for large g_d [compare (**a**) and (**b**)]. Figure reprinted from [22]

4 Correlations of Crystallized Dipolar Bosons

Bosons—the eeriest of the just two fundamental families of elementary particles in the universe—when cooled down to temperatures close to absolute zero only manifest their unearthly etiquette. Indeed, when several of them cuddle together they behave as if they were a single particle and not a noisy multitude; they metamorphose into a Bose-Einstein condensate [48–50].

As if this was not weird enough, they all behave—in such low temperatures—more like waves rather than like particles, thus underpinning their innate quantum mechanical nature. To perplex the situation a little, one needs to consider that any two particles of the system will interact: repel, attract or something in between, that can depend on the distance and orientation of the duo. Now there is the catch: if the interactions are weak enough the "condensate-like behavior" dominates and hence all atoms hardly feel the presence of each other. All move freely around and their motions are not correlated (anyone can go wherever). They do, in fact, flow that freely

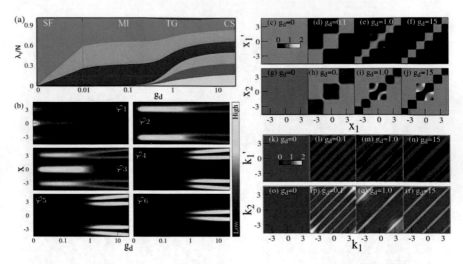

Fig. 2 *Tracing the crystallization of dipolar bosons in lattices* with eigenvalues (**a**) and eigenfunctions (**b**) of the one-body reduced density matrix, one-body Glauber correlation functions in real (**c**)–(**f**) and momentum (**k**)–(**n**) space, and two-body Glauber correlation functions in real (**g**)–(**j**) and momentum space (**o**)–(**r**). As the dipolar interactions, g_d, become stronger, the eigenvalues of the 1-RDM (**a**) transition from one (condensation, superfluidity, SF), through three (Mott insulator, MI) to six (crystal, CS) significant contributions. The transitions in (**a**) are accompanied by structural changes in the scaled eigenfunctions ($\lambda_i \varphi_i$) of the 1-RDM (**b**). As a key result we demonstrate that each quantum phase can be identified by its one- and two-body correlations in real ((**c**)–(**f**) and (**g**)–(**j**), respectively) and/or momentum space ((**k**)–(**n**) and (**o**)–(**r**), respectively]. Figure panels adapted from [21]

and rapidly that this state has the peculiar advantage of being called a "superfluid". The crucial piece of information here is this very connection between correlations (almost none) and phase of the system (superfluid).

Now consider stronger interactions—in particular ones that resemble the interactions between electric dipoles—and a periodic structure, which means little 'hills' and 'valleys' where the particles preferentially sit—that resembles a crystalline order of ions in a metal. In this scenario our—before—superfluid transitions into a different phase where it does not flow as freely anymore: a Mott insulator. For even stronger interactions, the system takes an ordered appearance and reaches a crystal state, i.e. a frozen configuration of atoms. These phases can show a wide variety of characteristics that differ drastically; as if it were a different material every time.

In the present work [21], we demonstrate how the transitions described above happen across the different phases of the system, see Fig. 2. The transitions are monitored as the interactions become stronger and stronger and their correlations are mapped out in a very accurate computational fashion using MCTDH-X. Quantum correlations undergird the phase transitions; this innate connection is the novelty of our work.

5 Cavity-Induced Phase Transitions of Bose-Einstein Condensates

At ultracold temperature, bosonic atoms will condense into the same state and form a so-called Bose-Einstein condensate (BEC) [48–50]. In a BEC, all the atoms become coherent and therefore the whole collection of atoms behaves like a unique entity - a superfluid. However, the superfluidity of the atoms can be significantly modified if they are placed in an optical cavity. In its simplest realization, such a cavity consists of two very reflective mirrors. Inside a cavity, photons are bouncing between the two mirrors. If the resonance frequency of the cavity matches the difference of energy levels inside the atoms, these photons can easily excite the atoms. By virtue of the reflection in the cavity the photons mediate an infinite-range interaction between any two atoms [72]. In the present setup, the interaction strength can be controlled by tuning the strength of a laser that irradiates the atoms. A strong enough interaction can drive the BEC into a self-organized state where the atoms gather into clusters with equal spacings between each other [73–75], and reduce or even eliminate the superfluidity in the system [76].

In the work [20], we use MCTDH-X [14] to investigate such an elimination of superfluidity stemming from the infinite-range interaction. As the interaction strength increases from zero, the atoms first cluster in a superlattice structure, where the super-fluid character is only partially still retained. As the interaction strength becomes even stronger, the atoms in different clusters start to lose information about each other. Eventually, the atoms from different clusters become completely uncorrelated and superfluidity is totally lost: the system is then said to exhibit a Mott insulating behavior. With MCTDH-X, the loss of superfluidity and the appearance of Mott insulation can be captured through many observables of the system (Fig. 3), such as the momentum space density distribution, Glauber one-body and two-body correlation functions [30], and natural orbital occupations [45].

6 Summary and Outlook

Our works [15–22] demonstrate how fruitful the combination of the MCTDHB theory and its software implementations [12–14] with the Hazel Hen Cray XC40 cluster has been and continues to be: novel, fundamental, and intriguing insights on the many-body physics of ultracold interacting atoms in cavities, with dipolar or contact interactions as well as the position and momentum space variances of many-body systems and the effects of the range of interactions have been unveiled in a multitude of setups. All the works in this report [15–22] and many other previous works with MCTDHB have investigated p-body RDMs and p-body correlation functions for $p \leq 2$ (see, for example, [29, 77–84]). We identify the computation and investigation of p-body RDMs and p-body Glauber correlation functions for $p > 2$ as one main

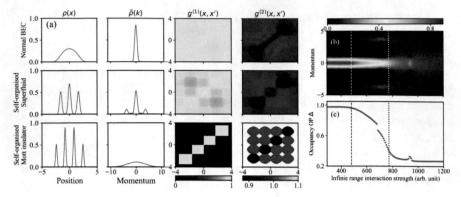

Fig. 3 *Phases of BECs with different correlation effects inside an optical cavity.* **a** The position space density distribution $\rho(x)$ (first column), momentum space density distribution $\tilde{\rho}(k)$ (second column), Glauber one-body correlation function $g^{(1)}(x, x')$ (third column), Glauber two-body correlation function $g^{(2)}(x, x')$ (fourth column) in a normal BEC state (first row), a self-organized superfluid state (second row) and a self-organized Mott insulator phase (third row). **b** The momentum density distribution and **c** the occupancy order parameter as a function of the infinite range interaction strength. Three different phases can be clearly seen according to these two quantities. The occupancy order parameter is defined by the MCTDHB orbital occupancies $\Delta = \sum_{i=1}^{M} \rho_i^2$ with ρ_i the occupancies of the orbitals [45]. The figure panels are reproduced from Ref. [20]

immediate goal. Other aims include the computation of full distribution functions of observables and classification strategies with single-shot images [45, 69, 88].

Acknowledgements Financial support by the Deutsche Forschungsgemeinschaft (DFG) is grate-fully acknowledged. OEA acknowledges funding by the Israel Science Foundation (Grant No. 600/15). We acknowledge financial support by the Austrian Science Foundation (FWF) under grants P32033 and M2653 as well as by the Wiener Wissenschafts- und TechnologieFonds (WWTF) project No. MA16-066. We acknowledge financial support from FAPESP, the Department of Science and Technology, Government of India under DST Inspire Faculty fellowship, a FAPESP (grant No. 2016/19622-0), a UFC fellowship, and financial support from the Swiss National Science Founda-tion and Mr. G. Anderheggen.

References

1. A.I. Streltsov, O.E. Alon, L.S. Cederbaum, Phys. Rev. A **73**, 063626 (2006)
2. A.I. Streltsov, O.E. Alon, L.S. Cederbaum, Phys. Rev. Lett. **99**, 030402 (2007)
3. O.E. Alon, A.I. Streltsov, L.S. Cederbaum, J. Chem. Phys. **127**, 154103 (2007)
4. O.E. Alon, A.I. Streltsov, L.S. Cederbaum, Phys. Rev. A **77**, 033613 (2008)
5. K. Sakmann, A.I. Streltsov, O.E. Alon, L.S. Cederbaum, Phys. Rev. Lett. **103**, 220601 (2009)
6. A.U.J. Lode, K. Sakmann, O.E. Alon, L.S. Cederbaum, A.I. Streltsov, Phys. Rev. A **86**, 063606 (2012)
7. A.U.J. Lode, Phys. Rev. A **93**, 063601 (2016)
8. *Multidimensional Quantum Dynamics: MCTDH Theory and Applications*, edited by H.-D. Meyer, F. Gatti, G.A. Worth (Wiley-VCH, Weinheim, 2009)

9. K. Sakmann, *Many-Body Schrödinger Dynamics of Bose-Einstein Condensates, Springer Theses* (Springer, Heidelberg, 2011)
10. *Quantum Gases: Finite Temperature and Non-equilibrium Dynamics*, edited by N. P. Proukakis, S.A. Gardiner, M.J. Davis, M.H. Szymanska, Cold Atoms Series, Vol. 1 (Imperial College Press, London, 2013)
11. A.U.J. Lode, *Tunneling Dynamics in Open Ultracold Bosonic Systems, Springer Theses* (Springer, Heidelberg, 2015)
12. A.I. Streltsov, L.S. Cederbaum, O.E. Alon, K. Sakmann, A.U.J. Lode, J. Grond, O.I. Streltsova, S. Klaiman, R. Beinke, *The Multiconfigurational Time-Dependent Hartree for Bosons Package*, version 3.x, http://mctdhb.org, Heidelberg/Kassel (2006-Present)
13. A.I. Streltsov, O.I. Streltsova, *The multiconfigurational time-dependent Hartree for bosons laboratory*, version 1.5, http://MCTDHB-lab.com
14. A.U.J. Lode, M.C. Tsatsos, S.E. Weiner, E. Fasshauer, R. Lin, L. Papariello, P. Molignini, C.Lévêque, MCTDH-X: The time-dependent multiconfigurational Hartree for indistinguishable particles software, http://ultracold.org (2019)
15. O.E. Alon, J. Phys.: Conf. Ser. **1206**, 012009 (2019)
16. S.K. Haldar, O.E. Alon, New J. Phys. **21**, 103037 (2019)
17. S.K. Haldar, O.E. Alon, J. Phys.: Conf. Ser. **1206**, 012010 (2019)
18. O.E. Alon, L.S. Cederbaum, Chem. Phys. **515**, 287 (2018)
19. O.E. Alon, Mol. Phys. (2019). https://doi.org/10.1080/00268976.2019.1587533
20. R. Lin, L. Papariello, P. Molignini, R. Chitra, A.U.J. Lode, Phys. Rev. A **100**, 013611 (2019)
21. B. Chatterjee, M.C. Tsatsos, A. U. J. Lode (2019). https://doi.org/10.1088/1367-2630/aafa93, New J. Phys. 21, 033030
22. S. Bera, B. Chakrabarti, A. Gammal, M.C. Tsatsos, M.L. Lekala, B. Chatterjee, C. Lévêque, A. U. J. Lode (2019). http://doi.org/10.1038/s41598-019-53179-1 (in press in Scientific Reports)
23. A.U.J. Lode, K. Sakmann, R.A. Doganov, J. Grond, O.E. Alon, A.I. Streltsov, L.S. Cederbaum, in High Performance Computing, in *Science and Engineering '13: Transactions of the High Performance Computing Center, Stuttgart (HLRS) 2013*, ed. by W.E. Nagel, D.H. Kröner, M.M. Resch (Springer, Heidelberg, 2013), pp. 81–92
24. S. Klaiman, A.U.J. Lode, K. Sakmann, O.I. Streltsova, O.E. Alon, L.S. Cederbaum, A.I. Streltsov, in High Performance Computing, in *Science and Engineering '14: Transactions of the High Performance Computing Center, Stuttgart (HLRS) 2014*, ed. by W.E. Nagel, D.H. Kröner, M.M. Resch (Springer, Heidelberg, 2015), pp. 63–86
25. O.E. Alon, V.S. Bagnato, R. Beinke, I. Brouzos, T. Calarco, T. Caneva, L.S. Cederbaum, M.A. Kasevich, S. Klaiman, A.U.J. Lode, S. Montangero, A. Negretti, R.S. Said, K. Sakmann, O.I. Streltsova, M. Theisen, M.C. Tsatsos, S.E. Weiner, T. Wells, A.I. Streltsov, in High Performance Computing, in *Science and Engineering '15: Transactions of the High Performance Computing Center, Stuttgart (HLRS) 2015*, ed. by W.E. Nagel, D.H. Kröner, M.M. Resch (Springer, Heidelberg, 2016), pp. 23–50
26. O.E. Alon, R. Beinke, L.S. Cederbaum, M.J. Edmonds, E. Fasshauer, M.A. Kasevich, S. Klaiman, A.U.J. Lode, N.G. Parker, K. Sakmann, M.C. Tsatsos, A.I. Streltsov, in High Performance Computing, in *Science and Engineering '16: Transactions of the High Performance Computing Center, Stuttgart (HLRS) 2016*, ed. by W.E. Nagel, D.H. Kröner, M.M. Resch (Springer, Heidelberg, 2016), pp. 79–96
27. O.E. Alon, R. Beinke, C. Bruder, L.S. Cederbaum, S. Klaiman, A.U.J. Lode, K. Sakmann, M. Theisen, M.C. Tsatsos, S.E. Weiner, A.I. Streltsov, in High Performance Computing, in *Science and Engineering '17: Transactions of the High Performance Computing Center, Stuttgart (HLRS) 2017*, ed. by W.E. Nagel, D.H. Kröner, M.M. Resch (Springer, Heidelberg, 2018), pp. 93–115
28. O. E. Alon, V. S. Bagnato, R. Beinke, S. Basu, L. S. Cederbaum, B. Chakrabarti, B. Chatterjee, R. Chitra, F. S. Diorico, S. Dutta, L. Exl, A. Gammal, S. K. Haldar, S. Klaiman, C. Lévêque, R. Lin, N. J. Mauser, P. Molignini, L. Papariello, R. Roy, K. Sakmann, A. I. Streltsov, G. D. Telles, M. C. Tsatsos, R. Wu, and A. U. J. Lode, *High Performance Computing in Science and Engineering '17: Transactions of the High Performance Computing Center, Stuttgart (HLRS)*

2018, edited by W. E. Nagel, D. H. Kröner, and M. M. Resch (Springer, Heidelberg, 2019), pp. XX-YY

29. K. Sakmann, A.I. Streltsov, O.E. Alon, L. S. Cederbaum Phys. Rev. A **78**, 023615 (2008)
30. R.J. Glauber, Phys. Rev. **130**, 2529 (1963)
31. O. Penrose, L. Onsager, Phys. Rev. **104**, 576 (1956)
32. O.E. Alon, L.S. Cederbaum, Phys. Rev. Lett. **95**, 140402 (2005)
33. R.W. Spekkens, J.E. Sipe, Phys. Rev. A **59**, 3868 (1999)
34. A.I. Streltsov, L.S. Cederbaum, N. Moiseyev, Phys. Rev. A **70**, 053607 (2004)
35. S. Zöllner, H.-D. Meyer, P. Schmelcher, Phys. Rev. A **74**, 063611 (2006)
36. P. Noziéres and D. Saint James, J. Phys. (Paris) **43**, 1133 (1982)
37. P. Noziéres, in *Bose-Einstein Condensation*, ed. by A. Griffin, D.W. Snoke, S. Stringari (Cambridge University Press, Cambridge, 1996), p. 15
38. E.J. Mueller, T.-L. Ho, M. Ueda, G. Baym, Phys. Rev. A **74**, 033612 (2006)
39. U.R. Fischer, P. Bader, Phys. Rev. A **82**, 013607 (2010)
40. Q. Zhou, X. Cui, Phys. Rev. Lett. **110**, 140407 (2013)
41. Y. Kawaguchi, Phys. Rev. A **89**, 033627 (2014)
42. S.-W. Song, Y.-C. Zhang, H. Zhao, X. Wang, W.-M. Liu, Phys. Rev. A **89**, 063613 (2014)
43. H.H. Jen, S.-K. Yip, Phys. Rev. A **91**, 063603 (2015)
44. A.R. Kolovsky, Phys. Rev. A **95**, 033622 (2017)
45. B. Chatterjee, A.U.J. Lode, Phys. Rev. A **98**, 053624 (2018)
46. R. Roy, A. Gammal, M.C. Tsatsos, B. Chatterjee, B. Chakrabarti, A.U.J. Lode, Phys. Rev. A **97**, 043625 (2018)
47. D. Jaksch, C. Bruder, J.I. Cirac, C.W. Gardiner, P. Zoller, Phys. Rev. Lett. **81**, 3108 (1998)
48. C.J. Pethick, H. Smith, *Bose-Einstein Condensation in Dilute Gases*, 2nd edn. (Cambridge University Press, Cambridge, 2008)
49. M.H. Anderson, J.R. Ensher, M.R. Matthews, C.E. Wieman, E.A. Cornell, Science **269**, 198 (1995)
50. K.B. Davis, M.-O. Mewes, M.R. Andrews, N.J. van Druten, D.S. Durfee, D.M. Kurn, W. Ketterle, Phys. Rev. Lett. **75**, 3969 (1995)
51. M. Girardeau, J. Math Phys. **1**, 516 (1960)
52. G. Zürn, F. Serwane, T. Lompe, A.N. Wenz, M.G. Ries, J.E. Bohn, and S. Jochim Phys. Rev. Lett. **108**, 075303 (2012)
53. T. Jacqmin, J. Armijo, T. Berrada, K.V. Kheruntsyan, and I. Bouchoule Phys. Rev. Lett. **106**, 230405 (2011)
54. O.E. Alon, L.S. Cederbaum, Phys. Rev. Lett. **95**, 140402 (2005)
55. S. Zöllner, H.-D. Meyer, P. Schmelcher, Phys. Rev. A **78**, 013629 (2008)
56. S. Zöllner, G.M. Bruun, C.J. Pethick, S.M. Reimann, Phys. Rev. Lett. **107**, 035301 (2011)
57. S. Zöllner, Phys. Rev. A **84**, 063619 (2011)
58. G.E. Astrakharchik, YuE Lozovik, Phys. Rev. A **77**, 013404 (2008)
59. A.S. Arkhipov, G.E. Astrakharchik, A.V. Belikov, Y.E. Lozovik, JETP Lett. **82**, 39 (2005)
60. G.E. Astrakharchik, G.E. Morigi, G. De Chiara, J. Boronat, Phys. Rev. A **78**, 063622 (2008)
61. F. Deuretzbacher, J.C. Cremon, S.M. Reimann, Phys. Rev. A **81**, 063616 (2010)
62. B. Chatterjee, I. Brouzos, L. Cao, P. Schmelcher, J. Phys. B **46**, 085304 (2013)
63. E. Fasshauer, A.U.J. Lode, Phys. Rev. A **93**, 033635 (2016)
64. S. Klaiman, A.U.J. Lode, A.I. Streltsov, L.S. Cederbaum, O.E. Alon, Phys. Rev. A **90**, 043620 (2014)
65. A.U.J. Lode, M.C. Tsatsos, J.L. Temp, Phys. **181**, 171 (2015)
66. T. Wells, A.U.J. Lode, V.S. Bagnato, M.C. Tsatsos, J.L. Temp, Phys. **180**, 133 (2015)
67. U.R. Fischer, A.U.J. Lode, B. Chatterjee, Phys. Rev. A **91**, 063621 (2015)
68. A.U.J. Lode, B. Chakrabarti, V.K.B. Kota, Phys. Rev. A **92**, 033622 (2015)
69. A.U.J. Lode, C. Bruder, Phys. Rev. Lett. **118**, 013603 (2017)
70. S.E. Weiner, M.C. Tsatsos, L.S. Cederbaum, A.U.J. Lode, Sci. Rep. **7**, 40122 (2017)
71. A.U.J. Lode, C. Bruder, Phys. Rev. A **94**, 013616 (2016)

72. R. Mottl, DOI: https://dx.doi.org/10.3929/ethz-a-010336231, *Roton-type mode softening in a dissipative quantum many-body system with cavity-mediated long-range interactions*, PhD Thesis, ETH Zürich (2014)
73. P. Domokos, H. Ritsch, Phys. Rev. Lett. **89**, 253003 (2002)
74. D. Nagy, G. Szirmai, P. Domokos, Eur. Phys. J. D **48**, 127 (2008)
75. K. Baumann, C. Guerlin, F. Brennecke, T. Esslinger, Nature **464**, 1301 (2010)
76. J. Klinder, H. Keßler, M. Bakhtiari, Reza, M. Thorwart, and A. Hemmerich, Phys. Rev. Lett. **115**, 230403 (2015)
77. A.U.J. Lode, S. Klaiman, O.E. Alon, A.I. Streltsov, L.S. Cederbaum, Phys. Rev. A **89**, 053620 (2014)
78. O.E. Alon, A.I. Streltsov, K. Sakmann, A.U.J. Lode, J. Grond, L.S. Cederbaum, Chem. Phys. **401**, 2 (2012)
79. I. Březinová, J. Burgdörfer, A.U.J. Lode, A.I. Streltsov, L.S. Cederbaum, O.E. Alon, L.A. Collins, B.I. Schneider, J. Phys.: Conf. Ser. **488**, 012032 (2014)
80. A.U.J. Lode, A.I. Streltsov, K. Sakmann, O.E. Alon, L.S. Cederbaum, Proc. Natl. Acad. Sci. **109**, 13521 (2012)
81. J. Grond, A.I. Streltsov, A.U.J. Lode, K. Sakmann, L.S. Cederbaum, and O. E. Alon Phys. Rev. A **88**, 023606 (2013)
82. J.H.V. Nguyen, M.C. Tsatsos, D. Luo, A.U.J. Lode, G.D. Telles, V.S. Bagnato, R.G. Hulet, Phys. Rev. X **9**, 011052 (2019)
83. P. Molignini, L. Papariello, A.U.J. Lode, R. Chitra, Phys. Rev. A **98**, 053620 (2018)
84. A.U.J. Lode, F.S. Diorico, R. Wu, P. Molignini, L. Papariello, R. Lin, C. Lévêque, L. Exl, M.C. Tsatsos, R. Chitra, N.J. Mauser, New J. Phys. **20**, 055006 (2018)
85. C. Lévêque and L. B. Madsen, DOI: https://doi.org/10.1088/1367-2630/aa6319, New J. Phys. **19**, 043007 (2017)
86. C. Lévêque and L. B. Madsen, DOI: https://doi.org/10.1088/1361-6455/aacac6, J. Phys. B **51**, 155302 (2018)
87. C. Lévêque, L.B. Madsen, *to be published in J* (Chem, Phys, 2019)
88. K. Sakmann, M. Kasevich, Nat. Phys. **12**, 451 (2016)

Hadronic Contributions to the Anomalous Magnetic Moment of the Muon from Lattice QCD

M. Cè, A. Gérardin, G. von Hippel, B. Hörz, H. B. Meyer, D. Mohler,
K. Ottnad, S. Schaefer, J. Wilhelm, and H. Wittig

Abstract The Standard Model of Particle Physics describes three of the four known fundamental interactions: the strong interaction between quarks and gluons, the electromagnetic interaction, and the weak interaction. While the Standard Model is extremely successful, we know that it is not a complete description of nature. One way to search for physics beyond the Standard Model lies in the measurement of precision observables. The anomalous magnetic moment of the muon $a_\mu \equiv \frac{1}{2}(g-2)_\mu$, quantifying the deviation of the gyromagnetic ratio from the exact value of 2 predicted by the Dirac equation, is one such precision observable. It exhibits a persistent discrepancy of 3.5 standard deviations between the direct measurement and its theoretical prediction based on the Standard Model. The total uncertainty of 0.5 ppm of this prediction is dominated by effects of the strong interaction, notably the contributions from hadronic vacuum polarisation and from hadronic light-by-light scattering. In this project we address some of the leading systematic uncertainties in a first principles determination of these contributions using Lattice Quantum Chromodynamics.

1 Introduction

The discovery of the Higgs boson at the LHC in 2012 has firmly established the Standard Model (SM) of Elementary Particle Physics. While the SM successfully describes all processes at high-energy colliders, it fails to answer some of the most pressing problems in particle physics, such as the nature of dark matter or the asymmetry between matter and antimatter. Therefore, efforts to search for physics beyond the SM have intensified in recent years, with the LHC experiments searching for new particles and interactions. An alternative approach seeks to detect deviations

M. Cè · A. Gérardin · G. von Hippel · B. Hörz · H. B. Meyer · D. Mohler (✉) · K. Ottnad ·
J. Wilhelm · H. Wittig
Institut für Kernphysik and Helmholtz Institut Mainz, Universität Mainz, Becher Weg 45, 55099
Mainz, Germany
e-mail: damohler@uni-mainz.de

A. Gérardin · S. Schaefer
NIC, DESY-Zeuthen, Platanenallee 6, 15738 Zeuthen, Germany

© Springer Nature Switzerland AG 2021
W. E. Nagel et al. (eds.), *High Performance Computing in Science and Engineering '19*,
https://doi.org/10.1007/978-3-030-66792-4_6

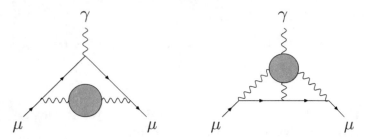

Fig. 1 Diagrams for the hadronic vacuum polarisation (left pane) and for the hadronic light-by-light scattering (right pane) contributions to a_μ

between measurements of precision observables, such as the neutron lifetime or the proton radius, and their theoretical predictions based on the SM. One very prominent example is the anomalous magnetic moment of the muon, $a_\mu = \frac{1}{2}(g-2)_\mu$, i.e. the deviation of its gyromagnetic ratio g_μ from the exact value of 2 predicted by the Dirac equation. This quantity exhibits a persistent discrepancy of 3.5 standard deviations between the direct measurement and its theoretical prediction based on the SM [1]:

$$a_\mu = \begin{cases} 116\,592\,080(54)(33) \cdot 10^{-11} & \text{(experiment)} \\ 116\,591\,828(43)(26)(2) \cdot 10^{-11} & \text{(SM prediction)}. \end{cases} \tag{1}$$

The total uncertainty of 0.5 ppm of the SM prediction is dominated by effects of the strong interaction, notably the contributions from hadronic vacuum polarisation (HVP), a_μ^{hvp} and from hadronic light-by-light scattering (HLbL), a_μ^{hlbl}. The contributions are shown in Fig. 1. The estimate for a_μ^{hvp} is typically obtained from dispersion theory using the experimentally determined cross section $e^+ e^- \to$ hadrons as input [2–4]. This implies that – even though the overall precision of 0.5% is quite high – the theory estimate is subject to experimental uncertainties. The hadronic light-by-light scattering contribution has so far only been determined via model estimates [4–7]. The relative uncertainty of 25% in a_μ^{hlbl} is large and difficult to quantify.

Two new experiments (E989 at Fermilab [9, 10] and E34 at J-PARC [11]) will reduce the error of the direct experimental determination of a_μ by a factor four in the years ahead. This has sparked a new activity, the *Muon g − 2 Theory Initiative*,[1] which represents a concerted effort between theorists and experimentalists, seeking to produce a more precise theoretical prediction for a_μ which can match the precision of the new experiments. The goal of our project, namely to provide highly precise estimates for a_μ^{hvp} and a_μ^{hlbl}, is a crucial ingredient in this endeavour.

The current state-of-affairs urgently calls for a first-principles approach. A suitable method is Lattice Quantum Chromodynamics (Lattice QCD), a formulation of the fundamental interaction between quarks and gluons on a Euclidean space-time grid (lattice). In Lattice QCD, physical observables of the discretised theory of the strong

---

[1] https://muon-gm2-theory.illinois.edu.

interaction are calculated by Monte-Carlo integration. The effects of the discretisation and the finite volume have to be controlled by calculations with different lattice spacings and different lattice volumes. Due to computational challenges, most past and current calculations are also performed at unphysical quark masses, extrapolating the results to the physical values. In order to obtain more precise estimates of a_μ^{hvp} and a_μ^{hlbl} it is necessary to avoid long extrapolations in the quark mass [8]. Therefore, obtaining precise results for a_μ^{hvp} and a_μ^{hlbl} at the physical value of the light (up and down) and strange quark masses is a crucial step towards a decisive test of the SM with much increased sensitivity.

In Sect. 2 we describe the observables we need to calculate, the major computational challenges, and the main features of the software setup used for our calculations. We provide details of our simulations on HazelHen and describe the current status of the gauge field generation for the ensemble at physical light-quark masses. The status of these simulations has previously been described in [12]. In Sect. 3 we summarise the status of our determination of the hadronic vacuum polarisation contribution to the anomalous magnetic moment of the muon. For more details the reader is referred to [13]. We conclude with a brief summary and outlook section.

2 Observables and Computational Setup

The long-term goal is the determination of a_μ^{hvp} with sub-percent precision, including quark-disconnected contributions, and effects from isospin breaking. In the current project our main focus is on the quark-connected part of a_μ^{hvp} at the physical pion mass and for small lattice spacings.

2.1 Observables

Our methodology is based on the expression for a_μ^{hvp} derived in the "time-momentum representation" (TMR) [24]

$$a_\mu^{\text{hvp}} = \left(\frac{\alpha}{\pi}\right)^2 \int_0^\infty dx_0\, \widetilde{K}(x_0) G(x_0), \quad G(x_0) = -\frac{a^3}{3} \sum_{k=1}^{3} \sum_{\mathbf{x}} \langle J_k(x_0, \mathbf{x}) J_k(0)\rangle,$$

$$(2)$$

where $G(x_0)$ denotes the summed correlator of the electromagnetic current $J_\mu = \frac{2}{3}\bar{u}\gamma_\mu u - \frac{1}{3}\bar{d}\gamma_\mu d - \frac{1}{3}\bar{s}\gamma_\mu s \cdots$, and $\widetilde{K}(x_0)$ is a known kernel function. To improve statistics, the TMR correlator in Eq. 2 is averaged over the three spatial directions. Unlike the more conventional method based on a four-dimensional Fourier transform

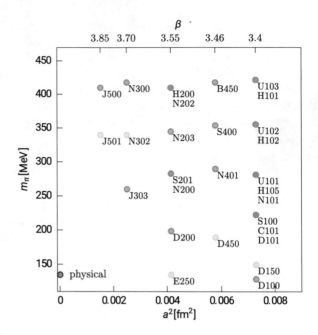

Fig. 2 Landscape of CLS 2+1-flavour ensembles along a trajectory with constant trace of the quark mass matrix Tr(M) = const. The lattice spacing squared is displayed on the x-axis, while the y-axis shows the pion mass. Ensembles still in production are shown in yellow. Multiple ensemble names next to a single dot indicate ensembles with different volumes at the (otherwise) same set of simulation parameters

of the current-current correlator [25, 26], the TMR formulation can be used both with periodic and open boundary conditions.

The importance of results computed directly at the physical pion mass has been emphasised recently in Ref. [8]. Since there is no firm theoretical guidance for the pion mass dependence of a_μ^{hvp}, the commonly used fit *ansätze* are at best phenomenological and can lead to ambiguities of the order of a few percent at the physical pion mass. In order to remove all doubts as to the reliability of the results, simulations must therefore be performed directly at or very near $m_\pi = m_\pi^{\mathrm{phys}}$, which we achieved by computing a_μ^{hvp} on ensemble E250 generated within this project.

The formulation of lattice QCD used in this project is the same as that used for all other ensembles with $2 + 1$ dynamical flavours by the *Coordinated Lattice Simulations* (CLS) consortium, i.e. the tree-level Symanzik-improved gauge action together with the non-perturbatively O(a) improved Wilson fermion action. An overview plot of the existing gauge ensembles is shown in Fig. 2. The simulations have been performed using the Hybrid Monte Carlo (HMC) algorithm as implemented in the openQCD code developed by M. Lüscher and S. Schaefer [14], which incorporates a number of significant algorithmic developments, including the use of

- twisted-mass preconditioning and determinant reweighting [14, 15] for the mass-degenerate light-quark doublet to avoid problems with the near-zero modes typical for Wilson fermions ("exceptional configurations"),
- the Rational Hybrid Monte Carlo (RHMC) algorithm including determinant splitting and reweighting [16] for the strange quark,

- a local deflation-accelerated solver [17, 18] to avoid slowing down in the chiral quark mass region,
- a hierarchical, multiple time-step integration scheme using the second and fourth-order integrators of Omelyan, Mryglod and Folk [19] to preserve high accuracy and achieve good acceptance rates.

While most of the CLS gauge field ensembles make use of open boundary conditions in the time direction [14, 20] to avoid getting trapped in topological sectors [21] as the continuum limit is approached, the calculations at physical pion and kaon masses described below use periodic boundary conditions to maximise the statistical resolution obtained.

In addition to the generation of gauge ensembles, our project relies on the calculation of quark propagators and their subsequent contractions to form suitable hadronic correlation functions. The performance depends crucially on an efficient solver for the inhomogeneous Dirac equation, i.e.

$$D_{\mathrm{w}}\psi = \eta, \tag{3}$$

where D_{w} is the O(a) improved Wilson-Dirac operator, and η some source vector. In our project we use the deflation-accelerated GCR solver described by Lüscher[17, 18], which is also used in the generation of the gauge ensembles.

The large system size of ensemble E250 implies that the numerical effort to compute the observables will be much larger than for the remaining set of CLS ensembles. For all ensembles, except E250, the correlation functions are computed using point sources, randomly distributed in space and in the centre of the lattice in the time direction. In order to achieve a high-statistics calculation of a_{μ}^{hvp} and of the pion transition form factor $\mathcal{F}_{\pi^0\gamma^*\gamma^*}$ at moderate cost, propagators on E250 are estimated using stochastic sources, with noise partitioning in spin, colour and time [27, 28]. Each source has support on a single, randomly chosen timeslice. Errors are estimated throughout the calculation using the jackknife procedure with blocking in order to take into account auto-correlation effects.

The solution to Eq. 3 represents the quark propagator $\psi(x) = S(x, x^{\mathrm{src}})$, where x^{src} denotes the position of the source vector η. Quark propagators are then contracted to form the quark-connected two- and three-point correlation functions

$$C_{\mu\nu}^{(2)}(\tau) = a^3 \sum_{\mathbf{x}} \left\langle J_{\mu}^{\mathrm{loc}}(x_0^{\mathrm{src}}, \mathbf{x}^{\mathrm{src}}) J_{\nu}^{\mathrm{ps}}(x_0, \mathbf{x}) \right\rangle, \quad \tau = x_0 - x_0^{\mathrm{src}}, \tag{4}$$

$$C_{\mu\nu}^{(3)}(\tau, \tau_\pi) = a^6 \sum_{\mathbf{x},\mathbf{y}} e^{-i\mathbf{q}\cdot\mathbf{x}} e^{i\mathbf{p}\cdot\mathbf{y}} \left\langle J_{\mu}^{\mathrm{loc}}(x_0^{\mathrm{src}}, \mathbf{x}^{\mathrm{src}}) J_{\nu}^{\mathrm{ps}}(x_0, \mathbf{x}) P^{\dagger}(y_0, \mathbf{y}) \right\rangle,$$

where $\tau_\pi = \min(x_0 - y_0, x_0^{\mathrm{src}} - y_0)$, and the superscripts "loc" and "ps" denote the local and conserved (point-split) vector currents, respectively. The calculation of the three-point function $C_{\mu\nu}^{(3)}(\tau, \tau_\pi)$ proceeds by computing the sequential propagator for which the source vector η in Eq. 3 is replaced by the solution vector ψ, multiplied by a suitable Dirac matrix and the phase factor $e^{i\mathbf{p}\cdot\mathbf{y}}$. The two-point function $C_{\mu\nu}^{(2)}$

is used to compute the vector correlator $G(x_0)$ that appears in the expression for a_μ^{hvp} of Eq. 2, while the three-point function $C_{\mu\nu}^{(3)}$ is required for the determination of the transition form factor $\mathcal{F}_{\pi^0\gamma^*\gamma^*}$ [22, 23]. Knowledge of the latter is required to determine the contribution to a_μ^{hlbl} from the pion pole, which is expected to be dominant.

2.2 Code Performance and Workflow

To achieve the target precision for the hadronic contributions to a_μ, generation of an ensemble of gauge configurations with physical pion and kaon masses is mandatory. This is the main computational task of the project. For the gauge field generation we use version 1.6 of the openQCD code [14] developed by Martin Lüscher and project contributor Stefan Schaefer, which is publicly available under the GPL license.

For the gauge field generation runs on HazelHen at HLRS we used two setups

A Local lattice volume of size $12 \times 8 \times 16 \times 16$ per MPI rank with 6912 MPI ranks on 288 nodes
B Local lattice volume of size $12 \times 8 \times 8 \times 16$ per MPI rank with 13824 MPI ranks on 576 nodes

In each case a hypercube of $2 \times 3 \times 2 \times 2$ processes was grouped onto a single node, to minimise off-node communication. This process setup was also determined to be optimal by the Cray profiler. For optimal run performance on HazelHen the following MPI parameters had to be used:

```
export MPICH_GNI_MAX_EAGER_MSG_SIZE=131072
export MPICH_GNI_MAX_VSHORT_MSG_SIZE=8000
export MPICH_USE_DMAPP_COLL=1
export MPICH_DMAPP_HW_CE=1
export MPICH_DMAPP_A2A_ARIES=1
export MPICH_GNI_ROUTING_MODE=ADAPTIVE_3
export MPICH_GNI_A2A_ROUTING_MODE=ADAPTIVE_3
```

Figure 3 shows the runtime history for a single HMC update with a trajectory length of 2 molecular dynamics units (MDU) compared to a previous low-statistics (thermalisation) run on MOGON II at Mainz University. Evidently, some slow runs have been observed on HazelHen, while no large runtime fluctuations were seen on MOGON II. This behaviour has in part been tracked down to machine-related interference from other jobs.

In addition to the gauge field generation, we also performed propagator runs on HazelHen with Setup C below:

C Local lattice volume of size $16 \times 16 \times 16 \times 8$ per MPI rank with 5184 MPI ranks on 216 nodes.

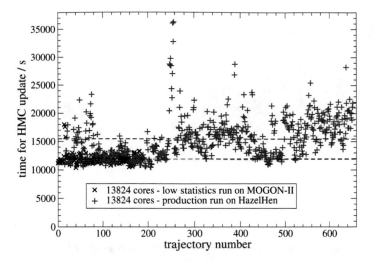

Fig. 3 Runtime history for completed HMC trajectories with 13824 MPI ranks on HazelHen (red pluses) compared to a previous thermalisation run on MOGON II (black crosses) at Mainz University. The average runtime is denoted by the dashed horizontal lines. In our proposal we estimated that HMC updates on these machines would have a similar runtime. While this is the case for the fastest runs, the runtime-variation on HazelHen is larger than on MOGON II

Figure 4 shows a histogram of typical propagator runtimes on HazelHen. For the reduced partition size the runs are much more stable, even though some outliers with larger runtime occurred.

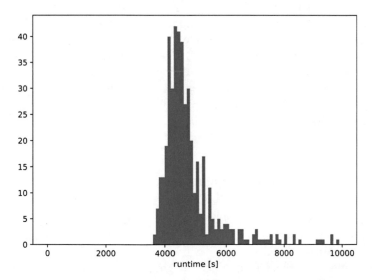

Fig. 4 Sample histogram of propagator runs on HazelHen. The mean is 4996 s and the standard deviation is 1679 s, i.e. 34% of the mean. There is an extended heavy tail of longer runs, up to 23940 s, i.e. 479% of the mean (not shown)

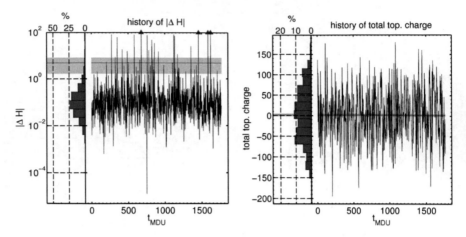

Fig. 5 Monte-Carlo histories of the Hamiltonian deficit ΔH (left) and the total topological charge (right) from the current E250 run performed on HazelHen

2.3 Status of the Gauge Field Generation

We now report on the status of gauge field generation for ensemble E250 with physical pion mass. Until the end of April 2019, roughly 440 gauge field configurations corresponding to 1760 MDU have been generated on HazelHen. The calculation of observables was performed in parallel on HazelHen and on compute clusters at JGU Mainz.

In the left pane of Fig. 5 we show the Monte Carlo history of the Hamiltonian deficit ΔH: the absence of any severe spikes demonstrates that the algorithmic setup is very stable, resulting in a high acceptance rate $(88.5 \pm 1.2)\%$ even for this demanding simulation at the physical pion mass. The right pane of Fig. 5 shows the Monte Carlo history of the topological charge: despite the fact that the lattice spacing is small, one clearly sees a high tunnelling frequency, which provides evidence that the algorithm correctly samples different topological sectors, if a chain of sufficient length is simulated.

3 The Hadronic Vacuum Polarisation a_μ^{hvp}

In this section we summarise the physics results, which are described in more detail in a recent publication [13], with a focus on the impact from the simulations with physical quark masses.

First results for the integrand of Eq. 2 (light-, strange-, and charm-quark contributions) on the ensemble with physical pion mass are shown in Fig. 6. The results

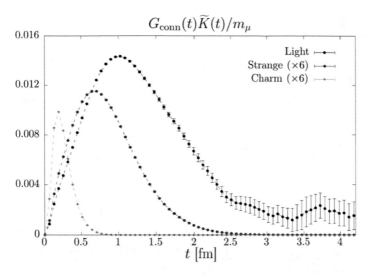

Fig. 6 The light-quark (black circles), strange-quark (blue circles), and charm-quark (green circles) integrands of Eq. 2 on ensemble E250. Note that the strange and charm-quark contributions are plotted enlarged by a factor 6. The displayed discretisation is the local-local one for the light and strange contributions, and the local-conserved one for the charm contribution

show that the statistical precision of $G(x_0)$ in the tail region is already quite good, and will further improve as statistics is being accumulated.

Figure 7 shows the pion mass dependence of the light-quark (top pane), strange-quark (middle pane), and charm-quark (bottom pane) contributions to a_μ^{hvp}. The chiral-continuum extrapolations of the data using two different discretisations of the electromagnetic current to check for discretisation effects yield the results at the physical point (black circles). For further details please refer to [13]. Comparing this result to the data with (close-to) physical pion and kaon masses from the newly generated ensemble E250 illustrates the importance of the E250 ensemble produced as part of project GCS-HQCD on HazelHen, providing a vital constraint to the chiral and continuum extrapolation shown in Fig. 7. We have achieved a statistical uncertainty of just over two percent on a_μ^{hvp} on the physical-mass ensemble E250, and of $1.0 - 1.2\%$ on all other ensembles. The result of our chiral-continuum extrapolation are in very good agreement with the values of $a_\mu^{\mathrm{hvp},l}$ directly obtained on ensemble E250. The statistical uncertainty on the final result is only 20% lower than the statistical uncertainties on E250, highlighting the importance of our simulations with physical pion and kaon masses on HazelHen.

Fig. 7 Results for the pion mass dependence of the light-quark (top pane), strange-quark (middle pane), and charm-quark (bottom pane) contribution to a_μ^{hvp} in units of 10^{-10}. The variable on the x-axis is the dimensionless ratio $\tilde{y} = \frac{m_\pi^2}{16\pi^2 f_\pi^2}$. The black curve represents the chiral dependence in the continuum, and the black point the final result at the physical pion mass. The triangles correspond to the lattice data points determined from correlation functions with local and conserved currents and the circles to results from correlation functions using only the local current

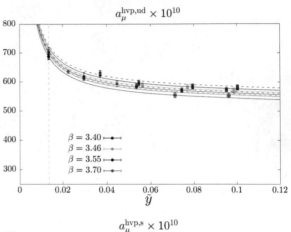

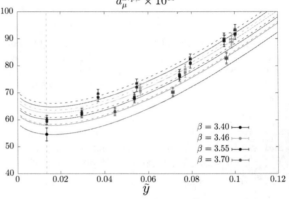

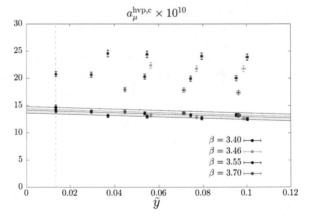

4 Summary and Outlook

In this paper we have reported on a recent calculation of the hadronic vacuum polarisation contribution to a_μ based on CLS gauge ensembles with $2+1$ dynamical flavours of $O(a)$ improved Wilson quarks, and highlighted the important role of QCD calculations with physical light-quark masses on supercomputing resources, specifically HazelHen at HLRS. A full description of our calculations for the HVP is presented in [13]. We obtain the result

$$a_\mu^{hvp} = (720.0 \pm 12.4_{stat} \pm 9.9_{syst}) \cdot 10^{-10}, \tag{5}$$

where the first error is statistical, and the second is an estimate of the total systematic uncertainty, which among other things accounts for the fact that the corrections due to isospin breaking have not yet been included in the central value. We thus find that the current overall error of our determination is 2.2%.

Future improvements will focus on calculations of the isospin-breaking contributions (see [27, 28] for progress towards this goal), on the improved determination of the quark-disconnected (or correspondingly the isoscalar) contribution, and on improved statistical accuracy at the physical pion mass.

Acknowledgements This work is partly supported by the Deutsche Forschungsgemeinschaft (DFG, German Research Foundation) grant HI 2048/1-1 and by the DFG-funded Collaborative Research Centre SFB 1044 *The low-energy frontier of the Standard Model*. The Mainz $(g-2)_\mu$ project is also supported by the Cluster of Excellence *Precision Physics, Fundamental Interactions, and Structure of Matter* (PRISMA+ EXC 2118/1) funded by the DFG within the German Excellence Strategy (Project ID 39083149) and by the European Research Council (ERC) under the European Union's Horizon 2020 research and innovation programme through grant agreement 771971-SIMDAMA. Calculations for this project were partly performed on the HPC clusters "Clover" and "HIMster II" at the Helmholtz-Institut Mainz and "MOGON II" at JGU Mainz. The authors also gratefully acknowledge the Gauss Centre for Supercomputing e.V. (www.gauss-centre. eu) for funding this project by providing computing time on the GCS Supercomputer HazelHen at Höchstleistungsrechenzentrum Stuttgart (www.hlrs.de) under project GCS-HQCD. We are grateful to our colleagues in the CLS initiative for sharing ensembles.

References

1. C. Patrignani et al., Chin. Phys. C **40**(10), 100001 (2016). https://doi.org/10.1088/1674-1137/40/10/100001
2. M. Davier, A. Hoecker, B. Malaescu, Z. Zhang, Eur. Phys. J. C **71**, 1515 (2011). https://doi.org/10.1140/epjc/s10052-012-1874-8; https://doi.org/10.1140/epjc/s10052-010-1515-z.[Erratum: Eur. Phys. J. C72 1874(2012)]
3. K. Hagiwara, R. Liao, A.D. Martin, D. Nomura, T. Teubner, J. Phys. **G38**, 085003 (2011). https://doi.org/10.1088/0954-3899/38/8/085003
4. T. Blum, A. Denig, I. Logashenko, E. de Rafael, B. Lee Roberts, T. Teubner, G. Venanzoni, arXiv:1311.2198 (2013)
5. F. Jegerlehner, A. Nyffeler, Phys. Rept. **477**, 1 (2009). https://doi.org/10.1016/j.physrep.2009.04.003

6. J. Prades, E. de Rafael, A. Vainshtein, Adv. Ser. Direct. High Energy Phys. **20**, 303 (2009). https://doi.org/10.1142/9789814271844_0009
7. J. Bijnens, EPJ Web Conf. **118**, 01002 (2016). https://doi.org/10.1051/epjconf/201611801002
8. M. Golterman, K. Maltman, S. Peris, Phys. Rev. D **95**(7), 074509 (2017). https://doi.org/10.1103/PhysRevD.95.074509
9. J. Kaspar, Nucl. Part. Phys. Proc. **260**, 243 (2015). https://doi.org/10.1016/j.nuclphysbps.2015.02.051
10. M. Fertl, Hyperfine Interact. **237**(1), 94 (2016). https://doi.org/10.1007/s10751-016-1304-7
11. M. Otani, J.P.S. Conf, Proc. **8**, 025010 (2015). https://doi.org/10.7566/JPSCP.8.025010
12. D. Mohler, S. Schaefer, J. Simeth, EPJ Web Conf. **175**, 02010 (2018). https://doi.org/10.1051/epjconf/201817502010
13. A. Gérardin, M. Cè, G. von Hippel, B. Hörz, H.B. Meyer, D. Mohler, K. Ottnad, J. Wilhelm, H. Wittig.: Phys. Rev. D **100**, 014510. https://doi.org/10.1103/PhysRevD.100.014510. arXiv:1904.03120 (2019)
14. M. Lüscher, S. Schaefer, Comput. Phys. Commun. **184**, 519 (2013). https://doi.org/10.1016/j.cpc.2012.10.003
15. M. Lüscher, F. Palombi, PoS **LATTICE2008**, 049 (2008). https://doi.org/10.22323/1.066.0049
16. M.A. Clark, A.D. Kennedy, Phys. Rev. Lett. **98**, 051601 (2007). https://doi.org/10.1103/PhysRevLett.98.051601
17. M. Lüscher, JHEP **07**, 081 (2007). https://doi.org/10.1088/1126-6708/2007/07/081
18. M. Lüscher, JHEP **0712**, 011 (2007). https://doi.org/10.1088/1126-6708/2007/12/011
19. I. Omelyan, I. Mryglod, R. Folk, Comput. Phys. Commun. **151**(3), 272 (2003). https://doi.org/10.1016/S0010-4655(02)00754-3; http://www.sciencedirect.com/science/article/pii/S0010465502007543
20. M. Lüscher, S. Schaefer, JHEP **1107**, 036 (2011). https://doi.org/10.1007/JHEP07(2011)036
21. S. Schaefer, R. Sommer, F. Virotta, Nucl. Phys. B **845**, 93 (2011). https://doi.org/10.1016/j.nuclphysb.2010.11.020
22. A. Gérardin, H.B. Meyer, A. Nyffeler, Phys. Rev. D **94**(7), 074507 (2016). https://doi.org/10.1103/PhysRevD.94.074507
23. A. Gérardin, H.B. Meyer, A. Nyffeler.: Phys. Rev. D **100**, 034520 https://doi.org/10.1103/PhysRevD.100.034520. arXiv:1903.09471 (2019)
24. D. Bernecker, H.B. Meyer, Eur. Phys. J. A **47**, 148 (2011). https://doi.org/10.1140/epja/i2011-11148-6
25. T. Blum, Phys. Rev. Lett. **91**, 052001 (2003). https://doi.org/10.1103/PhysRevLett.91.052001
26. M. Della Morte, B. Jäger, A. Jüttner, H. Wittig, JHEP **03**, 055 (2012). DOI https://doi.org/10.1007/JHEP03(2012)055
27. A. Risch, H. Wittig, EPJ Web Conf. **175**, 14019 (2018). https://doi.org/10.1051/epjconf/201817514019
28. A. Risch, H. Wittig, in *36th International Symposium on Lattice Field Theory (Lattice 2018) East Lansing, MI, United States, July 22–28, 2018* (2018)

Molecules, Interfaces, and Solids

Holger Fehske and Christoph van Wüllen

Also in this funding period the field of molecules, interfaces, and solids has profited enormously from the computational resources provided by the High Performance Computing Center Stuttgart and the Steinbuch Centre for Computing Karlsruhe. In this section, we selected only one-third of the broadly diversified projects in this area to demonstrate the impact of high performance computing in chemistry, electronic structure calculations, and biology inspired physics.[1]

We start with two smaller scale projects run at the SCC. In both cases, molecular dynamics (MD) simulations were performed using the GROMACS program, and typical runs use about 300 CPU cores. The group of Tallarek at the university of Marburg simulates the liquid chromatography process, which is important for the separation of a mixture of substances in a solvent. While the solution passes a solid stationary phase in a chromatography column, different substances in the solution move with different speed and can thus be separated at the end of the column. For the fixed phase, silica decorated with long alkyl chains, and as a solvent, a mixture of water and acetonitrile is used. The calculations show how the solvent becomes inhomogeneous near the decorated surface, where the less polar acetonitrile component enriches close to the nonpolar alkyl chains, and within these zones of increased acetonitrile content, some molecules exhibit must faster lateral mobility such that they can pass the columns at a higher speed. The results from this microscopic model have then been used as an input to mesoscopic and macroscopic simulations that finally model the overall chromatography process.

H. Göddeke and L. Schäfer from the university of Bochum use the same software to model ATP binding cassette transporters. These use the univeral biological energy source (ATP) to translocate substrate molecules through the cell membrane, that is, transporting material into and out of living cells. To this end, the transporter must actively perform large changes in its overall shape from an inward-facing to an outward-facing conformation. Such a large-amplitude motion of course implies

[1]Holger Fehske, Institut für Physik, Lehrstuhl Komplexe Quantensysteme, Ernst-Moritz-Arndt-Universität, Greifswald, Felix-Hausdorff-Str. 6, 17489 Greifswald, Germany, e-mail: fehske@physik.uni-greifswald.de, Christoph van Wüllen, Fachbereich Chemie, Technische Universität Kaiserslautern, Erwin-Schrödinger-Str. 52, 67663 Kaiserslautern, Germany, e-mail: vanwullen@chemie.uni-kl.de

that this is a relatively slow process. Using the computational resources of the SCC, the MD calculations could simulate trajectories for about 200 nano seconds within a (wall-clock) day of computing time, and thus they could run 100 MD simulations each covering 500 nsec. In six cases, a transition from the inward-facing to the outward-facing conformation occured spontaneously, and it was found that first the gate at the inner side of the membrane closes, yielding a very tight intermediate, and then the gate at the outer side of the membrane opens. In some further cases, this process went half-way, ending in the tight intermediate, but in most of the cases no large amplitude change was observed. This calls for extending the investigation toward larger simulation times.

An outstanding example is the report by M. Hummel, W. Verestek, and S. Schmauder from the IMWF at Stuttgart University, on the influence of time and length scales in MD simulations. After a brief but valuable review of the computational and code evolution of MD and its newer variants, such as the parallel replica method and the hyper-, respectively, temperature-accelerated dynamics, the research team addressed the gap between the strain rate of experiments and MD. Above all, the authors studied the numerical strain rate exploration, using the MD software LAMMPS and the EAM potential. Remarkably, an aluminium single crystal with 268 million atoms could be constructed, which is in the size of the lower limit of the experiments. This also applies to the strain rate of ten millions per second. Therefore, an extrapolation is not necessary anymore, which is surely a great advantage. The central message is that the MD simulations lead to the same results as the experiments for this kind of problems, provided they are carried out at the same scales. Nevertheless, the results are very sensitive on the materials, in particular their dimension has a strong influence. In any case one has to be careful when results were transferred from one scale to the other.

Freiburg's University and Fraunhofer material-physics group has performed extensive first-principles density functional theory (DFT) GGA+U calculations, with a focus on the interplay between point defects and electronic structure in mixed-valent perovskites utilized for solid oxide fuel and solid oxide electrolyzer cells. The simulations are of major importance to optimize the functionality of perovskite electrodes. D. Mutter, D. F. Urban, and C. Elsässer used VASP to determine the atomic configurations and electronic structure, whereby the valence electrons are represented by plane waves and the core electrons are described by PAW pseudo-potentials. In order to test the proposed calculation scheme and benchmark the computational performance $LaFeO_3$ was taken as a prototype material. The main result presented is the phase diagram with respect to the chemical potentials of La and Fe and the pressure of the oxygen atmosphere, which provides the basis for discussing the phase stability and point defect formation energies. While $LaFeO_3$, as well as $LaMnO_3$ and $CaMnO_3$, will be electronically insulating in a defect-free configuration, which forbids the use as electrodes, point defects act as intrinsic dopants, leading to free charge carriers at elevated temperatures. Another interesting point addressed are the oxygen vacancy formation energies in $La_{1-x}Ca_xFe_{1-y}Mn_yO_3$, which makes the compound promising for electrode materials.

A very exciting and innovative application of DFT-based simulations was provided by Maria Fyta's group from the Institute for Computational Physics at the University Stuttgart. With rather modest computational costs the authors demonstrated the bio-sensitivity of a device—made up of gold electrodes functionalized by memantine with an added thiol group—for distinguishing DNA nucleotides, whereby DNA rotation and water effects were taken into account. The DFT was implemented in SIESTA, using a Perdew-Burke-Enzerhof exchange-correlation functional. The influence of the water solvent is taken into account by quantum/molecular mechanics simulations in the CP2K framework. The calculations, performed *in vacuo*, revealed nucleotide-specific transmission peaks with the rotation, and the inclusion of water significantly increases the dynamic behavior of the whole system. Strong nucleotide-specific trends were also found in the conductance variation of the functionalized device. For certain, the results will be important detecting DNA by future nano-pore sequencers.

We wish to stress that almost all projects supported in this field during the last year are of high scientific quality, also those which could not be included in this chapter because of space limitations. Beyond that they demonstrated the strong need of supercomputing facilities in modern science.

Characteristics of Surface Diffusion and Effective Pore Diffusion in Reversed-Phase Liquid Chromatography from Molecular Dynamics Simulations

Julia Rybka, Alexandra Höltzel, Nicole Trebel, and Ulrich Tallarek

Abstract Through the use of molecular dynamics (MD) simulations, solute distribution and mass transport at solid-liquid interfaces can be elucidated on the molecular level. In reversed-phase liquid chromatography (RPLC), retained analyte molecules can diffuse faster in the interfacial region between a hydrophobic stationary phase and a water–acetonitrile (ACN) mobile phase. ACN accumulates on top of the hydrophobic, alkyl-modified stationary phase (a C_{18}- or C_8-modified silica support) and forms an ACN-rich layer (the "ACN ditch") on top of the bonded-phase chains. Because the high content of the organic compound is more conducive to analyte mobility, lateral (surface-parallel) analyte diffusivity goes through a maximum in the ACN ditch. In this project, we investigate the characteristics of surface diffusion in RPLC by MD simulations using GROMACS with respect to the influence of the applied mobile-phase composition, chain length and grafting density of the bonded phase on the lateral mobility gain for a set of four typical aromatic hydrocarbon analytes. The simulated spatially-dependent analyte mobilities serve as input parameters to calculate the effective macroscopic bed diffusivities in hierarchical macro–mesoporous structures.

1 Introduction

Mass transport at solid-liquid interfaces plays a key role in processes and applications such as chromatography, heterogeneous catalysis, adsorption, chemical sensors or the performance of electrodes. To optimize process efficiency, detailed knowledge of the spatiotemporal distribution and diffusivities of solvent and solute molecules in porous media (microporous, mesoporous or hierarchically structured materials) is absolutely necessary, but rarely accessible by experimental data. Therefore, computer simulations are often used to gain a molecular-level picture of the structural organization of solvent molecules as well as solute distribution and dynamics in porous media [1].

J. Rybka · A. Höltzel · N. Trebel · U. Tallarek (✉)
Hans-Meerwein-Strasse 4, 35032 Marburg, Germany
e-mail: ulrich.tallarek@staff.uni-marburg.de

© Springer Nature Switzerland AG 2021
W. E. Nagel et al. (eds.), *High Performance Computing in Science and Engineering '19*,
https://doi.org/10.1007/978-3-030-66792-4_7

In reversed-phase liquid chromatography (RPLC), the most important and widely used liquid chromatography technique, analyte molecules dissolved in an aqueous-organic mobile phase (usually water–acetonitrile (W/ACN) or W-methanol mixtures) move through a hierarchically structured, macro–mesoporous material (either a packing of silica particles or a silica monolith) by pressure-driven flow. RPLC is used for the separation and purification of a wide variety of analyte molecules. The separation of analytes is accomplished on the basis of their retention, which depends on analyte-specific interactions with the surface of the stationary phase. Typical stationary phases for RPLC are silica surfaces bearing chemically-bonded alkyl chains, the most popular being dimethyl octadecylsilane (C_{18}) chains [2]. RPLC characteristics on the molecular level were investigated by computer simulations, mostly by molecular dynamics (MD) [3–10] and Monte-Carlo (MC) simulations [1]. Whereas mass transport through the macroporous interparticle space is advection-dominated, mass transport through the mesoporous intraparticle space, where analyte retention takes place, is diffusion-limited. Molecules can be retained either by partitioning into or adsorption onto the bonded-phase chains [1]. The analyte molecules traverse the mesopore space of the chromatographic bed through a combination of pore diffusion (diffusion in the pore liquid) and surface diffusion (diffusion along the surface of the stationary phase) [11]. Experimental RPLC studies have shown that surface diffusion of retained molecules can be faster than their diffusivity in the bulk mobile phase [12, 13]. The surface diffusion coefficient was observed to increase with the retention of an analyte, suggesting a positive correlation between surface diffusion and retention.

Through MC-simulated solvent density profiles at alkyl-modified silica surfaces, Siepmann and coworkers established the presence of three different regions in the RPLC system: (I) the bonded-phase region, which is dominated by the bonded-phase chains, (II) the interfacial region, where the total solvent density increases from 10% to 90% of its bulk value, and (III) the bulk region, where bulk properties of the applied mobile phase composition are attained [14, 15]. In region II, the density of the organic modifier is increased while W density is depleted in comparison to the bulk density in region III. Recent simulations with W/ACN binary mixtures in our group elucidated the molecular-level picture of surface diffusion at a C_{18}-modified silica surface from MD simulations [7, 8]. In the interface region between bonded phase and mobile phase, ACN molecules accumulate around the alkyl chain ends of the bonded phase, forming an ACN-rich border layer, the "ACN ditch". In this high-mobility region, retained analyte molecules can diffuse faster along the stationary phase surface than in the corresponding W/ACN mobile phase because the high content of ACN in the ditch is more conducive to analyte mobility [8].

In this project, we aim at a detailed characterization of this unique "fast lane"-mechanism of surface diffusion along hydrophobic surfaces in RPLC from MD simulations to tailor this transport mechanism towards an increased separation efficiency. The evolution of the ACN ditch is primarily dependent on the employed mobile phase, as well as the polarity and flexibility of the bonded-phase chains. In the present work, we investigate the influence of solvent composition and bonded-phase chain length and density on the retentive and diffusive behavior of a set comprising four aromatic hydrocarbon analytes, which vary in size and polarity: Benzene and ethylbenzene (apolar molecules that vary in size), and acetophenone and benzyl alcohol (weakly to moderately polar molecules that vary in hydrogen-bonding capacity). In the first sub-project [16], the W/ACN mobile phase composition equilibrated at a 10 nm C_{18}-modified silica mesopore with slit-pore geometry was varied from 80/20 (v/v) W/ACN up to 10/90 (v/v) W/ACN to elucidate the relation between surface diffusion and retention (the latter being controlled by the elution strength of the mobile phase). The resulting properties (analyte distribution, surface and pore diffusion) derived from the MD simulations can be traced up to the macroscopic scale of the chromatographic bed. In a follow-up study, the obtained spatially-dependent analyte diffusivities and distribution from MD simulations in a C_{18}-modified silica mesopore with varied mobile phase compositions are combined with simulations in physical 3D reconstructions of mesopore and macropore spaces to arrive at the effective macroscopic analyte diffusion across the entire chromatographic bed [17].

In the second sub-project, we investigate the contribution of bonded-phase chain length and grafting density to surface diffusion and retention. Type and grafting density of the bonded phase assumingly have a great impact on the distribution of analyte molecules (adsorption on top of the alkyl chains in region II, and partitioning into the bonded phase in region I), as well as on the spatially-dependent diffusivities. Solvent and analyte behavior at the C_{18}-modified silica surface is compared to the solvent and analyte profiles simulated at two C_8-modified silica surfaces for two mobile phase compositions: (1) a C_8-modified surface with identical alkyl-chain grafting density as the C_{18}-modified surface, and (2) a "high-density" C_8-modified silica surface with an increased C_8 grafting density. The C_8 phase is chosen because it is another common and widely applied stationary phase. The dense C_8 bonded phase is highly structured and the chains are predominantly parallel to each other, which is known as the "picket fence"-structure in the literature [18, 19]. An increased grafting density will even further enhance the "picket fence"-structure leading to the retained molecules being located primarily in the ACN ditch and maximizing the effect of the "fast lane"-surface diffusion.

2 Methods

2.1 Simulation Systems

Simulations are carried out in GROMACS (Groningen Machine for Chemical Simulations). GROMACS is a fast program for MD simulations, well suited for the parallelization on processor clusters [20, 21]. GROMACS versions 5.1.1 and 2016.5 are used for the simulations in this project.

The initial simulation system for sub-project 1 consisted of a fully periodic rectangular simulation box with dimensions of 12.4 nm (x) × 13.2 nm (y) × 10.93 nm (z) and a central, three-layer silica slab (0.93 nm wide in z-direction). Due to the applied periodic boundary conditions, this setup equals a 10 nm slit-pore. The initial bare silica surface, prepared by an approach of Coasne et al. [22] from the (111) face of β-cristobalite, was modified with 3.11 μmol C_{18} chains/m^2 and 0.93 μmol trimethylsilyl (TMS) endcapping groups/m^2 leaving 3.42 μmol residual silanol groups/m^2. Surface modification and grafting densities reflect well the properties of alkyl-modified silica stationary phases used in RPLC because only up to 50% of the original silanols can be modified due to steric reasons [2]. The C_{18}-modified silica surface was equilibrated with mobile phases ranging from 80/20 to 10/90 (v/v) W/ACN for each of the four analytes, resulting in 32 simulation systems.

In the second sub-project, two C_8-modified silica surfaces, which vary in grafting density and endcapping, were prepared from the initial C_{18}-modified silica surface by (1) changing the long C_{18} chains to the shorter C_8 chains resulting in a silica surface modified with 3.11 μmol C_8 chains/m^2 and 0.93 μmol TMS groups/m^2 and (2) changing all endcapping groups of the first C_8-modified silica surface to additional C_8 chains resulting in a second C_8-modified silica surface with 4.04 μmol C_8 chains/m^2 and no endcapping. Simulations were carried out for a W-rich mobile phase (70/30 (v/v) W/ACN) and an ACN-rich mobile phase (20/80 (v/v) W/ACN) at each C_8-modified surface for each analyte molecule, resulting in 16 simulation systems in total for sub-project 2.

2.2   Force Fields and Simulation Parameters

Force-field parameters for the silica surface (Si, O, and H atoms) were taken from Gulmen and Thompson [23, 24]. The transferable potentials for phase equilibria united-atom (TraPPE-UA) force field was used for the C_8 and C_{18} chains [25–27], TMS endcapping [25–27] and ACN molecules [28]. W molecules were described by the simple point charge/extended (SPC/E) force field [29]. The four aromatic hydrocarbon analytes were represented by the explicit CHARMM general force field (CGenFF) [30, 31]. Simulations were carried out in a canonical NVT ensemble (constant number of molecules N, simulation box volume V, and temperature T) at a temperature of 300 K controlled by a Nosé-Hoover thermostat with a 0.25 ps coupling constant. Equations of motion were integrated with a 1 fs time step. Each simulation system was equilibrated for 60 ns simulation time. Afterwards, productive simulation runs were carried out for up to 1500 ns until the spatially-dependent analyte density and diffusion-coefficient profiles of each system showed no further changes with increasing simulation time. A more detailed description of the simulation parameters, choice of force fields and validation can be found in our previous publications [7, 8] and in our most recent publication [16], which also comprises the results of sub-project 1.

Table 1 Scaling behavior of GROMACS 2016.5 on the ForHLR I system at the Steinbuch Centre for Computing (SCC) of the KIT for two typical simulation systems as used in sub-project 1. Testing was performed with varied number of cores for 15 mins run time. Performance (ns/day for the GROMACS trajectory), efficiency, and relative speedup normalized to 20 cores are given

80/20 (v/v) W/ACN

#cores	Performance (ns/day)	Efficiency (%)	Speedup
20	5.484	100	1
40	10.654	97.14	1.94
80	20.654	94.16	3.77
160	35.351	80.58	6.45
320	63.403	72.26	11.56
640	97.453	55.53	17.77

10/90 (v/v) W/ACN

20	9.738	100	1
40	18.280	93.86	1.88
80	33.255	85.37	3.41
160	58.010	74.46	5.96
320	89.308	57.32	9.17
640	124.64	40.00	12.80

2.3 Performance on the ForHLR I

The MD simulations with GROMACS were carried out on the ForHLR I computer cluster at the Karlsruhe Institute of Technology (KIT) using typically 160 processors for the slit-pore simulation systems of the current sub-projects 1 and 2. Because of the desired long simulation times for each system, a splitting into a number of shorter consecutive simulation runs was necessary, so that the run time for each simulation did not exceed the maximum run time of 72 h on the computer cluster.

2.3.1 Sub-project 1

In the following Table 1 and plot, scaling and performance information is shown for exemplary simulation runs of a simulation system, which consists of a C_{18}-modified silica surface with an 80/20 (v/v) W/ACN (W-rich) mobile phase, and a simulation system, which consists of a C_{18}-modified silica surface with a 10/90 (v/v) W/ACN (ACN-rich) mobile phase using GROMACS version 2016.5.

The total number of atoms and united-atom groups in the simulation system varies with the applied mobile phase composition. With increasing ACN content of the mobile phase, the total number of atoms and united-atom groups in the system decreases from 128,767 at 80/20 (v/v) W/ACN to 97,104 at 10/90 (v/v) W/ACN. Therefore, performance (in ns/day) and scalability also differ for each mobile phase

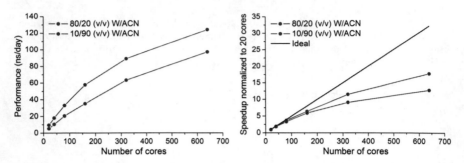

Fig. 1 Scaling behavior of GROMACS 2016.5 on the HPC system ForHLR I at the Steinbuch Centre for Computing (SCC) of the KIT for two typical RPLC simulation systems with varied solvent composition as used in sub-project 1

composition. Figure 1 shows that performance increases (and scalability decreases) with increasing ACN content of the mobile phase (due to the decreasing total number of atoms and united-atom groups in the system). The needed amount of trajectory varied strongly with the mobile phase composition and analyte molecule. For example, the apolar ethylbenzene is the most strongly retained compound of the analyte set. Because retention increases with the W content of the mobile phase, ethylbenzene remains either inside the bonded-phase region (partitioning) or on top of the apolar bonded-phase chains (adsorption) and rarely enters the bulk region of an 80/20 (v/v) W/ACN mobile phase. Thus, a simulation trajectory of almost 1500 ns was needed to receive the desired analyte profiles. In contrast, only a 410 ns trajectory was needed for ethylbenzene profiles with a 10/90 (v/v) W/ACN mobile phase. Compared to the strongly retained ethylbenzene, the generation of the benzyl alcohol profiles (the most weakly retained compound of the analyte set) needed only a 510 ns trajectory for the 80/20 (v/v) W/ACN mobile phase, and a 325 ns trajectory for the 10/90 (v/v) W/ACN mobile phase.

2.3.2   Sub-project 2

The shorter bonded-phase chains of the C_8-modified silica surfaces lead to a decrease of the bulky united-atom groups of the bonded-phase chains and to an increase in solvent molecules and thus, an increase in the total number of atoms and united-atom groups in the system. This results in a slightly increased scalability as shown in the following Table 2 and Fig. 2 for an exemplary simulation system (the C_8-modified silica surface with a grafting density of $3.11 \mu \text{mol/m}^2$ and 70/30 (v/v) W/ACN) as used in sub-project 2.

Table 2 Scaling behavior of GROMACS 2016.5 on the ForHLR I system at the SCC of the KIT for a typical simulation system as used in sub-project 2. Testing was performed with varied number of cores for 15 mins run time. Performance (ns/day for the GROMACS trajectory), efficiency, and relative speedup normalized to 20 cores are given

70/30 (v/v) W/ACN at C_8

#cores	Performance (ns/day)	Efficiency (%)	Speedup
20	5.342	100	1
40	10.208	95.55	1.91
80	20.364	95.30	3.81
160	35.298	82.60	6.61
320	62.321	72.91	11.67
640	98.480	57.61	18.44

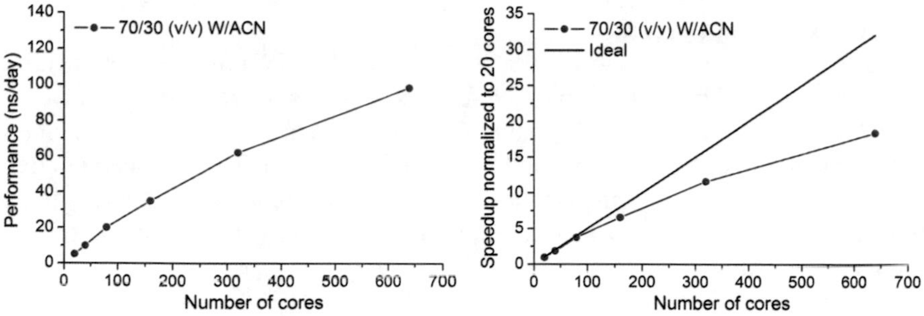

Fig. 2 Scaling behavior of GROMACS 2016.5 on the HPC system ForHLR I at the SCC of the KIT for a typical RPLC simulation system as used in sub-project 2

3 Results and Discussion

3.1 Variation of Mobile-Phase Composition (Sub-project 1)

The influence of the mobile phase composition on surface diffusion and retention was investigated for the four analyte molecules in a C_{18}-modified silica mesopore. The results of this sub-project are published in detail in reference 16. MD simulations of the C_{18}-modified silica mesopore confirmed that the ACN ditch is present in the interface region for mobile phases up to 80 vol% ACN. The ACN excess in the interfacial region decreases with increasing ACN content of the mobile phase [16]. In chromatographic practice, retention decreases with the polarity of an analyte and increases with the size of an analyte. For an analyte molecule, retention decreases with increasing ACN content of the mobile phase. In this study, the retention factor (k') gained from experiments is compared to simulated distribution coefficients (K), which correlate through $k' = K \times \beta$, where β is the phase ratio (the ratio of the stationary phase volume to mobile phase volume). The retention order (ethylbenzene

> benzene > acetophenone > benzyl alcohol) is well reproduced for all investigated mobile phase compositions.

The spatially-dependent diffusion-coefficient profiles of the four analytes show an increase in lateral mobility in the ACN ditch for mobile phases up to 20/80 (v/v) W/ACN. Like retention, the lateral mobility gain decreases with the ACN content in the mobile phase for the analytes ethylbenzene, benzene and acetophenone, whereas the lateral mobility gain of benzyl alcohol (the most polar compound of the analyte set) remains constant between 80/20 and 40/60 (v/v) W/ACN. Thus, the simulation data confirm the positive relation between analyte retention and lateral mobility gain from surface diffusion except for the most polar analyte. Though, this correlation does not extend to the analyte set as the lateral mobility gain of benzene benefits the most from surface diffusion, whereas ethylbenzene is the most strongly retained compound of the analyte set leading to the conclusion that retention and surface diffusion depend differently on analyte properties (e.g., molecule size, presence of side chains, or polarity) [16].

Additionally, this study showed that the bonded-phase chains have a lubricating effect on the lateral mobility gain of the retained analytes in the ACN ditch. Due to this lubrication, the lateral mobility gain can even exceed the lateral mobility increase that stems from the ACN enrichment in the interface region. This lubricating effect only occurs if analyte molecules in the ACN ditch have numerous contacts with the bonded-phase chains, which is the case only for W-rich solvent compositions (with less than 50 vol% ACN).

3.1.1 Multiscale Simulation of the Effective Bed Diffusion Coefficient

The approach to multiscale simulation of diffusion presented here connects the molecular-level picture of interfacial structure and dynamics obtained from MD simulations with mass transport in hierarchical porosity addressed by Brownian dynamics [17]. In this work, we target the overall, effective bed diffusion coefficients (D_{bed}) in macro–mesoporous materials used for adsorption, separation, and catalysis. These D_{bed}-values account for the solute dynamics at the interface between pore surface and pore liquid as well as for the actual hierarchical macro–mesoporous morphology of the material.

MD simulations carried out in the C_{18}-modified silica mesopore with mobile phase compositions ranging from 80/20 (v/v) W/ACN to 10/90 (v/v) W/ACN provide information on the interfacial solute dynamics via mesopore-level distributions of solute density and diffusion coefficient parallel to the pore surface (left panel of Fig. 3) under explicit consideration of surface chemistry, analyte properties, and mobile phase composition. This essential information is incorporated into Brownian dynamics simulations of the effective diffusion coefficient D_{meso} in the mesopore space morphology as physically reconstructed by scanning transmission electron microscopy (central panel of Fig. 3) [32]. The dependence of the simulated D_{meso}-values on analyte structure and composition of the mobile phase revealed nontrivial behavior regarding the contribution of partitioning and adsorption to solute reten-

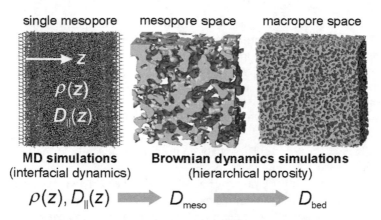

single mesopore   mesopore space   macropore space

MD simulations   **Brownian dynamics simulations**
(interfacial dynamics)   (hierarchical porosity)

$$\rho(z), D_{\parallel}(z) \implies D_{\mathrm{meso}} \implies D_{\mathrm{bed}}$$

Fig. 3 Multiscale simulation approach from the analyte density and diffusion coefficient profiles on the mesopore-level, as obtained by the MD simulations, via the effective diffusion coefficient D_{meso} in the physically-reconstructed mesopore space morphology, up to the macroscopic bed diffusion coefficient D_{bed} in the physically-reconstructed macropore space (reproduced from [17])

tion and effective diffusion in the mesopores. Mass transfer between pore space hierarchies is simulated using an effective homogeneous medium representation for the mesoporous domain in the macropore space morphology as physically reconstructed by confocal laser scanning microscopy (right panel of Fig. 3) [33], which finally allowed us to derive the macroscopic bed diffusion coefficients D_{bed}.

This multiscale simulation study has shown that the hierarchical macro–mesoporous structure can facilitate fast lateral equilibration of concentration gradients, which is beneficial to the separation efficiency of chromatographic columns and the plug-flow operation of fixed-bed reactors. Both D_{meso} and D_{bed} have immediate value as input to transport models of separation and catalytic reaction processes. It will expand the basis for predictive column modeling and help to eliminate ambiguity regarding adsorption and transport mechanisms in models that require specialized parameter values for use with particular solutes, support structures, and surface chemistries. This multiscale simulation approach to D_{meso} and D_{bed} is described in detail in our recent publication [17].

3.2 Influence of Surface Modification (Sub-project 2)

The influence of the surface modification with respect to chain length and density on surface diffusion and retention is investigated for the set of four analytes and two mobile phase compositions (70/30 (v/v) W/ACN and 20/80 (v/v) W/ACN) at a C_8-modified silica surface with endcapping, and at a high-density C_8-modified surface. From the C_{18}-modified surface to the C_8-modified surfaces, the position of the ACN ditch and the borders between the three regions shift closer to the silica surface as chain length decreases. The spatially-dependent analyte density profiles reveal that the contributions from adsorption and partitioning to retention cannot be distinguished as clearly as compared to the C_{18}-modified surface. For a 70/30 (v/v) W/ACN mobile phase, the ACN excess in the ditch increases from the C_8 surface to the C_{18} surface to the high-density C_8 surface. Concomitantly, the lateral mobility gain from surface diffusion increases from the C_8 surface to the C_{18} surface to the high-density C_8 surface. For 80/20 (v/v) W/ACN, the lateral mobility gain increases from the C_8-modified surface to the high-density C_8-modified silica surface to the C_{18}-modified silica surface. These data contribute significantly to the understanding of the bonded-phase influence to surface diffusion, which comprises the chain conformation and mobility, number of contacts with the analyte molecules residing in the ACN ditch, as well as the influence on ACN accumulation and orientation in the ditch. The experimentally-determined retention order of the analyte set is also preserved in the simulated distribution coefficients at the C_8-modified surfaces.

4 Conclusions

The influence of mobile phase composition, bonded-phase chain length and grafting density on analyte retention, surface diffusion and lateral mobility gain have been investigated in detail in the current sub-projects for alkyl-modified silica mesopores with slit-pore geometry. The MD data confirmed the presence of the ACN ditch for W/ACN compositions up to 80 vol% ACN at C_{18}-modified silica pores. At W-rich mobile phases, the bonded-phase causes a lubricating effect benefiting the lateral mobility gain from surface diffusion. The information gained from analyte diffusivity and distribution in a single mesopore was used in a multiscale simulation to access D_{meso} and D_{bed} in the hierarchically porous, physically-reconstructed structure of the chromatographic bed. This approach can be easily adapted to a wide variety of hierarchically porous materials and applications. The simulated D_{meso} and D_{bed} values function as important input parameters for fixed-bed models applied in catalysis and separation. MD simulations at C_8-modified silica surfaces showed the influence of the shorter alkyl chains and their "picket-fence" structure on analyte behavior. Though the C_8 chains are less flexible, ACN and analyte diffusivity equal or even exceed the lateral mobility at the more flexible C_{18}-modified silica surface due to a stronger accumulation of ACN molecules at the high-density C_8 silica surface. Performance tests of simulation systems with C_{18}- and C_8-modified silica slit pores showed that the scalability decreases with increasing ACN content of the mobile phase and increasing chain length (as the total number of atoms in the system decreases). The scalability allowed the parallelization with 160 processors and an efficiency of at least 75–80%.

Further simulation studies will comprise the influence of pore shape and curvature on bonded-phase and solvent structure, as well as on analyte surface diffusion and retention by investigating a set of cylindrical mesopores with varied pore diameter. The limited available space between the bonded-phase chains should hinder the partitioning of analyte molecules into the bonded-phase chains leading to an increased analyte density in the ACN ditch, which would benefit the mass transport through the mesopore. As larger simulation systems are needed for the investigation of cylindrical pore systems, scalability increases significantly, suggesting that a larger amount of processors can be used simultaneously.

Acknowledgements This work was supported by the Deutsche Forschungsgemeinschaft DFG (Bonn, Germany) under grant TA 268/11-1. Simulations were carried out on the ForHLR I cluster at the Steinbuch Centre for Computing (SCC) of the Karlsruhe Institute of Technology (Karlsruhe, Germany) under the project acronym RPLCMD.

References

1. R.K. Lindsey, J.L. Rafferty, B.L. Eggiman, J.I. Siepmann, M.R. Schure, J. Chromatogr. A **1287**, 60–82 (2013)
2. U.D. Neue, *HPLC Columns: Theory, Technology, and Practice* (Wiley-VCH, New York, 1997)
3. K.A. Lippa, L.C. Sander, R.D. Mountain, Anal. Chem. **77**, 7852–7861 (2005)
4. K.A. Lippa, L.C. Sander, R.D. Mountain, Anal. Chem. **77**, 7862–7871 (2005)
5. A. Fouqueau, M. Meuwly, R.J. Bemish, J. Phys. Chem. B **111**, 10208–10216 (2007)
6. J. Braun, A. Fouqueau, R.J. Bemish, M. Meuwly, Phys. Chem. Chem. Phys. **10**, 4765–4777 (2008)
7. J. Rybka, A. Höltzel, S.M. Melnikov, A. Seidel-Morgenstern, U. Tallarek, Fluid Phase Equilib. **407**, 177–187 (2016)
8. J. Rybka, A. Höltzel, U. Tallarek, J. Phys. Chem. C **121**, 17907–17920 (2017)
9. K.E. Hage, P.K. Gupta, R.J. Bemish, M. Meuwly, J. Phys. Chem. Lett. **8**, 4600–4607 (2017)
10. K.E. Hage, R.J. Bemish, M. Meuwly, Phys. Chem. Chem. Phys. **20**, 18610–18622 (2018)
11. F. Gritti, G. Guiochon, Anal. Chem. **78**, 5329–5347 (2006)
12. F. Gritti, G. Guiochon, AIChE J. **57**, 333–345 (2011)
13. F. Gritti, G. Guiochon, AIChE J. **57**, 346–358 (2011)
14. L. Zhang, J.L. Rafferty, J.I. Siepmann, B. Chen, M.R. Schure, J. Chromatogr. A **1126**, 219–231 (2006)
15. J.L. Rafferty, J.I. Siepmann, M.R. Schure, J. Chromatogr. A **1218**, 2203–2213 (2011)
16. J. Rybka, A. Höltzel, A. Steinhoff, U. Tallarek, J. Phys. Chem. C **123**, 3672–3681 (2019)
17. U. Tallarek, D. Hlushkou, J. Rybka, A. Höltzel, J. Phys. Chem. C **123**, 15099–15112 (2019)
18. C. Stella, S. Rudaz, J.L. Veuthey, A. Tchapla, Chromatographia **53**, 113–131 (2001)
19. W. Melander, Cs. Horváth, Reversed-phase chromatography, in *High-Performance Liquid Chromatography – Advances and Perspectives*, ed. by Cs. Horváth, vol. 2 (Academic Press, New York, 1980)
20. B. Hess, C. Kutzner, D. van der Spoel, E. Lindahl, J. Chem. Theory Comput. **4**, 435–447 (2008)
21. M.J. Abraham, T. Murtola, R. Schulz, S. Páll, J.C. Smith, B. Hess, E. Lindahl, SoftwareX **1–2**, 19–25 (2015)
22. B. Coasne, F. di Renzo, A. Galarneau, R.J.M. Pellenq, Langmuir **24**, 7285–7293 (2008)
23. T.S. Gulmen, W.H. Thompson, Langmuir **22**, 10919–10923 (2006)
24. K.G. Steenbergen, J.L. Kern, Z. Wang, W.H. Thompson, B.B. Laird, J. Phys. Chem. C **120**, 5010–5019 (2016)
25. M.G. Martin, J.I. Siepmann, J. Phys. Chem. B **102**, 2569–2577 (1998)
26. J.M. Stubbs, J.J. Potoff, J.I. Siepmann, J. Phys. Chem. B **108**, 17596–17605 (2004)
27. M.G. Martin, J.I. Siepmann, J. Phys. Chem. B **103**, 4508–4517 (1999)
28. C.D. Wick, J.M. Stubbs, N. Rai, J.I. Siepmann, J. Phys. Chem. B **109**, 18974–18982 (2005)
29. H.J.C. Berendsen, J.R. Grigera, T.P. Straatsma, J. Phys. Chem. **91**, 6269–6271 (1987)

30. K. Vanommeslaeghe, E. Hatcher, C. Acharya, S. Kundu, S. Zhong, J. Shim, E. Darian, O. Guvench, P. Lopes, I. Vorobyov et al., J. Comput. Chem. **31**, 671–690 (2010)
31. N.M. Fischer, P.J. van Maaren, J.C. Ditz, A. Yildirim, D. van der Spoel, J. Chem. Theory Comput. **11**, 2938–2944 (2015)
32. S.-J. Reich, A. Svidrytski, D. Hlushkou, D. Stoeckel, C. Kübel, A. Höltzel, U. Tallarek, Ind. Eng. Chem. Res. **57**, 3031–3042 (2018)
33. K. Hormann, U. Tallarek, J. Chromatogr. A **1312**, 26–36 (2013)

Atomistic Dynamics of Alternating Access Mechanism of an ABC Transporter

Hendrik Göddeke and Lars Schäfer

Abstract ATP-binding cassette (ABC) transporters are ATP-driven molecular machines. ATP binding and hydrolysis in the nucleotide-binding domains (NBDs) are coupled to large-scale conformational changes of the transmembrane domains (TMDs), which leads to the translocation of substrate molecules across biological membranes. The atomic details of the structural dynamics underlying the conformational transitions and the coupling of NBD and TMD motions remained largely *terra incognita*. Here, we used all-atom molecular dynamics (MD) simulations to characterize the conformational transitions underlying the working cycle of the heterodimeric ABC exporter TM287/288 from *Thermotoga maritima*. Multi-microsecond MD simulations reveal how ATP binding triggers a spontaneous conformational transition from the initial inward-facing (IF) conformation via an occluded (Occ) intermediate to an outward-facing (OF) conformation. ATP binding induces tightening of the NBD dimer, which involves closing and reorientation of the two NBD monomers. Simultaneous closure of the cytoplasmic (intracellular) TMD gate region leads to the Occ state. Subsequent wide opening of the periplasmic (extracellular) TMD gate yields the OF conformer. This distinct sequence of events imposes tight coupling of NBDs and TMDs and ensures that the cytoplasmic and periplasmic TMD gates are not open at the same time to both sides of the membrane.

1 Introduction

All-atom molecular dynamics (MD) simulations of large biomolecular systems are computationally expensive, limiting the accessible length scales and time scales to typically about 10^5 to 10^6 atoms and a few μs, respectively. Unfortunately, many biological processes occur on slower time scales and are thus inaccessible to con-

H. Göddeke · L. Schäfer (✉)
Theoretical Chemistry, Ruhr University Bochum, Bochum, Germany
e-mail: lars.schaefer@rub.de

© Springer Nature Switzerland AG 2021   117
W. E. Nagel et al. (eds.), *High Performance Computing in Science and Engineering '19*,
https://doi.org/10.1007/978-3-030-66792-4_8

Fig. 1 MD simulation
system of heterodimeric
ABC transporter TM287/288
embedded in a solvated
POPC lipid bilayer (ca.
132,000 atoms in total).
TM287 is shown in purple,
TM288 in blue. The
phosphocholine headgroups
of the lipids are shown as
orange spheres, and the lipid
hydrocarbon tails in yellow.
Water oxygen and hydrogen
atoms are shown in red and
white, respectively

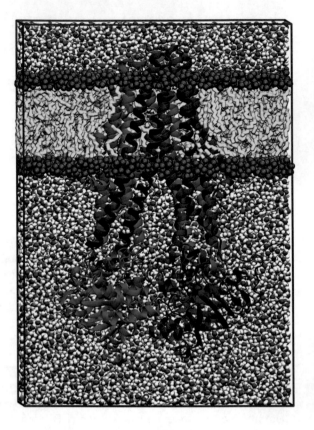

ventional MD simulations. One example for such a slow process is the large-scale
conformational transition of ATP-binding cassette (ABC) transporters. ABC transporters are membrane-embedded molecular machines that use the free energy stored
in ATP, the chemical energy-currency of the biological cell, to translocate a wide
range of transport substrates across biological membranes [1]. They are found in all
kingdoms of life and can be classified into importers, which are found exclusively
in prokaryotes, and exporters, which are present in every organism. ABC exporters
transport a broad range of chemically diverse molecules out of the cell, such as
lipids, peptides, and drugs, also including chemotherapeutics. Hence, these proteins
play a role in multidrug resistance of cancer cells and bacteria, and malfunction of
ABC transporters is linked to hereditary diseases such as cystic fibrosis and neonatal
diabetes [2].

ABC transporters share a common structural architecture. As shown in Fig. 1 for
the ABC exporter TM287/288 investigated here, a dimer of two nucleotide binding
domains (NBDs) is connected to a dimer of two transmembrane domains (TMDs).
The NBDs bind and hydrolyze ATP, while the TMDs form the substrate translocation
chamber. The free energy gained from ATP binding and/or hydrolysis in the NBDs is
transmitted to the TMDs, which undergo large-scale conformational changes to cycle

between an inward-facing (IF) and an outward-facing (OF) conformation, thereby providing access to both sides of the membrane in an alternating manner. This ultimately leads to substrate translocation. However, despite this common basic working principle, different ABC subfamilies can differ in terms of both structure and mechanism [3]. While bacterial ABC exporters are often homodimers, most eukaryotic exporters are heterodimers [4, 5]. Many heterodimers have one consensus and one degenerate nucleotide binding site. In the degenerate site, the catalytic residues that are essential for ATP hydrolysis [6] as well as residues in the ABC signature motif deviate from the consensus sequence, hence impairing catalytic activity.

The current mechanistic understanding of ABC transporters is largely based on high-resolution structures obtained by X-ray crystallography or cryo-electron microscopy which show static snapshots of different conformations that could play a role in the functional working cycle. However, although such structural information is of course extremely useful, it cannot provide the complete picture because the mechanism of ABC transporters is governed by dynamics. The nature of the coupled motions of the NBDs (which dimerize and partly dissociate upon nucleotide binding and unbinding, respectively) and the TMDs (which undergo large-scale conformational transitions between the IF and OF states, thereby providing access to one or the other side of the membrane) remain to be resolved.

In principle, all-atom MD simulation can provide the lacking atomic-level insights into the structural dynamics underlying the transport mechanism. However, due to the large computational effort MD simulations of complete ABC transporters in explicit membrane and water environment have been carried out only for a few systems, typically covering time scales up to a few hundred nanoseconds [7, 8]. These time scales are too short for complete conformational transitions between the IF and OF conformations to occur spontaneously. The aim of the present project was to characterize in full atomic detail the dynamics of the large-scale conformational transitions in the ABC exporter TM287/288 from the thermophilic bacterium *Thermotoga maritima*. TM287/288 is a bacterial homolog of the eukaryotic ABC exporters CFTR, which is linked to cystic fibrosis [9], and TAP, which is a key player in the immune system [10].

To overcome the above limitations in our simulations, we (i) used extensive all-atom MD sampling of more than 50 μs accumulated simulation time in total, and (ii) carried out our simulations at an elevated temperature of 375 K, which accelerates structural transitions. This was motivated by the fact that TM287/288 is from a hyperthermophile that lives under extreme conditions up to 90 °C [11], and hence the higher temperature appeared to be a natural choice.

Our MD simulations of TM287/288 are based on the X-ray structure from Hohl and coworkers [4, 5] that shows the transporter in an IF state, both for the apo protein and with the non-hydrolyzable ATP analog AMP-PNP bound to the degenerate NBD. In this work, we focus on the conformational transition, therefore we performed the MD simulations in the presence of ATP-Mg but without transport substrate. As many ABC transporters, TM287/288 shows a basal ATPase activity, i.e., the transporter cycles between different conformational states in a futile manner driven by ATP binding and hydrolysis alone [12].

The MD simulations reveal how TM287/288, when loaded with two ATP-Mg molecules, undergoes conformational transitions from the initial IF structure via an occluded intermediate to an OF conformation. The OF structures observed in our MD simulations were validated by EPR spectroscopy of membrane-reconstituted TM287/288. The distance distributions between spin-labeled residue pairs measured by EPR support the MD simulations and, furthermore, show that the lipid bilayer is essential for stabilizing the wide-open extracellular gate of the transporter in the OF conformation, as seen in the MD simulations. These results, which are described in Ref. [13], reveal the complete dynamic pathways of the large-scale conformational IF-to-OF transition that is underlying the working cycle of the ABC exporter. After this work was published, the group headed by Markus Seeger (Zurich University) obtained an X-ray crystal structure of the OF conformation of TM287/288 at 3.2 Å resolution (PDB 6QUZ), which very closely resembles the OF structure predicted by MD simulation [14].

2   Computational Details

2.1   Molecular Dynamics Simulations

The MD simulations were carried out with GROMACS version 5.1 [15]. The Amber99SB-ILDN protein force field [16, 17] was used together with the Berger parameters [18] for the lipids and the parameters of Meagher and coworkers [19] for ATP-Mg. The 2.6 Å resolution X-ray structure of TM287/288 in an inward-facing conformation [4, 5] with AMP-PNP bound to the degenerate site (PDB 4Q4A) was used as starting structure for the simulations, after converting AMP-PNP into ATP. A second ATP was docked into the consensus site, which was empty in the X-ray crystal structure. The resulting structure, loaded with 2 ATP-Mg molecules, was inserted into a POPC lipid bilayer [20] and solvated with water. The final simulation box contained 255 lipids, 33,370 TIP4P-2005 waters [21], and chloride ions to neutralize the periodic simulation box. The total system size was ca. 132,000 atoms. Before the MD, the system was energy-minimized with steepest descent and equilibrated for 10 ns in the NpT ensemble with harmonic position restraining potentials on all protein heavy atoms.

Short-range nonbonded interactions were treated with a buffered Verlet pair list [22] with potentials smoothly shifted to zero at a 1.0 nm cut-off. Long-range electrostatics were treated with the smooth particle mesh Ewald (PME) method [23] with a grid spacing of 0.12 nm. SETTLE [24] was used to keep the water molecules rigid, and LINCS [25] was used to constrain all other bond lengths. Virtual interaction sites were used for the hydrogen atoms [26], allowing 4 fs time steps for the integration of the equations of motion. The simulations were carried out in the NpT ensemble. Temperature was kept constant by coupling to the velocity rescaling thermostat of Bussi and coworkers [27]. For constant pressure, semi-isotropic p-coupling was applied. One hundred MD simulations, each of length 500 ns, were initiated with different random seeds for the initial velocity distribution.

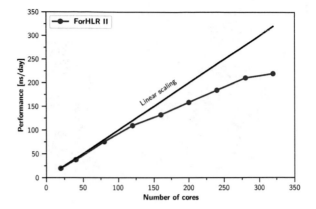

Fig. 2 Scaling of the MD simulations of the ABC transporter system shown in Fig. 1 on ForHLR II. Simulations were carried out with GROMACS version 5.1 and typically run on 200 or 280 CPU-cores (i.e., on 10 or 14 nodes) at a parallelization efficiency of 80% or 75%, respectively. The black diagonal line indicates ideal scaling

2.2 Computational Performance

The MD simulations were carried out at the Steinbuch Centre for Computing (SCC) in Karlsruhe on ForHLR II. For our simulation system comprising of ca. 132,000 atoms, a good performance was achieved on 10 nodes (200 cores) with a parallelization efficiency of 80% and an absolute performance of 159 ns/day (Fig. 2). The runtime of a typical single simulation of 500 ns was thus ca. 75 h. On 14 nodes (280 CPU-cores) the parallelization efficiency was still good (75%), and the absolute performance was 211 ns/day.

3 Results and Discussion

One hundred MD simulations, each of length 500 ns, were initiated from the inward-facing (IF) crystal structure of TM287/288 with ATP-Mg bound to the degenerate NBD [4, 5]. In all simulations, the transporter was inserted into an explicit POPC lipid bilayer and a second ATP-Mg was docked into the consensus NBD that was empty in the IF X-ray crystal structure, see Sect. 2.

In 6 MD simulations, a spontaneous large-scale conformational transition from the initial IF structure to an OF conformation was observed (Fig. 3). This complex conformational transition involves formation of a tight NBD dimer, TMD closure at the cytoplasmic gate, and finally TMD opening at the periplasmic gate. The transition to OF proceeds via an occluded (Occ) intermediate, in which the NBD dimer is tightened and the cytoplasmic gate is closed, but the TMDs did not yet open at the periplasmic gate. As both TMD gates (periplasmic and cytoplasmic gate) are closed, the Occ intermediate is sealed to both sides of the membrane, ensuring that the transporter is never a fully open channel in the membrane.

In the 6 simulations in which TM287/288 completely transitioned from IF to OF, the occluded state was visited only transiently. However, in 11 other simulations,

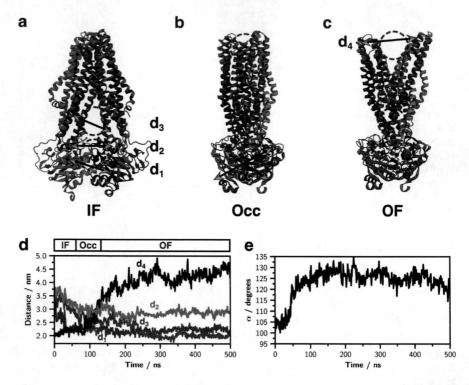

Fig. 3 MD simulation of large-scale alternating access conformational transition. (a–c) Initial IF, intermediate Occ, and final OF conformations observed during MD. The $C\alpha$–$C\alpha$ distances d_1, d_2, d_3 and d_4 indicate NBD dimerization, movement of the coupling helices, closing of the cytoplasmic TMD gate, and opening of the periplasmic TMD gate, respectively. The reorientation of the two NBDs with respect to each other is described by the NBD twisting angle α. (d, e) Time traces of the four indicated distances and of the NBD twisting angle α during representative IF-to-OF transition

the transporter transitioned to Occ and got trapped in that state for the remaining simulation time, i.e., up to 500 ns. In the remaining 83 simulations, the energy barrier towards Occ was not overcome during the 500 ns simulation time, and TM287/288 remained in the initial IF conformation.

Although the MD simulations did not draw on any structural information about a desired target structure, the final OF conformation obtained after IF-to-OF transition closely resembles the OF X-ray structure published later [14]. The low $C\alpha$-RMSD of ca. 0.3 nm for the entire transporter reflects thermal fluctuations around an average structure that is close to the X-ray crystal structure (Fig. 4).

The above mechanism is in line with the ATP switch model [28] in that ATP binding, and not hydrolysis, suffices to trigger the IF-to-OF conformational transition. After release of the transport substrate, ATP hydrolysis and unbinding of the reaction products ADP + P_i would thus merely be required to reset the conformational cycle back to the IF state that is again receptive for binding of transport substrate in the

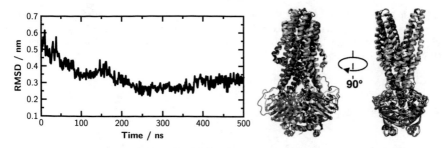

Fig. 4 The OF conformer predicted from MD simulations after IF-to-OF transition is compared to the X-ray structure of the OF conformer [14]. (Left panel) Time trace of Cα-RMSD of entire TM287/288 transporter with respect to the OF X-ray structure during representative IF-to-OF transition. The similarity of the structures is visualized by a superposition of the final snapshot after 500 ns of MD simulation (shown in color) with the X-ray structure (shown in grey)

TMDs and of ATP-Mg in the consensus NBD [6]. In a ABC heterodimer such as TM287/288, catalytic activity occurs predominantly in the consensus site. Thus, one ATP molecule would always remain bound to the degenerate site and mediate interactions between the two NBD monomers, precluding them from fully dissociating during the cycle.

4   Conclusions and Outlook

The conformational transitions between IF and OF states of the heterodimeric ABC exporter TM287/288 were studied by all-atom MD simulations in explicit membrane and water environment. The MD simulations reveal the conformational changes underlying the working cycle of the ABC transporter and provide an atomistic picture of the alternating access mechanism. First, ATP binding to both nucleotide binding domains triggers tightening of the NBD dimer by a concerted closing and twisting motion, that is, upon dimerization the two NBD monomers reorient to form a fully closed dimer. At the same time, the cytoplasmic TMD gate closes, which leads to an occluded intermediate. Finally, the periplasmic TMD gate opens to yield the OF conformation. The MD simulation results are supported by X-ray crystallography and EPR spectroscopy [13, 14].

To fully characterize the functional working cycle, substrate molecules should be explicitly considered. Due to the inherently dynamic nature of the process, mapping out the pathways along which substrates are shuttled through the ABC exporter is a formidable challenge that has not been achieved at atomic resolution up to now. The present results are a key step in this direction. As the conformational transitions occur spontaneously in our simulations, it might be possible to extract a small set of suitable collective variables for free energy simulations of the IF/OF transition in the presence of transport substrate molecules.

Acknowledgements The Steinbuch Centre for Computing (SCC) in Karlsruhe is acknowledged for providing computational resources. This work was funded by the Deutsche Forschungsgemeinschaft (DFG) through an Emmy Noether grant to L.S. (SCHA 1574/3-1) and Cluster of Excellence RESOLV (EXC-2033) project number 390677874.

References

1. A.L. Davidson, E. Dassa, C. Orelle, J. Chen, Microbiol. Mol. Biol. Rev. **72**(2), 317 (2008)
2. C.F. Higgins, Nature **446**, 749 (2007)
3. K.P. Locher, Nat. Struct. Mol. Biol. **23**, 487 (2016)
4. M. Hohl, C. Briand, M.G. Grütter, M.A. Seeger, Nat. Struct. Mol. Biol. **19**(4), 395 (2012)
5. M. Hohl, L.M. Hürlimann, S. Bohm, J. Schoppe, M.G. Grütter, E. Bordignon, M.A. Seeger, Proc. Natl. Acad. Sci. USA **111**, 11025 (2014)
6. M. Prieß, H. Göddeke, G. Groenhof, L.V. Schäfer, ACS Cent. Sci. **4**(10), 1334 (2018)
7. E. Lindahl, M.S.P. Sansom, Curr. Opin. Struct. Biol. **18**(4), 425 (2008)
8. J. Li, P.C. Wen, M. Moradi, E. Tajkhorshid, Curr. Opin. Struct. Biol. **31**, 96 (2015)
9. D.C. Gadsby, P. Vergani, L. Csanády, Nature **440**, 477 (2006)
10. D. Parcej, R. Tampé, Nat. Chem. Biol. **6**(8), 572 (2010)
11. R. Huber, T.A. Langworthy, H. König, M. Thomm, C.R. Woese, U.B. Sleytr, K.O. Stetter, Arch. Microbiol. **144**(4), 324 (1986)
12. M.H. Timachi, C.A. Hutter, M. Hohl, T. Assafa, S. Bohm, A. Mittal, M.A. Seeger, E. Bordignon,eLife **6**, e20236 (2017)
13. H. Göddeke, M.H. Timachi, C.A.J. Hutter, L. Galazzo, M.A. Seeger, M. Karttunen, E. Bordignon, L.V. Schäfer, J. Am. Chem. Soc. **140**(13), 4543 (2018)
14. C.A.J. Hutter, M.H. Timachi, L.M. Hürlimann, I. Zimmermann, P. Egloff, H. Göddeke, S. Kucher, S. Stefanic, M. Karttunen, L.V. Schäfer, E. Bordignon, M.A. Seeger, Nat. Commun. **10**, 2260 (2019)
15. M.J. Abraham, M. Murtola, R. Schulz, S. Páll, J.C. Smith, B. Hess, E. Lindahl, SoftwareX **1–2**, 19 (2015)
16. V. Hornak, R. Abel, A. Okur, B. Strockbine, A. Roitberg, C. Simmerling, Proteins **65**, 712 (2006)
17. K. Lindorff-Larsen, S. Piana, K. Palmo, P. Maragakis, J.L. Klepeis, R.O. Dror, Proteins **78**, 1950 (2010)
18. O. Berger, O. Edholm, F. Jähnig, Biophys. J. **72**, 2002 (1997)
19. K.L. Meagher, L.T. Redman, H.A. Carlson, J. Comput. Chem. **24**(9), 1016 (2003)
20. C. Kandt, W.L. Ash, D.P. Tieleman, Methods **41**, 475 (2007)
21. J.L.F. Abascal, C. Vega, J. Chem. Phys. **123**, 234505 (2005)
22. S. Páll, B. Hess, Comput. Phys. Commun. **184**(12), 2641 (2013)
23. U. Essmann, L. Perera, M.L. Berkowitz, T. Darden, H. Lee, L.G. Pedersen, J. Chem. Phys. **103**(19), 8577 (1995)
24. S. Miyamoto, P.A. Kollman, J. Comput. Chem. **13**, 952 (1993)
25. B. Hess, J. Chem. Theory Comput. **4**, 116 (2008)
26. K.A. Feenstra, B. Hess, H.J.C. Berendsen, J. Comput. Chem. **20**(8), 786 (1999)
27. G. Bussi, D. Donadio, M. Parrinello, J. Chem. Phys. **126**, 014101 (2007)
28. C.F. Higgins, K.J. Linton, Nat. Struct. Mol. Biol. **11**, 918 (2004)

Molecular Dynamics Simulations—A Time and Length Scale Investigation

Martin Hummel, Wolfgang Verestek, and Siegfried Schmauder

Molecular Dynamics simulation is often struggling with reproducing experimentally accessible dimensions in time and length. We present simulations connecting experiments and molecular dynamics simulations, without the explicit need of extrapolation.

1 Introduction

Molecular Dynamics (MD) simulation is a widely used and accepted method in materials science. Nevertheless, there are still questions arising about the reliability of the obtained results. This report shall provide a short overview over the development, which took place over the last decades. First in hardware, from theoretical concepts like the Turing machine until high performance clusters like "Hazel Hen" at the University of Stuttgart. Thereafter, software evolution is discussed, where there is on the one hand side optimization of available software codes and the modification of these simulation codes, in terms of the model itself, on the other hand. All three together,the improvement in hardware as well as software and modeling leads to a

M. Hummel (✉) · W. Verestek · S. Schmauder
Institut für Materialprüfung, Werkstoffkunde und Festigkeitslehre, Universität Stuttgart, Stuttgart, Germany
e-mail: Martin.Hummel@imwf.uni-stuttgart.de

W. Verestek
e-mail: Wolfgang.Verestek@imwf.uni-stuttgart.de

S. Schmauder
e-mail: Siegfried.Schmauder@imwf.uni-stuttgart.de

© Springer Nature Switzerland AG 2021   125
W. E. Nagel et al. (eds.), *High Performance Computing in Science and Engineering '19*,
https://doi.org/10.1007/978-3-030-66792-4_9

tremendous extension of the time and length scales, which are accessible by MD calculations. Although this tremendous improvement took place, MD simulations are still limited to small scales in dimension and time. This leads sometimes to discussions about the applicability of the results. A common approach is the extrapolation across several orders of magnitude, which is often argued against. Some strain rate convergence studies, available in literature, are presented here, both experimentally and numerically. Our numerical convergence study by MD shows an almost constant value for maximum tensile stresses in polycrystalline aluminum systems on the one hand, and ongoing decrease of maximum stresses in single crystalline aluminum samples on the other. Concerning length scales, an experimental size dependence analysis of Huang et al. [1] is presented and repeated numerically with molecular dynamics in similar size.

2   Computational Evolution

In 1936, Alan Turing suggested the Turing machine, which can be seen as the first ancestor of modern computers. To avoid rivalry and taking part in the discussion about which computer is the first, we head forward to the first supercomputers which were developed in the 1960s. Also in the 60s, more specific in 1965 Gordon Moore predicted the development that the number of components of integrated circuits will double each year [2], which is referred to as Moore's law. Ten years later, in 1975s he reduced the rate to a doubling rate every two years. Due to the improvement of the transistors themselves, it is often stated as a doubling in performance every 1.5 years. Being solely an historical observation and extrapolation, it is strongly impressive how accurate this prediction holds true, with some curtailment, until today. The graph in Fig. 1 shows the development of the peak speed of supercomputers and high performance clusters (HPCs) since 1936, which confirms Moore's law. In June 2018 the worlds fastest supercomputer is running at the Oak Ridge National Laboratory at 122.3 petaflops [3].

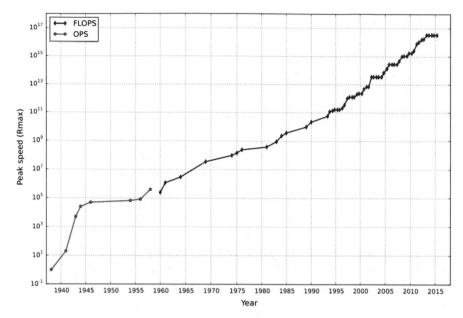

Fig. 1 Development of the peak speed in (FL)OPS ((floating point) operations per second) over the years [4]

3 Code Evolution

Besides the tremendous development of the hardware capabilities, it is also an ambition to improve the simulation software efficiency. This aim is difficult to achieve, since not only the speed up is important, but also flexibility of the software is requested. This necessary flexibility is due to the increased application of molecular dynamics (MD) simulations, partly caused by the improvement of numerically feasible system sizes, through hardware and software improvements. The developers of MD code LAMMPS [5] state on their webpage [6] that the performance of their code differs rarely more than 5–10 % between the current LAMMPS version, written in C++ and the older versions, written in Fortran F90 or F77. It strongly depends on the routines used and the machines and compilers, "sometimes the C++ version is a bit faster; sometimes a bit slower". For the special case of simulations of metals using EAM [7, 8] potentials, a speed up of 2–4 times faster, compared to the predecessor, is obtained by changing the way of implementation. A large share of the acceleration of MD simulation is done by optimizing the parallelization. With increasing cluster sizes, which are used for the numerical calculations, the amount of communication data increased. This optimization of the parallelization is often strongly dependent on the used computational cluster. A large variety of benchmarks of parallelization is available on the LAMMPS webpage [6].

4 Modified Molecular Dynamics

The earlier described evolutions of the software and especially the hardware components already increase the accessible field of application. Nevertheless the achievable
dimensions in time and length remain in the nano-regime. Very large simulations,
containing billions of atoms, are capable to access lengths up to one micrometer; an
aluminum cube of 1 μm side length would contain about 60 billion atoms. To reach
a very long simulated time, e.g. 1 microsecond, at a usual time step width of 1 fs for
metal simulations in MD, one would require 1 billion simulation steps. Accessing
the μ-scale in both, size and time, together would still require 47,248 core years on
a Cray XT5, based on the 1.49e-6 sec/atom/time step determined at the benchmark
on the LAMMPS webpage [6]. Other methods of accelerating the simulations need
to be developed. A review paper of Voter et al. [9] is presenting different options,
how the accessible time range in atomistic simulation can be extended. Some of the
different approaches will be briefly introduced. For a detailed description the review
paper of Voter et al., or the referred primary literature, is recommended. The solutions
presented here, are tailored for special circumstances.

4.1 Parallel Replica Method

As first example, the parallel replica method, see Fig. 2, is a useful approach for
simulations with rare, infrequent events, e.g. for diffusive processes. The four step
procedure is initiated by the replication of the simulated system to an arbitrary number
of replicas, indicated by (A) in Fig. 2. Following phase (A) is the dephasing stage
(B), where an elimination of the correlations between the replicas is fulfilled by
periodically randomizing the momenta. Through replication and independence, the
phase space is explored faster by the factor of the number of the replicas, pictured in
stage (C). In the final stage (D) the rare event happened and the simulation clock is
advanced to the sum of all individual trajectory times of all replicas.

4.2 Hyperdynamics

The idea of hyperdynamics is to add a bias potential $\Delta V(r)$ to the real potential $V(r)$
describing the interaction between the simulated atoms, see Fig. 3. The addition of
the bias potential activates the transitions between two states, e.g. C and A, according
to the transition state theory much faster, since the escape rate k_{CA} from state C to
A is proportional to the energy barrier between C and A.

$$k_{CA} \sim \exp[V(r)/k_B T] \tag{1}$$

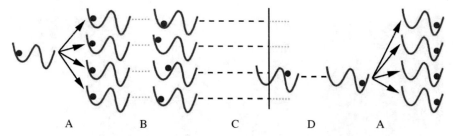

A   B   C   D   A

Fig. 2 Schematic illustration of the parallel replica method. The four steps are **a** replication of the system into M copies, **b** dephasing of the replicas, **c** independent trajectories until a transition is detected in any of the replicas, and **d** brief continuation of the transitioning trajectory to allow for correlated events such as recrossings or follow-on transitions to other states. The resulting configuration is replicated, beginning the process again [9]

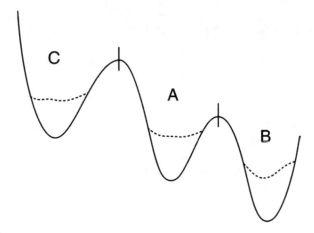

Fig. 3 Schematic illustration of the hyperdynamics method. The real interaction potential is amended by an artificial potential, the sum of both, the real and the artificial potential, is schematically depicted by the dashed line [9]

The speed up through hyperdynamics is given by the boost factor,

$$\text{boost(hyperdynamics)} = \frac{t_{hyper}}{t_{MD}} = \langle\exp[\delta V(r)/k_B T]\rangle_b \qquad (2)$$

divided by the extra computational cost due to the bias potential. The $\langle\rangle_b$ indicates an average over the trajectory on the biased potential. A boundary condition to the bias potential is given by the necessity to be zero at the maxima between two neighboring states. This is needed to preserve the trajectory. Throughout the whole hyperdynamics concept, it is assumed that there are no correlated events.

The hyperdynamics approach has been tested at our institute to reproduce the results of an annealing simulation of an amorphous Si-coating, which before was conventionally simulated without acceleration [10]. By reducing the depth of the

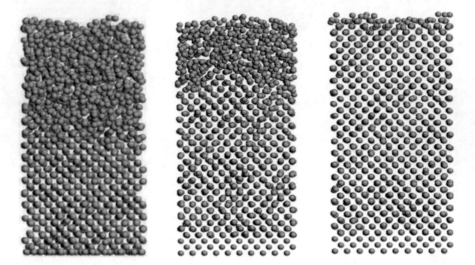

Fig. 4 Reproducing Fig. 9.6 of [10] showing the annealing of an amorphous Si-coating using hyperdynamics

minima of the Tersoff potential [11] to half the size, an effective speed up by a factor of 200 was obtained (Fig. 4).

The publication by Ghosh et al. [12] shows the applicability of hyperdynamics in the field of crack growth. It is shown, that the crack path is not essentially affected by the modulated potential.

4.3   Temperature-Accelerated Dynamics

Another displayed method of speeding up molecular dynamics by Voter et al. [9] is the temperature-accelerated dynamics (TAD). This method utilizes the fact of the potential $V(r)$ in Eq. 1 being strictly negative, and therefore an increase of the temperature T leads to an increased escape rate k. This method has to be handled with care since it might change the trajectories due to the different working temperature. A field where the TAD method is applicable is e.g. vapor-deposited crystal growth.

5   Strain Rate Convergence

In most research works, there is a gap concerning the strain rate of experiments and molecular dynamics simulations. The influencing effect has been investigated in

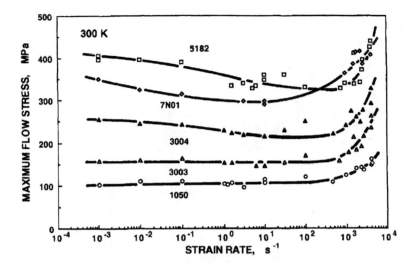

Fig. 5 The experimentally investigated relationship between the maximum flow stress and the strain rate for five different aluminum alloys [13]

both regimes and will be presented in the following. Moreover, special cases will be discussed, where these regimes can meet.

5.1 Experimental Strain Rate Exploration

In 1991 Higashi et al. [13] published a study of five different aluminum alloys at different strain rates reaching from 0.001 s^{-1} to 4,000 s^{-1}. Figure 3 of [13] is shown as Fig. 5 here, it shows the maximum flow stress in dependence of the strain rate for each alloy. The slow strain rates were accomplished with an instron machine, intermediate strain rates with a hydraulic tensile testing machine and the dynamic tensile tests with a split Hopkinson pressure bar. It is observed, that for the alloys with only little concentration of alloying elements 1050 (0.27 wt% Fe and 0.10 wt% Si) and 3003 (0.15 wt% Cu, 1.20 wt% Mn, 0.61 wt% Fe and 0.27 wt% Si), there is no strain rate dependence for slow and medium range strain rates. The alloys with higher concentration of alloying elements even show a decrease of the maximum flow stress with increasing strain rate. In contrast, the dynamic tensile tests with the split Hopkinson pressure bar show all an increase of maximum flow stress with increasing strain rate, regardless of the composition, whereat the 1050 alloy increases the least.

   In the work of Peng et al. [14] also a wide range of strain rates has been investigated, but specifically for the aluminum alloy 6082-T6. Tensile tests were performed in the quasi static regime up to dynamic tension as well. In total, the strain rates range from 0.001 s^{-1} to 3,400 s^{-1}. The dynamic tensile test, which are used to obtain the high strain rates, were performed by using the split Hopkinson pressure bar,

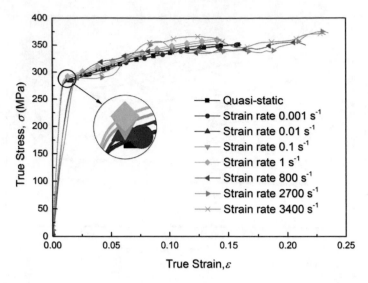

Fig. 6 Experimental stress-strain curves for tensile testing of an aluminum alloy 6082-T6 at different strain rates

like it is used in the experiments for Fig. 5. The resulting stress-strain-relations are depicted in Fig. 6. It can be seen that the overall behavior is quite similar to each other, independent of the used strain rate. In the elastic part it is observed, that the slope of the initial part of the stress-strain curve, namely the Young's modulus, is lower at higher strain rates and the transition into the plastic regime happens at higher strain. The stress level during the plastic phase is again very similar, although there are higher fluctuations observed at the experiments for high strain rates.

In comparison Figs. 5 and 6 show an identical behavior for slow strain rates, whereas the high strain rates show slight differences. Although both experiments have been carried out by the split Hopkins pressure bar test, the increase of the maximum stress, which has been observed in 1991, is not confirmed in the more recent experiments of 2016. This might be due to improvements in the experimental set up, but as well, it is strongly depending on the used alloy. For different other aluminum alloys, e.g. 7075 [15] almost constant maximum flow stresses have been measured, even for higher strain rates.

A different possibility of applying high strain rates is given by shock wave loading. Here the applied stress is compressive. The asymmetry of yield strength in tension and compression has been investigated in [16], where the compressive yield stress was found to be 20% higher for a very fine grained aluminum alloy AA1050. With increasing grain size the asymmetry is vanishing. Kanel et al. published in [17] a large series of shock wave impact experiments on different aluminum and aluminum alloy test specimens. Two different kinds of shock wave generators were used. First, an impact of flyer plates with an impact velocity of 600–700 ms^{-1}, which leads, together with the specimen thickness of 300–400 µm, to strain rates of $3 - 6 \times 10^5$ s^{-1}, was

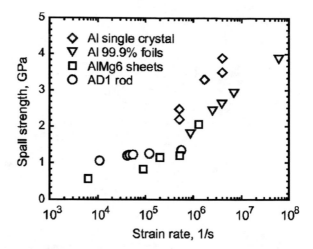

Fig. 7 Spall strength at room temperature for different aluminum and aluminum alloy compositions and geometries as a function of the strain rate [17]

used. Second, the pulsed high-power proton beam KALIF [18] was used on plates with thicknesses between 50 and 100 μm, resulting in strain rates of $1.5^{-3} \times 10^{6}\, s^{-1}$. Figure 7 is taken out of [17], where several results of high strain rate compression experiments are accumulated in one graph. The first impression might be a strongly increasing spall strength with increasing strain rate. A close look reveals that this impression holds true only for strain rates larger than $10^{6}\, s^{-1}$. For lower strain rates, the change is quite moderate. AD1 is an aluminum alloy with 99.3 wt. % aluminum. For the AD1 rod, the exact values are taken out of [19], the spall strength for the strain rate of $1.1 \times 10^{4}\, s^{-1}$ is 1.06 GPa on the lower end and 1.38 GPa at the high strain rate of $8.8 \times 10^{5}\, s^{-1}$. Therefore, over two decades it is varying by about 20 %. This shows the importance of the composition of the alloy as well as the geometrical properties of the test specimen.

5.2 Numerical Strain Rate Exploration

The previous section has shown an improvement in the experiments, especially concerning the accuracy of the measurements; still the regime accessible by simulation is way faster than very fast experiments. Koh and Lee published in 2006 [20] an investigation of face centered cubic (fcc) nanowires for different sizes and strain rates. The metals used in the nano wires are gold and platinum in three different sizes each. The wires were modeled with a diameter of 2, 4 and 6 nm and double the length each, and therefore comprise 890, 6,660 and 24,000 metal atoms. The applied strain rates were $4.0 \times 10^{8}\, s^{-1}$, $4.0 \times 10^{9}\, s^{-1}$ and $4.0 \times 10^{10}\, s^{-1}$. The results of the 18 simulations are depicted in Fig. 8. A very similar behavior is observed in the flow part for all different wire sizes. In contrast, the height and the strain of the first peak

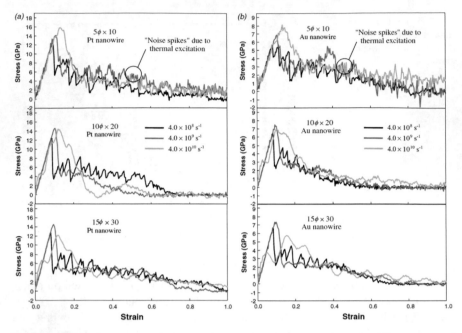

Fig. 8 Stress-strain response of 18 nanowires. **a** Pt and **b** Au, each for three different sizes and three different strain rates [20]

is different for different strain rates. Most of the time, but not always, the highest stress and highest strain at the first peak are obtained at the highest strain rate.

A strain rate comparison using a twelve-grained aluminum polycrystal with a size of $20 \times 20 \times 40$ nm^3, containing 960,000 aluminum atoms has been carried out. The range of strain rates reached from 5.0×10^6 s^{-1} to 5.0×10^{11} s^{-1} using IMD [21] as simulation software at a time step width of 2×10^{-15} s. The simulated temperature is 0 K and the used EAM potential is taken from Chen et al. [22]. The numbers in the legend designate the deformation per time step. Comparable fluctuations like observed in [20] and shown in Fig. 8, for the high strain rates are observed for the two highest strain rates in Fig. 9 as well. The obtained maxima are far too high, compared to the results obtained using the other strain rates. For strain rates of 5.0×10^9 s^{-1} and lower there is a maximum reached, followed by a stress decay in the plastic part of the deformation. The maximum stress is lowering with lower strain rates from 2.5 GPa for 5.0×10^9 s^{-1} over 2.1 GPa for 2.5×10^9 s^{-1} and 1.61 GPa for 2.5×10^8 s^{-1} to 1.58 GPa for 5.0×10^6 s^{-1}, whilst the strain at maximum is also decreasing.

Comparing the green and the blue stress-strain-curves in Fig. 10 it is seen that the reached stress values are almost the same, whereas the blue stress-strain-curve shows more details with the more jagged curve. Nevertheless the height of the maxima differs by only 30 MPa, which is less than 2% difference of the obtained value of 1,610 MPa for the strain rate of 2.5×10^8 s^{-1} (green) and 1,580 MPa for the strain rate of 5.0×10^6 s^{-1} (blue)

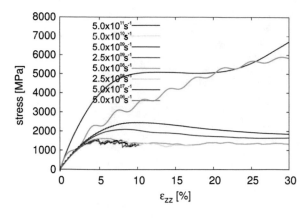

Fig. 9 Simulated stress-strain-curves for eight different strain rates. The numbers in the legend show the strain rate

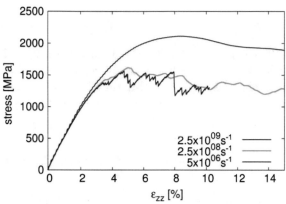

Fig. 10 Detailed view on the three stress-strain-curves of Fig. 9. The strain rate is given in the legend

Another long-term study of strain rates has been carried out on a polycrystalline cube with edge length 22 nm containing 642,000 aluminum atoms. The applied strain rates are $1.0 \times 10^5 \, s^{-1}$ to $1.0 \times 10^8 \, s^{-1}$, at a temperature of 300 K. Therefore, the, in molecular dynamics context, very low strain rate of $1.0 \times 10^5 \, s^{-1}$, represented by the red stress-strain-curve in Fig. 11, is only a factor of 25 higher compared to the experimental strain rate of $4,000 \, s^{-1}$. There have been 837,300,000 simulation steps calculated, which corresponds to $0.837 \mu s$, at a time step size of 1 fs. The downside of this strain rate is the enormous use of computational power. The simulation has been carried out using the molecular dynamics software LAMMPS [5] and the EAM potential for Al by Liu et al. [23]. The simulation took about 150 days on 144 computational cores in parallel, which is equivalent to 59.18 years on a single core processor. This high computational effort is not practicable for a broad use. The maxima of the different strain rates depicted in Fig. 11 are 1,686 MPa for the strain rate of $1.0 \times 10^8 \, s^{-1}$ (purple), 1,455 MPa for $1.0 \times 10^6 \, s^{-1}$ (blue), 1,346 MPa for $1.0 \times 10^6 \, s^{-1}$ (green) and 1,202 MPa for the strain rate of $1.0 \times 10^5 \, s^{-1}$ (red). Therefore, an influence of the strain rate is observed for this study, but the overall values are in very good agreement with the experimentally obtained spall stresses measured in

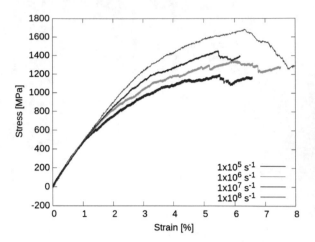

Fig. 11 Simulated stress-strain-curves for four strain rates. Using a cubic aluminum polycrystal with edge length of 22 nm, containing 642,000 atoms

[19] and also shown in Fig. 11. The experimental value corresponding to the green curve with 1,346 MPa at a strain rate of $1.0 \times 10^6 \, \text{s}^{-1}$ is 1,380 MPa at a strain rate of $8.8 \times 10^5 \, \text{s}^{-1}$ and 1,260 MPa at $1.2 \times 10^5 \, \text{s}^{-1}$ compared to the 1,202 MPa for the strain rate of $1.0 \times 10^5 \, \text{s}^{-1}$ depicted in the red stress-strain-curve.

5.3   Comparing Experimental and Numerical Strain Rates

The two subsections before have shown the effort in investigating the effect of strain rates in both, the experimental and the numerical, regimes. Almost constant values have been observed over a wide range of strain rates (six decades) in experiments, see Figs. 5 and 6. Furthermore, the convergence of the obtained maximum stress values has been shown for numerical simulations using molecular dynamics with almost constant maximum stress values for two decades in Fig. 9. Nevertheless, care has to be taken of the individual boundary conditions of each simulation series, which is seen in Fig. 11 where the convergence of the maximum stress is not yet reached even for the very low strain of $1.0 \times 10^5 \, \text{s}^{-1}$. There are many influences like the temperature or the choice of the interatomic potential, even for the same elements, which can change the convergence of the maximum stress at different strain rates. Both, the values of AD1 rod in Fig. 7 and the simulated values for aluminum in Fig. 11 are depicted in Fig. 12. It can be observed that the transition from experiment to simulation runs quite smooth. Remarkably the values obtained by the simulations are even lower than the experimental ones.

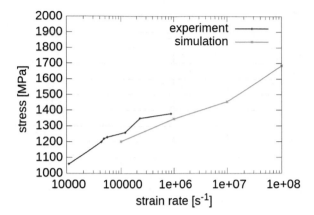

Fig. 12 Maximum stress values in dependence of the strain rate. Experimental values in red, simulated values in green

6 Size Effects

Besides the already listed effecting boundary conditions, like temperature and choice of potential, there is also an influence of the simulated system size, which is already shown in Fig. 8 in the work of Koh and Lee [20]. Additionally, it is intriguing, that the high strain rate experiments, resulting in high stress values, are derived from samples with small length scales. Therefore, the influence of the system size will be analyzed in the next subsections, first experimentally, then followed by numerical investigations.

6.1 Experimental Size Effects

Huang et al. [1] investigated the stress strain behavior of aluminum single crystals for different sizes. They fabricated samples for the mini-scale with length 1.5 mm, width 0.5 mm and thickness of 0.5 mm out of single crystalline aluminum blocks with a purity of 99.95%. The micro-scale sample, where a compression test was performed, is a rod with a length of 4.5 ± 0.5 μm and 2.0 ± 0.2 μm diameter. The sample has been prepared by using a focused ion beam (FIB) as the nano-samples as well. The schematic sketch and an SEM image of the nano-sized sample is given in Fig. 13. The size of block 2 in Fig. 13a is of the dimension 400 ± 100 nm in length, 200 ± 50 nm in width and 150 ± 50 nm in thickness. The orientation of the samples in all sizes is chosen to be the [111] or the [110] direction. For the different sizes the strain rate was chosen as 2×10^{-3} s^{-1} for the mini-sample, 7×10^{-3} s^{-1} for the micro-sample and 1×10^{-3} s^{-1} for the nano-sample.

The results of the deformation test are summarized in Table 1 for [111] oriented samples. A strong increase in the yield strength is observed with decreasing sample size. In the mini-scale the yield strength is reached at about 60 MPa, which is in good agreement with the literature value of 60–95 MPa for high purities. The reduction of

Table 1 Summary of the mechanical test results for the Al single crystal in all sizes obtained in [1] for the [111] oriented samples

	Size	Yield stress [MPa]	Elastic strain [%]
Mini scale	$1.5 \times 0.5 \times 0.5\,mm^3$	60 ± 3	0.1
Micro scale	$4.5 \pm 0.5 \times (2.0 \pm 0.2)^2\,\mu m^3$	180 ± 5	0.25
Nano scale	$400 \pm 100 \times 200 \pm 50 \times 150 \pm 50\,nm^3$	750 ± 10	1.0

the specimen size to the micro-sized sample shows an increase of the yield stress to 180 MPa. The further reduction of the sample size, by a factor of 10 in all dimensions, to obtain the nano-scale sample is leading to an even stronger increase of the yield stress to 750 ± 10 MPa. Simultaneously the elastic strain estimated from the unloading part is increasing from 0.1% for the mini-scale sample to 1.0% for the nano-scale sample.

6.2 Numerical Size Effects

The investigations of [1] are remodeled using molecular dynamics simulations. Therefore, an aluminum single crystal in the size of the lower limits of the experiment has been constructed with $300 \times 150 \times 100\,nm^3$, containing 268 million atoms. Therefore, the numerical effort is quite demanding and the strain rate was chosen to be $1 \times 10^7\,s^{-1}$. Periodic boundary conditions are applied in loading direction. The obtained stress-strain curve is depicted in Fig. 14 with the title '300 nm' in green. The maximum value of 1,535 MPa is reached at 3.6% strain. To investigate the influence of the size, the volume of the modeled system has been changed. The naming is according to the side length. The largest simulation named '450 nm' is of the size

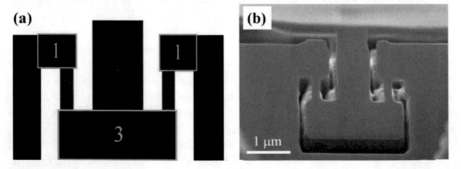

Fig. 13 Schematic drawing (**a**) and SEM micrograph (**b**) of the self-designed nano-tension specimen [1]

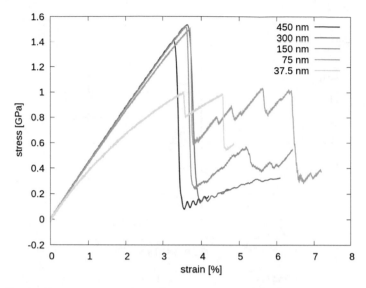

Fig. 14 Simulated stress-strain-curves for single crystalline aluminum at five different sizes

$450 \times 224 \times 153 nm^3$ and contains 909 million atoms. For the systems of the size of '75 nm' to '300 nm' the maximum stress is quite similar at 1.5 GPa. The reduction of the size to an eighth of the length leads to an increase of the maximum stress to a value of 1.2 GPa. This reduction of the maximum stress can be identified with the transition to the inverse Hall-Petch-regime which has been observed at 13.5 nm [24] for pure aluminum, which matches quite well with the side length of the wire with 11.5 and 18.6 nm. An increase of the flow stress in the plastic regime after the yield point is observed with reduction of the model size.

The difference between the obtained maximum stress values in the experiment (750 MPa) and simulation (1407 MPa) may be explained by extrapolating Fig. 12 with a square root fit, which is shown in Fig. 15. The maximum stress value of the wire at a strain rate of $1 \times 10^7 s^{-1}$ with 1407 MPa matches well to the value of the polycrystal with periodic boundary conditions (1455 MPa). The corresponding value

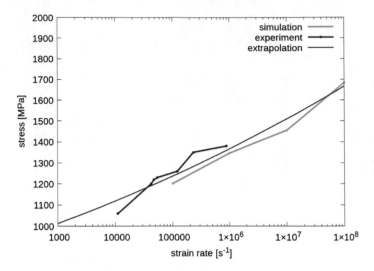

Fig. 15 Figure 12 with extrapolation via a fitted square root function

at a strain rate of 1×10^{-3} s^{-1} is determined to be 556 MPa, which is in quite good agreement with the experimentally obtained value of 750 MPa [1] and again, the simulated results lead to lower values compared to the experiments.

7 Conclusion

We made ends meet. The single crystalline simulation of aluminum is carried out at a strain rate, which is accessible by experiments, leading to similar results. Therefore, an extrapolation is not needed. Still it is shown, that the obtained results are strongly sensitive on the used material. The investigation concerning the dimension shows a huge influence. Much stronger than the influence of the strain rate. The experimental tensile test shows an increase of the maximum stress from the macro to the nano sample by a factor of 12.5 by decreasing the side length of the cube by the factor of 3750. Whilst the MD simulation of a same nano sized system leads to maximum value, which is, double as large as for the experiment. But the strain rate is with 1×10^{7} s^{-1} ten orders of magnitudes higher than the experimental one 1×10^{-3} s^{-1}. In summary, MD simulations lead to the same results as experiments, when they are carried out at the same scales. Nevertheless, one has to be careful when transferring results from one scale to another.

Acknowledgements The presented work was funded by the German Research Foundation (DFG) as part of the Special Research Field 716, sub-project B.7 and DFG project SCHM 746/154. This support is greatly appreciated. Special gratitude is dedicated to the HLRS for providing, updating and maintaining the infrastructures to run these simulations.

References

1. J.H. Wu, W.Y. Tsai, J.C. Huang, C.H. Hsieh, G.R. Huang, Mater. Sci. Eng.: A **662**, 296 (2016). https://doi.org/10.1016/j.msea.2016.03.076
2. G.E. Moore, IEEE Solid-State Circ. Soc. Newslett. **11**(3), 33 (2006). https://doi.org/10.1109/N-SSC.2006.4785860
3. The top500 project (2018). https://www.top500.org/lists/2018/06/
4. (2018). https://en.wikipedia.org/wiki/Supercomputer
5. S. Plimpton, J. Comput. Phy. (117), 1 (1995)
6. LAMMPS Molecular Dynamics Simulator (2018). http://lammps.sandia.gov
7. M.S. Daw, M.I. Baskes, Phy. Rev. B (29), 6443 (1984)
8. M.S. Daw, M.I. Baskes, Phy. Rev. B (50), 1285 (1983)
9. A.F. Voter, F. Montalenti, T.C. Germann, Ann. Rev. Mater. Res. **32**(1), 321 (2002). https://doi.org/10.1146/annurev.matsci.32.112601.141541
10. A.-P. Prskalo, *Molecular dynamics simulations of Si, SiC and SiN layered systems* (Südwest deutscher Verlag F, [S.l.], 2014). http://doi.org/10.18419/opus-2113
11. J. Tersoff, Phy. Rev. B **37**(12), 6991 (1988). https://doi.org/10.1103/PhysRevB.37.6991
12. S. Chakraborty, J. Zhang, S. Ghosh, Comput. Mater. Sci. **121**, 23 (2016). https://doi.org/10.1016/j.commatsci.2016.04.026
13. K. Higashi, T. Mukai, K. Kaizu, S. Tsuchida, S. Tanimura, Le Journal de Physique IV **01**(C3), C3 (1991). https://doi.org/10.1051/jp4:1991349
14. X. Chen, Y. Peng, S. Peng, S. Yao, C. Chen, P. Xu, PloS one **12**(7), e0181983 (2017). https://doi.org/10.1371/journal.pone.0181983
15. W.S. Lee, W.C. Sue, C.F. Lin, C.J. Wu, J. Mater. Process. Technol. **100**(1–3), 116 (2000). https://doi.org/10.1016/S0924-0136(99)00465-3
16. C.Y. Yu, P.L. Sun, P.W. Kao, C.P. Chang, Scripta Materialia **52**(5), 359 (2005). https://doi.org/10.1016/j.scriptamat.2004.10.035
17. G.I. Kanel, S.V. Razorenov, K. Baumung, J. Singer, J. Appl. Phy. **90**(1), 136 (2001). https://doi.org/10.1063/1.1374478
18. K. Baumung, H.J. Bluhm, B. Goel, P. Hoppé, H.U. Karow, D. Rusch, V.E. Fortov, G.I. Kanel, S.V. Razorenov, A.V. Utkin, O. Vorobjev, Laser Part. Beams **14**(02), 181 (1996). https://doi.org/10.1017/S0263034600009939
19. G.I. Kanel, S.V. Razorenov, A. Bogatch, A.V. Utkin, V.E. Fortov, D.E. Grady, J. Appl. Phy. **79**(11), 8310 (1996). https://doi.org/10.1063/1.362542
20. S.J.A. Koh, H.P. Lee, Nanotechnology **17**(14), 3451 (2006). https://doi.org/10.1088/0957-4484/17/14/018
21. J. Stadler, R. Mikulla, H.R. Trebin, Int. J. Mod. Phy. C **08**(05), 1131 (1997). https://doi.org/10.1142/S0129183197000990
22. Y.Q. Cheng, E. Ma, H.W. Sheng, Phy. Rev. Lett. **102**(24), 245501 (2009). https://doi.org/10.1103/PhysRevLett.102.245501
23. X.Y. Liu, P.P. Ohotnicky, J.B. Adams, C. Lane Rohrer, R.W. Hyland, Surf. Sci. (373), 357 (1997)
24. W.E. Nagel, D.H. Kröner, M.M. Resch (eds.), *High Performance Computing in Science and Engineering '14* (Springer International Publishing, Cham, 2015). https://doi.org/10.1007/978-3-319-10810-0

First-Principles Calculations of Phase Stability, Electronic Structure, and Defect Properties of Perovskites for SOFC/SOEC Electrodes

Daniel Mutter, Daniel F. Urban, and Christian Elsässer

Abstract Solid oxide fuel cells (SOFC) and solid oxide electrolyzer cells (SOEC), which transform chemical into electrical energy and vice versa, have the potential to make a significant contribution to the efforts of overcoming present problems of the energy economy in the near future. An optimal functionality of these devices requires a high catalytic activity at the electrodes, which strongly depends on point defect concentrations and on the capability of the material to allow for fast charge transfer reactions. Promising anode materials regarding these requirements are perovskite compounds (ABO_3), where the transition-metal ion on the B site can adopt different oxidation states by accepting and releasing electrons during the oxygen reactions at the SOEC/SOFC surfaces. For $LaFeO_3$, a typical representative of this material class, we present results regarding the phase stability and point defect formation energies derived by density functional theory GGA+U calculations. The influence of point defects on the electronic charge-carrier concentrations as a function of the oxygen partial pressure is studied and compared for the perovskite materials $LaFeO_3$, $LaMnO_3$ and $CaMnO_3$. In addition to the scientific results, the performance of the DFT calculations applied for these studies on the ForHLR I computer cluster is reported.

1 Introduction

The growing energy demand worldwide, along with a drastic consumption of fossil fuels, asks for innovative solutions in order to ensure a sustainable energy supply. Green technologies are on the rise, but the installation and maintenance of large facilities, which convert wind and solar energies into electricity still faces economical and technological challenges. It is therefore of major importance that the electricity

D. Mutter (✉) · D. F. Urban · C. Elsässer
Fraunhofer IWM, Wöhlerstraße 11, 79108 Freiburg, Germany
e-mail: daniel.mutter@iwm.fraunhofer.de

C. Elsässer
Albert-Ludwigs-Universität Freiburg, Freiburger Materialforschungszentrum (FMF),
Stefan-Meier-Straße 21, 79104 Freiburg, Germany

© Springer Nature Switzerland AG 2021
W. E. Nagel et al. (eds.), *High Performance Computing in Science and Engineering '19*,
https://doi.org/10.1007/978-3-030-66792-4_10

produced in these ways is used most effectively, even in times when the energy demand is low. An innovative approach to utilize the electric excess energy from green sources is the production of synthesis gas (H_2, CO) from abundant resources (H_2O and CO_2) by so-called high-temperature co-electrolysis in solid oxide electrolyser cells (SOECs) [3]. These gases can then be used either by the chemical industry, or to store energy which can again be set free by the reverse reaction in a solid oxide fuel cell (SOFC). SOFC and SOEC devices (SOCs) consist of stacks of cells with two electrodes (anode and cathode), at which the oxygen evolution and reduction reactions take place, separated by a solid electrolyte, through which the oxygen ions can move.

An optimal functionality of the electrodes requires a high catalytic activity at their surfaces, meaning e.g. the ability for chemisorption and dissociation of O_2 molecules, the charge transfer to O^{2-}, and the incorporation of O^{2-} ions into the lattice at vacant sites [16, 18]. Furthermore, suitable materials need to exhibit high electronic and ionic (O^{2-}) conductivities, as well as structural, chemical, and mechanical stability within a wide range of temperatures, both in contact with air on the one side and a solid electrolyte on the other side. Promising materials regarding these requirements are perovskites (ABO_3), with La, Ba or Sr ions on the A sites, and transition-metal ions (Mn, Fe, Co) on the B sites [6]. The B site ions can easily change their valence state (generally between +3 and +4), thereby accommodating the charge transfer taking place in the exchange of oxygen ions with the environment. The A site ions have fixed valence states +2 or +3, and provide, together with the oxygen anions, a stable structural framework. Due to the condition of overall charge neutrality, a deficiency at the A sites can lead to a reduced O content ($O_{3-\delta}$), which is necessary to a certain extent since vacant O sites enable the efficient incorporation of O into the structure as well as a high O diffusivity. The interplay between defect physics and electronic structure is of major importance for the functionality of the perovskite electrodes in SOC devices. In this study, this relation is explored in detail for the perovskites $(La_{1-x}Ca_x)(Fe_{1-y}Mn_y)O_{3-\delta}$ (with $0 \le x \le 1, 0 \le y \le 1$ and $\delta \ge 0$) by means of first-principles calculations based on the density functional theory (DFT).

2   Computational Methods

2.1   DFT Calculations

The DFT code *VASP* [7] was used to calculate the atomic configurations and the electronic structures of the materials of choice. The valence electrons are represented in a basis of plane waves, which is a computationally efficient choice when dealing with Bloch electrons in a periodic crystal modeled by infinitely repeated supercells. A plane wave cutoff energy of 520 eV was applied in this study. Core electrons were described by PAW pseudopotentials [8]. The generalized gradient approximation (GGA) of Perdew *et al.* was chosen for the exchange-correlation (xc) functional

[11]. In order to correct for the unphysical self-interaction of localized (strongly correlated) d-electrons in transition metal oxides, which is immanent in DFT calculations using approximate xc-functionals, the GGA+U method in the rotationally invariant formulation [2] was used, which describes the contribution of the metal d-electrons to the total energy functional by the Hubbard model. Following Wang *et al.* [17], U values for Fe (3.6 eV) and Mn (4.2 eV) were obtained within this study by finding the smallest deviation between experimental and computationally derived reaction energies between different Fe or Mn oxides. The magnetism of the transition metals Mn and Fe was treated by spin-polarized calculations. A $3 \times 4 \times 3$ sized k-point mesh centered at the Γ-point was applied for the Brillouin zone integrations of the orthorhombic perovskite structure in reciprocal space.

2.2 The Perovskite Structure

The perovskite compounds were set up in the orthorhombic crystal structure with space group *Pnma* (#62) and structurally relaxed with a BFGS algorithm [15] as implemented in *VASP*, until each force component had a value below 0.01 eV/Å. Volume relaxations were performed by total-energy minimization. Within this study, the orthorhombic structure was found to have a lower ground-state energy than the cubic (space group $Pm\bar{3}m$) and the trigonal ($R\bar{3}c$) perovskite structures. The orthorhombic supercells consisted of 80 atoms as shown in Fig. 1. Systems with mixed occupations on the A ($x = 0.5$) and/or B sublattices ($y = 0.5$) were represented by ordered configurations avoiding nearest neighbors of the same atomic types.

2.3 Point Defect Formation Energies

Knowledge about the occurrences, concentrations and charge states of point defects is essential for characterizing a perovskite material with respect to its catalytic activity in a SOC device. On the one hand, oxygen vacancies need to be present in the material since they facilitate the incorporation, migration and release of oxygen in contact with the solid electrolyte and the atmosphere. On the other hand, the concentrations of free charge carriers leading to the electronic conductivity of the electrode material depend on the position of the Fermi level, which is determined by the condition of charge neutrality in a system with differently charged point defects in various concentrations. At a temperature T, the defect concentration $x_{i,q}$ of a point defect i in the charge state q is related to the defect formation energy $\Delta E_{i,q}$ via:

$$x_{i,q} \propto \exp\left(-\frac{\Delta E_{i,q}}{k_B T}\right), \tag{1}$$

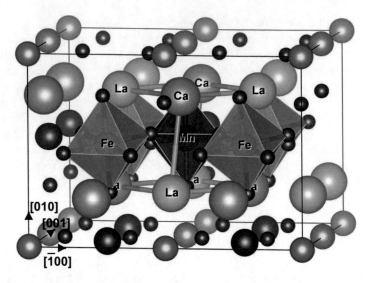

Fig. 1 Orthorhombic supercell consisting of 16 formula units of $La_{0.5}Ca_{0.5}Fe_{0.5}Mn_{0.5}O_3$ (80 atoms). The typical perovskite motive of an octahedrally oxygen-coordinated transition metal (B position, here occupied by Fe and Mn) in the center of a cube of cations (A position, La and Ca) is indicated, which is slightly distorted in the orthorhombic perovskite configuration compared to the cubic structure

with the Boltzmann constant k_B. In the supercell approach and under the assumption of dilute, i.e. non self-interacting defects, $\Delta E_{i,q}$ can be calculated using the well-known formula [19]

$$\Delta E_{i,q} = E_{i,q} - E_{bulk} + \sum_j n_j \underbrace{\left(\mu_j^{(0)} + \Delta\mu_j \right)}_{\mu_i} + q \underbrace{(\varepsilon_{VBM} + E_F)}_{\mu_{el}} + E_{i,q}^{corr}. \quad (2)$$

E_{bulk} stands for the energy of the defect-free bulk system, and $E_{i,q}$ denotes the energy of a system containing the point defect i in the charge state q. This defect is formed in the synthesis process of the material by exchanging n_j atoms of type j with an external reservoir of energy μ_j, or $\Delta\mu_j$ if referenced to the elemental ground state energy $\mu_j^{(0)}$. The sum runs over all atomic species j involved in the formation of the defect. Additionally, in the case of charged defects, electrons are exchanged with the electron reservoir of energy $\mu_{el} = \varepsilon_{VBM} + E_F$, which can be described by the energy of the valence band maximum (ε_{VBM}) and the position of the Fermi level relative to the VBM. An energy term $E_{i,q}^{corr}$ for correcting the artificial Coulomb interaction between a charged defect and its own periodic images, as well as for aligning the different energy zeros for charged and neutral cells is taken into account [4]. The charge corrections were performed applying an automatized scheme developed by Broberg and coworkers [1].

3 Results and Discussion

3.1 Phase Stability and Point Defect Calculations for LaFeO₃

LaFeO$_3$ (LFO) was taken as the prototype material in this study to test the calculation procedure and to benchmark the computational performance. The considered point defects include the vacancies on the La site ($\square_{La}$) in the charge states $q = 0, -1, -2, -3$, on the iron site ($\square_{Fe}$; $q = 0, -1, -2, -3$) and on the oxygen site ($\square_O$; $q = 0, +1, +2$), as well as atomic antisite defects Fe on La (Fe$_{La}$; $q = -1, 0, +1$) and La on Fe (La$_{Fe}$; $q = -1, 0, +1$). The choice of charge states results from the nature of the defect being acceptor-like ($\square_{La}$, $\square_{Fe}$) or donor-like ($\square_O$). Since the character of the antisites can not be defined with certainty in advance, a variety of positive and negative charge states was taken into account. In order to figure out which defects are dominant in LFO, the ranges of the chemical potentials $\Delta\mu_j$ introduced in Eq. (2) need to be determined. During the formation process, LFO is in equilibrium with the elemental reservoirs, therefore, the formation energy of LFO with respect to the elemental ground state phases can be expressed as:

$$\Delta E^{\text{form}} (\text{LFO}) = E^{\text{total}} (\text{LFO}) - (\mu_{La}^{(0)} + \mu_{Fe}^{(0)} + 3\mu_O^{(0)}) = \Delta\mu_{La} + \Delta\mu_{Fe} + 3\Delta\mu_O^{(0)}. \tag{3}$$

With the ground state references, hcp La and bcc Fe crystals and the O$_2$ molecule, and taking into consideration the correction energies for (a) the overbinding of O$_2$ in DFT [17] and (b) the mixture of total energies derived by DFT (elemental phases) and DFT+U (LFO) [5], ΔE^{form} (LFO) was calculated to 14.25 eV/f.u. for $U_{Fe} = 3.6$ eV. In addition to Eq. (3), the range of the chemical potentials is further confined by the formation energies of competing secondary phases such as Fe$_2$O$_3$ or La$_2$O$_3$, which are calculated analogously to Eq. (3), leading to the restrictions $2\Delta\mu_{Fe} + 3\Delta\mu_O < \Delta E^{\text{form}}$ (Fe$_2$O$_3$) and $2\Delta\mu_{La} + 3\Delta\mu_O < \Delta E^{\text{form}}$ (La$_2$O$_3$). To avoid also the formation of elemental phases, $\Delta\mu_j < 0$ must be valid for all atomic types j. If LFO is synthesized under an oxygen atmosphere, the corresponding chemical potential $\Delta\mu_O$ is directly related to the temperature and pressure of the atmospheric oxygen-gas phase [13]. The chemical potentials of the metal components can hardly be determined experimentally but are interpreted in the following way: the "poorer" ["richer"] the environment conditions of a component j are during the synthesis, the closer to zero [more negative] is the corresponding value of $\Delta\mu_j$. Figure 2 shows the phase diagram of LFO derived in this project for a typical synthesis temperature of the O$_2$ gas of 1400 K [9]. The formation energies for all the considered point defects and charge states described above were calculated within this phase diagram. It turned out, that for highly oxidizing, i.e. high pressure conditions ($p(O_2)$ above 10^5 atm) and also for highly reducing conditions (low oxygen pressure, $p(O_2)$ below 10^{-5} atm), some defect formation energies can adopt negative values. Based on Eq. (1), this would lead to very high defect concentrations in the range of 10^{23}/cm, for which no stable LFO crystal can be assumed. Also, such concentrations would be clearly above what can be regarded as the dilute limit which is the premise for

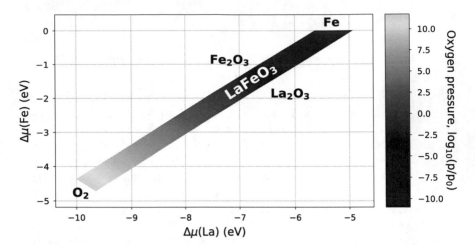

Fig. 2 Phase diagram of LFO with respect to the chemical potentials of elemental La and Fe, and the pressure of the oxygen atmosphere (color code) at a temperature of 1400 K ($p_0 = 1$ atm). The competing phases which form instead of LFO when crossing the respective borderlines are denoted

the validity of Eq. (2). Due to these difficulties in the interpretation, and since the corresponding oxygen pressures are hardly accessible experimentally, these extreme pressure regions are not further discussed.

In region of stability of the phase diagram, the formation energies for each defect in each considered charge state can now be obtained for the range of Fermi levels across the band gap. As a realistic example, the formation energies at a point being located in the center between the Fe_2O_3 and La_2O_3 lines and at a reasonable pressure ($\Delta\mu_{La} = -7.26$ eV, $\Delta\mu_{Fe} = -1.96$ eV, $p(O_2) = 0.2$ atm) [at $T(O_2) = 1400$ K] are shown in Fig. 3. A band-gap value of 2 eV was calculated for the choice of U_{Fe} in this work, which can be extracted from the density of states (see Fig. 4). For each considered defect, only the formation energy of the energetically lowest charge state at each Fermi level is shown in Fig. 3. The energies at the kinks where lines with different slopes cross are the transition levels between two charge states. Only in the region highlighted by the ellipse, there are only positive defect formation energies, so the Fermi level for a stable LFO crystal must lie in this range where the dominant defects can be identified as $\square_O^{q=+2}$, $\square_{La}^{q=-3}$, and $\square_{Fe}^{q=-3}$.

3.2 Electronic Structure of LaFeO₃, LaMnO₃ and CaMnO₃

The total electronic densities of states for LFO, $LaMnO_3$ (LMO) and $CaMnO_3$ (CMO) were calculated in the favorable antiferromagnetic spin configuration (Fig. 4). Each of these compounds exhibits a band gap, which for LFO (2 eV) and LMO (1.5 eV) is close to the experimental values (2.3 eV [14] and 1.8 eV [12], respectively),

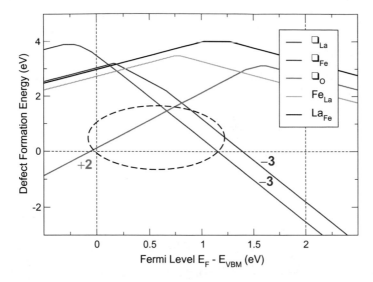

Fig. 3 Formation energies of charged defects as a function of the Fermi level at the point $(\Delta\mu_{La} = -7.26\ \text{eV},\ \Delta\mu_{Fe} = -1.96\ \text{eV},\ p(O_2) = 0.2\ \text{atm})$ [at $T(O_2) = 1400\ \text{K}$] in the phase diagram of LFO

while for CMO, a larger deviation between the computationally derived (1 eV) and the experimental band gap (3 eV) is found, as was also reported in the literature [10]. Due to these bandgap values, the considered perovskite compounds would be electronically isolating in a defect-free configuration, and therefore inapplicable as electrodes in SOC devices. However, the point defects act as intrinsic dopants which shift the Fermi level, leading to the existence of free charge carriers at elevated temperatures. The position of the Fermi level is determined by the condition of charge neutrality:

$$n + \sum_i \sum_{q<0} |q|\, x_{i,q} = p + \sum_i \sum_{q>0} q\, x_{i,q}, \tag{4}$$

with the concentrations of electrons in the conduction band, n, and electron holes in the valence band, p.

Using the formation energies for the defects in LFO and analogously obtained values for LMO and CMO, the Fermi-level dependent defect concentrations $x_{i,q}$ can be approximately calculated by Eq. (1). The change in vibrational entropy in the formation process of defects, which is part of the proportionality constant, was not yet taken into account at this stage of the project. In an automatized computation workflow, the Fermi level was changed stepwise and iteratively until zero net charge was reached. The charge carrier concentrations as functions of the oxygen partial pressure resulting at a processing temperature of 1400 K are shown in Fig. 5. The broad areas cover the whole range of the chemical potentials $\Delta\mu_{La}$, $\Delta\mu_{Ca}$, $\Delta\mu_{Fe}$ and

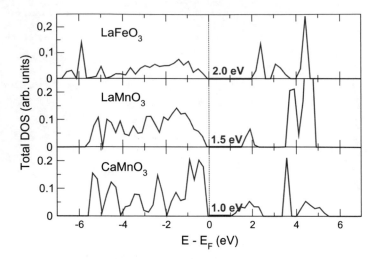

Fig. 4 Electronic densities of states of LaFeO₃, LaMnO₃ and CaMnO₃

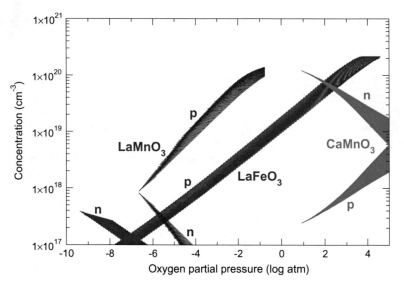

Fig. 5 Comparison of charge carrier concentrations in LFO, LMO and CMO as a function of the oxygen gas pressure at a temperature of 1400 K

$\Delta\mu_{Mn}$ within the respective phase diagrams. While the slopes between the different compounds are similar, strongly differing pressure regimes become visible for similar carrier concentrations. This is due to different dominant defects with strongly varying formation energies and transition levels in the three compounds.

3.3   Oxygen Vacancy Formation Energies of $La_{1-x}Ca_xFe_{1-y}Mn_yO_3$

Since the oxygen vacancy concentration is of major importance for the catalytic activity of the perovskite materials, the corresponding formation energy was calculated also for the compounds with mixed occupancies on the A- and B-sublattices, namely $x = 0.5$ and/or $y = 0.5$ in $La_{1-x}Ca_xFe_{1-y}Mn_yO_3$ in an ordered configuration as shown in Fig. 1. Only the neutral charge states were considered in this first stage, which already provides insight into the changes of vacancy concentration and hence catalytic behavior with composition. The results are shown in Fig. 6, again as before at an oxygen gas temperature of 1400 K and at pressure conditions of 0.1 atm, for which LFO has been successfully synthesized [9]. Since the metallic components are not involved in the formation of oxygen defects, no dependence on the respective chemical potentials in the phase diagrams needs to be taken into account. While compositional changes on the transition metal sublattice do not exhibit systematic trends, the oxygen vacancy formation energies decrease considerably with increasing Ca/La ratio.

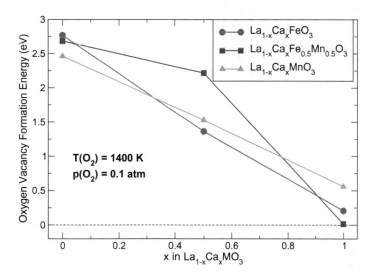

Fig. 6 Oxygen vacancy formation energy of $La_{1-x}Ca_xFe_{1-y}Mn_yO_3$ with $x = 0, 0.5, 1$ and $y = 0, 0.5, 1$

4 Performance of the Calculations on the ForHLR 1 Cluster

The accurate DFT calculations performed in this project require a large amount of computation time, mainly stemming from the self-consistent-field procedure to determine the electronic structure in DFT, and from the relaxation of the ions to their lowest energy (zero force) configuration. For example, finding a reasonably converged energy minimum of a La vacancy in a pre-converged perovskite ($LaFeO_3$) supercell of 40 atoms (8 formula units) needs about 52 CPU hours for a total of 110 electronic and ionic steps (performed on 160 cores in parallel). Increasing the size of the supercell by a factor of 4 (160 atoms) leads to a CPU time of about 20 times as long for the relaxation (1,000 CPU hours). As a compromise between saving computation time (smaller supercells) and increasing the accuracy by decreasing the artificial defect-defect interactions (larger supercells), the intermediate supercell size with 80 atoms was used for most of the calculations.

We apply the *VASP* code in this project, which is known to perform well on parallel computing clusters, since it is highly optimized and exhibits a good scaling behavior. *VASP* parallelization is based on MPI, and the calculations can be set up to utilize parallelization over k-points, energy bands, images in nudged elastic band calculations, and combinations of all those. In Fig. 7, the scaling behavior of *VASP* is illustrated for the relaxation of a La vacancy in a 40-atom LFO perovskite supercell. Using a number of cores between 80 and 160 appears to be an optimal choice for an efficient performance. The perfect scaling with the number of k-points, i.e. the density of the mesh of the reciprocal lattice needed for the Brillouin zone integrations,

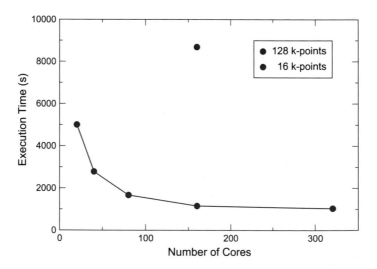

Fig. 7 Computation time of *VASP* with respect to the number of cores and the number of k-points

is also obvious: a benchmark calculation of a system with 8 times as many k-points took about 8 times longer.

Considering for each perovskite composition each reasonable point defect in all of the possible charge states requires a large amount of CPU time which can only be mastered by highly optimized and reliably running HPC computer clusters such as the ForHLR I of the Steinbuch Centre of Computing at the KIT.

5 Conclusions

The existence of intrinsic lattice defects enables the catalytic performance of undoped perovskite materials of the class $(La_{1-x}Ca_x)(Fe_{1-y}Mn_y)O_{3-\delta}$, making them highly promising candidates for electrode materials in solid oxide fuel and electrolyzer cells. Using the high performance computer cluster ForHLR I of the Steinbuch Centre of Computing at the KIT, we studied the complex interplay between point defects and electronic structures by applying accurate first-principles DFT calculations. We developed an automatized workflow, starting from confining the stability region in the phase diagram of the material of choice, followed by the determination of point defect formation energies and finally obtaining defect concentrations, the Fermi level position and charge carrier concentrations iteratively and self-consistently. This procedure, applied in this work exemplarily to $LaFeO_3$, $LaMnO_3$ and $CaMnO_3$, makes it possible to link performance determining quantities derived theoretically by atomistic simulations to experimentally accessible environment conditions during the material synthesis, which opens up and guides into novel directions for further materials design and optimization.

Acknowledgements This work was funded by the *Federal Ministry of Education and Research*, Germany, via the *Kopernikus* project *Power2X*. The DFT calculations were performed on the computational resource ForHLR I of the Steinbuch Centre of Computing at the KIT, funded by the *Ministry of Science, Research, and Arts* Baden-Württemberg and by the DFG ("Deutsche Forschungsgemeinschaft").

References

1. D. Broberg, B. Medasani, N.E. Zimmermann, G. Yu, A. Canning, M. Haranczyk, M. Asta, G. Hautier, PyCDT: A Python toolkit for modeling point defects in semiconductors and insulators. Comput. Phys. Commun. **226**, 165–179 (2018)
2. S.L. Dudarev, G.A. Botton, S.Y. Savrasov, C.J. Humphreys, A.P. Sutton, Electron-energy-loss spectra and the structural stability of nickel oxide: An LSDA+U study. Phys. Rev. B **57**, 1505–1509 (1998)
3. S. Foit, R.A. Eichel, I.C. Vinke, L.G.J. de Haart, Power-to-syngas - an enabling technology for the transition of the energy system? Angew. Chem. Int. Ed. **56**, 5402 (2017)
4. C. Freysoldt, J. Neugebauer, C.G. Van de Walle, Fully ab initio finite-size corrections for charged-defect supercell calculations. Phys. Rev. Lett. **102**, 016402 (2009)

5. A. Jain, G. Hautier, S.P. Ong, C.J. Moore, C.C. Fischer, K.A. Persson, G. Ceder, Formation enthalpies by mixing GGA and GGA+U calculations. Phys. Rev. B **84**, 045115 (2011)
6. E.A. Kotomin, R. Merkle, Y.A. Mastrikov, M.M. Kuklja, J. Maier, Chapter 6: Energy Conversion: Solid Oxide Fuel Cells, in *Computational Approaches to Energy Materials* pp. 149–186 (John Wiley & Sons, Ltd., 2013)
7. G. Kresse, J. Furthmüller, Efficient iterative schemes for ab initio total-energy calculations using a plane-wave basis set. Phys. Rev. B **54**, 11169 (1996)
8. G. Kresse, D. Joubert, From ultrasoft pseudopotentials to the projector augmented-wave method. Phys. Rev. B **59**, 1758–1775 (1999)
9. Kungl, H.: personal communication
10. M. Molinari, D.A. Tompsett, S.C. Parker, F. Azough, R. Freer, Structural, electronic and thermoelectric behaviour of $CaMnO_3$ and $CaMnO_{3-\delta}$. J. Mater. Chem. A **2**, 14109–14117 (2014)
11. J.P. Perdew, K. Burke, M. Ernzerhof, General gradient approximation made simple. Phys. Rev. Lett. **78**, 1396 (1997)
12. T. Saitoh, A.E. Bocquet, T. Mizokawa, H. Namatame, A. Fujimori, M. Abbate, Y. Takeda, M. Takano, Electronic structure of $La_{1-x}Sr_xMnO_3$ studied by photoemission and x-ray-absorption spectroscopy. Phys. Rev. B **51**, 13942–13951 (1995)
13. A. Samanta, E. Weinan, S.B. Zhang, Method for defect stability diagram from ab initio calculations: A case study of $SrTiO_3$. Phys. Rev. B **86**, 195107 (2012)
14. M.D. Scafetta, A.M. Cordi, J.M. Rondinelli, S.J. May, Band structure and optical transitions in $LaFeO_3$: theory and experiment. J. Physics: Condens Matter **26**, 505502 (2014)
15. D.F. Shanno, Conditioning of quasi-newton methods for function minimization. Math. Comp. **24**, 647–656 (1970)
16. J. Suntivich, K.J. May, H.A. Gasteiger, J.B. Goodenough, Y. Shao-Horn, A perovskite oxide optimized for oxygen evolution catalysis from molecular orbital principles. Science **334**, 1383 (2011)
17. L. Wang, T. Maxisch, G. Ceder, Oxidation energies of transition metal oxides within the GGA+U framework. Phys. Rev. B **73**, 195107 (2006)
18. L. Wang, R. Merkle, Y.A. Mastrikov, E.A. Kotomin, J. Maier, Oxygen exchange kinetics on solid oxide fuel cell cathode materials - general trends and their mechanistic interpretation. J. Mater. Res. **27**, 2000–2008 (2012)
19. S.B. Zhang, S.H. Wei, A. Zunger, H. Katayama-Yoshida, Defect physics of the $CuInSe_2$ chalcopyrite semiconductor. Phys. Rev. B **57**, 9642–9656 (1998)

Dynamics and Solvent Effects in Functionalized DNA Sensing Nanogaps

Frank C. Maier, Maofeng Dou, and Maria Fyta

Abstract The potential of reading-out DNA molecules, using functionalized electrodes embedded in nanopores, is discussed here. Focus is given on a functionalization using tiny diamond-like hydrogenated cages, the diamondoids. A derivative (known as memantine) of the smallest diamondoid is taken. This offers hydrogen bonding possibilities to DNA molecules. Accordingly, we place a small DNA unit, a nucleotide, within the functionalized nanogap. Based on quantum-mechanical calculations, we discuss the alteration in the electronic features of the electrodes by varying the relative arrangements of the nucleotides between these electrodes. Quantum transport calculations were also performed and have shown distinct features among the nucleotides revealing the high sensitivity of the electrodes in distinguishing the nucleotide identity. At a next step, the effect of an aqueous environment was taken into consideration through quantum-mechanical/molecular-mechanics simulations. This effect is discussed in view of the water structuring close to the electrodes and their influence on the device characteristics. These results would be highly relevant in detecting DNA using novel nanopore sequencers.

1 Introduction

Sensing biomolecules, such as DNA, is essential for detecting their sequences and is highly relevant to medical care and therapy design. One of the novel sequencers, the so called "next generation sequencing platforms" involves nanopores. The latter are nanometer-sized holes in materials and have the ability to detect DNA [2, 9, 37]. A nanopore can electrophoretically thread a biomolecule within a salt solution realizing single-molecule experiments [5]. In order to not only detect the passage of a DNA through the pore, but also read-out its exact sequence, different protocols have been proposed. One of these is based on the use of transverse tunneling currents [6, 40]. This read-out technique requires that metallic electrodes are embedded into

F. C. Maier · M. Dou · M. Fyta (✉)
Institute for Computational Physics, Universität Stuttgart, Allmandring 3, 70569 Stuttgart, Germany
e-mail: mfyta@icp.uni-stuttgart.de

© Springer Nature Switzerland AG 2021
W. E. Nagel et al. (eds.), *High Performance Computing in Science and Engineering '19*,
https://doi.org/10.1007/978-3-030-66792-4_11

155

the nanopore. Applying a voltage difference across these electrodes would generate an electronic current expected to identify the DNA nucleotides passing through the nanopore. Although this is a very promising technique, often the tunneling signals for the four nucleotides overlap, leading to a low signal-to-noise ratio [39]. Enhancing the nucleotide-specificity of these signals could be possible through a functionalization of the nanopores [2]. In functionalized nanopores, the metallic electrodes embedded in the nanopore are functionalized by molecules of the size of the nucleotides, interacting with the nucleotides through hydrogen bonds. This interaction is expected to give rise to highly nucleotide specific electronic signals. These signals are expected to also detect DNA mutations, that is distinguish "healthy" nucleotides from modified ones. Detection technologies have been improving along these lines as well [1, 20]. Being able to read-out mutations or modifications in DNA molecules can be a keystone in revolutionizing cancer therapies [21].

Along these lines, we have been very involved in identifying possible molecules able to efficiently functionalize a nanopore gap and increase the signal-to-noise ratio in the electronic current measurements. We have shown that the use of a small diamond-like molecule, known as diamondoid, can indeed enhance the read-out signals [30]. Diamondoids are nanoscale diamond-like carbon nanocages, which are terminated by hydrogen atoms, are thermodynamically very stable, have tunable properties, can be selectively tuned, and come in a variety of sizes and modifications [4]. For the functionalization of the electrodes, derivatives of diamondoids [38] can offer grafting possibilities to the metallic surface of the electrode, and additional donor/acceptor sites for binding to the DNA. Indeed, it has been shown that derivatives of diamondoids form hydrogen bonds to DNA nucleotides [3] and can enhance the optical properties of the single nucleotides [27]. We have also shown that it is possible to identify mutations through electronic transport measurements [31] and optical measurements [29]. We will further investigate the relevance for biosensing of one of the smallest diamondoid derivatives, known as memantine [22]. Memantine is bio-compatible and well known as a drug used in Alzheimer's disease [19, 25]. This molecule will be taken here as a potential candidate for detecting DNA in a functionalizing gold electrode gap. Our aim is to assess the influence of other important factors in the DNA detection process. These factors mainly involve the dynamics of the nucleotide in the pore and a surrounding fluidic environment.

2 Methodology

The main objective of the current study is to reveal the biosensitivity of a functionalized device for distinguishing DNA nucleotides by including dynamics and water effects. Accordingly, we consider a device made up of gold (111) electrodes. One of the electrodes is functionalized with a memantine, to which a thiol group was added, in order to provide an anchoring to the electrode. In the following, we will use the notation 'memS' for denoting the functionalizing molecule. Within the memS-functionalized gap, we place a DNA nucleotide. The initial placement assures

Fig. 1 The functionalized device used in this work. It includes the gold electrodes (yellow spheres on the right and left), the diamondoid molecule functionalizing the left electrode and an 2'-deoxyadenosine 5'-monophosphate (A) placed within the gap opened by the functionalized electrodes. The green lines mark the traces of a rotation of the nucleobase within the gap (see text). The hydrogen, carbon, nitrogen, oxygen, phosphorus, and thiol atoms are colored in white, gray, blue, red, pink, and purple, respectively. The sulfur atom is used to graft the diamondoid on the electrode. In this setup the electronic current can flow from the left to the right electrode though electron tunneling

that the amine group of the memantine forms hydrogen bonds to the nucleotide. For this, the four healthy DNA nucleotides (i.e. nucleobases with the sugar phosphate backbone), namely 2'-deoxyadenosine 5'-monophosphate, 2'-deoxythymidine 5'-monophosphate, 2'-deoxycytosine 5'-monophosphate, 2'-deoxyguanosine 5'-monophosphate and the 2'-Deoxy 5'-methylcytidine mutation are considered here. For simplicity, these are represented as 'A', 'T', 'C', 'G', and 'd5mC', respectively. The memS-functionalized electrodes with a nucleotide placed within these is depicted in Fig. 1.

In order to compute the electronic properties of this device, we have used the density functional theory (DFT), as implemented in SIESTA [32]. The gold electrodes are considered and represented by a single-zeta plus polarization basis set, the remaining atoms are represented by double-zeta plus polarization basis set. The exchange-correlation functional is approximated by the Perdew-Burke-Ernzerhof (PBE-GGA) functional [23]. The fineness of the real-space grid, the mesh-cutoff, is set to 200 Ry. During the structural relaxation, the integration over k-space is performed with $4 \times 4 \times 1$ k-points. For the calculation of the electronic properties and transport properties we used a k-space of $5 \times 5 \times 1$ k-points. The structural relaxation was performed until the maximum atomic forces reached less than 0.01 eV/Å.

The supercell of $14.8 \times 14.8 \times 39.7$ Å consists of two gold electrodes (111), each represented by 5×5 gold atoms, in 5 layers each. The diamondoid memantine and one nucleotide is placed in the 18Å wide gap between both electrodes in such a way, that the phosphate group of the nucleotide is oriented close to the right electrode and the diamondoid is attached to the left electrode by a sulfur atom. For the investigation of DNA dynamics effects, the nucleotide is rotated at increments of 20° with respect to the diamondoid and along the axis connecting both electrodes (see green traces in Fig. 1). At each rotation, both the diamondoid and the nucleotide are relaxed. The transport calculations were performed using the nonequilibrium Green's function method (NEGF), implemented in TRANSIESTA [33]. For post-processing and the calculation of the transmission spectra, we have used TBTrans with a finer k-point sampling of $10 \times 10 \times 1$ k-points. The electronic transmission was calculated at zero bias. All calculations are performed at the Γ-point.

The influence of a water environment is taken into account through Quantum Mechanics/Molecular Mechanics (QM/MM) simulations as implemented in CP2K [14, 34, 35]. Within this scheme, the Au electrodes are calculated at the MM level. The respective interactions are modeled using the embedded atom model (EAM) potential [8] obtained from the LAMMPS database [24]. The DNA nucleotides, water, and memS are calculated at the QM level using DFT. The interactions between the Au electrodes and water, DNA nucleotides, and memS are modelled using Lennard-Jones potentials with parameters from the literature [26]. The core electrons are described by norm-conserving Goedeker-Teter-Hutter (GTH) pseudopotentials [10, 13]. The valence electrons are represented using double-ζ valence plus polarization (DZVP) basis sets of a Molopt-type [36]. The exchange-correlation interactions are described by the Perdew-Burke-Ernzerhof (PBE) functional [23], while van der Waals interactions are accounted for through the Grimme D3 dispersion correlation [11]. The energy cut off for the auxiliary plane waves is set to 300 Ry. A vacuum layer with a length of 15 Å along the direction normal to the electrode surfaces is added to avoid spurious image charge interactions. 30 ps simulations with a time step of 0.5 fs are first performed in the canonical (NVT) ensemble with a velocity-rescaling thermostat and a time constant of 10 fs at 300 K. This equilibration step is followed by a 10 ps (NVT) simulation using the thermostat with a time constant of 100 fs. The latter results were used for the analysis. During the simulation, the S atom, one carbon atom in the memS, and one carbon atom on the backbone of the DNA nucleotide are kept fixed in order to constrain the memS and DNA nucleotide close to the Au electrodes. The water density was set to 0.03 Å^{-3}, close to the liquid water density. Additional details can be found elsewhere [7].

2.1 Utilization of Computational Resources

The SIESTA 4.0 version was used for the simulations. The SIESTA code is available online and free of charge for academic purposes [32]. It is written in Fortran 95 and memory is allocated dynamically. Our current compilation supports NetCDF (network Common Data Form) and is a parallel version using Intel MPI Libraries. In order to provide insight onto the required computational resources, we provide a short analysis on DNA systems. The nanogap, diamondoid, nucleotide system, in the case of A, includes 320 atoms. Each single relaxation calculation was performed on 40 threads (2 nodes, 20 processors per node each (ppn)) for a simulation time of approximately four days of CPU time. The first part of the transport calculations were performed using the corresponding TRANSIESTA version [33], running on one node with 20 threads for three hours. For the second part, we used the SIESTA utility TBTrans, which ran on three nodes with 20 ppn each and 60 threads in total for a walltime of approximately 30 hours. The MPI version of CP2K (5.1) was used for the QM/MM calculations. The nanopore, diamondoid, water, and nucleotide system, in the case of A, includes 557 atoms. From these, the diamondoid, water, and nucleotide (307 atoms) are described at the QM level and the Au electrodes (250 atoms) are described at the MM level. A typical calculation of a trajectory was performed on 160 processors (8 nodes and 20 processors per node) and required a CPU time of 8-10 seconds for each time-step.

3   Discussion

In order to discuss the results, we first begin with the influence of a nucleotide rotation within the nanogap and reveal the electronic characteristics of the device. These results refer to the *in vacuo* case. We next move on with the same setup and include the water solvent. For these calculations, a single relative configuration of the DNA molecule, the nucleotide, and the functionalizing diamondoid was taken.

3.1   Influence of DNA Rotation

We include some dynamics effects by simply rotating the nucleotide along the transport direction across the functionalized nanogap, while preserving the hydrogen bond associated with the diamondoid. Taking A as a representative nucleotide, this is rotated within the nanogap as indicated by the green traces in Fig. 1. We first analyze the electronic features related to this rotation and focus on the variation of the electronic energy levels. Mainly the levels around the Fermi level are considered. Specifically, we focus on the first and second highest occupied and lowest unoccupied molecular orbitals, HOMO and LUMO respectively. A voltage bias was also applied in order to evaluate the electronic current that tunnels across the electrodes. The corresponding results are shown in Fig. 2 for two healthy nucleotides, A and C. As a first observation, small variations of the electronic energy levels can be observed. For both nucleotides, the energy levels lie lower for very small and very large angles and for intermediate angles (140–200°). The HOMO levels shown are closer to each other, while the two LUMO levels are further away, but follow the exact same trends.

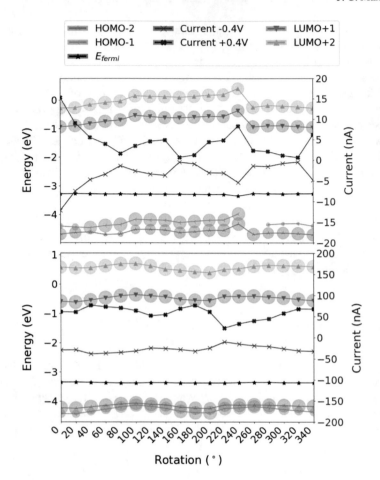

Fig. 2 Electronic energy levels for the case of A (top) and C (bottom) nucleotide placed within the gap. The graphs depict the electronic levels around the Fermi energy (E_{fermi}, black lines), starting from the HOMO-2 up to the LUMO+2 and their variation with the rotation angle. In order to highlight a possible correlation between these levels and the electronic transmission, we show the electronic current at an applied bias of $+0.4V$ and $-0.4V$ in dark and light red, respectively, as denoted by the legends

When a bias voltage is applied, the corresponding tunneling current is higher for a positive bias, than for a negative one. The respective currents do not follow very specific trends with respect to the rotation angle. The fact that the electronic current in the case of A (top panel in Fig. 2) shows more rich dynamics is probably related to the larger size of A compared to T. Accordingly, the larger molecule has more degrees of freedom, thus mirrors richer dynamics. Note, that the current in the case of A is one magnitude less than the current of the system including C. Overall, the electronic current generated by a voltage bias V_{bias} applied across the electrodes can

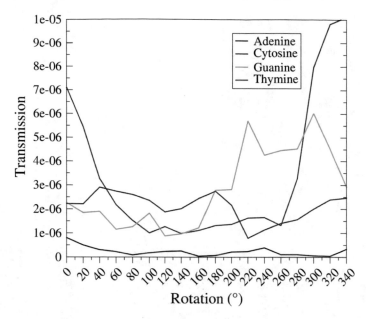

Fig. 3 The electronic transmission across the functionalized device for the case of the four canonical nucleotides. The dependence of the transmission at the fermi level with respect to the rotation angle of the nucleotides within the nanogap is depicted

be obtained from the integral over the electronic transmission curves in the range $\left[-\frac{V_{bias}}{2}, +\frac{V_{bias}}{2}\right]$.

The full electronic transmission curves for each rotation step and all nucleotides were also evaluated (not shown). An excerpt of these calculations, the transmission at the Fermi level (at no gating), is depicted in Fig. 3. These results show that the transmission is rotation dependent and in the case of A the transmission level at the Fermi level correlates with the current generated by a voltage bias, as shown in Fig. 2. On the other hand, C exhibits a significant change in the transmission at the Fermi level, at the highest and lowest rotation angles, while the current remains fairly steady. The significant differences of almost one order of magnitude, across different rotation angles, between the currents for A and C indicate that even at zero voltage bias nucleotide-specific features are retained in the nanogap. The variation in the transmission at the Fermi level point out to a high complexity and highlight the necessity for a more in depth analysis of the interplay between DNA and the functionalizing diamondoid. Preliminary results show that clear resonance peaks remain throughout the rotation and slightly shift relative to the energy range of the resonance peak for no rotation, i.e. zero degrees. This indicates again a possible tuning by means of a gating voltage.

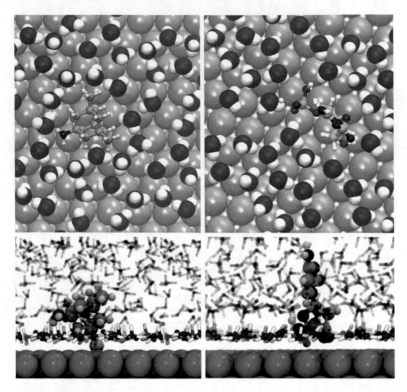

Fig. 4 A typical snapshot of the memS-functionalized electrodes, including C and water molecules. The latter are depicted in white (hydrogen) and red (oxygen). On the left and right are shown the left and right diamondoid-functionalized electrode, respectively. The left electrode is functionalized by the diamondoid, the right is closer to the nucleotide. The top and bottom panels depict a top and side view of the electrodes, respectively

3.2 Influence of a Water Solvent

At a next step, we include the influence of the solvent and related thermal fluctuations on the characteristics of the functionalized electrodes through the QM/MM calculations, as discussed above. A snapshot of the system is depicted in Fig. 4 for the case of C within the functionalized nanogap. Overall, the distance between memS-C increases to 2.1 Å when water is included. Note, that, unless otherwise stated, the hydrogen bond between the diamondoid and the nucleotide is defined as the distance between the N atom of the diamondoid and the closest H atom of the nucleotide. These correspond to the acceptor and donor of the hydrogen bond between the two molecules, respectively. An exception was found only for the case of A, for which the shortest hydrogen bond length was found between the H atom of the diamondoid and the closest N atom of the nucleotide throughout. The simulations revealed, that the water molecules tend to move towards the gold electrodes. Due to the reorientation of the water molecules arising from the polarization induced by the image charges, some hydrogen atoms are pointing towards the gold surfaces. The water molecules are weakly bonded to the backbone of the nucleotide. In the case of memS, this shows a more hydrophobic behavior and tends to repel water. From the figure, a water layer close to both electrodes forms, in which most O atoms of the water are equidistant to the Au atoms of the electrodes, as has been previously observed [12, 15, 18]. This relatively large distance is a consequence of the weak binding of water monomers to the metal surface. It can also be seen, that the water molecules lie either flat close to the surfaces or with the H atoms pointing away from the surfaces.

This water structuring, specifically around the molecules (memS and nucleotide), is strongly influencing their hydrogen bonding. In order to monitor this influence, we have followed the temporal evolution of the memS-C hydrogen bond. In reality, more hydrogen bonds may occur for certain configurations, but for simplicity we have monitored the dynamics of the shortest bond length, as this approach would lead to the largest conductance (see below), thus possibly to measurable values. In fact, it has been previously reported, that the distance between molecules in a gap can be directly correlated to the conductance across the nanogap [16, 17] through the formula

$$G = G_0 \exp(-\beta d_{hb}), \tag{1}$$

where G is the electron conductance, G_0 is a reference conductance, and β is the electronic decay constant extracted from hydrogen bond characteristics of diamondoid-nucleotides complexes [7, 28]. The reference conductance is given through $G_0 = g_0 T(E_F)$, where $g_0 = 2e^2/h$ is the conductance quantum and $T(E_F)$ is the electronic transmission at the Fermi level. Note, that as we use only one recognition molecule on the left electrode, but refer to the conductance across two electrodes, we need to add to the diamondoid-nucleotide hydrogen bond distance the distance from the nucleotide to the right electrode. The value of the resulting distance enters the equation through the d_{hb} and is depicted for all nucleotides in Fig. 5(top). Based on this distance and according to Eq. 1, the corresponding dynamics in the conductance across the device is summarized for all healthy and the mutated nucleotide in the same figure. The fact that the conductance remains relatively constant with time, when water is included, indicates that even in this case the nucleotides can be identified by time averaging of the measurements. Most importantly, the conductance values are nucleotide-specific and reveal again the strong potential of the functionalized nanogap for detecting DNA. Even d5mC shows deviations from its healthy counterpart C, showing the potential for reading-out mutations, as well.

3.3 Conclusions

In this work, we have used quantum-mechanical calculations, both static and dynamic, as well as quantum transport simulations in order to reveal the biosensing potential of diamondoid-functionalized gold nanogaps. We have focused on the

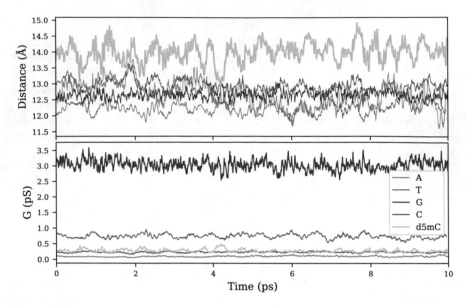

Fig. 5 The dynamics of the total distance between memS-nucleotide taking into account also the distance to the electrodes (see text) with time (top). The respective conductance is shown on the lower panel. In both panels, the results are shown for all nucleotides, as denoted by the legends

influence of rotations of DNA nucleotides within the nanogap and have observed nucleotide-specific electronic currents and rotation-dependent distinct changes in the transmission at the Fermi level. As all calculations so far served a proof-of-principle concept, these were performed in vacuo. Following those, we have added a solvent—at a first approximation—through water molecules. We could show that the inclusion of water significantly increases the dynamic behavior of the whole system. Based on the variation of the hydrogen bond distance with time, we could extract the time evolution of the conductance variation of the functionalized device. Strong nucleotide-specific trends were found here, as well. In view of novel nanopore sequencing techniques, this work demonstrates the potential of functionalized electrode embedded in a nanopore as efficient DNA detectors. In order to further evaluate all the aspects of these nanopores and their biosensitivity, further studies should be directed towards a more detailed investigation of the dynamics and solvent effects. The influence of translations of the nucleotide in the nanogap (as an additional degree of freedom in our model), certain salt solutions, as well as extrapolating to longer DNA lengths, and tuning the nanopore-DNA interactions would be necessary. These would further assist in evaluating the errors in the measurements used to eventually detect the sequence of a DNA passing through a functionalized nanopore. In the end, we hope, that we will be able to reveal the potential of diamondoid-functionalized electrodes in distinguishing among DNA nucleotides even by including more realistic effects.

Acknowledgements Financial support from the collaborative network SFB 716 "Dynamic simulations of systems with large particle numbers" funded by the German Funding Agency (Deutsche Forschungsgemeinschaft-DFG) is acknowledged. This work was performed on the computational resourcew ForHLR Phase I funded by the Ministry of Science, Research and the Arts Baden-Württemberg and DFG.

References

1. E.M Boon, D.M Ceres, T.G Drummond, M.G. Hill, J.K. Barton. Mutation detection by electrocatalysis at dna-modified electrodes. Nat. Biotechnol. **18**(10), 1096–1100 (2000)
2. D. Branton, D.W Deamer, A. Marziali, H. Bayley, S.A. Benner, T. Butler, M. Di Ventra, S. Garaj, A. Hibbs, X. Huang, et al. The potential and challenges of nanopore sequencing. Nat. Biotechnol. **26**(10), 146 (2008)
3. F.C. Maier, G. Sivaraman, M. Fyta, The role of a diamondoid as a hydrogen donor or acceptor in probing dna nucleobases. Eur. Phys. J. E **37**(10), 1 (2014)
4. W.A. Clay, J.E.P. Dahl, R.M.K. Carlson, N.A. Melosh, Z.-X. Shen, Physical properties of materials derived from diamondoid molecules. Rep. Progress Phys. **78**(1), 016501 (2015)
5. Cees Dekker, Solid-state nanopores. Nat. Nanotechnol. **2**(4), 209–215 (2007)
6. M. Di Ventra and Masateru Taniguchi. Decoding dna, rna and peptides with quantum tunnelling. Nat. Nanotechnol. **11**(2), 117 (2016). Review
7. M. Dou, F.C. Maier, M. Fyta, The influence of a solvent on the electronic transport across diamondoid-functionalized biosensing electrodes. Nanoscale (2019)
8. S.M. Foiles, M.I. Baskes, M.S. Daw, Embedded-atom-method functions for the fcc metals Cu, Ag, Au, Ni, Pd, Pt, and their alloys. Phys. Rev. B **33**(12), 7983–7991 (1986)
9. Maria Fyta, Threading dna through nanopores for biosensing applications. Journal of physics: condensed Matter **27**(27), 273101 (2015)
10. S. Goedecker, M. Teter, J. Hutter, Separable dual-space Gaussian pseudopotentials. Phys. Rev. B **54**(3), 1703–1710 (1996)
11. S. Grimme, J. Antony, S. Ehrlich, H. Krieg, A consistent and accurate ab initio parametrization of density functional dispersion correction (DFT-D) for the 94 elements H-Pu. J. Chem. Phys. **132**(15), 154104 (2010)
12. Axel Groß, Florian Gossenberger, Xiaohang Lin, Maryam Naderian, Sung Sakong, Tanglaw Roman, Water structures at metal electrodes studied by ab initio molecular dynamics simulations. J. Electrochem. Soc. **161**(8), E3015–E3020 (2014)
13. C. Hartwigsen, S. Goedecker, J. Hutter, Relativistic separable dual-space Gaussian pseudopotentials from H to Rn. Phys. Rev. B **58**(7), 3641–3662 (1998)
14. Jürg Hutter, Marcella Iannuzzi, Florian Schiffmann, Joost VandeVondele, Cp2k: atomistic simulations of condensed matter systems. Wiley Interdisciplinary Rev.: Comput. Molecular Sci. **4**(1), 15–25 (2014)
15. A. Huzayyin, J.H. Chang, K. Lian, F. Dawson, Interaction of water molecule with au(111) and au(110) surfaces under the influence of an external electric field. J. Phys. Chem. C **118**(7), 3459–3470 (2014)
16. P. Krstić, B. Ashcroft, S. Lindsay, Physical model for recognition tunneling. Nanotechnology **26**(8), 084001 (2015)
17. M.H. Lee, O.F. Sankey, Theory of tunneling across hydrogen-bonded base pairs for dna recognition and sequencing. Phys. Rev. E **79**(5), 051911 (2009)
18. X. Lin, A. Groß. First-principles study of the water structure on flat and stepped gold surfaces. Surf Sci. **606**(11-12), 886–891 (2012)
19. S. Matsunaga, T. Kishi, N. Iwata, Memantine monotherapy for alzheimer's disease: a systematic review and meta-analysis. Plos One **10**(4), e0123289 (2015)

20. C. Merstorf, B. Cressiot, M. Pastoriza-Gallego, A. Oukhaled, J.-M. Betton, L. Auvray, J. Pelta, Wild type, mutant protein unfolding and phase transition detected by single-nanopore recording. ACS Chem. Biol. **7**(4), 652–658 (2012)

21. M. Murtaza, S.-J. Dawson, D.W.Y. Tsui, D. Gale, T. Forshew, A.M. Piskorz, C. Parkinson, S.-F. Chin, Z. Kingsbury, A.S.C. Wong et al., Non-invasive analysis of acquired resistance to cancer therapy by sequencing of plasma dna. Nature **497**(7447), 108 (2013)

22. A. Nazem, G.A. Mansoori, Nanotechnology building blocks for intervention with alzheimer's disease pathology: Implications in disease modifying strategies. J. Bioanal. Biomed. **6**(2), 9 (2014)

23. J.P. Perdew, K. Burke, M. Ernzerhof, Generalized gradient approximation made simple. Phys. Rev. Lett. **77**, 3865 (1996)

24. Steve Plimpton, Fast parallel algorithms for short-range molecular dynamics. J. Comput. Phys. **117**(1), 1–19 (1995). Mar

25. B. Reisberg, R. Doody, A. Stöffler, F. Schmitt, S. Ferris, H. Möbius, Memantine in moderate-to-severe alzheimer's disease. New England J. Med. **348**(14),1333–1341 (2003)

26. Nadja Sändig, Francesco Zerbetto, Molecules on gold. Chem. Commun. **46**(5), 667–676 (2010)

27. C.S. Sarap, P. Partovi-Azar, M. Fyta, Optoelectronic properties of diamondoid-dna complexes. ACS Appl. Bio Mater. **1**(1), 59–69 (2018)

28. C.S. Sarap, P. Partovi-Azar, M. Fyta, Optoelectronic properties of diamondoid-DNA complexes. ACS Appl. Bio. Mater. **1**(1), 59–69 (2018)

29. C.S. Sarap, P. Partovi-Azar, M. Fyta, Enhancing the optical detection of mutants from healthy dna with diamondoids. J. Mater. Chem. B (2019)

30. G. Sivaraman, R.G. Amorim, R.H. Scheicher, M. Fyta, Diamondoid-functionalized gold nanogaps as sensors for natural, mutated, and epigenetically modified dna nucleotides. Nanoscale **8**(19), 10105–10112 (2016)

31. G. Sivaraman, R.G. Amorim, R.H. Scheicher, M. Fyta, Insights into the detection of mutations and epigenetic markers using diamondoid-functionalized sensors. RSC Adv. **7**(68), 43064–43072 (2017)

32. J.M. Soler, E. Artacho, J.D. Gale, A. Garcia, J. Junquera, P. Ordejon, D. Sinchez-Portal, The siesta method for ab initio order- n materials simulation. J. Phys.: Condensed Matter **14**(11), 2745 (2002)

33. K. Stokbro, J. Taylor, M. Brandbyge, O. Pablo, Transiesta: a spice for molecular electronics. Ann N. Y. Acad. Sci. **1006**(1), 212–226 (2003)

34. T. Laino, F. Mohamed, A. Laio, M. Parrinello, An efficient real space multigrid QM/MM electrostatic coupling. J. Chem. Theory Comput. **1**, 1176–1184 (2005)

35. T. Laino, F. Mohamed, A. Laio, M. Parrinello, An efficient linear-scaling electrostatic coupling for treating periodic boundary conditions in QM/MM simulations. J. Chem. Theory Comput. **2**, 1370–1378 (2006)

36. J. VandeVondele, J. Hutter, Gaussian basis sets for accurate calculations on molecular systems in gas and condensed phases. J. Chem. Phys. **127**(11), 114105 (2007)

37. B.M. Venkatesan, R. Bashir, Nanopore sensors for nucleic acid analysis. Nat. Nanotechnol. **6**(10), 615–624 (2011)

38. Y. Zhou, A.D. Brittain, D. Kong, M. Xiao, Y. Meng, L. Sun, Derivatization of diamondoids for functional applications. J. Mater. Chem. C **3**, 6947 (2015)

39. M. Zwolak, M. Di Ventra, Electronic signature of dna nucleotides via transverse transport. Nano Lett. **5**(3), 421–424 (2005)

40. M. Zwolak, M. Di Ventra, Colloquium: Physical approaches to dna sequencing and detection. Rev Modern Phys. **80**(1), 141 (2008)

Materials Science

Johannes Hötzer and Britta Nestler

This chapter presents the results of computational research activities in the field of Material Science conducted on the high performance computers at the High Performance Computing Center Stuttgart and the Steinbuch Centre for Computing Karlsruhe. Interdisciplinary activities in the field are selected and composed demonstrating the successful usage of massive parallel HPC systems to improve material properties for future applications in the field of mobility and magnets.[1]

New multicomponent permanent magnetic materials with reduced content of rare earth elements to be used in energy converters are investigated by Robert Bosch GmbH within the project SimMag. For their ab initio methods using density functional theory (DFT) simulations, the scientists applied the Vienna Ab initio Simulation Package (VASP) on the ForHLR-I using up to 64 CPUs on 8 nodes. Based on experimental findings, the thermodynamic stability, the hard magnetic properties and temperature dependencies of new high potential phases of a rare-earth-free compound, $(Fe, Mn)_2(P, Si)$, and two rare-earth containing phases, $(Y, Sm)(Fe, Ti)_12$ and $Y(Fe,V)_12$ are studied in simulations. Saturation magnetization as well as anisotropy constants were determined to predict the permanent magnetization properties of the new materials.

In a further Bosch project, Intercon, the long-term behavior of the nickel-based anode of a fully ceramic solid oxide fuel cells (SOFC) is studied. To investigate aging mechanisms in SOFCs as a result of the microstructural evolution of the anode material, phase-field simulations as well as experiments are conducted. Ab initio DFT simulations are employed on the high performance clusters in order to determine nickel surface and grain boundary energies as well as to provide self-diffusion coefficients. These quantities are required as input parameters of the phase-field simulations of microstructure evolution and hence contribute to design new anode materials. The DFT simulations are run with VASP on the ForHLR-I.

The team around Prof. Dr. Blügel from the Institute for Advanced Simulation and Peter Grünberg Institut at the Forschungszentrum Jülich developed a unique large-

---

[1]Johannes Hötzer, Institute for Applied Materials - Computational Materials Science (IAM-CMS), Karlsruhe Institute of Technology (KIT), Karlsruhe, Germany, Britta Nestler, Institute for Applied Materials - Computational Materials Science (IAM-CMS), Karlsruhe Institute of Technology (KIT), Karlsruhe, Germany.

scale electronic structure Density Functional Theory code KKRnano to study magnetic B20 compounds. In their code, the Green's function for the Kohn-Sham equation is solved using an extended Korringa-Kohn-Rostoker (KKR) scheme. The method scales linearly with the number of atoms, so that system sizes with a large number of atoms can be realized. Improved algorithms further enable to compute complex non-collinear magnetic structures. Due to the decomposition of the space into non-overlapping cells around the atoms, the calculations are separated into single-cell parts enabling large-scale simulations. The KKRnano code is parallelized with MPI and OpenMP. By applying KKRnano to the material B20-MnGe, performance runs for a supercell of 1728 atoms on high-clocked Intel Xeon processors and up to 576 nodes were performed on the Hazel Hen supercomputer.

Atomistic Simulation of Permanent Magnets With Reduced Rare-Earth Content

Anika Marusczyk, Holger Wüst, and Tim Kolb

Abstract The goal of 'Perfekt', a publicly funded joint project with several partners from science and industry, is to detect, characterize and industrialize new (multi-component) permanent magnetic phases that can be used for energy converters in a cost efficient way. The desired compositions contain no, or at least a drastically reduced (70%) content of, rare earth metals so that the economical as well as technological dependence on expensive rare earth elements can be decreased. The potential energy densities of the anticipated new materials should ideally be as high as those of state of the art magnets, $(BH)max > 350 \, kJ/m^3$. However, materials that can close the gap between ferrites and Fe-Nd-B on the energy density-cost-landscape (saturation polarization > 1 T, anisotropy constant $> 1 \, MJ/m^3$) are also desirable for applications such as smaller engines and sensors. The experimental works on the assessment of new high potential phases should be guided by simulative studies, for which ab-initio methods using density functional theory were used. The associated research focused on the determination of the thermodynamic stabilities of relevant structures at 0 K and for the most promising phases, temperature dependencies were analyzed taking into account phonon contributions to the free energy. In addition to determining the saturation magnetization, the anisotropy constants of these phases were calculated, enabling an assessment of their permanent magnetic qualities. Within the reporting period and scope of 'SimMag', we investigated three compounds that have been experimentally predicted as promising. These include one rare-earth-free compound, $(Fe,Mn)_2(P,Si)$, and two rare-earth containing phases, $(Y,Sm)(Fe,Ti)_{12}$ and $Y(Fe,V)_{12}$. The derived results will be presented in the following.

A. Marusczyk (✉) · H. Wüst · T. Kolb
Corporate Sector Research and Advance Engineering, Rober Bosch GmbH,
Robert-Bosch-Campus 1, 71272 Renningen, Germany
e-mail: Anika.Marusczyk@email.address; Anika.Marusczyk@de.bosch.com

© Springer Nature Switzerland AG 2021
W. E. Nagel et al. (eds.), *High Performance Computing in Science and Engineering '19*,
https://doi.org/10.1007/978-3-030-66792-4_12

1 Compuational Methods

In this work, all density functional theory calculations were performed with the Vienna Ab initio Simulation Package (VASP, version 5.4.1 on ForHLRI) [4–7] using the Projector Augmented Wave Method (PAW) [1, 8] while treating the Mn as well as the Fe semi-core p electrons as valence states. The valence state configuration of Sm was varied up to Sm^{3+}, as will be described in the results section. To describe the effects of exchange and correlation, we used the Perdew-Burke-Ernzerhof parametrization of the generalized gradient approximation [15, 16]. Phases containing rare-earth elements were additionally investigated using the Perdew-Wang (PW91) [17] parametrization of the generalized gradient approximation, the details of which will be described in the results section. All structures were fully relaxed including spin-polarization. For the determination of the anisotropy constants we conducted static calculations taking into account spin-orbit coupling.

Hexagonal Fe_2P was modeled using a 9-atom supercell including 3 formula units (FUs) as well as a 18-atom supercell whenever necessary. The orthorhombic Fe_2P structure contains 36 atoms. After careful convergence parameter investigation for the hexagonal and orthorhombic Fe_2P host structures, a reciprocal k-point density between 40 and 100 k-points per $\mathring{A}^3$ and a planewave cutoff between 520 eV and 560 eV were chosen. The lower parameters were used for relaxations and energy calculations and the higher ones for the determination of the vibrational entropy contributions. For the calculation of the anisotropy constants, highly converged energies are needed and the used parameters are explicitly discussed below. The phases containing rare-earth elements were modeled using a tetragonal 26-atom structure, for which a gamma centered k-mesh of 3x3x7 and a plane-wave cutoff of 560 eV were used.

For the work flow management, the Automated Interactive Infrastructure and Database for Computational Science (AiiDA) [18] software was used in combination with Custodian [14] and the Atomistic Simulation Environment (ASE) [10] as well as the Python Materials Genomics (pymatgen) [14] were employed for the pre- and post-processing. The structures were visualized using VESTA [12].

2 Rare-Earth Free Phase: (Fe, Mn)₂(P, Si)

Within the project 'Perfekt', (Fe, Mn)₂(P, Si) was experimentally identified as a promising hard magnetic phase and in order to guide the experimental efforts to design the best possible candidate within the quaternary configuration space, our work focused on the full characterization of the system in terms of stability at 0 K and finite temperatures as well as hard magnetic properties. The details of our calculations will be published elsewhere and we will therefore give a summary of the most important findings in the following.

There are three competing phases of the Fe_2p host structure, hexagonal, orthorhombic and trigonal, which were all considered. The latter was found to be unstable at
the concentrations of interest and was therefore not the focus of our proceeding
investigation. We determined the quaternary phase diagram of the Fe_2P host structure by substituting various amounts of Mn and/or Si for Fe and P respectively into
the hexagonal and orthorhombic phases to model different concentrations. The site
preference occupancy was modeled by sequential substitution of the atoms. This
means that after having found the ground state of one concentration, the next atom
is substituted into this configuration without considering all possible combinations.
For more efficiency, this concept has been automated by use of the greedy algorithm
[19], which proved to give reliable results for our system. The validation was conducted by calculating all possible combinations for the substitution of up to 3 Mn
and/or Si atoms into the 9-atom Fe_2P host structure for comparison.

It should be noted that these extensive stability calculations were conducted with
lowered cutoffs while the most relevant phases were considered with more accurate
parameters. For these pre-relaxed structures, we used 4 nodes with 8 tasks per node
with a runtime ranging from a few minutes for simple and small pre-relaxed configurations to 10h for larger structures containing more elements. We controlled the
parallelization of VASP by defining the number of bands that are treated in parallel
(NPAR-tag) and used NPAR = 4.

In order to characterize the stability of the quaternary $(Fe_{1-x}Mn_x)_2P_{1-y}Si_y$ system, all competing binary and ternary phases were taken into account and the convex
hull was constructed. We carefully analyzed the regions of stability as well as the
saturation magnetization as predicted by our spin-polarized calculations with respect
to various concentrations in $(Fe_{1-x}Mn_x)_2P_{1-y}Si_y$. The orthorhombic and hexagonal
phases were found to exhibit the highest stability at around 50% Mn and low Si
contents. Fortunately, these concentrations almost coincide with the compositions
of maximal saturation magnetization that are reached at near 50% Mn and all Si
contents.

In order to fully assess which of the considered phases exhibit superior magnetic
properties, the anisotropy constant, K_1, was calculated from the energetic difference
between structures whose spin is oriented into the z-direction, E_z, and the x-direction,
E_x,

$$K_1 = \frac{E_z - E_x}{V}, \tag{1}$$

where V is the volume.

This approach requires at least two fully relativistic, non-collinear density functional theory calculations that take into account spin-orbit coupling. Since even small
changes in the total energies of E_z and E_x (≈ 0.001 eV) will lead to unacceptable
($\approx 100\%$) deviations, highly converged results are indispensable. In this work, we
used an approach in which we first accurately and fully relaxed the crystal structure and then performed a non-self-consistent, static, non-collinear calculation for
the different spin orientations. As even slight adjustments of the parameters lead to

noticeable changes in the anisotropy constants, we carefully investigated its convergence with respect to all relevant influencing factors. These include the smearing used for the relaxation of the crystal structure, the k-mesh for the relaxation as well as the static non-collinear calculations, the plane-wave cutoff, the cell size and a change in spin orientation from z to $-z$ or x to y.

Even with our carefully chosen parameters, numerical errors cannot be excluded and we observed an oscillating behavior of the value up to large k-point densities. For all our calculations, a k-point density of 4000 k-points/atom was chosen, since the total energy would usually be considered converged for this value. Within our convergence analysis, we found that the anisotropy constant is less sensitive to the chosen plane-wave cutoff and a low value of 300 eV was chosen for a first screening, which was then increased to 520 eV for the most relevant phases. We used the Methfessel-Paxton method with a smearing width of 0.05 eV for the structural optimizations and the tetrahedron method for the static, non-collinear calculations. The effects of spin-orientation and cell size did not play a significant role. It should be noted that we conducted our anisotropy constant calculations for the thermodynamically most stable phases only and did not evaluate the effect of less stable site occupancies.

Within the determined scatter, satisfactory agreement with the experimental anisotropy constant value for Fe_2P [2] could be achieved. We found that while the phase with the overall highest saturation magnetization exhibits a negligible and therefore unfavourable anisotropy, the ternary FeMnP phase, whose saturation magnetization is only slightly lower, exhibits a promising anisotropy constant value (Table 1). In addition to this, a quaternary phase with low Mn content leads to satisfactory intrinsic hard magnetic properties.

For the described analysis, we used 32 or 64 CPUs (4 or 8 nodes and 8 tasks per node) and a parallelization for the bands of NPAR = 8 with the non-collinear version of Vasp.5.4.1 on the ForHLRI. The runtime deviated depending on the used parameters (e.g. k-mesh, cell size), but was generally in the range of a few hours. It should be noted that we frequently experienced memory related issues and had to increase the request to 15 GB to avoid an abrupt interruption of the job in these cases.

Based on our current findings, FeMnP constitutes the most promising phase within the considered configuration space. Even though Si substitution drastically decreases the anisotropy constant (while increasing the saturation magnetization), there might be other, chemically similar, elements with a more promising effect on the magnetic properties of FeMnP. To investigate this issue, S, Al and Se, as well as N and B, which have been shown to have a beneficial effect on the magnetic properties in some cases [9], were substituted into the host structure and the magnetic properties of the resulting phases were determined. As the intent at this point was to determine a general trend, only full substitution of P was considered so that the site preferences did not have to be taken into account. The results are summarized in Table 1.

Among the considered elements, Al is the only one that increases the saturation magnetization compared to FeMnP. However, since the magneto-crystalline anisotropy becomes planar, it does not constitute a promising alloying element. Even

Table 1 Calculated saturation magnetization and anisotropy constant of the indicated phases

Phase	J_s [T]	K_1 MJ/m^3
FeMnP	1.36	2.2
FeMnS	1.07	2.0
FeMnAl	1.46	−0.7
FeMnSe4	1.01	1.2
FeMnN	−0.01	–
FeMnB	0.01	–

though S and Se somewhat decrease the saturation magnetization, they do exhibit positive anisotropy constants and might therefore lead to a promising candidate for a permanent magnet. The considered compositions are, however, inferior to FeMnP and further work should be conducted to identify potentially more promising compositions in ternary Fe-Mn-S and Fe-Mn-Se as well as quaternary Fe-Mn-S-P and Fe-Mn-Se-P systems. In addition to this, the stability of the corresponding phases needs to be clarified.

It should be noted that since N and B substitution leads to zero saturation magnetization, the anisotropy constants for these materials were not calculated. In previous works [9], these elements have been interstitially substituted into the host structures, leading to an increase of the anisotropy. Therefore, the effect of small amounts of interstitial B or N as opposed to full substitution in promising host structures should be investigated in future work. For this part, 32 CPUs were used with 4 nodes and 8 tasks per node with an average runtime of a few hours per job.

3 Rare-Earth Containing Phases: $(Y, Sm)(Fe, Ti)_{12}$ and $Y(Fe, V)_{12}$

$Sm_{0.4}Y_{0.6}Fe_{10.7}Ti_{1.3}$ and $YFe_{11}V$ were experimentally identified as promising hard magnetic phases and in order to generate an enhanced understanding and guide the experimental design within the project 'Perfekt', they were investigated using density functional theory methods. Opposed the previously presented results, the focus was not on the full assessment of phase stabilities of the ternary and quaternary configuration space, but on the analysis of the proposed compositions in order to generate an understanding of the influence of the substitutional elements. This was also motivated by the fact that the stability of the tetragonal 1:12 phases is well characterized in the literature and the stabilizing effect of the substation of Fe by other transition metals, such as V and Ti, is well described [11, 13].

We first modeled the host structures, tetragonal Sm_2Fe_{24} and Y_2Fe_{24}, and ensured their correct description by convergence analysis regarding the k-mesh and the plane wave cutoff. These calculations typically ran on 32 CPUs, using 4 nodes and 8

Table 2 Calculated saturation magnetization and anisotropy constant of the indicated phases

Phase	J_s [T]	K_1 MJ/m^3
Sm_2Fe_{24}	1.68	1.19
$Sm_2Fe_{23}Ti$	1.52	1.69
$Sm_2Fe_{22}Ti_2$	1.37	1.73
Y_2Fe_{24}	1.69	1.73
$Y_2Fe_{23}V$	1.54	2.26
$Y_2Fe_{22}V_2$	1.40	1.89
$YSmFe_{24}$	1.68	1.45
$YSmFe_{23}Ti$	1.52	1.88
$YsmFe_{22}Ti_2$	1.38	1.84

tasks per node, and a runtime of a few hours. To model the desired phases, we then successively substituted the relevant elements into the structures until we obtained compositions close to the experimental ones at feasible cell sizes (24 atoms). In a first step, we considered all possible non-equivalent occupancies of 1 and 2 Ti atoms in Sm_2Fe_{24} to determine the ground state of $Sm_2Fe_23Ti_1$,$Sm_2Fe_{22}Ti_2$ as well as $SmYFe_{22}Ti_2$ by substitution of one Y atom. The resulting site occupancy preference of 8i > 8j > 8f is in agreement with the literature [13]. Based on this knowledge, we determined the site occupancy preference of 1 and 2 V atoms on an Fe site in YFe_{12} with a reduced number of calculations and found it to coincide with our Ti results.

The calculated quantities of interest, the saturation magnetization and the anisotropy constant, are given in Table 2. As expected, substitution of Fe-atoms by the non-magnetic elements Ti and V decreases the saturation magnetization while the anisotropy constant increases in most cases. Therefore, the substitution of Fe by another transition metal cannot only stabilize the desired tetragonal phase, but also has a beneficial effect on the anisotropy constant. A correlation between the latter and geometrical aspects such as the c/a-ratio could not be observed. It is unclear at this point why the anisotropy constant of $Y_2Fe_{23}V$ does not follow the expected trend and increases significantly compared to the other compositions.

For this part, we used 32 or 64 CPUs (4 or 8 nodes and 8 tasks per node), a parallelization for the bands of NPAR=8 with the non-collinear version of Vasp.5.4.1 on the ForHLRI for the calculation of the anisotropy constants and a runtime of a few hours depending on the explicit structure. We frequently experienced memory related issues when performing non-collinear calculations and had to increase the request to 15 GB to avoid an abrupt interruption of the job in these cases.

It should be noted that the calculated values for the Y-Fe-V system are in satisfactory agreement with experimental ones measured within the project (J_s=1.3 T, $K_1 = 1.8$ MJ/m^3 for $YFe_{11}V$ at room temperature) as well as literature data, demonstrating similar saturation magnetizations [11] and anisotropy constants for YFe_{12} [3]. Opposed to this, we observed significant quantitative deviations for the Sm-Y-Fe-Ti phases. While the saturation magnetization of 1.26 T, that was measured for

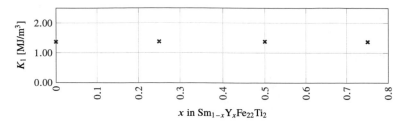

Fig. 1 Calculated anisotropy constant with x $Sm_{1-x}Y_xFe_{22}Ti_2$

Table 3 Calculated total magnetization, anisotropy constant and equilibrium lattice parameters of $SmFe_{12}$ determined with the indicated GGA parametrization and valence configuration

GGA and valence state	Total mag [μ_B]	Anisotropy constant [MJ/m^3]	a,b,c [Å]
PBE Sm^{3+}	45.17	0.69	8.54, 8.46, 4.74
PW91 Sm	–	–	–
PBE Sm^{2+}	37.01	1.69	8.37, 8.36, 4.68
PBE Sm^{3+}	40.03	1.62	8.49, 8.49, 4.70
PBE Sm^{3+}, no relaxation	43.23	1.16	8.50, 8.50, 4.71
PW91 Sm^{3+}, no relaxation	40.26	1.62	8.50, 8.50, 4.71
PW91 Sm^{2+}	41.63	1.02	8.50, 8.54, 4.70
PW91 Sm^{3+}, DFT+U	41.13	1.23	8.57, 8.59, 4.71

$Sm_{0.4}Y_{0.6}Fe_{10.7}Ti_{1.3}$ at room temperature within the project, exhibits good agreement with our results, the found anisotropy constant of $K_1 = 3.2$ MJ/m^3 is significantly underestimated by our calculations. This is even more pronounced for the Y-free systems, whose anisotropy constant has been determined to 7.6 MJ/m^3 at 4.3 K [20]. At the same time, the saturation magnetization in the reference is in good agreement with our values. Finally, the expected trend of a decrease in anisotropy with Sm-content is not met (Fig. 1), indicating an even qualitative disagreement with experiment.

Therefore, the reliability of results derived in such a way should be questioned. One possible source of error might be that, opposed to Y, Sm contains f-electrons that are known to be difficult to describe with density functional theory methods. To deal with this, one often uses a frozen core approach and pseudopotentials that treat the problematic f-electrons as part of the effective nucleus potential. By doing so, a valence state of Sm^{3+} is often assumed, which is not always justified. It is therefore questionable, whether a model using such approximations can reliably predict quantities such as anisotropy constants.

To investigate some of these issues, we varied the used pseudopotential as well as the parametrization of the generalized gradient approximation (GGA) exchange-correlation functional and observed significant differences in the results (Table 3).

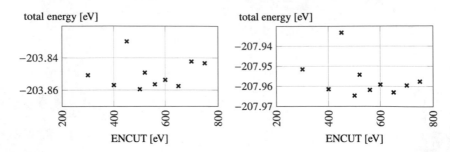

Fig. 2 Convergence of the total energy with respect to the plane-wave cutoff for $SmFe_{12}$ (left) and YFe_{12} (right) when using a PW91 exchange-correlation functional

These can likely be correlated with the geometry of the optimized structures as e.g. the use of a PBE exchange-correlation functional for the structural optimization resulted in a de-tetragonalization and therefore a very low anisotropy constant. To confirm this, we also calculated the anisotropy constant with the same parameters, but without previous structural relaxation and observed a significant increase in K_1. In addition to this, we also investigated the effect of placing an-onsite repulsion onto the Sm f-states (DFT+U, U = 6.87–0.76 eV), which did not lead to results in better agreement with experiment. It should be noted that we were not able to obtain converged results when using pseudopotentials with a Sm valence state.

The final values shown in Table 2 were derived by use of the PW91 parametrization of the GGA exchange-correlation functional assuming a Sm^{3+} valence state. It should be noted that even though these values are in better agreement with experiment than the PBE results, a large error due to the made assumptions in the frozen core approximation cannot be excluded. This is underlined by our plane wave cutoff convergence analysis, exemplary shown for PW91 and both of our systems (Fig. 2), which demonstrates a very strong oscillation of the total energy. At the same time, values such as the volume or the saturation magnetization do converge at reasonable cutoffs (not shown here). The used CPUs and average runtime are comparable to the previously described work packages.

4   Summary and Outlook

In order to complement and guide the experimental works conducted within the publicly funded project 'Perfekt', the research conducted within the project 'Simmag' focused on the determination of thermodynamic stabilities and hard magnetic properties of relevant phases by use of density functional theory methods. Within the reporting period, we investigated three compounds that were previously experimentally predicted as promising candidates. These include one rare-earth-free compound, $(Fe, Mn)_2(P, Si)$, and two rare-earth containing compounds, $(Y, Sm)(Fe, Ti)_{12}$ and $Y(Fe, V)_{12}$.

The rare-earth free system, $(Fe_{1-x}Mn_x)_2P_{1-y}Si_y$, was fully characterized in terms of phase stabilities as well as hard magnetic properties. While ternary Fe-P-Si does not constitute a promising permanent magnetic candidate, the highly favorable hexagonal phase can be stabilized by Mn substitution. It should be noted that while Si addition increases the saturation magnetization, it also leads to unfavorable anisotropy constants. Consequently, the most favorable candidates stem from the ternary Fe-Mn-P system with the best candidate being FeMnP. The details of our investigations are to be published elsewhere.

Based on the preferred FeMnP phase, a screening for other promising compositions was performed as well. Among the considered structures, FeMnS and FeMnSe exhibit the best hard magnetic properties, but are still inferior to FeMnP. Since the effects of composition have so far been neglected within the screening investigations, further work should be conducted to identify potentially more promising compositions in the ternary Fe-Mn-S and Fe-Mn-Se as well as the quaternary Fe-Mn-S-P and Fe-Mn-Se-P systems. Furthermore, the effect of interstitial substitution of small elements, such as B or N, should be clarified for the most promising phases.

In addition to the rare-earths-free system, we also investigated compounds containing relatively cheap rare-earth elements such as Sm and Y. During the experimental high-throughput efforts within the project 'Perfekt', $Sm_{0.4}Y_{0.6}Fe_{10.7}Ti_{1.3}$ and $YFe_{11}V$ were identified as promising hard magnetic phases and investigated using density functional theory within 'SimMag' during the reporting period. Since the stability of these systems is well described in the literature, we focused on the analysis of the hard magnetic properties of the phases in order to generate an understanding of the influence of the individual elements to guide the experimental design.

To this end, we successively substituted atoms into the known host structures and calculated the saturation magnetization and anisotropy constants of the determined ground states. While our obtained results for phases without Sm were in sufficient qualitative agreement with experimental data, the addition of the f-electrons containing Sm led to results that even qualitatively deviate from known values. After a careful analysis of our used parameters and methods within the standard framework of VASP, we came to the conclusion that more work should be conducted on the model development for the calculation of anisotropy constants of Sm-containing phases. Generally speaking, the accurate determination of anisotropy constants using standard density functional theory methods is highly non-trivial and further work should be conducted to develop more precise methods.

Acknowledgements This work was performed on the computational resource ForHLR I funded by the Ministry of Science, Research and the Arts Baden-Württemberg and DFG ("Deutsche Forschungsgemeinschaft"). Financial support from the Federal Ministry of Education and Research (BMBF) via the PERFEKT project (No. 03XP0023A) is gratefully acknowledged.

References

1. P.E. Blöchl, Projector augmented-wave method. Phys. Rev. B **50**, 17953–17979 (1994)
2. H. Fuji, T. Hokabe, T. Kamigaichi, T. Okamoto, Magnetic Properties of Fe_2P Single Crystal. J. Phys. Soc. Jpn. **43**, (1977)
3. L. Ke, D.D. Johnson, Intrinsic magnetic properties in $r(Fe_{1-x}Co_x)_{11}Tiz(r = Y and Ce; z = H, C, and N)$. Phys. Rev. B **94**, 024423 (2016)
4. G. Kresse, J. Furthmüller, Efficiency of ab-initio total energy calculations for metals and semiconductors using a plane-wave basis set. Comput. Mater. Sci. **6**(1), 15–50 (1996)
5. G. Kresse, J. Furthmüller, Efficient iterative schemes for ab initio total-energy calculations using a plane-wave basis set. Phys. Rev. B **54**, 11169–11186 (1996)
6. G. Kresse, J. Hafner, Ab initio molecular dynamics for liquid metals. Phys. Rev. B **47**, 558–561 (1993)
7. G. Kresse, J. Hafner, Ab initio molecular-dynamics simulation of the liquid-metal-amorphous-semiconductor transition in germanium. Phys. Rev. B **49**, 14251–14269 (1994)
8. G. Kresse, D. Joubert, From ultrasoft pseudopotentials to the projector augmented-wave method. Phys. Rev. B **59**, 1758–1775 (1999)
9. W. KAurner, G. Krugel, C. Elsässer. Theoretical screening of intermetallic thmn12-type phases for new hard-magnetic compounds with low rare earth content. Sci. Rep. **8**, 24686 (2018)
10. A.H. Larsen, J.J. Mortensen, J. Blomqvist, I.E. Castelli, R. Christensen, M. Dulak, J. Friis, M.N. Groves, B. Hammer, C. Hargus, E.D. Hermes, P.C. Jennings, P.B. Jensen, J. Kermode, J.R. Kitchin, E.L. Kolsbjerg, J. Kubal, K. Kaasbjerg, S. Lysgaard, J.B. Maronsson, T. Maxson, T. Olsen, L. Pastewka, A. Peterson, C. Rostgaard, J. Schiotz, O. Schütt, M. Strange, K.S. Thygesen, T. Vegge, L. Vilhelmsen, M. Walter, Z. Zeng, K.W. Jacobsen, The atomic simulation environment a python library for working with atoms. J. Phys.: Condensed Matter **29**(27), 273002 (2017)
11. H. Ma, Z. Huang, B. Chen, W. Qiang, G. Sun, The stabilization effect of the substituted atoms and the magnetism for intermetallic compounds $YFe_{12x}V_x$. Sci. China Phys. Mech. Astron. **53**(7), 1239–1243 (2010)
12. K. Momma, F. Izumi, VESTA3 for three-dimensional visualization of crystal, volumetric and morphology data. J. Appl. Crystallography **44**(6), 1272–1276 (2011)
13. C. Nan-xian, H. Shi-qiang, W. Yu, S. Jiang, Phase stability and site preference of Sm(Fe, T)₁2. J. Mag. Mag. Mater. **233**(3), 169–180 (2001)
14. S.P. Ong, W.D. Richards, A. Jain, G. Hautier, M. Kocher, S. Cholia, D. Gunter, V.L. Chevrier, K.A. Persson, G. Ceder, Python materials genomics (pymatgen): A robust, open-source python library for materials analysis. Comput. Mater. Sci. **68**, 314–319 (2013)
15. J.P. Perdew, K. Burke, M. Ernzerhof, Generalized gradient approximation made simple. Phys. Rev. Lett. **77**(18), 3865–3868 (1996)
16. J. P. Perdew, K. Burke, and M. Ernzerhof. Generalized gradient approximation made simple [Phys. Rev. Lett. 77, 3865 (1996)]. Phys. Rev. Lett. **78**(7), 1396–1396 (1997)
17. J.P. Perdew, Y. Wang, Accurate and simple analytic representation of the electron-gas correlation energy. Phys. Rev. B **45**, 13244–13249 (1992)
18. G. Pizzi, A. Cepellotti, R. Sabatini, N. Marzari, B. Kozinsky, Aiida: automated interactive infrastructure and database for computational science. Comput. Mater. Sci. **111**, 218–230 (2016)
19. K. Terayama, T. Yamashita, T. Oguchi, K. Tsuda. Fine-grained optimization method for crystal structure prediction. npj Comput. Mater. **4** (2018)
20. I. Tereshina, N. Pankratov, E. Tereshina Chitrova, V. Verbetski, A. Salamova, T. Mydlarz, W. Suski. Magnetocrystalline anisotropy of hydrogenated and nitrogenated rare-earth intermetallic compound smfe11ti. Phys. Metals Metallography **99**, S46–S49, 01 (2005)

Atomistic Simulation of Nickel Surface and Interface Properties

Anika Marusczyk, Senja Ramakers, Matthias Kappeler, Patricia Haremski, Matthias Wieler, and Piero Lupetin

Abstract This report describes the research conducted by use of the ForHLRI within the publicly funded project 'Kersolife100', in which the long-term performance of a fully ceramic solid oxide fuel cell (SOFC) concept is investigated. The project aims at modeling and understanding the dominant degradation mechanisms in SOFCs so that the required lifetime of the SOFC-stacks can be ensured. One major cause of ageing is the unfavourable microstructural evolution of the nickel-based anode that occurs upon SOFC operation. The associated mechanisms are modeled by use of phase field methods within 'Kersolife100'. For a successful outcome, the availability of accurate material parameters is crucial, but until now not given. Complementary to experimental efforts, the ab-initio research activities of this year therefore focused on the determination of the relevant nickel surface and interface properties. By combining experimental and simulation results, a deeper understanding of the anode aging mechanism can be generated and the identification of counter measures can be guided. In this report, the current results of the ab-initio activities are summarized.

1 Compuational Methods

In this work, all density functional theory calculations were performed with the Vienna Ab initio Simulation Package (VASP, version 5.4.1 on ForHLR I) [14–17] using the Projector Augmented Wave Method (PAW) [1, 18] while treating the nickel p-states as valence states. During structural optimization, the Methfessel-Paxton smearing method with $\sigma=0.01$ was used to determine the partial orbital occupancies. To obtain highly converged results, a static energy calculation using the tetrahedron method with Bloechl correction was then conducted. The slab and grain boundary structures were constructed from fully relaxed bulk structures including spin-polarization and only ionic relaxation was allowed for the total energy calculation of the slabs. The convergence criterion was held constant for all calculations

A. Marusczyk (✉) · S. Ramakers · M. Kappeler · P. Haremski · M. Wieler · P. Lupetin
Corporate Sector Research and Advance Engineering, Rober Bosch GmbH,
Robert-Bosch-Campus 1, 71272 Renningen, Germany
e-mail: Anika.Marusczyk@de.bosch.com

© Springer Nature Switzerland AG 2021
W. E. Nagel et al. (eds.), *High Performance Computing in Science and Engineering '19*,
https://doi.org/10.1007/978-3-030-66792-4_13

at 10^{-8} eV for the electronic cycle and 10^{-6} eV for the ionic cycle. To achieve highly accurate results, careful convergence analyses regarding cell size, relaxation methods, exchange-correlation functional, plane-wave cutoff and gamma-centered k-mesh were conducted, the results of which will be described in the following sections. For the job management, the Automated Interactive Infrastructure and Database for Computational Science (AiiDa) [30] software was used in combination with Custodian [26]. The crystal structures were generated with the Atomic Simulation Environment (ASE) [19] as well as the Python Materials Genomics (pymatgen) [26] and Vesta [25] was used for the visualization. The empirical potential calculations were performed with enlarged $(2 \times 2$ and $3 \times 3)$ cells using LAMMPS [31] with a convergence criterion of $\epsilon = 10^{-4}$ eV.

2   Nickel Surface Energies and Anisotropies

Nickel surface energies and anisotropies are of crucial importance as input parameters for the phase field simulations as they are likely to play a dominant role in the microstructural evolution of the Ni-based anode upon SOFC operation. At the same time, their experimental determination is highly challenging and the expected scatter due to the measurement methods is significant (compare e.g. Refs. [2, 5, 20, 32, 37, 40, 41]). At the same time, the determination of accurate metallic surface energies with ab-initio methods is highly non-trivial and careful convergence as well as exchange-correlation functional evaluations were conducted to obtain reliable energy values. The computational details of our efforts to determine these energies will be described in the next sections.

2.1   Convergence Parameters and Technical Details

The surface energy γ is defined as the difference between the energy of the surface structure E_{surf} and the bulk structure E_{bulk} normalized to the area of the surface A [40]:

$$\gamma = \frac{1}{2A}(E_{\text{surf}} - N \cdot E_{\text{bulk}}). \tag{1}$$

To determine E_{surf}, the slab method was used and the surface was modeled as a perfect fcc nickel phase with two interfaces to the vacuum due to the periodic boundary conditions. As a consequence, additional convergence parameters to the plane-wave cutoff and the k-mesh were considered for the slab calculations. These include the size of the vacuum and the number of atomic layers in the slab to ensure minimal interaction between the two surfaces, as well as the number of layers to be relaxed upon each optimization to account for the deceased tight binding of surface atoms

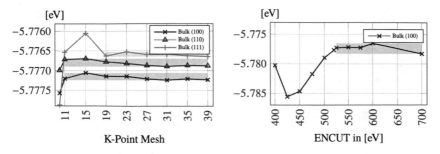

Fig. 1 Convergence of the total energy of bulk Ni fcc with respect to the k-mesh (left) and plane-wave cutoff (right)

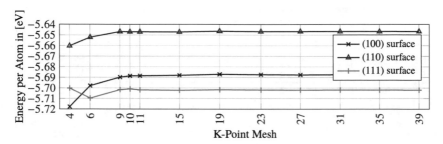

Fig. 2 Convergence of the total energy with respect to the vacuum size for low Miller index orientations of the surface

compared to those in the bulk. We calculated the ground state energies of differently oriented surfaces characterized by their Miller indices and to achieve high accuracy, the bulk energies E_{bulk} for low Miller index ($<=1$) structures were calculated for the corresponding orientations of the bulk as well.

Convergence analyses were conducted for the three chosen exchange-correlation functionals (local density approximation (LDA) [4], generalized gradient approximation under the Perdew-Burker-Ernzerhof parametrization (PBE) [27, 28], generalized gradient approximation for solids and surfaces (PBEsol) [29]) and representative results for the plane-wave cutoff and the k-mesh are depicted in Fig. 1. These fast calculations for the small bulk structures were used as a guideline for the convergence evaluation of the computationally more demanding slab structures. Representative results for the low Miller index orientation surfaces with respect to the the vacuum slab, number of relaxed layers and slab size can be found in Figs. 2, 3 and 4. Based on these calculations, we chose a k-mesh of $23 \times 23 \times 1$, a plane-wave cutoff of 550 eV, a vacuum slab of 14 Å, a cell size of 13–17 layers and relaxed 2–4 layers.

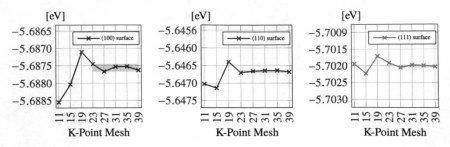

Fig. 3 Convergence of the total energy with respect to the number of relaxed layers

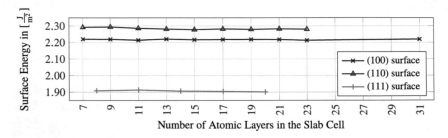

Fig. 4 Convergence of the surface energy with respect of the total atomic layers in the slab

2.2 Surface Energies

The results of our surface energy calculations have been submitted to Acta Materialia and are described in detail in the corresponding article. The surface energies, γ_{hkl}, for low Miller index structures were determined with the three exchange-correlation functionals, all of which lead to excellent agreement within the expected experimental scatter. However, LDA is known to overestimate [3] and GGA-PBE to underestimate surface energies [42] while the GGA-PBEsol value, which was in between the other two, is expected to be most reliable. It was therefore used for the calculation of structures with higher Miller indices.

To obtain these results, calculations on 80 or 120 CPUs (depending on the cell size) with an average runtime of 18h were conducted. Surfaces with higher Miller index orientations require larger unit cells and therefore the average runtime on 120 CPUs amounted to 140 h. Due to the requirement of high accuracy, we chose strict numerical parameters and cutoffs for our calculations. This led to frequent convergence issues, especially for larger cell sizes, that were solved by increasing the number of cores and runtime. It should be noted that the parallelization of VASP can be controlled by defining the number of bands that are treated in parallel (NPAR-tag). We made use of this option for the slab calculations, requiring at least 80 CPUs, and fixed the parameter to 8, as recommended by the VASP manual (NPAR $\approx \sqrt{\text{\#cores}}$).

Clearly, the calculation of accurate surface energies using ab-initio methods is computationally very demanding. Empirical potentials constitute a much less expen-

Table 1 Empirical potentials and their abbreviations used in this work

Abbreviation	Potential	Source
EAM_fcc	Embedded-atom-method functions for the fcc metals	[10, 11]
EAM_TCS	EAM Angelo Moody Baskes modified by Tehranchi Curtin Song	[8, 9, 38, 39]
EAM_DMF	EAM Dynamo Mishin Farkas	[23, 24]
EAM_DM	EAM Dynamo Mendelev	[21, 22]
Pair_Morse	Pair Morse Shifted GirifalcoWeizer HighCutoff	[6, 7, 12]
EMT	EMT Asap Standard Jacobsen Stoltze Norskov	[13, 35, 36]

sive method and if they provided reliable results, their use would therefore be highly desirable. To investigate this issue, we benchmarked a set of six available empirical potentials (Table 1) from the OpenKim database [38] and compared them to our accurate density functional theory values. The chosen potentials differ in their model or fitting method. While the embedded atom method potential EAM_fcc has been fitted to the sublimation energy, equilibrium lattice constant, elastic constants, vacancy-formation energies of the pure metals and the heat of solution of the binary alloys, EAM_DMF has been fitted to experimental and ab initio data for monoatomic nickel and EAM_DM to liquid and gas properties. Pair_Morse is a Pair Morse potential fitted to the energy of vaporization, the lattice constant and the compressibility and EMT is an Effective Medium Theory potential fitted to bonding in metallic systems. EAM_TCS is the only embedded atom method potential that also includes an interaction between Ni and H.

We found EAM_TCS to give the most reliable weighted surface energies within 10 % accuracy. However, none of the used empirical potentials correctly reproduced the energetic hierarchy determined by density functional theory. In view of first phase field simulations conducted at the Hochschule Karlsruhe that revealed the importance of anisotropy to correctly describe the microstructural evolution of the Ni-based anode, the observed discrepancy in predicting the relative energies between the accurate PBEsol and empirical potential method prohibits the use of latter for our purposes. While we will still refer to empirical potentials in future work packages whenever possible, density functional theory calculations will continue to be the focus of this project. For this work package, direct calculation on the visualization nodes was feasible.

3 Nickel Grain Boundary Energies

Nickel grain boundary energies are important material parameters to be used as input for the phase field simulations and their determination constitutes an ongoing work package. The grain boundary energy γ_{GB} is calculated as the difference in energy between structures with and without a grain boundary

$$\gamma_{GB} = \frac{E_{GB,tot} - N \cdot E_{bulk}}{2A_{GB}}, \tag{2}$$

where $E_{GB,tot}$ is the total energy of the system with two grain boundaries due to periodic boundary conditions, N is the number of atoms in the system, E_{bulk} the energy of the bulk per atom, and A_{GB} the surface area of the grain boundary.

In a first step, the necessary parameters to achieve high accuracy were evaluated and using our research on nickel surface energies as a guideline, we conducted a reduced convergence analysis regarding the planewave cutoff and k-mesh for a $\Sigma 3(111)60°$ twist boundary that is expected to be highly stable [3]. For further calculations, we chose a k-mesh of $12x\frac{12}{b}x\frac{12}{c}$ ($\approx 3.5\mathring{A}$) and 550 eV as a planewave cutoff, consistent with the surface calculations. The convergence analysis was conducted for a 24-atoms cell, using 16 CPUs with an average runtime of 15 min for low k-mesh and planewave cutoff values and an average of 1.5h for high values.

When determining grain boundary energies with density functional theory, convergence with respect to the cell size needs to be ensured. Due to the periodic boundary conditions, each structure has two grain boundary plains and to achieve high accuracy, their interaction should be minimized by increasing the distance between them. After having confirmed the dominance of the c-lattice parameter, we investigated the energy convergence with respect to c for the stable $\Sigma 3$ grain boundary while varying the relaxation method between full, ionic only and fully static. Full relaxation of the structure was found to be needed for accurate results at feasible cell sizes and convergence was reached at a cell length of 50 $\mathring{A}$, in agreement with similar density functional theory calculations for bcc Tungsten [34]. For larger cell sizes, we increased the number of CPUs to 32 for cells up to 72 atoms (average runtime 5h) and 64 for cells up to 144 atoms (average runtime 25h). To ensure efficient parallelization, we made use of the NCORE-tag available in VASP and chose 8 cores per orbital, as recommended in the VASP-manual. The previously used NPAR-tag can be related to NCORE by NCORE=CPUs/NPAR.

Due to the large number of possible grain boundaries in fcc nickel, we are currently investigating the effect of cell size and relative energies with the less computationally expensive empirical potentials. The results will then be benchmarked to those obtained from highly accurate density functional theory calculations for selected structures, whose choice was based on our previous surface energy calculations as well as literature results. Consequently, we constructed symmetric tilt boundaries around the 110-axis for the four most stable surface orientations as previously determined. Our current results and computational details are shown in Table 2, demonstrating good agreement with values known from the literature [33]. For these highly

Table 2 Grain boundary energy results and computational details of the four most stable grain boundaries

Sigma	# atoms	Plane	Twist	Tilt	γ_{GB} mJ/m^2	CPUh	# CPUs
3	48	111	rot. axis: (111) angle: 60/180	rot. axis: (110) angle: 109.5	58.5	2.36	32
3	20	211	rot. axis: (211) angle: 180	rot. axis: (110) angle: 70.5	1283.1	5.39	32
9	68	221	rot. axis: (221) angle: 180	rot. axis: (110) angle: 141.1	1211.1	29.83	96
11	88	311	rot. axis: (311) angle: 180	rot. axis: (110) angle: 50.5	517.0	9	160

converged calculations, we increased the number of CPUs to 96 for large cells while keeping NCORE at 8. The results generated in this workpackage will be published in a scientific journal.

4 Nickel Surface Self-Diffusion Constants

Based on an internal evaluation of experimental results, nickel surface diffusion is expected to play a dominant role in the microstructural evolution of the anode. It is expected to involve the migration of adsorbed nickel atoms and therefore knowledge of stable nickel adsorption sights on the nickel surfaces, including the atomistic models, is needed in a first step. Our derived results are described in detail in the submitted paper (compare Sect. 2). The relevant energy barriers for the migration of a nickel atom on the surface can then be calculated using the Nudged Elastic Band method (NEB) and by considering differently oriented surfaces, the effect of anisotropy on the diffusion constants can be assessed. In addition to this, the influence of the expectedly adsorbed hydrogen atoms on the diffusion constant should be investigated, which is why we also considered the adsorption of hydrogen on nickel surfaces. It should be noted that the derived diffusion constants should be evaluated in close comparison to available diffusion constant values. Depending on the outcome, an extension or reduction of the methodology (e.g. explicit calculation of vibrational entropies or the use of empirical potentials) should be considered.

The adsorption energies of nickel on the expected stable sites of the low Miller index surfaces were calculated using density functional theory as well as the most promising empirical potential EAM_TCS (compare Table 1). We found an excellent agreement between both methods, which is why the computationally much less expensive empirical potential constitutes an attractive alternative for similar future research questions. Opposed to this, the adsorption of hydrogen on Ni surfaces using empirical potentials did not give reliable results and density functional theory calculations, predicting a relative stability of hydrogen adsorption sites in agreement with the literature, are indispensable. The derived atomistic models will be used to investigate the influence of hydrogen on the nickel surface self-diffusion constants. The density functional theory calculations were performed on 80 or 120 CPUs depending on the cell size with an average calculation time of 120 h.

In order to determine the desired diffusion constants as input for the phase field simulations, first energy barrier calculations for the simplest case, an interstitial hopping mechanism, were conducted using the Nudged Elastic Band method. In a first step, the thermodynamically stable (111) surface was considered and the minimal energy path was guessed by interpolation. To be feasible, the k-mesh was reduced to $9 \times 5 \times 1$ for a $2 \times 1 \times 1$ supercell and the effect of the number of images as well as the number of fixed atoms was investigated. First calculations led to unconverged results and underestimated energy barriers and the most promising numerical parameters and methods are currently under investigation. The NEB calculations were conducted on 64 or 144 CPUs depending on the number of images with an average runtime of 15–50 h.

5   Summary and Outlook

The goal of this project is the determination of the most relevant material parameters of a Ni-based SOFC anode to be used as input parameters for phase field simulations conducted within the publicly funded project 'Kersolife100'. These strive to model the microstructural evolution of the anode under SOFC operating conditions in order to make understanding driven predictions of its long-term performance. Within the reporting period, our research activities focused on the determination of the relevant nickel surface energies, nickel grain boundary energies and nickel surface self-diffusion constants using density functional theory. After having conducted extensive convergence analyses to obtain highly accurate nickel surface energies, we considered different surface orientations up to a Miller index of 3 and achieved excellent agreement with known literature values within the expected scatter. First phase field simulations revealed the importance of anisotropy for the accurate prediction of the microstructural evolution of the Ni-based anode, excluding the possibility to use the computationally much less demanding empirical potentials that have so far not lead to a satisfactory description of the anisotropy.

Based on our surface results, we were able to obtain first converged nickel grain boundary energies with a reduced set of convergence tests for selected structures. The latter were chosen based on our knowledge of the stable surfaces as well as literature data and will be used for further analysis of the most reliable methodology. This includes the finalization of our preliminary investigation of cell size convergence and a comparison of our accurate density functional theory results with those derived from the computationally much less demanding empirical potentials. Using the latter, we aim to generate an enhanced understanding of the cell size influence, which is not feasible with density functional theory due to cell size limitations. In addition to this, we can obtain a qualitative understanding of the relative stability of the immense number of possible grain boundary structures based on which we will perform highly accurate density functional theory calculations for selected structures to be used as input for phase field simulations.

Once pure nickel grain boundaries have been satisfactorily described, the more complicated nickel-electrolyte grain boundaries can be considered. In a first step, the atomistic models for the electrolyte, an Y-stabilized Zr-oxide, must be developed and a reliable description of the bulk oxide must be ensured. The mixed grain boundaries will then be constructed based on our knowledge of pure surfaces and grain boundaries and their determined energies will be transferred to the phase field simulation. Since the mixed structures are expected to become extremely large, it might be necessary to decrease our strict numerical parameters and rely on less accurate results. Since there is currently no data available on mixed nickel-YSZ grain boundaries, this is still expected to be a highly valuable result.

In addition to interfacial energies, diffusion constants are another important input parameter for the phase field simulations. In a first step, we considered nickel surface self-diffusion, which is expected to play a dominant role based on an internal evaluation of experimental results. Since we expect the migration of adsorbed nickel atoms on the surface to be of crucial importance, we determined the stable adsorption sites of nickel on differently oriented stable nickel surfaces and derived the corresponding atomistic models. In addition to this, we also considered the adsorption of hydrogen on the surface in order to investigate its effect on the diffusion constant. Using the derived atomistic models, we conducted first Nudged Elastic Band calculations to determine the relevant energy barriers. Further research is, however, needed to obtain reliable results, which will be part of this year's activities. It should be noted that the derived diffusion constants will be evaluated in close comparison to available experimental values. Depending on the outcome, an extension or reduction of the methodology (e.g.. explicit calculation of vibrational entropies or use of empirical potentials) should be considered. In view of expected timing and resources, grain boundary diffusion constants will likely not be considered within this project.

Acknowledgements This work was performed on the computational resource ForHLR I funded by the Ministry of Science, Research and the Arts Baden-Württemberg and DFG ("Deutsche Forschungsgemeinschaft") and conducted within the project "Kersolife100", funded by the Federal Ministry for Economic Affairs and Energy (03ET6101A).

References

1. P.E. Blöchl, Projector augmented-wave method. Phys. Rev. B **50**, 17953–17979 (1994)
2. E.A. Clark, R. Yeske, H.K. Birnbaum, The effect of hydrogen on the surface energy of nickel. Metall. Trans. A **11**(11), 1903–1908 (1980)
3. G.I. Csonka, J.P. Perdew, A. Ruzsinszky, P.H.T. Philipsen, S. Lebègue, J. Paier, O.A. Vydrov, J.G. Ángyán, Assessing the performance of recent density functionals for bulk solids. Phys. Rev. B **79**, 155107 (2009)

4. S. De Waele, K. Lejaeghere, M. Sluydts, S. Cottenier, Error estimates for density-functional theory predictions of surface energy and work function. Phys. Rev. B **94**, 235418 (2016)
5. R. Digilov, S. Zadumkin, V. Kumykov, K. Khokonov, Measurement of surface tension of refractory metals in solid state. Fiz. Met. Metalloved. **41**, 979–982 (1976)
6. R.S. Elliott, This is a model driver for the morse pair potential shifted to zero energy at cutoff separation (2014), Online Accessed 07 Sept 2018
7. R.S. Elliott, This is a Ni morse model parameterization by girifalco and weizer using a high accuracy cutoff distance (2014) Online Accessed 07 Sept 2018
8. R.S. Elliott, EAM model driver with hermite cubic spline interpolation (2018) Online Accessed 28 Aug 2018
9. R.S. Elliott, E.B. Tadmor, Knowledgebase of interatomic models application programming interface (2011) Online; Accessed 28 Aug 2018
10. S.M. Foiles, M.I. Baskes, M.S. Daw, Embedded-atom-method functions for the fcc metals Cu, Ag, Au, Ni, Pd, Pt, and their alloys. Phys. Rev. B **33**, 7983–7991 (1986)
11. S.M. Foiles, M.I. Baskes, M.S. Daw, Erratum: Embedded-atom-method functions for the fcc metals Cu, Ag, Au, Ni, Pd, Pt, and their alloys. Phys. Rev. B **37**, 10378–10378 (1988)
12. L.A. Girifalco, V.G. Weizer, Application of the Morse potential function to cubic metals. Phys. Rev. **114**, 687–690 (1959)
13. K.W. Jacobsen, P. Stoltze, J.K. Nørskov, A semi-empirical effective medium theory for metals and alloys. Surf. Sci. **366**(2), 394–402 (1996)
14. G. Kresse, J. Furthmüller, Efficient iterative schemes for ab initio total-energy calculations using a plane-wave basis set. Phys. Rev. B **54**, 11169–11186 (1996)
15. G. Kresse, J. Furthmller, Efficiency of ab-initio total energy calculations for metals and semi-conductors using a plane-wave basis set. Comput. Mater. Sci. **6**(1), 15–50 (1996)
16. G. Kresse, J. Hafner, Ab initio molecular dynamics for liquid metals. Phys. Rev. B **47**, 558–561 (1993)
17. G. Kresse, J. Hafner, Ab initio molecular-dynamics simulation of the liquid-metal-amorphous-semiconductor transition in germanium. Phys. Rev. B **49**, 14251–14269 (1994)
18. G. Kresse, D. Joubert, From ultrasoft pseudopotentials to the projector augmented-wave method. Phys. Rev. B **59**, 1758–1775 (1999)
19. A.H. Larsen, J.J. Mortensen, J. Blomqvist, I.E. Castelli, R. Christensen, M. DuAĆak, J. Friis, M.N. Groves, B. Hammer, C. Hargus, E.D. Hermes, P.C. Jennings, P.B. Jensen, J. Kermode, J.R. Kitchin, E.L. Kolsbjerg, J. Kubal, K. Kaasbjerg, S. Lysgaard, J.B. Maronsson, T. Maxson, T. Olsen, L. Pastewka, A. Peterson, C. Rostgaard, J. Schiotz, O. Schaijtt, M. Strange, K.S. Thygesen, T. Vegge, L. Vilhelmsen, M. Walter, Z. Zeng, K.W. Jacobsen, The atomic simulation environment—a python library for working with atoms. J. Phys.: Condens. Matter **29**(27), 273002 (2017)
20. P.S. Maiya, J.M. Blakely, Surface self a diffusion and surface energy of nickel. J. Appl. Phys. **38**(2), 698–704 (1967)
21. M. Mendelev, M. Kramer, S. Hao, K. Ho, C. Wang, Development of interatomic potentials appropriate for simulation of liquid and glass properties of NiZr2 alloy. Phil. Mag. **92**(35), 4454–4469 (2012)
22. M.I. Mendelev, Finnis-Sinclair potential for the Ni-Zr system developed by Mendelev et al. (2012) (2018) Online; Accessed 07 Sept2018
23. Y. Mishin, EAM Ni potential (2018) Online; Accessed 07 Sept 2018
24. Y. Mishin, D. Farkas, M.J. Mehl, D.A. Papaconstantopoulos, Interatomic potentials for monoatomic metals from experimental data and ab initio calculations. Phys. Rev. B **59**, 3393–3407 (1999)
25. K. Momma, F. Izumi, VESTA3 for three-dimensional visualization of crystal, volumetric and morphology data. J. Appl. Crystallogr. **44**(6), 1272–1276 (2011)
26. S.P. Ong, W.D. Richards, A. Jain, G. Hautier, M. Kocher, S. Cholia, D. Gunter, V.L. Chevrier, K.A. Persson, G. Ceder, Python materials genomics (pymatgen): A robust, open-source python library for materials analysis. Comput. Mater. Sci. **68**, 314–319 (2013)

27. J.P. Perdew, K. Burke, M. Ernzerhof, Generalized Gradient Approximation Made Simple. Phys. Rev. Lett. **77**(18), 3865–3868 (1996)
28. J.P. Perdew, K. Burke, M. Ernzerhof, Generalized Gradient approximation made simple [Phys. Rev. Lett. 77, 3865 (1996)]. Phys. Rev. Lett. **78**(7), 1396–1396 (1997)
29. J.P. Perdew, A. Ruzsinszky, G.I. Csonka, O.A. Vydrov, G.E. Scuseria, L.A. Constantin, X. Zhou, K. Burke, Restoring the density-gradient expansion for exchange in solids and surfaces. Phys. Rev. Lett. **100**, 136406 (2008)
30. G. Pizzi, A. Cepellotti, R. Sabatini, N. Marzari, B. Kozinsky, Aiida: automated interactive infrastructure and database for computational science. Comput. Mater. Sci. **111**, 218–230 (2016)
31. S. Plimpton, Fast parallel algorithms for short-range molecular dynamics. J. Comput. Phys. **117**(1), 1–19 (1995)
32. T. Roth, The surface and grain boundary energies of iron, cobalt and nickel. Mater. Sci. Eng. **18**(2), 183–192 (1975)
33. M.D. Sangid, H. Sehitoglu, H.J. Maier, T. Niendorf, Grain boundary characterization and energetics of superalloys. Mater. Sci. Eng., A **527**(26), 7115–7125 (2010)
34. D. Scheiber, R. Pippan, P. Puschnig, L. Romaner, Ab initiocalculations of grain boundaries in bcc metals. Modell. Simul. Mater. Sci. Eng. **24**(3), 035013 (2016)
35. J. Schiotz, Effective medium theory as implemented in the ase/asap code (2015). Online; Accessed 07 Sept 2018
36. J. Schiotz, Standard effective medium theory potential for face-centered cubic metals as implemented in ase/asap (2015) Online; Accessed 07 Sept 2018
37. D.R. Stickle, J.P. Hirth, G. Meyrick, R. Speiser, A new technique for measuring the effects of oxygen activity on surface energies: Application to nickel. Metall. Trans. A **7**(1), 71–74 (1976)
38. E.B. Tadmor, R.S. Elliott, J.P. Sethna, R.E. Miller, C.A. Becker, The potential of atomistic simulations and the Knowledgebase of Interatomic Models. JOM **63**(7), 17–77 (2011)
39. A. Tehranchi, A modification of the angelo et al. ni-h potential which enhances the binding energies of h atoms to the gbs in nickel (2018) Online; Accessed 28 Aug 2018
40. R. Tran, Z. Xu, B. Radhakrishnan, D. Winston, W. Sun, K.A. Persson, S.P. Ong, Surface energies of elemental crystals. Sci. Data **3** (2016)
41. W. Tyson, W. Miller, Surface free energies of solid metals: Estimation from liquid surface tension measurements. Surf. Sci. **62**(1), 267–276 (1977)
42. L. Vitos, A. Ruban, H. Skriver, J. Kollar, The surface energy of metals. Surf. Sci. **411**(1), 186–202 (1998)

KKRnano: Quantum Description of Skyrmions in Chiral B20 Magnets

Marcel Bornemann, Paul F. Baumeister, Rudolf Zeller, and Stefan Blügel

Abstract We present the linear-scaling electronic-structure code KKRnano which is useful to perform Density-Functional-Theory (DFT) calculations for systems containing a large number of atoms. The code is based on a modified Korringa-Kohn-Rostoker (KKR) Green-function scheme and allows us to treat systems with up to thousands of atoms per unit cell. The capability of the code for the treatment of non-collinear alignments of atomic spins is used for the investigation of nanometer-sized magnetic textures in the germanide B20-MnGe, a material that is considered to play an important role in future spintronic devices. A performance analysis of KKRnano on Hazel Hen shows an efficient scaling behaviour with increasing system size and demonstrates an extensive integration of highly optimized libraries.

1 Introduction

We have developed a unique electronic structure code, KKRnano [1–3], specifically designed for petaFLOP computing. Our method scales linearly with the number of atoms, so that we can realize system sizes of up to half a million atoms in a unit cell if necessary. Recently, we implemented a relativistic generalization of our

M. Bornemann (✉) · S. Blügel
Peter Grünberg Institute and (PGI-1) and Institute for Advanced Simulation (IAS-1),
Forschungszentrum Jülich, 52425 Jülich, Germany
e-mail: m.bornemann@fz-juelich.de

S. Blügel
e-mail: s.bluegel@fz-juelich.de

M. Bornemann
RWTH Aachen University, 52056 Aachen, Germany

P. F. Baumeister
Jülich Supercomputing Centre, Forschungszentrum Jülich, 52425 Jülich, Germany
e-mail: p.baumeister@fz-juelich.de

R. Zeller
Institute for Advanced Simulation (IAS-1), Forschungszentrum Jülich, 52425 Jülich, Germany
e-mail: ru.zeller@fz-juelich.de

© Springer Nature Switzerland AG 2021   191
W. E. Nagel et al. (eds.), *High Performance Computing in Science and Engineering '19*,
https://doi.org/10.1007/978-3-030-66792-4_14

algorithm enabling us to calculate complex non-collinear magnetic structures, such as skyrmions, in real space. Skyrmions are two-dimensional magnetization solitons, i.e., two-dimensional magnetic structures localized in space, topologically protected by a non-trivial magnetization texture, which has particle-like properties. The focus of our work is on the germanide MnGe that is particularly interesting among the chiral magnetic B20 compounds, as it exhibits a three-dimensional magnetic structure that is not yet understood [4, 5].

This report is structured as follows: In Sect. 2 we give a brief introduction on the Korringa-Kohn-Rostoker (KKR) method and explain how linear scaling is achieved. Our results on B20-MnGe are presented in Sect. 3. In Sect. 4 and Sect. 5 the focus is on code-specific issues and in Sect. 6 on improvements made for the electrostatic solver. The parallelization scheme of KKRnano is described in detail and a performance analysis of our code on Hazel Hen is conducted.

2 Numerical Methods and Algorithms

Contrary to other Density Functional Theory (DFT) codes, which determine the Kohn-Sham orbitals by solving the Kohn-Sham differential wave equation, in KKR-nano the Green function for the Kohn-Sham equation is obtained by solving an integral equation in real space. In this so called Korringa-Kohn-Rostoker (KKR) scheme, space is divided into non-overlapping cells around the atoms, and the calculations are separated into single-cell parts where only the potential within the individual cell enters, and a large complex linear matrix equation for the matrix elements $G_{LL}^{nn'}(\epsilon)$ of the Green function $G(r, r, \epsilon)$:

$$G_{LL'}^{nn'}(\epsilon) = G_{LL'}^{r,nn'}(\epsilon) + \sum_{n''} \sum_{L''} G_{LL''}^{r,nn''}(\epsilon) \sum_{L'''} \Delta t_{L''L'''}^{n''}(\epsilon) G_{L'''L'}^{n''n'}(\epsilon). \qquad (1)$$

In the following we refer to $\Delta t_{L''L'''}^{n''}(\epsilon)$ as the Δt-matrices and $G_{LL''}^{r,nn''}(\epsilon)$ as the reference Green functions. The problem given in Eq. (1) is identical to solving a complex linear matrix equation of the form

$$\mathbf{Ax} = \mathbf{b}. \qquad (2)$$

As is well known, this problem scales as $O(N^3)$, where N denotes the dimension of the $N \times N$ matrix $\mathbf{A}$ when standard solving schemes are used. However, by choosing a reference system of repulsive potentials, $\mathbf{A}$ can be made sparse and this can be exploited by using customized iterative algorithms for solving linear sparse matrix equations. Our method uses the transpose-free quasi-minimal-residual (TFQMR) algorithm [6] which enables us to achieve $O(N^2)$ scaling. Convergence of the TFQMR iterations is achieved by working in the complex energy plane. Both concepts, the repulsive reference system and application of complex energy, were

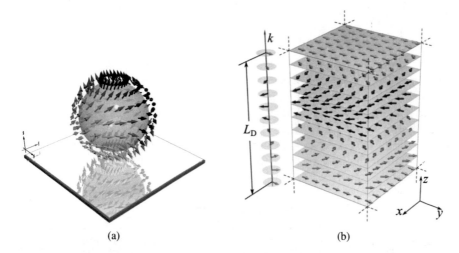

Fig. 1 Magnetic textures that are found experimentally in B20-MnGe: **a** Magnetic anti-hedgehog texture that is wrapped around a singularity at the center. Published with kind permission of Nikolai Kiselev 2018. All Rights Reserved. **b** Helical spin spiral that propagates in (001) direction. Reprinted from [5] and licensed under CC BY 3.0

introduced by us into the KKR method [7, 8] and are used worldwide for many years. By truncating inter-atomic interactions above a certain distance, an $O(N)$ scaling behavior can be realized, and this feature should be used in any calculation involving more than 1000 atoms. In most applications the TFQMR solver accounts for the major part of the computational work. For super-large-scale calculations (above 100,000 atoms) the electrostatics solver was identified to require a non-negligible amount of computing resources [9], a drawback which could be alleviated considerably by recent algorithmic improvements (see Sect. 6).

3 Complex Magnetic Textures in B20-MnGe

B20-MnGe is a good candidate material to be investigated with the new version of KKRnano which now contains the feature of non-collinear magnetism and spin-orbit coupling. In a recent study [4], it was found by transmission electron microscopy that 3D magnetic objects exist in B20-MnGe. The authors of [4] came to the conclusion that their data indicates a cubic lattice of skyrmionic hedgehogs and anti-hedgehogs (see Fig. 1a) with a lattice constant of about 3–6 nm. The singularity at the center of the texture exists only in the micromagnetic description since the atomic magnetic moment of the atom, which is located at the center, remains finite [10]. Findings by Kanazawa et al. suggest that the lattice is set up by a superposition of three orthogonal helical structures, also referred to as 3Q state [11]. Here, the local magnetization is given by

$$\mathbf{M}(\mathbf{r}) = \begin{pmatrix} \sin qy + \cos qz \\ \sin qz + \cos qx \\ \sin qx + \cos qy \end{pmatrix}, \tag{3}$$

where $q = \frac{2\pi}{\lambda}$ is the wavenumber given in terms of the helical wavelength λ and x, y and z are the spatial coordinates within the unit cell. Note, that $\mathbf{M}(\mathbf{r})$ is not normalized. In contrast to other systems exhibiting a similar magnetic phase, the rather short helical wavelength in B20-MnGe allows one to perform Density Functional Theory (DFT) calculations with KKRnano. B20-MnGe is currently the subject of extensive investigation [4, 12–15]. This is mainly inspired by the discovery of skyrmions as small information-carrying particles that could potentially be used in spintronic devices [16]. At present there is a lack of a convincing explanation of what is observed in experiment. Research in the framework of micromagnetic models identified both magnetic frustration (RKKY interaction) as well as spin-orbit coupling induced Dzyaloshinskii-Moriya (DM) interaction as potentially crucial to a better understanding [17, 18]. While the 3Q state certainly constitutes the most interesting non-trivial magnetic texture in B20-MnGe, there are also reports that the magnetic ground state in this system is actually a helical spiral (see Fig. 1b) [19] which was observed up to a temperature of 170 K [20]. These two observations are clearly contradictory and it has not yet been explained how both can coexist within the same material. The helical spin spiral in B20-MnGe, the so-called 1Q state, forms along the (001) direction and is described by the relation

$$\mathbf{M}(\mathbf{r}) = \begin{pmatrix} \cos qz \\ -\sin qz \\ 0 \end{pmatrix}. \tag{4}$$

We started our investigations by considering a $6 \times 6 \times 6$ B20-MnGe supercell (1728 atoms), where we use PBEsol as exchange-correlation functional and include only a single k-point, i.e., the Γ-point. In this initial comparison of ferromagnetic (FM), 1Q and 3Q state the respective states are imposed on the system by forcing the atomic exchange-correlation B-fields to point into specific directions. For the equilibrium lattice constant $a = 4.80$ Å, the ferromagnet is the state with the lowest total energy. As this contradicts experimental observations, we take into consideration that in experiment the crystal structure might inadvertently differ from the ideal structure. Such discrepancies can for instance be caused by strain that originates from the manufacturing process of the sample. Therefore, it is reasonable to check how the results change, when the lattice constant is varied. In the upper part of Fig. 2 the calculated total energy is shown as function of the lattice constant for FM, 1Q and 3Q state. Clearly, neither the 1Q nor the 3Q state constitutes the ground state, when the experimental lattice constant is assumed. Yet, by increasing or decreasing the lattice constant the energetic difference can be made smaller. We focus on an increase of the lattice constant rather than a decrease since the system goes into the low-spin state below $a = 4.65$ Å [21] and according to experiment the non-trivial textures exist in the high-spin regime. A crucial transition point is found around $a = 5.0$ Å,

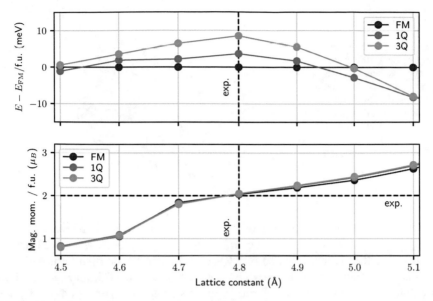

Fig. 2 Comparison of Ferromagnetic (FM), helical spiral (1Q) and hedgehog lattice (3Q) state with KKRnano. Top: Difference of total energies with the FM state as reference state for different lattice constants. The experimental lattice constant is $a = 4.80$ Å. 1Q and 3Q state are energetically preferable for $a \geq 5.0$ Å. Bottom: Magnetic moment per Mn atom increases with lattice constant. High-spin/Low-spin transition is clearly visible between $a = 4.60$ and $a = 4.70$ Å. Experimentally, the magnetic moment is measured to be $\approx 2\mu_B$. Published with kind permission of Marcel Bornemann 2018. All Rights Reserved

where by imposing the 1Q or 3Q state the energy can be made smaller than for the ferromagnetic state. In general, for $a \geq 5.0$ Å both helical states are favored over the ferromagnetic one. Obviously, an artificial increase of the lattice constant by 0.2 Å ($\approx 4\%$) or more is fairly large. However, probes in experiment are seldom if ever perfectly clean and impurities in the sample need to be considered as a source of error in the final analysis. One potential effect of impurities is chemical pressure that causes a spatial expansion of the lattice structure. An example of the possible effects of positive chemical pressure can be found in Co-doped B20-FeGe [22]. Here, it was experimentally observed that doping can increase the melting temperature and change the magnetic properties of a B20 alloy. In the lower part of Fig. 2, the evolution of the magnetic moment with varying lattice constant is tracked. The resulting magnetic moment for the experimental lattice constant nicely agrees with the magnetic moment of approximately $2\mu_B$/f.u. reported by experimentalists [19]. Furthermore, the high-spin/low-spin transition is recognizable between $a = 4.60$ and $a = 4.70$ Å. It can also be observed that the magnetic moment increases, when the lattice constant is increased. This is a common behaviour often observed in metallic systems. For larger lattice constants the magnetic moments of the three different magnetic textures differ more than for smaller lattice constants. This might be connected to the observation of the differences in the total energy.

Fig. 3 Illustration of a
Bloch point in real space
with all magnetic moments
pointing out of the center of
the Bloch sphere. We use the
same magnetic configuration
but invert the spin direction
so that all moments point
into the center. Published
with kind permission of
Nikolai Kiselev 2018. All
Rights Reserved

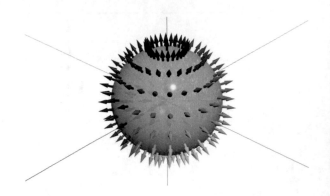

We continued our investigation of B20-MnGe by studying the influence of spin-orbit coupling (SOC) enhancement on the magnetic energy landscape. An advantage of the way SOC is implemented in KKRnano is that its contribution is added to the scalar-relativistic potential in a perturbation-like manner. This allows to scale its strength and make it artificially stronger. In this context we perform a comparison of the Bloch point (BP) state and the ferromagnetic (FM) state. The BP texture is defined by means of the four spherical parameters ϕ, θ, Φ and Θ. ϕ and θ designate the position of an **individual atom** in the unit cell which is described by the radius r and the common polar and azimuthal angle

$$\phi = \arctan(y/x) \tag{5}$$

and

$$\theta = \arccos\left(\frac{z}{\sqrt{x^2 + y^2 + z^2}}\right). \tag{6}$$

Here, we define the origin of the coordinate system, i.e., the tuple ($x = 0$, $y = 0$, $z = 0$), to be at the center of the unit cell. In this frame of reference, all atoms that lay in an x-y-plane that intersects with the center are described by $\theta = \pi/2$. The orientation of the **individual atomic magnetic moments** for a BP texture is then defined by the polar angle

$$\Phi = \phi + \phi_1 \tag{7}$$

and the azimuthal angle

$$\Theta = 2\arctan\left(\cot\frac{\theta}{2}\right), \tag{8}$$

where the angles designating the atomic position enter as arguments. ϕ_1 is a phase factor. An illustration of a BP is given in Fig. 3. Note, that in contrast to that illustration we conduct our investigation for a BP with $\phi_1 = \pi$, where magnetic moments are inverted, i.e., all moments point into instead of out of the center.

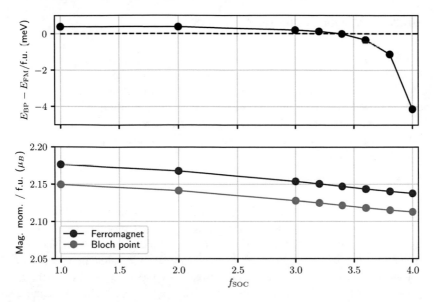

Fig. 4 Effect of increased SOC on B20-MnGe in a $6 \times 6 \times 6$ supercell. Top: Total energy difference between (relaxed) Bloch point and (relaxed) ferromagnet. Bottom: Magnetic moment of ferromagnet and Bloch point state. Published with kind permission of Marcel Bornemann 2018. All Rights Reserved

For our calculations we again use a $6 \times 6 \times 6$ supercell but this time with a $2 \times 2 \times 2$ k-point-mesh and LDA as exchange-correlation functional. Here, we choose LDA because it has been used extensively in all KKR codes in the past and we want to eliminate the possibility of numerical problems that could occur when SOC is artificially enhanced. The magnetic moments are allowed to relax during the convergence process. This leads to a small canting of the moments which is a known effect in B20 materials [23]. A series of calculations is conducted ranging from the physical value of the SOC to an enhancement of it by a factor of $f_{SOC} = 4.0$ (see Fig. 4). As could be expected from the investigation of 1Q and 3Q state before, the BP state is energetically not preferred over the FM state for a small scaling of SOC. However, when SOC is scaled further up to $f_{SOC} = 3.5$, both states are energetically more or less equivalent. Above $f_{SOC} = 3.5$ the BP state is clearly preferred over the FM state with an energy difference of up to 4 meV/f.u. Within the parameter range that we checked in this study, the most beneficial scaling value for the BP is found to be $f_{SOC} = 4.0$. The effect of SOC scaling on the magnetic moment can be deemed negligible. Over the whole range it decreases by 0.07 μ_B for each of the two states (see again Fig. 4).

4 Parallelization Scheme of KKRnano

In order to use the full potential of modern supercomputers, KKRnano features both MPI and OpenMP parallelization. The parallelization concept is sketched in Fig. 5. There are three MPI levels to parallelize over atoms, spins and energy points. OpenMP is used in various parts of the code, mainly to parallelize important loops. The single-cell solver for calculations involving non-collinear magnetism and the TFQMR solver that solves the Dyson equation in Eq. (1) are the parts that potentially benefit the most from OpenMP.

MPI parallelization over atoms

The most crucial MPI level in KKRnano is the one for atoms since the iterative KKR formalism allows us to solve the multiple-scattering problem locally for each atom, if the Green functions of the reference system and the Δt-matrices of the other scattering centers are provided. These quantities are calculated by each task for the atoms it is responsible for and then broadcasted to the other tasks that require them. Afterwards, the Dyson equation can be solved independently by each task. The charge density in each atomic cell and the local charge moments can also be calculated independently since only the diagonal nn-elements of the Green function are needed for this. In order to subsequently obtain the potential the local charge moments are broadcasted to the other tasks that require them for the solution of the Poisson equation.

MPI parallelization over spin channels

If the system of interest is a collinear magnet, the two spin channels can be handled by two distinct MPI tasks since the magnetic Kohn-Sham equations are separable. Due to the relatively small additional MPI communication effort, this yields an almost ideal speed-up by a factor of two. It should be noted that in the non-collinear KKR formalism such a separation of spin channels is no longer possible because all operations involving the Δt-matrices and Green functions need to be performed in full spin space, i.e., $\{\uparrow\uparrow, \uparrow\downarrow, \downarrow\uparrow, \downarrow\downarrow\}$. Thus, the spin parallelization can only be used in connection with a collinear calculation.

MPI parallelization over energy points

The requirement to calculate the Green function at different energy points offers another possibility for parallelization since the values of the Green function $G(\epsilon_i)$ at energy points ϵ_i can be obtained independently. Since the number of TFQMR iterations considerably depends on the imaginary part of ϵ_i, a good load balancing is achieved by grouping the energy points into two or three groups such that the sum of the walltime used in each group is approximately equal. The grouping is dynamically updated during the self-consistency cycle by using the times measured for each energy point in each self-consistency step.

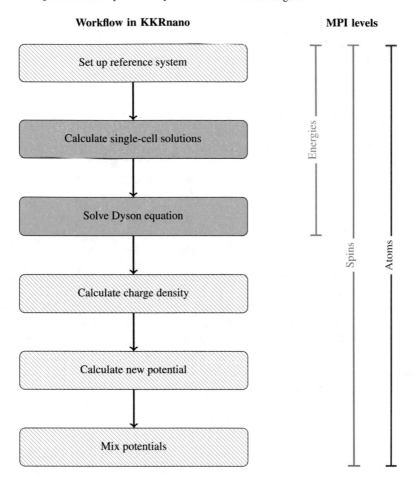

Fig. 5 Schematic representation of MPI and OpenMP parallelization in KKRnano. The most important steps in the KKR workflow are depicted on the left side and the three MPI regions over atoms, spins and energy points are indicated on the right side. Parts filled with blue comprise routines where OpenMP is used and where this can be of high importance to the overall performance while in the striped blue parts OpenMP is used but is less significant. Published with kind permission of Marcel Bornemann 2018. All Rights Reserved

OpenMP parallelization

KKRnano can be compiled either with or without support for OpenMP. If OpenMP is enabled, loops primarily in the TFQMR solver but also in the routines that calculate the single-cell solutions are executed using parallel threads. This is particularly useful on architectures that support simultaneous multithreading (SMT). However, we try to use BLAS (Basic Linear Algebra Subprograms) library routines for arithmetic operations, e.g., matrix-matrix multiplications, throughout our code. BLAS libraries usually have their own built-in SMT support. Therefore, the available SMT threads must be partitioned between the explicit OpenMP parallel regions in the code and the implicit parallelization of the BLAS library. Here, the optimal partitioning is highly architecture-dependent and a general recommendation cannot be given. On Hazel Hen the best performance is achieved, when all threads are used by the BLAS library.

Multiple Atoms per MPI Task

Assigning multiple atoms to one MPI task is beneficial, if large systems are calculated on a comparatively small allocation of compute nodes. KKRnano can treat multiple atoms per atomic MPI task by using the following algebraic scheme: We rewrite the Dyson equation (see Eq. 1 and Eq. (2)) with indices as

$$\sum_{\mu}\sum_{L'} A_{LL'}^{\nu\mu} X_{L'L''}^{\mu} = b_{LL''}^{\nu}, \tag{9}$$

where ν and μ indicate atomic indices and L, L' and L'' denote the angular momentum components. In order to describe the dimensions of the constituents of the equation above, we declare the following parameters: N_{cl} is the number of atoms in the reference cluster that defines the reference Green function. N_{tr} is the number of atoms in the truncation cluster. Only interactions with atoms that lay within the truncation cluster of an atom are considered. This is essential to achieve the linear-scaling of the multiple-scattering problem. All quantities are expanded in spherical harmonics up to l_{max}, which is usually fixed to $l_{max} = 3$. Then A in Eq. (9) is of dimension $N_{cl}(l_{max} + 1)^2 \times N_{tr}(l_{max} + 1)^2$ and X of dimension $N_{tr}(l_{max} + 1)^2 \times (l_{max} + 1)^2$. For more than a single atom per task, X is a matrix of dimension $N_{tr}(l_{max} + 1)^2 \times N_{loc}(l_{max} + 1)^2$, where N_{loc} is the number of atoms treated by one MPI task of the atom parallelization level. The corresponding linear system (with the index γ running over N_{loc}) is then

$$\sum_{\mu}\sum_{L'} A_{LL'}^{\nu\mu} X_{L'L''}^{\mu\gamma} = b_{LL''}^{\nu\gamma}. \tag{10}$$

5   Performance of KKRnano on Hazel Hen

In this section we present performance results of KKRnano that were obtained on Hazel Hen. A runtime analysis of the parts that are related to the specific physical problems is given which is followed by an overview of how much time is spent in distinct classes of routines, e.g., library calls and MPI communication. The weak-scaling benchmarks on Hazel Hen are performed for B20-MnGe with the magnetic moments being treated as non-collinear. For this, KKRnano is compiled with the Intel Fortran compiler with optimization level O_2 and linked to the Intel Math Kernel (MKL) library which handles all BLAS calls. The runs are conducted with a task

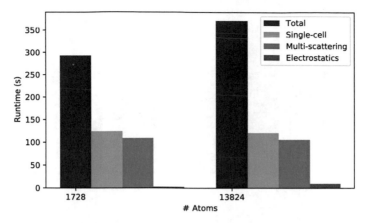

Fig. 6 Runtime of a single DFT self-consistency step for different B20-MnGe supercell sizes with KKRnano. The total runtime and the individual runtimes for the single-cell solver, the multiple-scattering solver and the electrostatics solver are given. Published with kind permission of Marcel Bornemann 2018. All Rights Reserved

distribution of 24 MPI processes per node. The OpenMP threads are reserved for the multi-threaded MKL library. We choose two supercells containing 1728 and 13824 atoms and perform one self-consistency step for each system. We only activate the MPI parallelization over atoms and not the ones over spins and energies. Then, 72 and 576 nodes are needed for the two benchmark calculations, since each MPI task is assigned to one atom. Each MPI process reads four shared binary direct-access files before the self-consistency step. Two of them are index-files of the size of a few hundred kByte while the other two can grow with system size up to several GByte. The combined size of the files is roughly $N \times 0.5$ MBytes, where N is the number of atoms. After the single self-consistency step the potential is written out. This and the read-in of the four files mentioned above accounts for most of the time spent in I/O. Fig. 6 shows the total runtimes of a single self-consistency iteration and how much time is spent on solving the individual physical problems for the two supercells containing 1728 and 13824 atoms. The remaining time can be mainly attributed to Fortran direct-access I/O which does not scale well on a Lustre file system. Its increase for the larger system probably arises from the amount of file accesses which is proportional to the number of MPI processes. The implementation of a more suitable I/O library (e.g., MPI I/O or SIONlib) is likely to solve this issue. The single-cell and multiple-scattering solvers are expected to account for most of the computational work in KKRnano which is why linear scaling is of particular importance in these parts of the code. Our results show that both parts scale indeed linearly, i.e., the runtime stays constant as long as the ratio of CPU cores to atoms remains constant. Furthermore, it should be noted that the walltime needed for one self-consistency step of a system with 1728 atoms is less than five minutes which is a good value, especially when considering that the magnetic moments are allowed to be non-collinear. For larger systems the electrostatic Poisson solver contributes

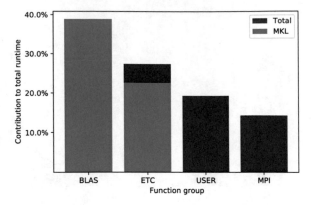

Fig. 7 Runtime results from Cray Performance Measurement Tool for a single KKRnano self-consistency step for a 6 × 6x6 supercell of B20-MnGe (1728 atoms). 72 Hazel Hen nodes are used with 1728 MPI processes. The runtimes are divided into four subgroups which are described in the text. The fraction of the total runtime in each subgroup that is spent in routines of Intel's highly-optimized MKL library is highlighted in blue. Published with kind permission of Marcel Bornemann 2018. All Rights Reserved

considerably to the overall runtime [9], while its impact for the mid-sized systems investigated here is negligible. With the recent algorithmic improvements (see Sect. 6) its impact also for very large systems is expected to be small enough.

Next, we conduct a performance analysis on the level of individual routines. The Cray Performance Tools package is helpful for this purpose. It allows to sample the time that is spent within each routine and therefore gives a good impression of which routines are performance-critical and should be considered for further optimization. Fig. 7 shows the sampling results for a 6 × 6 × 6 supercell calculation using 1728 MPI tasks on 72 nodes. The routines are grouped according to their origin. The *BLAS* group comprises the kernels of library-based linear algebra operations. All external routines for which no source file was given at compile time are subsumed in *ETC*. Such routines are, e.g., Fortran I/O routines and MKL helper routines that are indispensable for BLAS but not directly related to the MKL kernel. The *USER* group accounts for all routines from the KKRnano source code, i.e., our self-written routines. Finally, *MPI* contains all routines needed for communication and synchronization between the MPI tasks. Performing linear algebra operations accounts for almost 40% of the total runtime. Prior tests have shown that Intel's MKL outperforms Cray's LibSci and therefore KKRnano is linked to the former. The routines that fall into the *ETC* category consume 28% of the total runtime. The MKL routines dominate this group and it can be stated that the ratio of MKL in the total runtime is 62%. This is remarkable as it indicates that more than half of the runtime is spent in routines that are already highly optimized. The only group of routines that can benefit from source code tuning is the *USER* group. However, a hypothetical performance improvement by a factor of two would reduce the overall runtime by merely 10% because this group only needs 20% of the total runtime. MPI routines are not crucial to performance as they amount to only 15% of the total computational effort.

6 Electrostatic Solver

In KKRnano the electrostatic potential in each cell i is separated into a part, which arises from the density $\rho^i(\mathbf{r})$ in the same cell, and into a part, which arises from the density $\rho^j(\mathbf{r})$ in all other cells. The intercell part is expanded in spherical harmonics as $V^i(\mathbf{r}) = \sum_{lm} a^i_{lm} r^l Y_{lm}(\mathbf{r})$ with

$$a^i_{lm} = \sum_{j \neq i} \frac{4\pi}{2l+1} \int_{\text{cell } j} d\mathbf{r}' \rho^j(\mathbf{r}') \frac{Y_{lm}(\mathbf{r}' + \mathbf{R}^j - \mathbf{R}^i)}{|\mathbf{r}' + \mathbf{R}^j - \mathbf{R}^i|^{l+1}} \tag{11}$$

where $\mathbf{R}^i$ and $\mathbf{R}^j$ denote the cell centers and the sum is over all atoms in the supercell and in the periodic images of the supercell. The sum over the periodic images, which extremely slowly converges, is transformed by Ewald's method into two exponentially converging sums over real and reciprocal lattice vectors. The efficiency of Ewald's method depends on the optimal choice of the Ewald splitting parameter η and the optimal numbers N_R and N_G of real and reciprocal lattice vectors. This aspect was not addressed in [9] and for the results shown in Fig. 7. Recently, the behaviour of the electrostatic solver on Hazel Hen was thoroughly investigated for large systems containing 32 to 131072 atomic sites. The systems studied contained one empty site describing a vacancy while all other sites were occupied by Cu atoms. The density and the potential for these systems were determined self-consistently and the smallest numbers N_R and N_G were chosen which preserve averaged accuracies of 10^{-7} Ry for the potential. The main time is used for the setup of Madelung constants before the self-consistency cycle. These setup times, i.e., CPU times needed using one MPI task for each atomic site, are shown in Table 1. An unfavorable choice of η as $\eta_{\text{cell}} = \sqrt{\pi}/a$, where a is the edge length of the basic four-atom cube used to construct the supercells, leads to the times in the first row, which are large and increase approximately as $N_{\text{sites}}^{1.6}$. In the literature, often $N_R \approx N_G$ is assumed which gives the *optimal* η as $\eta_{\text{lit}} = \sqrt{\pi}/\Omega^{1/3}$, where Ω is the volume of the supercell. This leads to the times in the second row, which are considerably smaller and increase only proportionally to N_{sites}. Another improvement, shown in the third row, is obtained by omitting contributions in the real sum from neighboring supercells if $|\mathbf{R}^i - \mathbf{R}^j|$ is larger than a cutoff value and by eliminating every second term in the reciprocal sum with the help of the parity of spherical harmonics. Finally, by choosing $N_R < N_G$, to take into account that terms in the reciprocal sum can be calculated faster, η can be tuned to gain another factor of four (see times in the last row) making calculations with more than 100000 atoms feasible.

Table 1 Setup times for the Madelung constants (in seconds) for different number of atomic sites N_{sites} and different choice of the Ewald splitting parameter

N_{sites}	256	2048	16384	131072
η_{cell}	0.349	23.027	643.528	>10000
η_{lit}	0.119	0.990	7.918	61.933
η_{lit}	0.099	0.807	6.338	49.454
η_{opt}	0.027	0.192	1.665	13.173

7 Conclusions

In this report we introduced an extended Korringa-Kohn-Rostoker method as the foundation of our large-scale Density Functional Theory application KKRnano. Consecutively, we presented results that were obtained on Hazel Hen by applying KKR-nano to the material B20-MnGe. These results, which are partly published in a recent paper [24], are promising and suggest further investigation of complex magnetic textures in this particular compound. The high-clocked Intel Xeon processors of Hazel Hen make a single DFT self-consistency step for a supercell of 1728 atoms feasible within less than five minutes and hence allows one to obtain results in a timely manner.

During the *Optimization of Scaling and Node-level Performance on Hazel Hen* workshop at HLRS in 2018 we were given the chance to further optimize our code and were able to identify a computational bottleneck. The performance of KKRnano on Hazel Hen was significantly improved by resolving it. The encountered problem was connected to unnecessary copying of a large complex-valued array, when the magnetic moments are allowed to be non-collinear. All results in Sect. 5 were obtained with this optimized version. Further optimization for standard calculations with a few thousand atoms is difficult since only 20% of the runtime is spent in hand-written code. The remaining 80% is spent in library calls that are already highly tuned. These routines mainly deal with MKL-BLAS and MPI communication.

Acknowledgements The authors gratefully acknowledge the Gauss Centre for Supercomputing e.V. (www.gauss-centre.eu) for funding this project by providing computing time through the project GCS-KKRN on the GCS Supercomputer Hazel Hen at Hchstleistungsrechenzentrum Stuttgart (HLRS).

References

1. R. Zeller, J. Phys.: Condens. Matter **20**, 294215 (2008)
2. A. Thiess, R. Zeller, M. Bolten, P.H. Dederichs, S. Blügel, Phy. Rev. B **85**, 235103 (2012)
3. M. Bornemann, Large-scale Investigations of Non-trivial Magnetic Textures in Chiral Magnets with Density Functional Theory. PhD thesis, RWTH Aachen University, Aachen. Submitted

4. T. Tanigaki, K. Shibata, N. Kanazawa, X. Yu, Y. Onose, H.S. Park, D. Shindo, Y. Tokura, Nano Lett. **15**, 5438 (2015)
5. F.N. Rybakov, A.B. Borisov, S. Blügel, N.S. Kiselev, New J. Phys. **18**, 045002 (2016)
6. R.W. Freund, N.M. Nachtigal, Numer. Math. **60**, 315 (1991)
7. R. Zeller, J. Deutz, P. Dederichs, Solid State Commun. **44**, 993 (1982)
8. R. Zeller, P.H. Dederichs, B. Újfalussy, L. Szunyogh, P. Weinberger, Phy. Rev. B **52**, 8807 (1995)
9. D. Brömmel, W. Frings, B.J.N. Wylie, JUQUEEN Extreme Scaling Workshop 2017. Tech. rep. (2017)
10. E. Feldtkeller, IEEE Trans. Magn. **53**, 1 (2017)
11. N. Kanazawa, S. Seki, Y. Tokura, Adv. Mater. **29**, 1603227 (2017)
12. N. Kanazawa, Y. Onose, T. Arima, D. Okuyama, K. Ohoyama, S. Wakimoto, K. Kakurai, S. Ishiwata, Y. Tokura, Phys. Rev. Lett. **106**, 156603 (2011)
13. N. Kanazawa, J.H. Kim, D.S. Inosov, J.S. White, N. Egetenmeyer, J.L. Gavilano, S. Ishiwata, Y. Onose, T. Arima, B. Keimer, Y. Tokura, Phy. Rev. B **86**, 134425 (2012)
14. S.V. Grigoriev, N.M. Potapova, S.A. Siegfried, V.A. Dyadkin, E.V. Moskvin, V. Dmitriev, D. Menzel, C.D. Dewhurst, D. Chernyshov, R.A. Sadykov, L.N. Fomicheva, A.V. Tsvyashchenko, Phys. Rev. Lett. **110**, 207201 (2013)
15. N. Martin, M. Deutsch, F. Bert, D. Andreica, A. Amato, P. Bonfà, R. De Renzi, U.K. Röler, P. Bonville, L.N. Fomicheva, A.V. Tsvyashchenko, I. Mirebeau, Phy. Rev. B **93**, 174405 (2016)
16. A. Fert, N. Reyren, V. Cros, Nat. Rev. Mater. **2**, 17031 (2017)
17. E. Altynbaev, S.A. Siegfried, E. Moskvin, D. Menzel, C. Dewhurst, A. Heinemann, A. Feoktystov, L. Fomicheva, A. Tsvyashchenko, S. Grigoriev, Phys. Rev. B **94**, 174403 (2016)
18. T. Koretsune, N. Nagaosa, R. Arita, Sci. Rep. **5**, 13302 (2015)
19. A. Yaouanc, P. Dalmas de Réotier, A. Maisuradze, B. Roessli, Phy. Rev. B **95**, 174422 (2017)
20. O.L. Makarova, A.V. Tsvyashchenko, G. Andrc, F. Porcher, L.N. Fomicheva, N. Rey, I. Mirebeau, Phys. Re. B **85**, 205205 (2012)
21. U.K. Röler, J. Phys: Conf. Ser. **391**, 012104 (2012)
22. M.J. Stolt, X. Sigelko, N. Mathur, S. Jin, Chem. Mater. **30**, 1146 (2018)
23. V.A. Chizhikov, V.E. Dmitrienko, Phys. Rev. B **88**, 214402 (2013)
24. M. Bornemann, S. Grytsiuk, P.F. Baumeister, M. dos Santos Dias, R. Zeller, S. Lounis, S. Blügel, J. Phy.: Condensed Matter **31**, 485801 (2019)

Reactive Flows

Dietmar Kröner

The chapter about Reactive flows consists of two contributions:
"Towards full resolution of spray breakup in flash atomization conditions using DNS"
by D. Loureiro, J. Reutzsch, A. Kronenburg, B. Weigand, K. Vogiatzaki
and
"Enhancing OpenFOAM's Performance on HPC Systems"
by Thorsten Zirwes, Feichi Zhang, Jordan A. Denev, Peter Habisreuther, Henning Bockhorn, and Dimosthenis Trimis
These papers are related to the DNS for droplet formations and some guidelines for improving the performance of OpenFOAM, respectively.[1]

In the first contribution the authors consider the ignition of rocket thrusters in orbits. This requires injection of cryognic propellants into the combustion chamber. The chamber's initial very low pressure leads to flash boiling that will then determine the dynamics of the spray breakup, the mixing of fuel and oxidizer, the reliability of the ignition and the subsequent combustion process. As details of the spray breakup process of cryogenic liquids under flash boiling conditions are not yet well understood, the authors use direct numerical simulations (DNS) to simulate the growth, coalescence and bursting of vapor bubbles in the superheated liquid that leads to the primary breakup of the liquid oxygen jet. Considering the main breakup patterns and droplet formation mechanisms for a range of conditions, the authors evaluate the effectiveness of the volume of fluid (VoF) method together with the continuum surface stress (CSS) model to capture the breakup of thin lamellae formed at high Weber numbers between the merging bubbles. A grid-refinement study indicates convergence of the mass averaged droplet sizes towards an a priori estimated droplet diameter. The order of magnitude of this diameter can be estimated based on thermodynamic conditions.

In the second contribution general guidelines for improving OpenFOAM's performance on HPC clusters are given. A comparison of the serial performance for different compilers shows that the Intel compiler generally generates the fastest

---

[1] Dietmar Kröner, Abteilung für Angewandte Mathematik, Universität Freiburg, Freiburg, Germany,
e-mail: dietmar.kroener@googlemail.com

executables for different standard applications. More aggressive compiler optimization options beyond O3 yield performance increases of about 5 to improvements of up to 25 not lead to a performance gain. The parallel scaling behavior of reactive flow solvers shows an optimum at 5000 cells per MPI rank in the tested cases, where caching effects counterbalance communication overhead, leading to super linear scaling. In addition, two self-developed means of improving performance are presented: the first one targets OpenFOAM's most accurate discretization scheme "cubic". In this scheme, some polynomials are unnecessarily re-evaluated during the simulation. A simple change in the code can reuse the results and achieve performance gains of about 5Score-P/Vampir and load imbalances due to the computation of the chemical reaction rates are identified. A dynamic-adaptive load-balancing approach has been implemented for OpenFOAM's reacting flow solvers which can decrease computation times by 40. Both projects have been supported by the Deutsche Forschungsgemeinschaft.

Towards Full Resolution of Spray Breakup in Flash Atomization Conditions Using DNS

D. D. Loureiro, J. Reutzsch, A. Kronenburg, B. Weigand, and K. Vogiatzaki

Abstract Ignition of rocket thrusters in orbit requires injection of cryogenic propellants into the combustion chamber. The chamber's initial very low pressure leads to flash boiling that will then determine the dynamics of the spray breakup, the mixing of fuel and oxidizer, the reliability of the ignition and the subsequent combustion process. As details of the spray breakup process of cryogenic liquids under flash boiling conditions are not yet well understood, we use direct numerical simulations (DNS) to simulate the growth, coalescence and bursting of vapour bubbles in the superheated liquid that leads to the primary breakup of the liquid oxygen jet. Considering the main breakup patterns and droplet formation mechanisms for a range of conditions, we evaluate the effectiveness of the volume of fluid (VOF) method together with the continuum surface stress (CSS) model to capture the breakup of thin lamellae formed at high Weber numbers between the merging bubbles. A grid refinement study indicates convergence of the mass averaged droplet sizes towards an a priori estimated droplet diameter. The order of magnitude of this diameter can be estimated based on thermodynamic conditions.

D. D. Loureiro (✉) · A. Kronenburg
Institute for Combustion Technology, University Stuttgart, Stuttgart, Germany
e-mail: daniel.dias-loureiro@itv.uni-stuttgart.de

A. Kronenburg
e-mail: kronenburg@itv.uni-stuttgart.de

J. Reutzsch · B. Weigand
Institute of Aerospace Thermodynamics, University Stuttgart, Stuttgart, Germany
e-mail: jonathan.reutzsch@itlr.uni-stuttgart.de

B. Weigand
e-mail: bernhard.weigand@itlr.uni-stuttgart.de

K. Vogiatzaki
Advanced Engineering Centre, University of Brighton, Brighton, UK
e-mail: k.vogiatzaki@brighton.ac.uk

© Springer Nature Switzerland AG 2021   209
W. E. Nagel et al. (eds.), *High Performance Computing in Science and Engineering '19*,
https://doi.org/10.1007/978-3-030-66792-4_15

1 Introduction

Flash boiling is an important spray disintegration and mixing process in upper stage rocket engines and orbital maneuvering systems where liquid propellants are injected into the reaction chamber that is initially at very low pressure. Due to the low pressure microscopic vapour bubbles spontaneously nucleate within the continuous liquid phase and grow while the liquid continues to evaporate. The dynamics of this process will then determine the spray breakup and the mixing of fuel and oxidizer—the latter being important for ignition and the subsequent combustion process. Sher et al. [20] and Prosperetti [14] provide a quite comprehensive overview of the underlying physics and the dynamics of bubble growth under superheated conditions.

The key characteristics of a jet injected into a combustion chamber are spreading angle, jet penetration and the resulting droplet size distribution after breakup. At the institute of rocket propulsion at the German Aerospace Center (DLR) Lampoldshausen high-speed shadowgraphy measurements are performed to correlate spreading angle and penetration of cryogenic nitrogen and oxygen jets with the injection conditions such as the level of superheat and the mass flow rate. Additional Phase Doppler Anneometry (PDA) shall be used to measure droplet size and velocity distributions. The temperatures for the cryogenic liquids can range from 80K to 120K and initial chamber pressures are as low as 40 mbar. Experimental conditions thus favour all breakup regimes from aerodynamic breakup to a fully flashing spray [7]. The latter conditions are used as a reference in this work where we focus on the microscopic processes leading to the primary breakup of a fully-flashing liquid oxygen jet.

Methods that resolve the entire flashing chamber are usually based on Reynolds averages or filtered values for the solution of the flow and mixing fields. These methods focus on the macroscopic characteristics of the jet, they are, however, not able to fully resolve the small scales of the breakup process and require closure models for the unresolved scales. Studies using Eulerian-Lagrangian methods (cf. [4, 18]) rely on empirical models for the initial droplet size distribution and various additional assumptions on droplet shapes and relative velocities. One fluid approaches [9, 11] do not resolve individual bubbles and rely on mass-transfer models (such as the homogeneous relaxation model) which require calibration. As calibration data is scarce and most relaxation models are calibrated by using data from water flows in a channel independent of the material system they are applied to, it seems necessary to provide detailed (microscopic) information about size distribution of the generated droplets, their surface area and relative velocities for cryogenic flashing sprays. Some of these data shall now be obtained from direct numerical simulation (DNS) under the relevant conditions. The DNS are performed on a small domain, representative of the conditions that would be found near the exit of the injector nozzle. With fully resolved bubbles and the introduction of phase change at the interface, we simulate the growth, deformation and coalescence of multiple bubbles, leading to the formation of ligaments and liquid films that breakup and burst into small droplets that constitute the spray.

In this work, we focus on the dynamics of the breakup of thin sheets or lamellae that form in between the bubbles while they grow and interact. The bursting of lamellae will then introduce the smallest droplets present after breakup (some being physical, some being artificial and related to the mesh size), and a resolution criterion for DNS is now sought such that the artificial droplets related to the mesh size are insignificant in terms of mass and area compared to the real droplet size distribution.

2 Numerical Method

We use the in-house code Free-Surface 3D—FS3D [5] for the DNS of the atomization process. The code uses a finite volume method to discretize the incompressible Navier-Stokes equations, while capturing a fully resolved liquid-vapour interface with phase change and surface tension using the Volume of Fluid (VOF) method with PLIC reconstruction. The incompressible treatment seems justified as the low temperatures and the near vacuum pressure typical for the flashing of cryogenic liquids ensures sub-critical conditions. Here, the gas and liquid phases have a well defined interface for each individual bubble, the interface velocities are relatively low and these conditions are well suited for DNS using FS3D.

The Navier-Stokes equations are solved in a one-field formulation for a continuous velocity, $\mathbf{u}$, and pressure field, p, yielding

$$\frac{\partial}{\partial t} (\rho \mathbf{u}) + \nabla \cdot [\rho \mathbf{u}\mathbf{u}] = \nabla \cdot \mu \left[\nabla \mathbf{u} + \nabla (\mathbf{u})^T \right] - \nabla p + \mathbf{f}_\sigma, \tag{1}$$

where $\mathbf{u}\mathbf{u}$ is a dyadic product. Buoyancy forces have been neglected, and $\mathbf{f}_\sigma$ denotes the effects of surface tension. The latter is non-zero only in the vicinity of the interface and modeled by the continuum surface stress (CSS) model [10]. In the VOF method an additional variable is transported representing the volume fraction of liquid in the cell, f, with $f = 1$ in the liquid phase, $f = 0$ in the gas phase and $0 < f < 1$ in cells containing the interface. Volume-averaged properties are then defined as $\rho = \rho_\ell f + \rho_v (1 - f)$ for the density and $\mu = \mu_\ell f + \mu_v (1 - f)$ for viscosity, where the subscripts ℓ and v denote the liquid and vapour phases. Constant temperature and incompressibility are assumed within each phase thus ρ_ℓ and ρ_v are constants, as well as μ_ℓ and μ_v. The f transport equation can be written as [17]

$$\frac{\partial f}{\partial t} + \nabla \cdot (f \mathbf{u}_\Gamma) = \frac{\dot{m}'''}{\rho_\ell}, \tag{2}$$

where $\mathbf{u}_\Gamma$ is the interface velocity and $\dot{m}'''$ is the liquid evaporation rate at the interface. The latter can be defined as $\dot{m}''' = a_\Gamma \dot{m}''$, using the evaporation mass flux, $\dot{m}''$, and the interface density, a_Γ, which represents the interface area per unit of volume. A PLIC (piecewise linear interface calculation) scheme [15] is used to determine the interface plane for each finite volume cell, and the exact position of the interface within the cell is determined by matching the volume bound by the plane with the volume fraction f. The PLIC reconstruction coupled with the CSS model for surface tension is a flexible approach that requires only moderate resolution when modeling bubble coalescence, liquid breakup and droplet collisions. However, during bubble merging thin lamellae will form, and unphysical dynamics of the interactions between the two interfaces of a thin lamella may be predicted if the lamella thickness is smaller than 4 computational cells [12]. This corresponds to the width of the stencil around the interface used to calculate $\mathbf{f}_\sigma$ and is responsible for some of the mesh dependent effects analysed in this work. The pressure field, p, is obtained implicitly by solving the pressure Poisson equation

$$\nabla \cdot \left[\frac{1}{\rho} \nabla p \right] = \frac{\nabla \cdot \mathbf{u}}{\Delta t}, \tag{3}$$

using an efficient multi-grid method [16]. Continuity is implied in the velocity divergence term, with $\nabla \cdot \mathbf{u} = 0$ except in interface cells (with $0 < f < 1$). There, $\nabla \cdot \mathbf{u}$ is determined to account for phase change with large density ratio as a function of the evaporation rate, $\dot{m}'''$, using the method detailed in [17] which ensures mass conservation in spite of a volume averaged density, ρ, being used. This naturally introduces the jump condition in the momentum conservation equation through the pressure field, p.

While DNS resolves all hydrodynamically important scales, it cannot capture the molecular processes at the interface that determine the evaporation rate. Phase change continues to require modeling. Here, we model the unknown $\dot{m}''$ according to the growth rate of a spherical bubble,

$$\dot{m}'' = \rho_v \dot{R}, \tag{4}$$

where $\dot{R}$ acts as an imposed interface velocity and is obtained from a numerical solution as presented by Lee and Merte [11]. They coupled the Rayleigh-Plesset equation with heat transfer at the bubble interface. This solution can provide evaporation and growth rates during all stages of the bubble growth while analytical solutions [19] that were favoured in the past, cover the heat diffusion driven stage only [2]. It should be noted that for the conditions analysed in this work the interface velocities are in the order of 10 m/s which ensures a low Mach number in the gas. However, due to the effect of varying vapour pressure in the early stages of bubble growth, the vapour density can vary substantially which is not captured with the current numerical approach. However, both $\dot{R}$ and ρ_v are calibrated to the bubble size at which the bubble merging occurs, as detailed in Sect. 3, which accounts the for the influence of interface cooling and vapour compressibility effects in the bubble growth up to that point. This provides an adequate approximation for the volume of vapour being generated at the interface, while the dynamics of the subsequent bubble breakup are captured by DNS with the assumption of constant fluid properties.

3   Computational Setup

As detailed in [13], the simulation domain represents a small volume of continuous liquid near the exit of the injector nozzle when flash boiling starts. For the current setup, no assumptions are necessary regarding bubble distribution or the type of interface instability leading to breakup. Nonetheless, regular arrays of equally spaced bubbles are used here for simplicity of setup, for repeatability of results and for a reduction in system parameters that allow for a better systematic comparison of different setup conditions (for a general impression compare e.g. Fig. 2 top left).

The initial relative velocity between bubbles and liquid is zero emulating a fluid element moving with bulk velocity. The liquid is free to expand through the use of continuity boundary conditions (outflow) and large buffer zones are used to prevent the interaction of the liquid-vapour interface with the boundaries. Finally, symmetry conditions are used to reduce the computational cost. The reference liquid temperature and ambient pressure conditions for the DNS, T_∞ and p_∞, correspond to values given in the experiments at DLR. The initial spacing between the bubbles is assumed to be uniform and should be related to the nucleation rate J and the mass flow rate. Since here a mass flow rate is not defined, such dependence is avoided and the bubble spacing is treated as a free geometric parameter and can represent different nucleation rates triggered by different levels of disturbances in the flow. As such, we define the parameter R_f as the final bubble radius at which the bubbles are expected to touch and start merging. Considering p_∞ and T_∞, we use the critical radius as a reference value and normalize R_f, $R_f^* = R_f/R_{\mathrm{crit}}$. The normalized quantity represents a growth factor since nucleation with R_{crit} given by $R_{\mathrm{crit}} = 2\sigma/(p_{\mathrm{sat}}(T_\infty) - p_\infty)$. Here, σ denotes the surface tension at T_∞. With the parameters p_∞ and T_∞, we use the Lee and Merte solution [11] for single bubble growth to determine the growth rate, $\dot{R}$, vapour density ρ_v, surface tension, σ, and vapour viscosity, μ_v, as functions of R_f^*. The density and viscosity of the liquid, ρ_ℓ and μ_ℓ, are functions of p_∞ and T_∞ and obtained from the equation of state library CoolProp [3].

3.1   Characterization of the Flow

Droplet dynamics are usually characterized by the Weber (We) and Ohnesorge (Oh) numbers. For a corresponding definition applicable to bubble dynamics we use the interface velocity, $\dot{R}$, and the final bubble diameter, $2R_f$, as characteristics scales and define

$$\mathrm{We}_b = \frac{\rho_\ell \dot{R}^2 2R_f}{\sigma}, \tag{5}$$

and

$$\mathrm{Oh}_b = \frac{\mu_\ell}{\sqrt{\rho_\ell \sigma 2R_f}}, \tag{6}$$

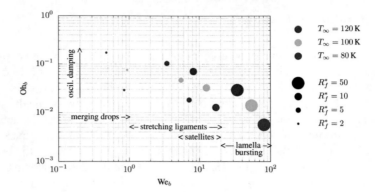

Fig. 1 Weber-Ohnesorge diagram to characterize the type of breakup

with subscript b indicating the characteristic numbers for bubbles. The Weber-Ohnesorge diagram (Fig. 1) shows the characteristic quantities for all DNS simulations that were part of the study.

Loureiro et al. [13] identified very distinct types of breakup that are also indicated in the diagram. These types are primarily determined by the We_b number while Ohnesorge number is typically below 1, but the larger Oh_b the stronger the damping of droplet oscillations due to larger viscous effects.

For cases located in the small Weber number regime ($We_b < 5$), surface tension forces dominate the bubble dynamics during merging and the interstitial liquid between the bubbles does not break into smaller droplets but coalesces generating relatively large droplets relative to the bubble size (cf. Fig. 2a). Cases with $We_b \sim \mathcal{O}(10)$ result in a binary droplet distribution of main droplets and satellites, where the size of the main droplets matches the interstitial volume $D \approx 1.94 R_f$ (Fig. 2b and c). In this regime the larger Oh_b (Fig. 2b) leads to more stable main droplets and smaller or no satellites. Finally, as depicted in Fig. 2d and e, cases with $We_b > 20$ show large deformation of the bubbles prior to merging, forming thin and flat liquid lamella between them. As the bubble continues to grow, the lamellae stretc.h and thin until they can no longer be properly resolved by the computational mesh, potentially generating a large number of artificial droplets.

3.2 Test Cases

For the mesh sensitivity studies we focus on two cases with $T_\infty = 120K$ (case A) and $T_\infty = 80K$ (case B) and $p_\infty = 1000Pa$. The two cases represent the two extrema of injection temperatures at the minimum vacuum pressure as realised in the experiments at DLR. The Weber numbers are between 30 and 100 and the test cases are therefore located in the "lamella bursting" regime indicated in Fig. 1. Each setup simulates the dynamics of 3^3 bubbles with details of the computational parameters

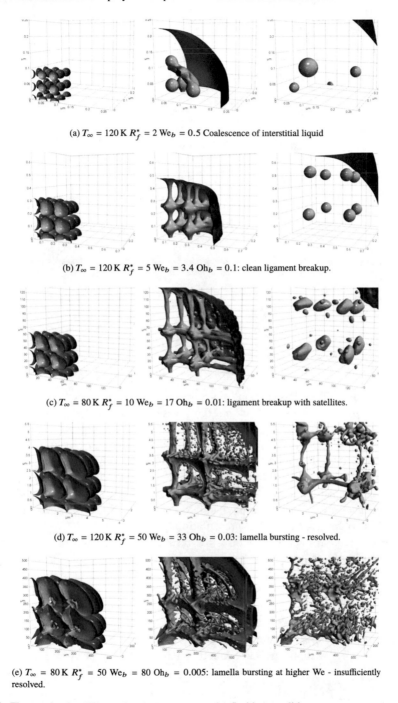

(a) $T_\infty = 120\,\mathrm{K}$ $R_f^* = 2$ $We_b = 0.5$ Coalescence of interstitial liquid

(b) $T_\infty = 120\,\mathrm{K}$ $R_f^* = 5$ $We_b = 3.4$ $Oh_b = 0.1$: clean ligament breakup.

(c) $T_\infty = 80\,\mathrm{K}$ $R_f^* = 10$ $We_b = 17$ $Oh_b = 0.01$: ligament breakup with satellites.

(d) $T_\infty = 120\,\mathrm{K}$ $R_f^* = 50$ $We_b = 33$ $Oh_b = 0.03$: lamella bursting - resolved.

(e) $T_\infty = 80\,\mathrm{K}$ $R_f^* = 50$ $We_b = 80$ $Oh_b = 0.005$: lamella bursting at higher We - insufficiently resolved.

Fig. 2 Time series for different breakup patterns under flashing conditions

Table 1 Simulation parameters and grid refinement levels

Case	A	B	C (Control)
T_∞	120K	80K	120K
p_∞	1000Pa	1000Pa	1000Pa
ΔT	58.71K	18.71K	58.71K
R_p	1022	30.12	1022
R_f^*	50	50	175
R_{crit}	$0.012\mu m$	$1.01\mu m$	$0.012\mu m$
We_b	33	80	80
Oh_b	2.9×10^{-2}	5.66×10^{-3}	1.5×10^{-2}
D_{ref}	$0.285\mu m$	$10.77\mu m$	$0.418\mu m$
128^3 grid			
$\Delta x = R_f/14$	$4.2 \times 10^{-8} m$	$3.8 \times 10^{-6} m$	$1.5 \times 10^{-7} m$
256^3 grid:			
$\Delta x = R_f/28$	$2.1 \times 10^{-8} m$	$1.9 \times 10^{-6} m$	$7.4 \times 10^{-1} m$
512^3 grid:			
$\Delta x = R_f/57$	$1.1 \times 10^{-8} m$	$9.5 \times 10^{-7} m$	$3.7 \times 10^{-8} m$
1024^3 grid:			
$\Delta x = R_f/114$	$5.3 \times 10^{-9} m$	–	–

given in Table 1. The physical conditions for flash atomization are typically characterized by p_∞, T_∞, the superheat and the pressure ratio. The level of superheat is
defined by $\Delta T = T_\infty - T_{sat}(p_\infty)$, or in terms of pressure as $R_p = p_{sat}(T_\infty)/p_\infty$.

As the Weber numbers of cases A and B are similar, we can expect the same type
of breakup pattern, but length scales are vastly different as a result of the difference
in T_∞. The table includes a third case, labeled case C, that matches We_b of case B but
uses a higher injection temperature, $T_\infty = 120K$. The case corroborates the results'
independence of Oh_b, the thermodynamic conditions and length scales and results
are not shown here for brevity of presentation.

One key issue for spray breakup is the resolution requirement, i.e. the expected
minimum droplet that needs to be resolved after spray breakup. The literature on
DNS of droplet break up has shown that as the Weber number is increased more
"smaller" droplets are expected and thus a higher resolution is required. However,
if DNS is expected to act in a predictive manner, an a priori criterion needs to be
established in order to define the necessary resolution. Such a criterion is currently
missing from existing literature. Here, we equate the droplet's volumetric surface
energy, $4\sigma/D_{ref}$, with the available kinetic energy in the liquid, $\rho_\ell \dot{R}^2/2$, yielding

$$D_{ref} \approx \mathcal{O}\left(\frac{8\sigma}{\rho_\ell \dot{R}^2}\right). \tag{7}$$

This value shall be understood as an order of magnitude estimate for the prediction of mesh resolution requirements and not as an exact formula for the smallest droplet size. The real minimum droplet diameter may deviate since $\dot{R}$ considers a local interface velocity of a single bubble only and bubble interactions and convective effects may alter the local (relevant) kinetic energy that leads to breakup.

4   Results

The spray breakup includes bubble growth, merging of the bubbles, formation of thin lamellae and shedding of droplets from the thin lamellae. The later stages of the lamellae formation and droplet shedding are shown in Figs. 3 and 4 for cases A and B, respectively. The "open" area between the liquid is where the bubbles have already merged and the liquid between the bubbles is now being pushed back and retracting due to continuous growth of the bubbles and movement of the outer surface. Due to the inertia of the liquid between the bubbles, the fluid does not fully retract immediately, but relatively flat thin liquid layers, the lamellae, persist. The retracting velocity of the lamellae is relatively fast and instability at the rim of the lamella leads to the shedding of droplets.

The subfigures in Figs. 3 and 4 show the effect of grid refinement. For all simulations, the estimated minimum droplet size D_{ref} is resolved by a minimum of 6 (case A) and 3 (case B) CFD cells. Some effects of grid resolution on breakup dynamics and droplet size can be observed. For case A (Fig. 3) the lamella seems to be more stable with increasing resolution, leading to a delay in the droplet formation. Most notably, the lower resolutions promote the ideas of breakup of a liquid rim due to longitudinal instability [1]. "Liquid finger" formation with subsequent droplet breakup seems to be the dominant breakup mechanism. However, finer resolutions seem to dampen the finger formation and the rim itself detaches from the lamella. For case B (Fig. 4) the droplets generated by lamella breakup seem to be well resolved and stable with increasing resolution. However, their number and size still changes between the 256^3 grid and 512^3 grid and further computations with 1024^3 grid nodes may be needed for a last conclusive assessment.

4.1   Statistical Analysis

It has been questioned in the past whether mesh independence can ever be achieved by simple mesh refinement [12] for droplet collision problems. It may be more important to achieve convergence of the mass weighted droplet size distribution, i.e. a decrease of the minimum droplet size with increased mesh resolution is accepted but these smallest droplets shall not significantly affect the mass balance and can thus be ignored.

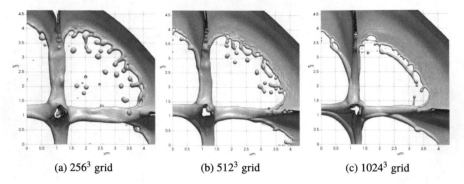

(a) 256^3 grid   (b) 512^3 grid   (c) 1024^3 grid

Fig. 3 Case A breakup patterns for different grid refinements at simulation time $t = 0.1\mu$s. The lamellae are represented by the iso-surface $f = 0.1$

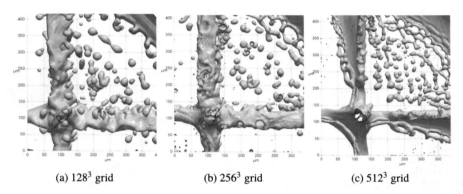

(a) 128^3 grid   (b) 256^3 grid   (c) 512^3 grid

Fig. 4 Case B breakup patterns for different grid refinements at simulation time $t = 0.58\mu$s. The lamellae are represented by the iso-surface $f = 0.1$

To sample the statistics a filter is applied such that only fully atomized (stable) droplets are counted and large ligaments or any residual droplets that fall below the resolution criterion and are not fully resolved are excluded. Representative diameters such as the arithmetic mean D_{10}, the Sauter mean D_{32} or the De Brouckere mean diameter $D_{43} = \sum_{i \in j}^{N_j} D_i^4 / \sum_{i \in j}^{N_j} D_i^3$, the latter corresponds to the mass-weighted mean of the spray, are then computed from these droplets. The dependence of these representative diameters on the mesh refinement is shown in Fig. 5. Both cases indicate convergence of the Sauter and De Brouckere means, and a resolution with 512^3 nodes seems sufficient to capture the droplet size of the droplets holding most of the liquid mass after breakup. The percentages indicate the deviation of D_{43} from the corresponding value at the finest mesh and should be understood as order of magnitude analysis as data are extracted from one instant in time and the exact values are likely to change somewhat during the course of the breakup. The larger sensitivity of D_{10} is due to the generation of a large number of artificial, small droplets. This is expected and has been documented in the archival literature. The figure also indicates the estimate for the minimum droplet size, D_{ref} (Eq. 7) . For case B the size of the

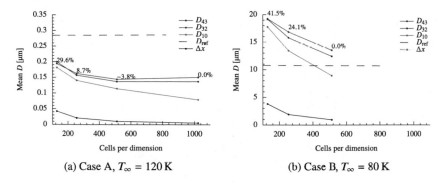

(a) Case A, $T_\infty = 120\,\mathrm{K}$ (b) Case B, $T_\infty = 80\,\mathrm{K}$

Fig. 5 Dependency of the D_{10} (arithmetic), D_{32} (Sauter) and D_{43} (De Brouckere) mean diameters with the mesh refinement for both cases studied and comparison with the cell size Δx and estimated droplet size D_{ref} (Eq. 7). For the D_{43} the percentage of the variation relative to the most refined case is shown

main droplet group seems to converge towards this value and for case A, the resulting droplets are—on average—half the size of D_{ref}. The good agreement for case B may be somewhat fortuitous, but both cases indicate that D_{ref} may provide a suitable first estimate for the mesh size needed for adequate resolution.

The normalized droplet size distributions shown in Fig. 6 provide a more quantitative measure and demonstrate more clearly the dependence of the droplet sizes on grid resolution with the presence of smaller droplets for the cases with higher resolution. Also, a bimodal droplet distribution is observed, with a clear segregation between one group of larger droplets that represents "physical" droplets after breakup and the much smaller, artificial satellite droplets. However, Fig. 6 should be seen in conjunction with Fig. 7 that shows the distribution of mass-weighted droplet diameters. It is apparent that the distribution of volume fraction (here equivalent to the mass weighted average) does not change much with resolution for case A (256^3 grid nodes gives a similar picture and only a coarser mesh with 128^3 grid nodes provide notably larger droplets) while case B seems to converge to the estimated D_{ref} for 512^3 grid nodes and a resolution requirement of $\Delta x \approx D_{\mathrm{ref}}/10$ may be a reasonable criterion for the prediction of the smallest (but relevant) droplet sizes being generated during a flash boiling process.

It should be noted that the convergence of the macroscopic averaged quantities has been shown e.g. for splashing by Gomaa et al. [8] in 2008. So the here obtained results are perfectly in line with this investigation.

4.2 Computational Performance

The general performance of FS3D on the Cray XC40 Hazel-Hen super computer located at the High Performance Computing Center Stuttgart (HLRS) has previously

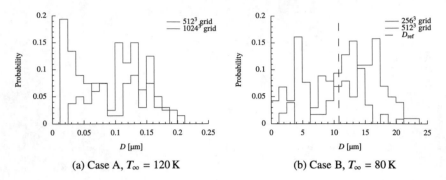

Fig. 6 Discrete probability density functions (PDF) of the equivalent droplet diameter D_i for cases A and B at the two highest levels of mesh refinement for each case

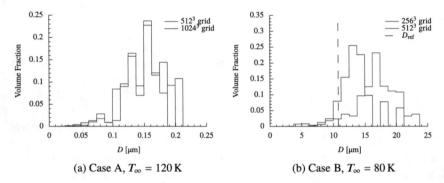

Fig. 7 Discrete volume fraction distribution function of the equivalent droplet diameter D_i for cases A and B at the two highest levels of mesh refinement for each case

been analysed in various works such as such as [6]. However, specific case configurations can affect the parallelization efficiency and the issue of efficient implementation needs to be revisited. Also note that the code allows for hybrid parallelization using MPI and OpenMP, but only MPI has been used for this study.

The largest case with 1024^3 grid nodes is the maximum mesh size that should be used on Hazel-Hen according to weak-scaling limitations of FS3D reported in [6]. Therefore, only the strong scaling efficiency is tested here. The test is based on case A with 1024^3 nodes while the number of processors has been varied. The domain is initialized with an array of 5^3 bubbles that are partially overlapping. This scenario is similar to the conditions when bubbles start to coalescence. The fluid properties and the bubble size/spacing is taken from case A as given in Table 1. The resolution of the bubbles at time of merging is about 100 cells per bubble diameter.

Memory requirements and significant initialization times limit the range for the number of processors, N_{proc}, that shall be used: here N_{proc} varies from 512 to 8192 distributed on compute nodes with 24 processors each. All relevant details are given in Table 2. The computational speed (in cycles per hour) is determined by the total

Fig. 8 Results for
strong-scaling on the for the
1024^3 grid: Compute speed
compared to ideal linear
scaling and scaling efficiency
(annotations)

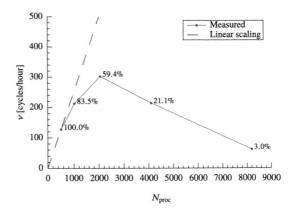

Fig. 8 Results for
strong-scaling on the for the
1024^3 grid: Compute speed
compared to ideal linear
scaling and scaling efficiency
(annotations)

computation time (wall time) minus the initialization time divided by a fixed number
of time-steps (cycles), viz.

$$\nu = \frac{t_{\text{total}} - t_{\text{init}}}{N_{\Delta t}}. \tag{8}$$

The number of time-steps, $N_{\Delta t}$, was set to 20 for all the cases and care has been taken
that the number of iterations of the pressure solver for each time step are comparable.
The strong-scaling efficiency is given by the ratio of the total computational cost per
time step relative to the reference value using 512 processors,

$$\eta = \frac{N_{\text{proc}}/\nu}{512/\nu_{512}} \times 100\%. \tag{9}$$

Figure 8 shows the computational speed, ν, as a function of N_{proc}. The annotated
percentages indicate the corresponding efficiency η for each case. The peak performance is measured for $N_{\text{proc}} = 2048$ which corresponds to a load of $64 \times 64 \times 128$
cells per processor. This only marginally differs from earlier results [6] that report
peak performances for 64^3 cells/processor on a 512^3 grid, and results are consistent
with a reported 20% weak-scaling efficiency at $N_{\text{proc}} = 4096$.

Computational requirements for the cases as reported in Sect. 3 can now be given:
case A with 1204^3 grid nodes requires 16600 cycles for a physical simulation time
of $0.14\mu s$. With $N_{\text{proc}} = 2048$, the simulation time may be as low as 55 h with a
total cost of 1.1×10^5 processor-hours. However, the number of iterations of the
pressure solver can vary significantly, especially upon topological changes and the
formation of complex interface structures. For case A the number of iterations varies
by a mere 20% and computational costs are therefore very moderate. In contrast, the
complexity of breakup is much larger for case B and the number of iterations by the
pressure solver can be up to two orders of magnitude higher. This effectively halts
simulations when breakup occurs and explains the omission of results of case B with
1024^3 nodes. Larger computations are to be conducted in future but may require
some improvement of the iterative pressure solver for the specific time steps when
bubble merging occurs.

Table 2 strong-scaling cases and results for 1024^3 grid

N_{proc}	Nodes	Cells/proc.	v (cycles/hour)	η (efficiency)
512	22	128x128x128 = 2097152	128	100%
1024	43	64x128x128 = 1048576	213	83%
2048	86	64x 64x128 = 524288	303	59%
4096	171	64x 64x 64 = 262144	215	21%
8192	342	32x 64x 64 = 131072	62	3%

5 Conclusions

The dynamics of breakup and droplet formation under flashing conditions have been investigated. The relevant parameter range as defined in corresponding experiments carried out at the DLR institute of rocket propulsion Lampoldshausen has been covered, but the focus of this work has been on the high Weber number cases as they present the most challenging conditions on grid resolution for capturing all the dynamic processes that determine the final droplet size distribution.

The smallest scales are expected to be located between the bubbles when they merge. There, thin lamellae are formed and droplets are formed at the rim of these lamellae by either "finger"-formation and subsequent droplet pinch-off or by break-off of the entire rim. This disintegration of the lamellae is expected to yield the smallest droplets, let them be of physical or artificial (mesh resolution induced) size. A mesh refinement study corroborates existing studies inasmuch the minimum droplet size decreases with increasing resolution and seems to be mesh dependent. However, the mass weighted droplet diameter tends to converge towards a given size that can be estimated by analysis of surface and kinetic energy acting on the bubble walls. The exact average droplet diameter may deviate from this estimate by a factor of two, however, an increase in resolution also demonstrates that the overall liquid mass associated with lamellae breakup approaches single digit percentage values and the highest resolution may not be needed for adequate approximation of the entire spray breakup process.

The dynamics of the lamella puncture and retraction are exposed, as well as their dependence on the mesh resolution. This can be used as a criterion to determine the significance of different size groups of the final droplet size distributions or to define a minimum resolution requirement to capture the bursting of lamellae in the high Weber number range. The current DNS therefore demonstrates that (1) the dynamics of the breakup influence the characteristics of droplet formation and droplet sizes, and (2) the current DNS implementation can be used for simulations of much larger domains that will then capture the complete spectrum of the droplet size distributions resulting from a flashing jet.

Acknowledgements The simulations presented in this work were performed on the CRAY XC40 Hazel Hen of the High Performance Computing Center Stuttgart (HLRS). This work is part of the HAoS-ITN project and has received funding from the European Union's Horizon 2020 research and innovation programme under the Marie Sklodowska-Curie grant agreement No 675676 (DL). We also acknowledge funding by DFG through the Collaborative Research Center SFB-TRR75 (JR, AK, BW) and by the UK's Engineering and Physical Science Research Council support through the grant EP/P012744/1 (KV).

References

1. G. Agbaglah, C. Josserand, S. Zaleski, Longitudinal instability of a liquid rim. Phys. Fluids **25**(2), 022103 (2013). https://doi.org/10.1063/1.4789971
2. R. Bardia, M.F. Trujillo, Assessing the physical validity of highly-resolved simulation benchmark tests for flows undergoing phase change. Int. J. Multiphase Flow **112** (2018). https://doi.org/10.1016/j.ijmultiphaseflow.2018.11.018
3. I.H. Bell, J. Wronski, S. Quoilin, V. Lemort, Pure and pseudo-pure fluid thermophysical property evaluation and the open-source thermophysical property library coolprop. Ind. Eng. Chem. Res. **53**, 2498–2508 (2014). https://doi.org/10.1021/ie4033999
4. R. Calay, A. Holdo, Modelling the dispersion of flashing jets using cfd. J. Hazard. Mater. **154**(1–3), 1198–1209 (2008)
5. K. Eisenschmidt, M. Ertl, H. Gomaa, C. Kieffer-Roth, C. Meister, P. Rauschenberger, M. Reitzle, K. Schlottke, B. Weigand, Direct numerical simulations for multiphase flows: an overview of the multiphase code FS3D. J. Appl. Math. Comput. **272**(2), 508–517 (2016)
6. M. Ertl, J. Reutzsch, A. Nägel, G. Wittum, B. Weigand, Towards the implementation of a new multigrid solver in the dns code fs3d for simulations of shear-thinning jet break-up at higher reynolds numbers, in *High Performance Computing in Science and Engineering ' 17* Springer International Publishing (2018), pp. 269–287
7. J.W. Gaertner, A. Rees, A. Kronenburg, J. Sender, M. Oschwald, D. Loureiro, Large eddy simulation of flashing cryogenic liquid with a compressible volume of fluid solver, in *ILASS–Europe 2019, 29th Conference on Liquid Atomization and Spray Systems* (2019)
8. H. Gomaa, B. Weigand, M. Haas, C. Munz, Direct Numerical Simulation (DNS) on the influence of grid refinement for the process of splashing, in *High Performance Computing in Science and Engineering '08 Transactions of the High Performance Computing Center, Stuttgart (HLRS)* (2008), pp. 241–255
9. I. Karathanassis, P. Koukouvinis, M. Gavaises, Comparative evaluation of phase-change mechanisms for the prediction of flashing flows. Int. J. Multiph. Flow **95**, 257–270 (2017)
10. B. Lafaurie, C. Nardone, R. Scardovelli, S. Zaleski, G. Zanetti, Modelling merging and fragmentation in multiphase flows with surfer. J. Comput. Phys. **113**(1), 134–147 (1994)
11. H.S. Lee, H. Merte, Spherical vapor bubble growth in uniformly superheated liquids. Int. J. Heat Mass Transf. **39**(12), 2427–2447 (1996)
12. M. Liu, D. Bothe: Numerical study of head-on droplet collisions at high weber numbers. J. Fluid Mech. **789**, 785–805 (2016). https://doi.org/10.1017/jfm.2015.725
13. D. Loureiro, J. Reutzsch, D. Dietzel, A. Kronenburg, B. Weigand, K. Vogiatzaki, DNS of multiple bubble growth and droplet formation in superheated liquids, in *14th International Conference on Liquid Atomization and Spray Systems* ,Chicago, IL, USA (2018)
14. A. Prosperetti, Vapor bubbles. Ann. Rev. Fluid Mech. **49** (2017)
15. W.J. Rider, D.B. Kothe, Reconstructing volume tracking. J. Comput. Phys. **141**(2), 112–152 (1998)
16. M. Rieber, Numerische modellierung der dynamik freier grenzflachen in zweiphasenstromungen. Ph.D. thesis, Universitat Stuttgart (2004)

17. J. Schlottke, B. Weigand, Direct numerical simulation of evaporating droplets. J. Comput. Phys. **227**, 5215–5237 (2008)
18. R. Schmehl, J. Steelant, Computational analysis of the oxidizer preflow in an upper-stage rocket engine. J. Propul. Power **25**(3), 771–782 (2009)
19. L. Scriven, On the dynamics of phase growth. Chem. Eng. Sci. **10**(1), 1 – 13 (1959). https://doi.org/10.1016/0009-2509(59)80019-1
20. E. Sher, T. Bar-Kohany, A. Rashkovan, Flash-boiling atomization. Prog. Energy Combust. Sci. **34**, 417–439 (2008)

Enhancing OpenFOAM's Performance on HPC Systems

Thorsten Zirwes, Feichi Zhang, Jordan A. Denev, Peter Habisreuther, Henning Bockhorn, and Dimosthenis Trimis

Abstract OpenFOAM is one of the most popular open source tools for CFD simulations of engineering applications. It is therefore also often used on supercomputers to perform large eddy simulations or even direct numerical simulations of complex cases. In this work, general guidelines for improving OpenFOAM's performance on HPC clusters are given. A comparison of the serial performance for different compilers shows that the Intel compiler generally generates the fastest executables for different standard applications. More aggressive compiler optimization options beyond O3 yield performance increases of about 5 % for simple cases and can lead to improvements of up to 25 % for reactive flow cases. Link-time optimization does not lead to a performance gain. The parallel scaling behavior of reactive flow solvers shows an optimum at 5000 cells per MPI rank in the tested cases, where caching effects counterbalance communication overhead, leading to super linear scaling. In addition, two self-developed means of improving performance are presented: the first one targets OpenFOAM's most accurate discretization scheme "cubic". In this scheme, some polynomials are unnecessarily reevaluated during the simulation. A simple change in the code can reuse the results and achieve performance gains of about 5 %. Secondly, the performance of reactive flow solvers is investigated with Score-P/Vampir and load imbalances due to the computation of the chemical reaction rates are identified. A dynamic-adaptive load balancing approach has been implemented for OpenFOAM's reacting flow solvers which can decrease computation times by 40 % and increases the utilization of the HPC hardware. This load balancing approach utilizes the special feature of the reaction rate computation, that no information of neighboring cells are required, allowing to implement the load balancing efficiently.

T. Zirwes (✉) · J. A. Denev
Steinbuch Centre for Computing, Karlsruhe Institute of Technology,
Hermann-von-Helmholtz-Platz 1, Karlsruhe, Germany
e-mail: thorsten.zirwes@kit.edu

F. Zhang · P. Habisreuther · H. Bockhorn · D. Trimis
Engler-Bunte-Institute, Karlsruhe Institute of Technology, Engler-Bunte-Ring 7, Karlsruhe, Germany
e-mail: feichi.zhang@kit.edu

© Springer Nature Switzerland AG 2021   225
W. E. Nagel et al. (eds.), *High Performance Computing in Science and Engineering '19*,
https://doi.org/10.1007/978-3-030-66792-4_16

Keywords OpenFOAM · Load balancing · Reactive flows · Performance optimization · Combustion

1 Introduction

Computational Fluid Dynamics (CFD) has proven to be an effective tool in engineering to aid the development of new devices or for gaining a deeper understanding of physical phenomena and their mutual interactions. One of the most popular open-source CFD tools is OpenFOAM [1]. It provides many tools for simulating different phenomena: from simple, incompressible flows to multi-physics cases of multi-phase reactive flows. In the case of combustion simulations, the interaction of chemical reactions and turbulent flow takes place on a large range of time and length scales [2, 3]. Therefore, supercomputers have to be employed to resolve all the multi-scale interactions and to gain deeper insight into the combustion dynamics by performing highly-resolved simulations.

In order to utilize the hardware provided by High Performance Computing (HPC) clusters efficiently, it is not sufficient to install OpenFOAM with its default settings. In this work, a general guideline on running OpenFOAM efficiently on supercomputers is given. Section 2 reviews the performance of OpenFOAM depending on the choice of compiler and the optimization settings as well as link-time optimization. Sections 2 and 5 give examples of the parallel scaling behavior. In the case of reactive flows, an optimal number of cells per MPI rank is discussed.

In addition to these guidelines, two self-developed methods for speeding up general OpenFOAM applications are discussed. Since the OpenFOAM applications on supercomputers are typically highly-resolved direct numerical simulations, it is likely that users will choose OpenFOAM's "cubic" discretization for spatial derivatives since it is the most accurate scheme OpenFOAM offers. In the implementation of this scheme, some polynomials are computed repeatedly which can be avoided. The necessary code changes and performance benefits are shown in Sect. 3 and the accuracy of the scheme is demonstrated with the well-known Taylor-Green vortex case [4].

The second self-developed method for improving OpenFOAM's performance improves the load balancing of reactive flow solvers. Since chemical reaction rates depend exponentially on temperature and are closely coupled with each other, they tend to be numerically stiff. This requires special treatment for the computation of the chemical reaction rates which can lead to large load imbalances. Section 4 presents the load balancing method which is done dynamically during the simulation. An application of this is given in Sect. 5, where a flame evolving in the Taylor-Green vortex case from Sect. 3 is studied.

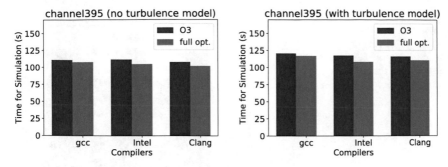

Fig. 1 Time required to compute the tutorial case channel395 depending on the compiler and optimization settings. Simulation without turbulence model (left) and with turbulence model (right)

2 Compilers and Optimization Settings

If OpenFOAM is compiled without changing any settings, it will be compiled with the gcc compiler. The only optimization flags given to the compiler is O3. This raises the question if OpenFOAM can benefit from using different compilers and more aggressive optimization options since O3 does not generate code specifically optimized for the CPU type of the cluster. In this section, two different cases are considered:

- incompressible flow: the standard tutorial case channel395 computed with OpenFOAM's pimpleFoam solver
- compressible reactive flow: the standard tutorial case counterflowFlame 2D_GRI computed with OpenFOAM's reactingFoam solver

These two cases differ in their physical complexity. For the incompressible flow case, only two conservation equations have to be solved (mass and momentum). Also, all fluid properties are assumed to be constant. The reactive flow case on the other hand simulates the chemical reactions of 52 different species. Therefore, conservation equations have to be solved for mass, momentum, energy and 51 chemical species. In addition to this, all fluid properties like density or heat capacity are computed as a function of gas mixture composition, temperature and pressure. Lastly, a large part of the simulation time stems from computing the chemical reaction rates based on the GRI3.0 [5] reaction mechanism.

Figure 1 shows the simulation time required to compute 100 s of OpenFOAM's tutorial case channel395 with OpenFOAM's pimpleFoam solver. The figure on the left shows the case without turbulence model and the case with turbulence model, which means that additional transport equations for the turbulence model are solved. Three different compilers are investigated: gcc 8, Intel2018 and Clang 7. Additionally, OpenFOAM was compiled using the default optimization setting O3 and more aggressive optimizations (full opt.). For gcc and Clang, the full optimization settings are -Ofast -ffast-math -fno-rounding-math- -march=native -mtune=native and for the Intel compiler -O3 fp-

Fig. 2 Time required to compute the first 0.003 s of the tutorial case `counterflowFlame2D_GRI` depending on the compiler and optimization settings

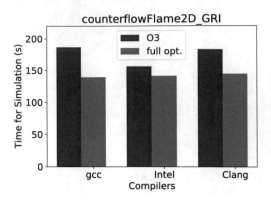

model fast=2 -fast-transcendentals -xHost. All timings have been recorded on the ForHLR II cluster at KIT [6] with OpenFOAM 5.x in serial.

The difference between the three compilers is approximately 3 %. The fastest code using only O3 as optimization option is generated by Clang, while the fastest code using the full optimization options is generated by the Intel compiler. Changing from the standard O3 to the full optimization options yields a performance increase of about 5 % on average. Including the turbulence model does not change the timings significantly because most of the time is spent in the pressure correction step.

For the reactive flow case however (shown in Fig. 2), the difference between the O3 and the fully optimized build becomes significant. For the gcc compiler, switching from O3 to the full optimizations increases the performance by about 25 %. The Intel compiler generates efficient code even with O3, so that the full optimizations increase the performance by about 10 % for the Intel compiler. Although there are large differences between the compilers with O3, they perform approximately the same using the full optimizations.

Using the full optimization settings as described before has not shown any drawbacks in accuracy of the simulation results. Final temperature profiles for the reactive flow case for example agreed within 0.1 % between the O3 build and the fully optimized build. It should be noted however, that other unsafe optimizations like strict aliasing has caused OpenFOAM to randomly crash. Therefore, using additional options to the ones shown above should be used with caution. It has also been found that link-time optimization (LTO) has no effect on performance at all. This is probably due to OpenFOAM being compiled into many different shared objects so that LTO cannot be very effective.

3 Optimizing the Cubic Discretization

OpenFOAM offers a large number of spatial discretization schemes to discretize spatial gradients for the finite volume method. Since CFD simulations on super-

Original Implementation	Optimized Implementation

```
const surfaceScalarField kSc
(
lambda*(scalar(1) - lambda*(scalar(3)
 - scalar(2)*lambda))
);

const surfaceScalarField kVecP
(
  sqr(scalar(1) - lambda)*lambda
);
const surfaceScalarField kVecN
(
  sqr(lambda)*(lambda - scalar(1))
);
```

```
if (kScGlob.get() == nullptr || mesh.changing())
{
  kScGlob.reset( new surfaceScalarField(
    lambda*(scalar(1) - lambda*(scalar(3) -
    scalar(2)*lambda))));
  kVecPGlob.reset( new surfaceScalarField(
    sqr(scalar(1) - lambda)*lambda) );
  kVecNGlob.reset( new surfaceScalarField(
    sqr(lambda)*(lambda - scalar(1))));
}

surfaceScalarField& kSc = *kScGlob;
surfaceScalarField& kVecP = *kVecPGlob;
surfaceScalarField& kVecN = *kVecNGlob;
```

Fig. 3 Original code for computing the cubic discretization scheme from `cubic.H` from Open-
FOAM (left) and optimized code version (right)

computers are usually highly-resolved, discretization schemes with high accuracy
should be employed. The most accurate scheme OpenFOAM offers is the `cubic`
scheme. It interpolates a quantity ϕ from the cell centers of the current cell C and
the neighboring cell N to the cell face f using a third order polynomial:

$$\phi_f = \lambda\phi_C + (1 - \lambda)\phi_N + A(\phi_N - \phi_C) + B\nabla\phi_C + C\nabla\phi_N \tag{1}$$

Here, λ is a function of the distance between the cell centers. The coefficients A, B, C
are themselves polynomials of λ:

$$A = 2\lambda^3 - 2\lambda^2 + \lambda \tag{2}$$
$$B = -\lambda^3 + \lambda^2 \tag{3}$$
$$C = -\lambda^3 + 2\lambda^2 - \lambda \tag{4}$$

Since these polynomials only depend on the distances between the cell centers,
they only have to be computed once for each simulation if the computational mesh
does not change. In OpenFOAM's implementation however, these polynomials are
computed at every time step and for every transport equation. So for example if the
conservation of momentum and energy use the `cubic` scheme, these polynomials
are computed twice per time step. In order to avoid this, OpenFOAM's code of the
cubic discretization can easily be modified so that these polynomials are computed
only once at the beginning of the simulation.

Figure 3 on the left shows the code from OpenFOAM which computes the three
polynomials. The code on the right is the modification. It starts by checking if the
polynomials have already been computed. If not, they are computed only once. They
are also re-computed if the mesh is changing during the simulation. If the polynomials
have already been computed, references to the precomputed values are used on the
subsequent computations.

The performance gain of this new implementation is tested with the well-known
Taylor-Green vortex [4] using OpenFOAM's `pimpleFoam` solver. The case consists
of a cube with side length $2\pi L$ where all boundary conditions are periodic. The

Fig. 4 Vorticity iso-surface of $0.1\,\mathrm{s}^{-1}$ at the start of the simulation (left) and after $20\,\mathrm{s}$ (right)

velocity profile at the start of the simulation is set to:

$$u_x = u_0 \sin\left(\frac{2\pi x}{L}\right) \cos\left(\frac{2\pi y}{L}\right) \cos\left(\frac{2\pi z}{L}\right) \tag{5}$$

$$u_y = -u_0 \cos\left(\frac{2\pi x}{L}\right) \sin\left(\frac{2\pi y}{L}\right) \cos\left(\frac{2\pi z}{L}\right) \tag{6}$$

$$u_z = 0 \tag{7}$$

This places counter-rotating vortices at each corner. The flow field then starts to become turbulent and decays over time. This is shown in Fig. 4.

By using the new implementation for the cubic scheme, which requires only minimal code changes, the total simulation time was decreased by 5 %.

The accuracy of the cubic scheme is shown in Fig. 5 on the left. The normalized, volume averaged dissipation rate over time computed with OpenFOAM is compared with a spectral DNS code [4].

$$\varepsilon = \frac{1}{V} \int_V S : S \, \mathrm{d}V \tag{8}$$

where S is the strain rate tensor. Both simulations use a grid with 512^3 cells; the OpenFOAM simulation was performed on Hazel Hen at the HLRS [7]. The maximum deviation is below 1 %, demonstrating that OpenFOAM's cubic scheme is able to accurately simulate the decaying turbulent flow.

On the right of Fig. 5 strong scaling results for this case using 384^3 cells are shown. The numbers above the markers show the parallel scaling efficiency. The tests have been performed with OpenFOAM 1812 using the standard pimpleFoam solver.

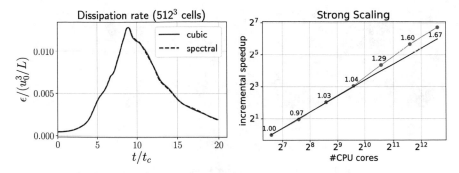

Fig. 5 Integral dissipation rate over time compared with a spectral DNS code (right) and strong scaling results (right)

4 Dynamic Load Balancing for Reactive Flows

OpenFOAM's parallelization strategy is based on domain decomposition together with MPI. By default, OpenFOAM uses third-party libraries like scotch or metis to decompose the computational domain into a number of sub-domains with approximately the same number of cells. This is the ideal strategy for most solvers and applications. For reactive flow simulations however, the chemical reaction rates have to be computed. Since the chemical reaction rates generally depend exponentially on temperature and are coupled to the rate of change of all chemical species, they are numerically very stiff and require very small time steps. In order to avoid using very small time steps in the simulation, an operator splitting approach is usually used. In this approach, the chemical reaction rates are computed by solving a system of ordinary differential equations (ODE) in each cell with special ODE solvers for stiff systems. These solvers take adaptive sub-time steps for integrating the chemical reaction rates over the CFD time step (see [8, 9] for a more detailed description of the operator splitting approach).

The size of the sub-time steps depends on the current local conditions. The lower the chemical time scales are, i.e. the faster the chemical reactions are, the more sub-time steps have to be taken. Consider for example the simulation of a turbulent flame in Fig. 6 using Sundials [10] as the ODE integrator. The light blue lines show the sub-domains into which the computational domain has been divided for the parallel simulation. It can easily be seen that the sub domains in the top and bottom row do not contain any cells where chemical reactions take place. Therefore, the computational effort for sub-domains in the middle where the flame is located is much higher. This leads to load imbalances during the simulation. In general, the position of the flame might change over time so that it is not possible to decompose the mesh once in an optimal way. Therefore, this section presents an approach to overcome load imbalances caused by chemical reaction computations that is done dynamically during runtime.

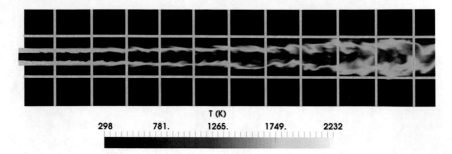

Fig. 6 Temperature field from the simulation of a turbulent flame. Blue lines show the sub-domains for the parallel simulation

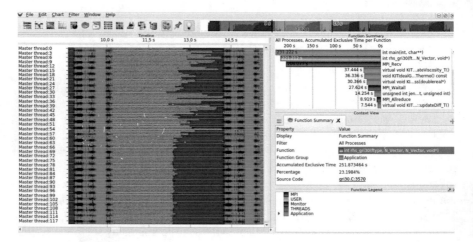

Fig. 7 Performance measurements with Score-P [11], visualized with Vampir [12]

Figure 7 shows performance measurements of a combustion simulation [9] with OpenFOAM on 120 MPI ranks recorded with Score-P [11] and visualized with Vampir [12]. The solver is a self-developed DNS extension for OpenFOAM [13]. The load imbalances can be seen on the left: depicted is one time step during the simulation. Each horizontal line represents one MPI rank. The x-axis represents time. The green areas are useful work performed by OpenFOAM, like solving the conservation equations. The gray regions show the computation of chemical reaction rates. The red areas show communication overhead. In this case, a small number of processes requires about twice as much time to compute the chemical reaction rates than most other processes. Therefore, most of the processes spend half of the simulation time waiting on the few slow processes. This of course wastes large amounts of computational resources. The more complex the chemical reaction mechanisms are (i.e. the more reactions and chemical species are considered), the more severe these load imbalances become.

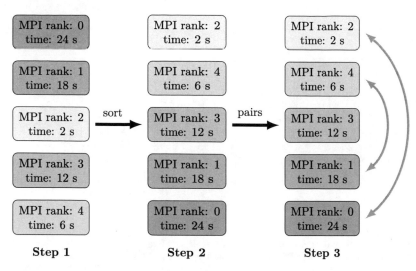

Fig. 8 Schematic drawing of the load balancing approach: Step 1: Measure time for computing chemical reaction rates. Step 2: Sort all processes by computing time. Step 3: Form pairs of processes that share their workload during the following time steps

In order to overcome this, a load balancing algorithm has been implemented which can be used with any reactive flow solver in OpenFOAM and achieves dynamic load balancing during runtime. It also exploits the fact, that the computation of chemical reaction rates does not require information from neighboring cells. This means, that the work load can be freely distributed among the processes without taking any connectivity information of the computational mesh into account, which makes the algorithm flexible and efficient.

The load balancing approach is based on the idea of forming pairs of processes that share their workload. Figure 8 shows an example of this idea: At the beginning of the simulation after the very first time step, each MPI rank measures the time it takes to compute the chemical reaction rates on its own sub-domain. The timing results are then sorted and communicated to all MPI ranks. In the example from Fig. 8, MPI rank 2 has the lowest computing time of 2 s and is therefore the first entry in the list. MPI rank 0 is the slowest with 24 s and therefore is the last entry in the list. The last step is to form pairs of processes: The fastest MPI rank forms a pair with the slowest. The second fastest forms a pair with the second slowest, and so on. The two processes in each pair will share their workload in the subsequent time steps of the simulation. Of course, communicating the timing results from all MPI ranks to all other MPI ranks is an expensive operation. If the position of the flame is rapidly changing, it might be a good idea to repeat this communication step every few time steps to make sure that each pair of processes consists of a slow and a fast process. Usually it is sufficient to check the global imbalances every few thousand time steps.

After the pairs have been formed, the slow process in the pair sends some of its cells to the fast process. After the fast process has computed the chemical reaction

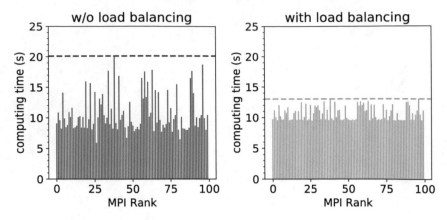

Fig. 9 Time for computing the chemical reaction rates on each MPI rank. The slowest rank determines the overall time for the simulation, shown as dashed line

rates on its own sub-domain, it also computes the reaction rates in the cells of the partner process and sends back the reaction rates in these cells along with the time it took to compute the rates. Based on this time, the number of cells that is sent between the processes is adjusted in the next time step. Due to this, the load balancing approach can adapt to changing conditions during the simulation. Because the calculation of chemical reaction rates does not depend on information in neighboring cells, the cells can be freely shared among the processes. For more information of this approach, see [9].

Figure 9 shows measurements for the time spent on computing chemical reaction rates for a case similar to the one in Fig. 7, using the self-developed DNS solver [13] and the GRI3.0 reaction mechanism, computed with 100 MPI ranks. It can be seen on the left, that some processes take much longer to compute the chemical reaction rates than most other processes. Since all processes have to wait until the last one finishes, most processes spend half the time waiting. Since the slowest process determines the overall time for computing the chemical reaction rates, 20 s per time step are needed for computing the chemical reaction rates in this simulation. Running the same simulation with the load balancing approach from above (Fig. 9 on the right) shows that the difference between the fastest and slowest processes has been drastically decreased. The ratio of the slowest to the fastest process in terms of computing time is about 4 without load balancing and about 1.3 with load balancing. The time per time step that is spent on chemical reaction rates reduces to about 13 s with load balancing, as shown by the dashed line. Often, the computation of chemical reaction rates is the most time consuming part of the simulation, so that the simulation time is reduced from 20 to 13 s per time step, saving about 40 % of the total simulation time.

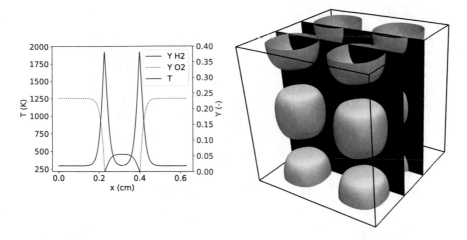

Fig. 10 Left: Initial profiles of temperature, hydrogen mass fraction and oxygen mass fraction along the centerline. Right: Iso-surface of the magnitude of vorticity of $4000\,s^{-1}$ and iso-surface of temperature at 1800 K colored in red

5 Reactive Taylor-Green Vortex

In this section, an application for reactive flow solvers is shown which uses both presented optimization methods: the improved cubic discretization scheme from Sect. 3 and the new load balancing approach from Sect. 4. In total, these led to a reduction of total computing time of about 20 %. This application uses the Taylor-Green vortex from Sect. 3 but embeds a flame into the flow field [14]. This setup allows to study the interaction of different types of flames with a well-defined turbulent flow field.

The initial conditions for the flow are the same as for the standard Taylor-Green vortex case (see Eqs. 5 and 6). Additionally, there are now profiles for the temperature and mass fractions of hydrogen and oxygen present. The gas properties and chemical reactions are taken from a reaction mechanism for hydrogen combustion [15]. Figure 10 on the left shows these profiles along the centerline. On the right, a 3D view of the initial conditions in terms of flow vorticity and temperature can be seen. As soon as the simulation starts, the flow field begins to decay into a turbulent flow while the flame burns the hydrogen and oxygen in the domain, leading to a complex interaction between the flow field and the flame. Figure 11 shows the 1800 K iso-surface colored by OH mass fraction and iso-surfaces of vorticity of the flow field during three points in time during the simulation.

The case as shown in Fig. 11 was computed on an equidistant grid with 256^3 cells. This case was used to perform a strong scaling test (Fig. 12) both with activated and deactivated computation of chemical reaction rates. The scaling tests are performed with up to 12288 CPU cores. When using up to about 1000 CPU cores or 15 000 cells per core, the scaling is slightly super linear. At 1536 cores, parallel efficiency is

Fig. 11 Iso-surface of
temperature at $T = 1800$ K
colored by the mass fraction
of the OH radical and
iso-surfaces of the vorticity
magnitude of $4000\,\mathrm{s}^{-1}$
(shown in white) at different
times: $t = 0.5\,\mathrm{ms}$ (top),
$t = 1.0\,\mathrm{ms}$ (middle),
$t = 2.0\,\mathrm{ms}$ (bottom)

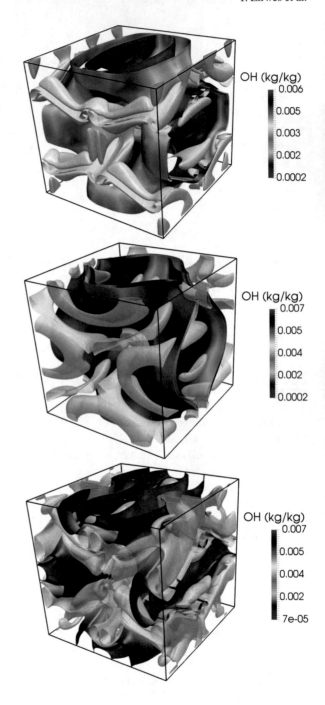

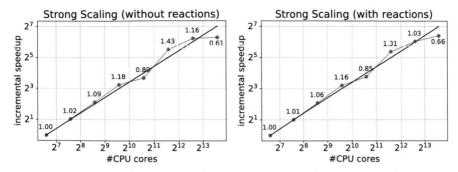

Fig. 12 Strong scaling for the reactive Taylor-Green vortex setup. Left: without chemical reactions. Right: With chemical reactions. Numbers above the markers show the parallel scaling efficiency

reduced to about 0.8 for both the reactive and non-reactive simulation. The optimum lies at 3072 CPU cores or about 5000 cells per core, where the parallel efficiency reaches 1.3 to 1.4. For larger fuels or larger reaction mechanisms, the optimal number of cells might be even lower. Due to the low number of cells, all sub-domains fit into the L3 cache of the CPU. Running the simulation with twice as many cores yields another good scaling result. But beyond this point (more than 12 000 CPU cores or less than 1500 cells per core) the scaling efficiency degrades rapidly. In conclusion, there is an optimum in terms of efficiency at about 5000 cells per core for this simulation using a relatively small reaction mechanism for hydrogen combustion [15]. Reducing the number of cells to about 2500 per core is still possible, but further reduction does not benefit the simulation time.

6 Summary

In order to utilize HPC hardware efficiently with OpenFOAM, the following recommendations are given:

- In order to maximize performance, OpenFOAM should be compiled with more aggressive optimization options like -Ofast -ffast-math -march=native -mtune=native for gcc and Clang and -O3 -fp-model fast=2 -fast-transcendentals -xHost for the Intel compiler. The effect of this on the accuracy of the simulation results was found to be negligible. The performance gain of using the full optimization options can be as large as 25 % for complex simulations and about 5 % for simple simulations with incompressible flows.
- In general, the Intel compiler generates the most efficient code, but the difference between the compilers are negligible if the full optimization options are used. Using only O3 as optimization option shows that the Intel compiler generally outperforms gcc and Clang.

- using even more aggressive optimization options like strict aliasing should be avoided since they have led to random crashes of OpenFOAM.
- link-time optimization (LTO) was found to have no effect on performance.

For reactive flow simulations, an optimal value in terms of cells per process was found to be about 5000. This case was run with a relatively small reaction mechanism so that this number may be lower for more complex reaction mechanisms including more chemical species and reactions.

OpenFOAM's most accurate spatial discretization scheme is its `cubic` scheme. It is based on a third order polynomial interpolation from cell centers to cell faces. A comparison with a spectral DNS code has shown that this scheme is able to predict the correct dissipation rate during turbulent decay of the Taylor-Green vortex. The performance of this discretization scheme can be increased by reusing the results of the interpolation polynomials as shown in the code modification in Sect. 3. This can reduce simulation times by an additional 5%.

For simulations using reactive flow solvers, load imbalances due to the computation of chemical reaction rates can lead to large amounts of wasted HPC resources. This is demonstrated by doing a performance measurement with Score-P/Vampir. A self-developed load balancing approach, which is specifically made for the imbalances caused by the computation of chemical reaction rates, is shown to reduce overall simulation times by up to 40%. This approach can be used with any reactive flow solver. The workload is adaptively shared between pairs of processes between a slow and a fast process. This can be combined with the automatic code generation for the chemical reaction rates as described in a previous work [8] to achieve even larger performance gains.

Acknowledgements This work was supported by the Helmholtz Association of German Research Centres (HGF) through the Research Unit EMR, Topic 4 Gasification (34.14.02). This work was performed on the national supercomputer Cray XC40 Hazel Hen at the High Performance Computing Center Stuttgart (HLRS) and on the computational resource ForHLR II with the acronym DNSbomb funded by the Ministry of Science, Research and the Arts Baden-Württemberg and DFG ("Deutsche Forschungsgemeinschaft").

References

1. OpenCFD, OpenFOAM: The Open Source CFD Toolbox. User Guide Version 1.4, OpenCFD Limited. Reading UK (2007)
2. T. Poinsot, D. Veynante, *Theoretical and Numerical Combustion* (R.T, Edwards, 2001)
3. R. Kee, M. Coltrin, P. Glarborg, *Chemically reacting flow: theory and practice* (John Wiley & Sons, 2005)
4. G.I. Taylor, A.E. Green, Mechanism of the production of small eddies from large ones, *Proceedings of the Royal Society of London Series A-Mathematical and Physical Sciences*, vol. 158, no. 895, pp. 499–521 (1937)
5. G. Smith, D. Golden, M. Frenklach, N. Moriarty, B. Eiteneer, M. Goldenberg et al., Gri 3.0 reaction mechanism
6. Karlsruhe institute of technology (2018), www.scc.kit.edu/dienste/forhlr2.php

7. High performance computing center stuttgart (2018) www.hlrs.de/systems/cray-xc40-hazel-hen
8. T. Zirwes, F. Zhang, J. Denev, P. Habisreuther, H. Bockhorn, Automated code generation for maximizing performance of detailed chemistry calculations in OpenFOAM, in *High Performance Computing in Science and Engineering '17*, ed. by W. Nagel, D. Kröner, M. Resch (Springer, 2017) pp. 189–204
9. T. Zirwes, F. Zhang, P. Habisreuther, J. Denev, H. Bockhorn, D. Trimis, Optimizing load balancing of reacting flow solvers in openfoam for high performance computing. ESI (2018)
10. Suite of nonlinear and differential/algebraic equation solvers http://computation.llnl.gov/casc/sundials
11. Score-p tracing tool, http://www.vi-hps.org/tools/score-p.html
12. Vampir visualization tool, http://www.paratools.com/vampir/
13. T. Zirwes, F. Zhang, P. Habisreuther, M. Hansinger, H. Bockhorn, M. Pfitzner, D. Trimis, *Quasi-DNS dataset of a piloted flame with inhomogeneous inlet conditions* (Turb. and Combust, Flow, 2019)
14. H. Zhou, J. You, S. Xiong, Y. Yang, D. Thévenin, S. Chen, Interactions between the premixed flame front and the three-dimensional taylor-green vortex. Proc. Combust. Instit. **37**(2), 2461–2468 (2019)
15. P. Boivin, *Reduced-kinetic mechanisms for hydrogen and syngas combustion including autoignition* (Universidad Carlos III, Madrid, Spain, Disseration, 2011)

Computational Fluid Dynamics

Ewald Krämer

The following chapter presents a selection of research projects conducted in the field of Computational Fluid Dynamics (CFD) during the reporting period. Numerical simulations were performed both on the CRAY XC40 Hazel Hen of the High Performance Computing Center Stuttgart (HLRS) and on the ForHLR II of the Steinbuch Centre for Computing (SCC) in Karlsruhe. As in the years before, CFD had the strongest request for supercomputing resources among all disciplines.[1]

This year, 41 annual reports were submitted and underwent a peer-review process. Research groups from various institutions across Germany took advantage from the opportunity to utilize the supercomputer resources offered by the two centres. Many important results and new scientific findings have been achieved and notable progress has been made in the field of CFD. Research has focused on the simulation of fluid dynamic phenomena being detectable only through an extremely fine temporal and spatial discretization, often in combination with high-order methods, as well as on the optimization of the employed numerical codes with respect to computational performance.

Fifteen contributions were selected for publication in this book, which means that, again, not all highly qualified reports could be considered. The spectrum of projects presented in the following covers fundamental research as well as application-oriented problems of industrial relevance. Challenging fluid dynamic problems are addressed in the fields of combustion, film cooling, jet impingement, aircraft, and rotorcraft aerodynamics, just to name a few. Different numerical methods, such as Finite Difference, Finite Volume, Smoothed Particle Hydrodynamics, or Discontinuous Galerkin methods, were employed using in-house or open-source codes. The first 5 articles describe the results of so-called Gauss Large Scale Projects, which means that these projects were granted more than 35 million core-hours per year from the Gauss Centre for Supercomputing (GCS).

Inanc and Kempf from the Institute of Combustion and Gasdynamics at the University of Duisburg-Essen performed fully resolved auto-igniting transient jet flame simulations by LES and DNS with direct chemistry using their in-house code PsiPhi.

---

[1]Ewald Krämer, Institute of Aerodynamics and Gas Dynamics, University of Stuttgart, Stuttgart, Germany, e-mail: kraemer@iag.uni-stuttgart.de

Their simulations reproduce two experiments conducted at the German Aerospace Center (DLR) in Stuttgart. The first one is a statistically steady reference case, and the second one is a pulsed case, where a high-velocity cold methane jet emanates into a laminar flow of hot exhaust products of a lean hydrogen/air mixture. With their simulations, the authors aimed at quantifying the growth of the initial ignition kernel using fully resolved flow fields and species concentrations. Furthermore, a reliable benchmark case of a pulsed jet flame was to be established for further use. A perfect agreement between experiment and simulations could be achieved for the steady jet. For the pulsed case, analyses are still ongoing, but the estimated delay times and the location of the auto-ignition match experimental observations. LES and DNS show similar results. 96,000 parallel compute cores were used on Hazel Hen for the DNS computations. In total, 53M core hours were spent within the project.

The contribution of *Peter and Kloker* from the Institute of Aerodynamics and Gas Dynamics, University of Stuttgart, deals with a DNS of film cooling at supersonic flow conditions. Target application is the thermal protection of rocket engine nozzles against the hot exhaust gases. For this purpose, a cool secondary gas (Helium) is blown at Mach 1.8 tangentially into the hot gas turbulent boundary layer of the main Mach 3.3 flow through a backward-facing step. In their work, the authors have studied the complex interaction between both flows using high-order DNS to gain further understanding of the mixing process. This extremely complex flow case requires a particularly tailored flow solver with high spatial and temporal accuracy. Their in-house code NS3D has been developed for such applications over years. Furthermore, it has continuously been improved with respect to an optimal performance on supercomputers. This task has benefitted, among others, from the biannually offered HLRS workshop "Optimization of Scaling and Node-level Performance", during which the cluster experts at HRLS help the users optimize their simulation codes.

At the Institute for Combustion Technology, RWTH Aachen University, *Davidovic, Bode, and Pitsch* have investigated parallelization strategies for Multiple Representative Interactive Flamelet (MRIF) combustion models. The use of regeneratively produced synthetic fuels has great potential to reduce greenhouse gas emissions. However, the effect of the individual fuel properties on the combustion and pollutant formation process is not entirely understood. Numerical simulations that help gain better insight into these processes require high computational power implying excellent parallel code performance. While the very good scaling behavior of the generic flow solver within the developed CIAO simulation framework had already been proven, the effect of different parallelization strategies of the MRIF combustion model had not been studied before. The authors reveal interesting aspects on this subject.

Against the backdrop of the on-going climate change, a strong focus in aeronautical research is on the reduction of CO_2 emissions. One means to achieve this goal is to decrease the aircraft's friction drag. A broad variety of passive and active measures has been investigated during the last decades. One technique, which has proven to reduce the viscous drag in a turbulent boundary layer, is the introduction of spanwise traveling transversal surface waves. In their contribution, *Albers, Meinke, and Schröder* from the Institute of Aerodynamics, RWTH Aachen University, pursue the question, whether this active drag reduction method is successful not only in zero-pressure gradient canonical flows but also in applications relevant for aviation, i.e., along curved surfaces with non-zero pressure gradients. For that purpose, they performed highly resolved implicit LES simulations around an aircraft wing section at a high freestream Reynolds number for incompressible inflow. About three-quarters of the surface underwent a transversal spanwise wave motion. A significant reduction in the overall drag (sum of viscous and pressure drag) was achieved even together with a slight increase in lift. They employed their in-house LES code using the MILES approach to model the smallest turbulent scales. The simulations were run on 4,800 cores with a grid size of 408 million cells.

Fröhlich, Meinke, and Schröder from the same institute have studied in detail the dynamics of particles in an isotropic turbulent flow. Their focus was on validating the standard two-way coupling approach, where the solid phase is approximated by point particles, by a direct particle-fluid simulation with all turbulent scales and the complete flow field in the vicinity of the particles being resolved. Non-spherical particles of Kolmogorov length-scale size with various aspect ratios have been considered. Notable deviations between the point-particle approach and the fully resolved reference solution could be found. Two billion grid cells were required to resolve the near-particle flow fields, and 48,000 cores were used in parallel on Hazel Hen.

Steigerwald, Reutzsch, Ibach, Baggio, Seck, Haus, and Weigand from the Institute of Aerospace Thermodynamics of the University of Stuttgart performed DNS of a wind-generated water wave with their in-house code Free Surface 3D (FS3D). The code is based on the Volume-of-Fluid (VOF) method and uses a Piecewise Linear Calculation (PLIC) algorithm. In their highly resolved simulations, they can show that wind-generated water waves develop starting from a quiescent water surface driven solely by the turbulent airflow. In the considered regime, both restoring forces that are acting to return the displaced water surface to its equilibrium position, i.e., the tension and the gravity force, are in the same order of magnitude. Such waves are called capillary-gravity waves. Different aspects of the capillary-gravity wave characteristic, such as topology, phase velocity, and temporal development of the wave energy, have been evaluated. A comparison with experimental data obtained in a wave tank of the University of Miami shows good agreement. In addition, the performance of the code was further improved by replacing large pointer arrays with allocatable arrays, leading to a significantly higher weak and strong scaling efficiency.

Turbulent flows over porous media were simulated *by Chu, Müller, and Weigand* at the same institute using DNS. The high-order open-source spectral/hp element framework Nektar++ was adopted for solving the incompressible Navier-Stokes equations. The computational domain is a channel partially filled with a porous medium, and the effect of this porous structure on the turbulent flow has been investigated using up to one billion grid points. The solver has an excellent parallel efficiency up to 96,000 cores on Hazel Hen. Using this highly resolved DNS, the authors intend to focus on two major physical phenomena, the turbulence modulation including drag reduction and flow control, and the turbulent/non-turbulent interface close to the porous surface.

In their article, *Beck, Dürrwächter, Kuhn, Meyer, Munz, and Rohde* from the Institute of Aerodynamics and Gas Dynamics and the Institute of Applied Analysis and Numerical Simulations at the University of Stuttgart, respectively, present their ongoing development of a highly efficient framework for the quantification of uncertainties (UQ) in computational fluid dynamics and computational aeroacoustics. Their baseline CFD-solver is the high-order Discontinuous Galerkin code FLEXI. Two uncertainty propagation environments have been developed. The intrusive one is based on the Stochastic Galerkin method, which was implemented into FLEXI. Validation test cases are shown and details about its implementation and its parallel performance are given. Besides, a new framework for non-intrusive methods, called POUNCE, was built-up, which includes a spectral projection method as well as a multilevel Monte Carlo method for the stochastic post-processing. For the latter, high stochastic accuracy is achieved by a very large number of simulations on a coarse grid, with each sample being computationally cheap, and only a few simulations on the finest grid. As application case, the aeroacoustics of an open cavity flow has been considered. To capture all acoustically relevant effects, 3D high-order zonal LES simulations have been performed at several points in the stochastic space and the stochastic noise spectrum for an uncertain inflow Reynolds number has been computed using the spectral projection method of the new UQ framework.

The topic of the next article, written by *Föll, Hitz, Keim, and Munz* from the Institute of Aerodynamics and Gas Dynamics, University of Stuttgart, is the use of modern data structures on high performance computers for high-fidelity compressible multi-phase simulations. Since in the homogeneous equilibrium, such simulations are usually based on real equations of state (EOS), and the direct evaluation of these EOS is too expensive, look-up tables are used. Applying specific data structures reduces computational time significantly, but, on the other hand, increases the memory requirements. The authors present a shared memory parallelization of look-up tables for binary mixtures based on the MPI 3.0 standard. They can show that, depending on the hardware architecture (Hazel Hen and its successor Hawk), up to 128 times more memory could be accessed using shared memory trees in contrast to the former non-shared memory approach. A detailed analysis of benefits and drawbacks of the new data structure is made. Application case was a shear layer simulation with two species. In addition, 3D simulations of colliding droplets were carried out using a parabolic relaxation model of the diffuse interface model of the isothermal Navier-Stokes-Korteweg equations implemented in the in-house code FLEXI. The influence of different model parameters on the coalescence behavior is shown.

Another application of a CFD-solver based on the Discontinuous Galerkin method is presented by *Genuit, Keßler, and Krämer* from the same institute. Their long-term aim is to develop a DG-code for the simulation of helicopter rotor flow phenomena. Results are shown for a rotor in hover and in untrimmed forward flight. The former show an excellent agreement with experimental data provided by Caradonna and Tung for a two-bladed model rotor. For the latter case, the functionality of the code for predicting typical phenomenological effects, such as blade-vortex-interaction, could be demonstrated. A very good scaling behaviur is shown for all orders of accuracy, but is currently limited to about 1,000 cores.

At the Institute of Nuclear Technology and Energy Systems, University of Stuttgart, *Pandey and Laurin* have investigated the effect of non-uniform body force on an incompressible turbulent channel flow by means of DNS. Depending on the strength of the axial body forces, which varied in wall-normal direction, different features occur. For increasing body forces, the flow relaminarises. When the force is increased further, a recovery of the turbulent structures takes place. The main motivation of this work is to study the effect of gravity on the vertical flow of supercritical CO_2. Several open-source codes have been applied in the project, the present one being the high-order spectral/hp element framework Nektar++. For the current set-up of 60 million degrees of freedom, strong scaling shows a linear speed-up up to 1920 cores, but for increasing problem sizes, which will occur when higher Reynolds numbers will be investigated in the future, the linear regime is extended to higher core numbers.

Camerlengo and Sesterhenn from the Institute of Fluid Dynamics and Technical Acoustics of the TU Berlin present DNS results for a compressible impinging jet flow. New results for a turbulent jet are compared to a previously published case with laminar inlet conditions. The analysis focuses on the heat transfer at the impingement wall and its spatial distribution, whose shape is determined by the vortex dynamics in the proximity of the wall. The CFD code used is the group's well-known NSF solver, which demonstrates a satisfactory parallel efficiency on Hazel Hen.

A study of the influence of atmospheric turbulence on aircraft is the subject of the contribution of *Müller, Ehrle, Lutz, and Krämer* from the Institute of Aerodynamics and Gas Dynamics at the University of Stuttgart. For that purpose, they performed numerical simulations on an extruded airfoil. Two different strategies to account for the broadband turbulent inflow are compared. In the first one, the turbulence is fed into the flow field via a momentum source term at some distance ahead of the leading edge and propagated towards the airfoil, which requires a high resolution of the computational grid and a spatially high-order numerical scheme. This computationally expensive method, called Resolved Atmosphere Approach, RAA, is compared with a simplified method, the so-called Disturbance Velocity Approach (DVA). In the latter, the atmospheric disturbances are not propagated through the flow field, but simply added to the flux balance by superposition. The authors show that, depending on the wave number, DVA is capable of covering the main effects of the atmospheric turbulence on the lift and local pressure spectra in the leading edge and suction peak region. Further downstream, the agreement worsens. For the pitching moment, the DVA results are acceptable only for small wave numbers. The simulations were performed with two different flow solvers, FLOWer and TAU from the German Aerospace Center, since the DVA method is only available in the latter code. Although the DVA approach is computationally cheaper, its implementation presently contains non-scaling routines, which deteriorate the parallel performance of the TAU code and are subject to future optimizations.

The helicopter group at the Institute of Aerodynamics and Gas Dynamics has for a long time been successfully engaged in high performance computations of rotorcraft aerodynamics. The well-established flow solver FLOWer has continuously been improved to increase its functionality as well as its efficiency on massive parallel supercomputers. One of the recent achievements, *Öhrle, Letzgus, Keßler, and Krämer* report on in their article, is the implementation of an Adaptive Mesh Refinement technique, which is shown to nearly halve the computational time for the simulation of a complete compound helicopter in fast forward flight without any loss of accuracy. Further topics of the article deal with a new shielding approach for DES simulations, which applies a sophisticated communication method to exchange flow variables in wall-normal direction, and with the turbulent structures in a highly resolved rotor wake in hover. Strong and weak scaling plots show the parallel efficiency of the FLOWer code for these types of applications. Good performance is achieved down to 16,000 grid cells per core, which means that a typical 500 million cells mesh can be run effectively on 32,000 cores.

In contrast to the afore-mentioned projects, *Chaussonnet, Dauch, Keller, Okraschevski, Ates, Schwitzke, Koch, and Bauer* from the Institute of Thermal Turbomachinery at the Karlsruhe Institute of Technology use a meshless, particle-oriented method for their simulations. For such methods, a proper load balancing, which significantly influences the run-time of the code, is more challenging compared to grid-based methods, since the number of elements (particles) inside a given physical volume depends on the local thermodynamical state of the fluid and, thus, might change during the simulation. In their contribution, the authors study the main factors influencing the load balancing for their Smoothed Particle Hydrodynamics (SPH) method and present the algorithmic strategies of the implementation. The evolution of the load balancing during a simulation is shown for a practical case of liquid atomization, where the domain contains more than one billion particles in average. Up to 15,000 cores were used on the tier-2 supercomputer For HLR II at the SCC in Karlsruhe.

The presented selection of projects reflects the continuous progress in supercomputer simulation in the field of CFD. It becomes obvious that for many real-world applications, numerical simulations on high performance computers are the only practicable way to get insight into the often complex, time-dependent flow physics that characterizes the problem under investigation. Some of these flow phenomena only take place on very small scales and can only be brought to light by an extremely fine discretization of the physical domains. To make such simulations possible within a reasonable time frame, the most powerful high performance computers are crucial. But also in industrial design processes, the use of high-fidelity numerical methods on modern supercomputers becomes more and more vital in order to increase the reliability and the efficiency of the simulations, mitigating development time, risks and costs and, thus, increasing industrial competitiveness. Engineers and researchers, who want to fully exploit the potential offered by high performance computing, must not focus solely on their physical problem. Moreover, they have to think about appropriate numerical schemes as well as about how to optimize their codes with respect to the specific hardware architecture they are going to use. Usually, this exceeds the capabilities and skills of one individual and, thus, different experts in the respective fields should work together. The very fruitful collaboration of researchers with experts from the HLRS and the SCC in most of the projects shows how this can work. Gratitude is owed in this respect to the staff of both computing centres for their individual and tailored support. Without their efforts, it would not have been possible to sustain the high scientific quality we see in the projects.

In 2020, both the HLRS and the SCC will install next-generation tier-1 and tier-2 supercomputers, respectively, offering their users once more a significant increase in computing performance. This will pave the way for even more advanced research. For sure, the new hardware architectures will require further adaptations and optimizations of the codes, which, again, can only succeed in a close cooperation of the researches and the experts at the computing centres.

Fully Resolved Auto-Igniting Transient Jet Flame Simulation

Eray Inanc and Andreas M. Kempf

Abstract This work investigates an auto-igniting impulsively started jet flame issuing into hot and vitiated co-flow by large eddy simulation (LES) with direct chemistry. The experiment from German Aerospace Center is reproduced. The direct chemistry model uses an augmented reduced mechanism that consists of 19 transported species. The targets are first to quantify the growth of the initial ignition kernel using fully resolved flow-fields and species concentrations, then to establish a reliable benchmark case for further studies. The grid study has shown that employed resolution is sufficient to describe the ignition chemistry since the ignition kernel appears at low velocities and fuel-lean conditions. Initial comparisons showed a perfect agreement between the simulations and the experiments for the statistically steady jet. For the transient part of this work, two injection cycles with higher-resolution and six injection cycles with lower-resolution LES have been performed. The estimated delay times and the location of the auto-ignition matched experimental observations. Most importantly, both the low and high-resolution LES show quite similar results, which implies that the resolved flow-fields are statistically converged, independent from the grid resolution. Only after rigorous validation, this simulation is planned to be an ideal benchmark case for such studies of pulsed jets.

1 Introduction

The initial burst of jet tears the environment, forms a vortex ring and evolves into a high-momentum structure. This vortex ring then pinches off and rapidly diffuses into the surrounding medium resulting in enhanced mixing. Since those steps are a function of time, such studies require either high-speed cameras or time-accurate simulations. Most favourable approach for the numerical investigations for these type of cases are direct numerical simulations (DNS) or highly-resolved large eddy simulations (LES) [12], but, these come with a high price of computational resource.

E. Inanc (✉) · A. M. Kempf
Chair of Fluid Dynamics, Institute for Combustion and Gasdynamics (IVG), University of Duisburg-Essen, Carl-Benz-Str. 199, Duisburg 47057, Germany
e-mail: eray.inanc@uni-due.de

© Springer Nature Switzerland AG 2021
W. E. Nagel et al. (eds.), *High Performance Computing in Science and Engineering '19*,
https://doi.org/10.1007/978-3-030-66792-4_17

Recently, the transient jets have received focus [6, 7, 12, 24], however, only for non-reactive cases.

As the chemical state of the flame is linked to the gas-mixing, the dynamics of the transient jet during a pulse affect the ignition, hence, the combustion. The relation between transient jet dynamics and chemistry due to auto-ignition needs to be understood.

Previous experiments studied the ignition at various temperatures and pressures for different fuels in laminar and turbulent conditions, such in shock tubes [26], rapid compression engines [5] or in combustion bombs and flames [17]. A focus for auto-ignition (AI) computations is also given in the context of Large Eddy Simulation (LES) [8] and DNS [19]. The detailed overview of studies on AI is omitted for brevity and are given in the review paper by Mastorakos [18].

Recent experiments in DLR (German Aerospace Institute, Stuttgart) successfully determined the delay time and the location of AI, and the subsequent development of the flame front of a transient jet by advanced high-speed imaging techniques [2–4, 21, 22]. In the present work, the experiments in DLR is reproduced using LES combined with direct chemistry approach, where the steady injection part of this experiment is used for additional validation purposes. The agreement would then yield a continuation for the transient injections by pulse-cycles. The target is to create a database of time-resolved evolution of an ignition kernel in a transient jet, as to be used for further researchers. Such a study of a highly resolved LES with direct chemistry of a pulsed jet in an immense computational domain has, in our knowledge, not been performed before.

2   Experiment

In the DLR experiments [2–4, 21, 22], a high-velocity cold methane jet at 290 K ($U_i = 177$ m/s) is emanated into a laminar co-flow of hot exhaust products of a lean hydrogen/air mixture at 1490 K ($U_o = 4$ m/s). The configuration is at atmospheric pressure. The injector tube has a diameter of $D = 1.5$ mm, leading to a jet-Reynolds number of 16,000. The co-flow mixture is obtained from the products of a lean-hydrogen flame ($\phi = 0.465$). The co-flow consists of oxygen ($X_{O_2} = 0.102$), nitrogen ($X_{N_2} = 0.72$) and gas phase water ($X_{H_2O} = 0.178$). The stoichiometric mixture fraction of this methane and co-flow mixture is 0.297.

There are two versions of the same experiment—a (statistically) steady reference case and the pulsed case. In the latter, four phases are repeated; i. opening the inlet-valve and ramping up of the injection rate; ii. steady injection of the fuel at a constant mass flow rate; iii. ramping-down of the fuel's velocity, and closing of the inlet-valve; and iv. preparation for the next cycle. Experimentalists have measured the mass fractions of methane, hydroxyl and temperature fields for each pulse using high-speed cameras for the statistical part, and the probability of the ignition delay times and heights above the burner for the pulsed part.

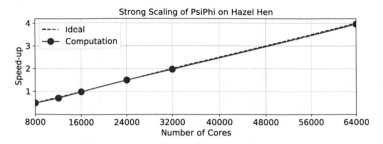

Fig. 1 Strong scaling behavior of PsiPhi on HPC Hazel Hen at Stuttgart

3 Numerical Methods and Algorithms

All simulations use group's in-house LES/DNS code PsiPhi [12, 14, 23] that is available in both density- and pressure-based formulations. The parallelisation performance of this code is given in Fig. 1 by strong scaling tests performed on HPC Hazel Hen at Stuttgart. Favre-filtered governing equations for mass, momentum, a total of 19 different species mass fractions and total enthalpy including the chemical part (without the kinetic part) are solved with a low-storage explicit third-order Runge–Kutta scheme in pressure-based formulation.

The simulations use a direct chemistry approach that employs an augmented reduced mechanism (ARM) developed by Lu and Law [16] that consists of 19 transported species, 11 quasi-steady-state species and 15 lumped reaction steps. This ARM mechanism is based on the GRI3.0 kinetics library [25], whereas the agreement between the reduced and the detailed mechanisms on the ignition delay times and propagation speeds along with its associated thermodynamics and transport databases has been found identical [16]. Hence, an estimation of a factor of two performance gain is obtained by employing this reduced mechanism compared to a detailed one without compensating on accurate AI dynamics. This study assumes a unity Lewis number.

The convective fluxes of scalars are interpolated using a total variation diminishing (TVD) scheme with the CHARM limiter [28]. A second-order central differencing scheme for the convective fluxes of momentum is used. A seventh-order Power-law velocity profile is imposed at the inlet. The inflow turbulence is generated by the method developed by Klein et al. [15] in an efficient implementation by Kempf et al. [13]. The integral length scale l_t and the turbulence intensity I_t are chosen as $0.205 D$ and 5% of the bulk velocity, respectively.

During the LES, the unresolved fluxes in momentum and scalars are estimated from eddy-viscosity and eddy-diffusivity approaches at a turbulent Schmidt number Sc_t of 0.7. Turbulent viscosity ν_t is determined by Nicoud's σ-model [20].

An analytical expression is used to estimate the scalar dissipation rate (SDR) of the mixture fraction, as given in Eq. (1). The sub-grid part of the filtered SDR $\tilde{\chi}$ is modeled with a simplified version of the method by Girimaji and Zhou [10] by adding turbulent diffusivity $D_t = \nu_t/Sc_t$ to the molecular diffusivity D.

$$\tilde{\chi} = 2D\nabla\tilde{Z}\nabla\tilde{Z} \tag{1}$$

For the pulsed part, a linear function is used for both ramping up and down of the fuel mass flow over 0.5 ms. Then, a continuous injection is performed for 10 ms, where the waiting time between the pulses is 39 ms. The waiting time corresponds to around 1.5 co-flow advection times, hence, no interference is observed between the cycles. Therewith, a cycle lasts 50 ms.

4   Computational Resources

The low-resolution steady (SL) and transient (TL) LES use an equidistant Cartesian grid with $\Delta = 0.1$ mm spacing and 80 million cells with dimensions of $65 \times 35 \times 35$ mm^3. These coarse computations use a CFL criterion of 0.7 for the temporal resolution (or around $1.8 \mu s$ depending on the flow conditions). The computational domain includes the final 3 mm of the fuel injector tube. In other words, the axial distance of the computational domain has been forty nozzle diameter long, which is exceptionally long for standard studies, but necessary to capture the possible ignition events that occurred far downstream. This low-resolution transient (TL) LES requires 660,000 CPUh per pulse cycle, where a total of six cycles are performed with a cost of four (4) Million compute unit hours (CPUh), using 11,016 compute-cores (459 nodes) in parallel through message passing interface (MPI).

For the high-resolution transient LES (TH), the grid size is reduced to $\Delta = 0.05$ mm, which requires 674 Million grid points in the $65 \times 35 \times 35$ mm^3 domain at a CFL criterion of 0.3. The final 3 mm of the fuel injector tube is also considered. This smaller CFL condition is a consequence for the stability of the code due to the stiffer gradients that are resolved at this resolution. This high-resolution LES requires 22.5 Million CPUh per pulse cycle, where a total of two cycles are performed with a cost of 45 Million CPUh, using 93,600 compute-cores (3900 nodes) that continuously run over 20 d. All of these computations are summarized in Table 1.

In the TL case, a total of five pulse cycles are phase-averaged to obtain ignition statistics, requiring to simulate a real-time of 311 ms. Only two cycles are performed with the computationally demanding TH case, where the results are cross-compared between the experiments and the simulations. The TH is stopped after 11 ms prior to the start of the second pulse cycle—since the third cycle is not planned. The first cycles are discarded from the statistics for both TL and TH cases.

A priori analysis of one-dimensional laminar flame investigations indicates that the thermal flame thickness should not be smaller than 0.15 mm, considering the highest possible strain rate before a laminar flame quenches. Thus, the flame in our simulations is resolved with at least three grid-points. However, it should be noted that the flame only ignites on far downstream locations, where the strain rates are relaxed, hence, the number of grid points in the flame front should be exponentially higher. It is worth noting that this resolution would be sufficient for a DNS of the

Table 1 Details of the low-resolution steady (SL), low-resolution transient (TL) and high-resolution transient (TH) LES with direct chemistry and their performance figures. The wall-time for each run is 24 h and the size of the computational domain is $65 \times 35 \times 35$ mm^3. The TL case is switched to inert by turning off the chemistry solver during the waiting time between the pulses

Name	Type	$\sum$Cells (M)	Δ/mm	Sim. time (ms)	Runs	#Steps/Run	#CPU/Run	$\sum$CPUh (M)
SL	cont.	80	0.10	75	15	18,000	11,016	4
TL	6-cycle	80	0.10	311	15	18,000	11,016	4
TH	2-cycle	**675**	**0.05**	61	20	11,000	**93,600**	45
								$\sum$ **53**

flame (away from the nozzle exit) as the Kolmogorov scale in the flame front is estimated as 0.17 mm.

During the waiting time of 39 ms between the cycles, the chemistry solver is turned off to hasten the simulation by a factor of ten, therefore, the results after the continuous injection of 10 ms to the next cycle are considered as inert (species source terms are set to zero). Only at the second-cycle of the TH case, the chemistry solver is turned on to study the details of the flow field during the ramp-down and closing of the valve phases. Therefore during this waiting time, the CFL criterion has been replaced with the diffusive time step limit [27] with a model constant of one over twelve due to the overwhelmed diffusion fluxes over the convection ones resulted from the highly-stiff chemistry and laminar flow velocities.

5 Results

The following results are produced from the simulations that have been carried out on the HPC Hazel Hen, with a grant no. of 44141 GCS-JFLA. These results costed a total of 53 Million CPUh.

5.1 Steady Flame

The simulations are validated with the experimental data [1, 9] during the (statistically) steady phase of the jet. The numerical data is taken from the SL case. The sampling starts after 25 ms or around one and a half co-flow's advection time later, which yields a sampling duration of 50 ms. Hence, the sample rate is 0.1 ms, which is much longer than the integral time scale $l_t/\tilde{U}'$ of 0.031 ms (acquired a total of 500 samples).

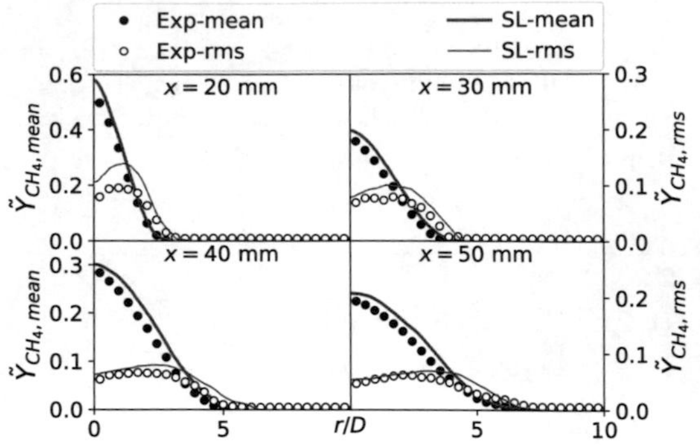

Fig. 2 Radial profiles of
mean and RMS of Mean
methane mass fraction $\tilde{Y}_{CH_4}$
from SL case at four axial
locations on the middle plane

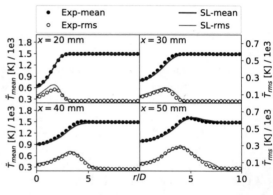

Fig. 3 Radial profiles of
mean and RMS of
temperature $\tilde{T}$ from SL case
at four axial locations on the
middle plane

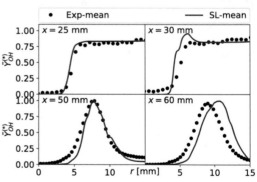

Fig. 4 Radial profiles of
mean normalized hydroxyl
$\tilde{Y}_{OH}^{(*)}$ from SL case at four
axial locations on the middle
plane

The radial profiles at four axial locations for mean and root-mean-square (RMS) of methane mass fraction $\tilde{Y}_{CH_4}$, temperature $\tilde{T}$ and normalized hydroxyl mass fraction $\tilde{Y}_{OH}^{(*)}$ are given in Figs. 2, 3 and 4, respectively. Upstream at 20 mm, the simulations over-predict the jet-penetration length as in the experiments, as shown in Fig. 2. The mass fraction fluctuations upstream are well-predicted, even though no a priori information about the velocity fluctuations at the inlet has been known. Further downstream at 30–50 mm, the simulations predict a slightly wider jet, but, the

Fig. 5 Temporal evolution of the jet with snapshots of mixture fraction $\tilde{Z}$ from the second pulse cycle of TH case

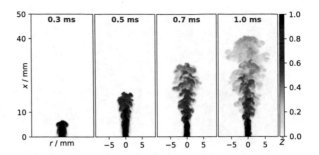

marginal difference between these results indicate that the mixing-dynamics are well described.

Figure 3 shows that the temperature predictions are in almost perfect agreement with the experiments for both mean and the RMS values. This outstanding agreement between the lift-off heights of these results confirms the quality of the simulations.

By the time the results are obtained from the experimentalists [1], the hydroxyl measurements have not been corrected. Therefore, it is agreed to normalize the mass fractions by their maximum values to compare at least the radial position and the thickness of the flame front. In Fig. 4, the location of the flame front is almost identical between the measurements and the calculations for locations up to 50 mm downstream. After 60 mm downstream, a slight shift of the flame towards the hot co-flow side is observed, where it should be noted that this location is very close to the outlet boundary. Overall, even the LES simulations agree remarkably well with the experiments that brought enough evidence to proceed with the pulsed case studies.

5.2 Pulsed Flame

This part uses the transient version of the experimental data available from DLR [2–4, 21, 22]. The TH case is estimated to be almost one order of magnitude more costly than the TL case, as mentioned earlier in Sect. 4. Whilst, a total of six cycles are sampled—five for TL and one for TH.

The temporal evolution of the jet during the second cycle of the TH case is illustrated in Fig. 5. The common vortex ring dynamics of a starting jet, such as the disintegration of the pinched vortex ring to the hot co-flow, are observable. Compared to the cold jets, this vortex ring dissipates to the surrounding co-flow much faster due to the high co-flow temperature. The potential core of the jet remains as short as 8 mm. For more physical insight into the behavior of pulsed jets (reactive and non-reactive), the reader is kindly referred to our recent papers [11, 12].

The formation of an AI kernel over time is presented with contours in Fig. 6 and with iso-surfaces in Fig. 7 by snapshots of temperature for the second pulse of the TH case, where ignition have happened around 1.9 ms. The horizontal cross-section

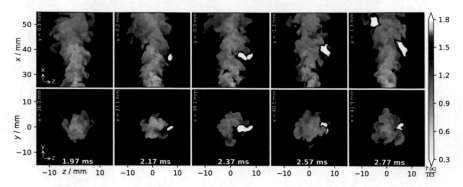

Fig. 6 Snapshots of temperature $\tilde{T}$ to represent the ignition event during the second pulse cycle of TH. The contour-planes are sliced on the position where the AI is occurring. Two view-perspectives of the jet, side view (top row) and top view (bottom row) are given for precise AI location

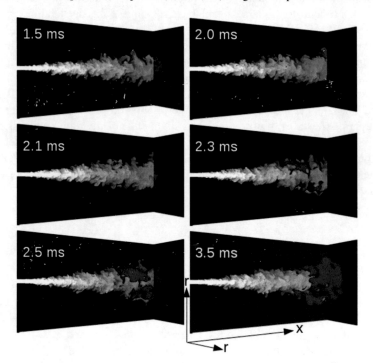

Fig. 7 Iso-surface of ignition temperature $\tilde{T}_{AI}$ from the second pulse cycle of TH case between 1.5 and 3.5 ms. Two slice-perspectives of the jet, middle-plane to the radial and last-plane to the axial directions are given for a better interpretation of the figure

in Fig. 6 is moved with the flow to track the AI kernel. A 15% temperature increase from $T_o = 1490$ K is defined as the outer limit of the AI kernel, which is denoted as the ignition temperature $\tilde{T}_{AI}$.

Figure 6 shows the initial ignition kernel at 1.97 ms. This AI kernel at $(x, y, z) = [36.3, 0.3, 7.8]$ mm emerges on the lean side of the jet's mixing layer, then rapidly grows and ignites the flame. Meanwhile, another kernel have formed slightly towards the downstream location at a later time of 2.57 ms. As the injection continues, several ignition spots appear in a quasi-random manner. The availability of the time-resolved three-dimensional fields of the simulated quantities in this simulations yields a precise information of the ignition dynamics, as represented in Fig. 7 with the ignition temperature iso-surfaces.

The initial ignition spot that appears on the relaxed flow fields can be correlated to the SDR $\tilde{\chi}$. Figure 8 shows the iso-surface of the critical SDR $\tilde{\chi}^*$. This critical SDR corresponds to the highest SDR value that a steadily burning one-dimensional laminar counter-flow flame can achieve. The iso-surface of the ignition temperature is also included to Fig. 8 with a plain red color to indicate the location of the ignition kernel.

Qualitatively, the first ignition spot appears on the SDR smaller than the critical SDR, which can be interpreted from the top sub-figure in Fig. 8 by the missing red iso-surface of the ignition temperature (the red iso-surface remains inside the critical SDR iso-surface, presumably the SDR there is pretty low). This spot also corresponds to a lean mixture fraction of 0.01, where this mixture fraction is named as most reactive mixture fraction. This ignition kernel remains on the lean-side during its expansion, where a small part of this kernel aligns with the stoichiometric mixing layer later on. As evident in the literature [18], the ignition spots should appear in pseudo-random locations, but are correlated to the mixing and other parameters, such as local SDR.

5.3 Pulse Cycle Deviations

The volume integrated temperature $\tilde{T}$ and the absolute value of the heat release rate $|\tilde{q}|$ over a cycle time are presented in Fig. 9. During the ramping up the fuel mass flow, the integrated temperature decreases due to the cold fuel. This is followed by the ignition and flame stabilization that increase the integrated temperatures. It is observed that several slight temperature drops are visible, which can be related to the turbulence that quenches the ignition kernel. Later on, the temperature increases again until the stably burning flame is established.

At around similar times where the temperatures start rising, a monotonic increase of integrated absolute heat release rate is seen until the flame becomes stabilized (Fig. 9b). Interestingly, the integrated heat release rate decreases slightly after it reaches its peak at around 3 ms, which is also the time where the gradient of the integrated temperature after the ignition reaches its maximum value. All of the cycles are almost identical before the ignition. A variation of the temperature in the flame

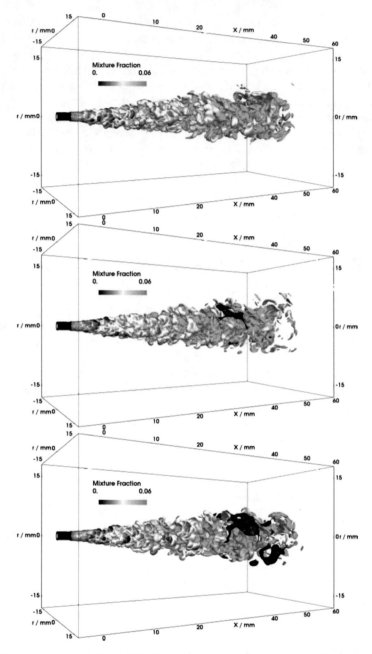

Fig. 8 Two Iso-surfaces of critical SDR $\tilde{\chi}^* = 20\,\mathrm{s}^{-1}$ and ignition temperature $\tilde{T}_{AI}$ from the second pulse cycle of the TH case at 2.0 ms (top), 2.2 ms (middle) and 2.4 ms (bottom). The critical SDR iso-surface is colored with the instantaneous mixture fraction $\tilde{Z}$ and the ignition temperature iso-surface is colored with plain red

Fig. 9 Volume integrated temperature $\tilde{T}$ (**a**) and absolute value of the heat release rate $|\tilde{q}|$ (**b**), together with the velocity pulses (dashed) from the TL and TH cases. The error bars represent the cycle-deviations of TL case

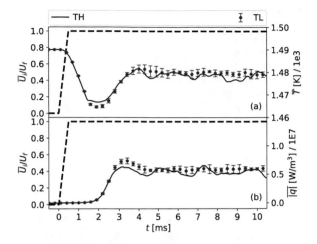

Fig. 10 Maximum temperature $\tilde{T}$ (**a**), mass fractions of formaldehyde $\tilde{Y}_{CH_2O}$ (**b**) and hydroxyl $\tilde{Y}_{OH}$ (**c**), together with the velocity pulses (dashed) from the TL and TH cases. The error bars represent the cycle-deviations of TL case

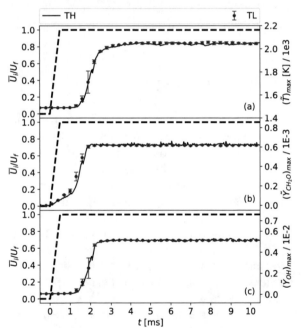

front is observed during these five cycles of TL case. Most of the deviations of the integrated temperature and heat release rate are seen after the flame stabilizes. This deviations can be an evidence for broken reaction zones of the flame front due to the turbulence.

The maximum of temperature, mass fractions of formaldehyde Y_{CH_2O} and hydroxyl Y_{OH} over the duration of a cycle are presented in Fig. 10. The non-monotonic shape of the temperature curve in Fig. 10a happens because of the failed

Fig. 11 Normalized PDFs
of the ignition delay times
(left) and ignition heights
above the burner (right) from
the TL case with five cycles
and the experiments. The
criteria for the ignition is
given in the text

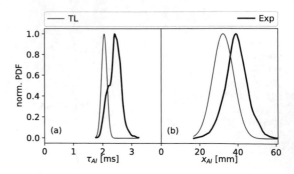

pre-ignition events. A typical failed ignition happens due to the turbulent eddies destroying the ignition spots. It seems that the ignition related species formaldehyde starts forming even just at the start of a cycle, then slowly built until the ignition happens. The failed AI events are more profoundly visible for the formaldehyde mass fraction (Fig. 10b) at 1.5 ms.

5.4 Ignition Statistics

The numerically estimated AI delay times are compared to the experimental ones by their probability density functions (PDF) in Fig. 11, where a Gaussian PDF is fitted to the means standard deviations of the five cycles from TL cases, whereas the experiments used 300 cycles. The AI delay time τ_{AI} are estimated by the time it takes for the temperature values to reach the ignition temperature $\tilde{T}_{AI}$ of 1700 K from to the time when the fuel just leaves the nozzle exit; a similar definition is used in the experiments [4].

The ignition happens between 1.8 and 2.3 ms, which is distributed to a small range. The computed mean delay time is 2.12 ms, which is 0.2 ms earlier than the experimental peak one, however, still in the range of the experimental certainty. Obviously, more samples for the delay times could yield a better agreement to the experimental figures.

The PDF of the height of the ignition kernel to the nozzle X_{AI} are shown in Fig. 11. This height is defined as the axial distance of the initial ignition spot from the burner exit. Similarly, a Gaussian PDF is fitted to the means standard deviations to the predictions. The ignition heights are spread to a wider range than the delay times, which means the discrepancy between the cycle is quite large. The closest ignition height happens at 23 mm, on the other hand, the farthest one happens at 36 mm above the burner exit. The mean predicted height is slightly close to the burner than the mean of the measurements.

6   Conclusions

Highly-resolved Large-eddy simulation (LES) of the pulsed auto-igniting (AI) jet flame in a vitiated co-flow experiment from DLR [2–4, 21, 22] was conducted using direct chemistry. The 19 species augmented reduced mechanism (ARM) for methane combustion developed by Lu and Law [16] was employed to describe the ignition and the further flame stabilization. The computational domain was large enough to cover up to 40 nozzle diameters above the burner exit in order to recover the ignition spots that could occur most downstream locations. A highly-resolved LES with direct chemistry of a pulsed jet in such a large computational domain had, in our knowledge, not been performed before. The aim was to generate a reliable data-set of a pulsed jet flame that could be further analyzed or be used as a benchmark for other researchers.

It was shown that a very good agreement was obtained between the computations and the experiments for the (statistically) steady jet: especially for the mixing dynamics, as well as the temperature and minor species' mass fraction profiles. The predicted flame lift-off heights were almost identical to the experimental evidence. A low-resolution LES was run over six pulses, whereas high-resolution LES was continued for two. The access to the temporally resolved three dimensional scalars of the simulations would definitely help for further understanding of the ongoing turbulence to ignition and turbulence to flame interactions. The predicted ignition delay times and heights were in the experimental range, however, more samples should be performed in order to complete the cycle-to-cycle variation statistics.

These simulations were performed in the HPC Hazel Hen in Stuttgart/Germany (grant no. of 44141 GCS-JFLA) using a total of 53 million CPUh. For the largest simulation, 93.600 CPUs were parallelly run over twenty days. This work successfully demonstrated the necessity of the HPC systems to understand the complex problems found in industrial systems.

Acknowledgements The authors are grateful for the financial support by DFG (Proj. No.: 393710272, KE 1751/13-1) and the Gauss Center High-Performance computing grant on Hazel Hen, Stuttgart (44141 GCS-JFLA).

References

1. C.M. Arndt, Personal communication, 26 November 2016 (2016)
2. C.M. Arndt, J.D. Gounder, W. Meier, M. Aigner, High-speed imaging of auto-ignition of pulsed methane jets in a hot vitiated co-flow. Appl. Phys. B **108**(2), 407–417 (2012)
3. C.M. Arndt, R. Schießl, J.D. Gounder, W. Meier, M. Aigner, Flame stabilization and auto-ignition of pulsed methane jets in a hot coflow: influence of temperature. Proc. Combust. Inst. **34**(1), 1483–1490 (2013)
4. C.M. Arndt, M.J. Papageorge, F. Fuest, J.A. Sutton, W. Meier, M. Aigner, The role of temperature, mixture fraction, and scalar dissipation rate on transient methane injection and auto-ignition in a jet in hot coflow burner. Combust. Flame **167**, 60–71 (2016)

5. G. Cho, D. Jeong, G. Moon, C. Bae, Controlled auto-ignition characteristics of methane-air mixture in a rapid intake compression and expansion machine. Energy **35**(10), 4184–4191 (2010)
6. J. Craske, IMA J. App. Math. **82**(2), 305–333 (2016)
7. J. Craske, M. van Reeuwijk, J. Fluid Mech. **792**, 1013–1052 (2016)
8. P. Domingo, L. Vervisch, D. Veynante, Large-eddy simulation of a lifted methane jet flame in a vitiated coflow. Combust. Flame **152**(3), 415–432 (2008)
9. A. Fiolitakis, P. Ess, P. Gerlinger, M. Aigner, Anwendung eines transportgleichungs-PDF-verfahrens zur berechnung der selbstzündung eines methan-freistrahles, in *Proceedings of 27. Deutscher Flammentag* (2015), pp. 617–628
10. S.S. Girimaji, Y. Zhou, Analysis and modeling of subgrid scalar mixing using numerical data. Phys. Fluids **8**, 1224–1236 (1996)
11. E. Inanc, A.M. Kempf, Numerical study of a pulsed auto-igniting jet flame with detailed tabulated chemistry. Fuel **252**, 408–416 (2019)
12. E. Inanc, M.T. Nguyen, S. Kaiser, A.M. Kempf, High-resolution LES of a starting jet. Comput. Fluids **140**, 435–449 (2016)
13. A.M. Kempf, M. Klein, J. Janicka, Efficient generation of initial-and inflow-conditions for transient turbulent flows in arbitrary geometries. Flow Turbul. Combust. **74**(1), 67–84 (2005)
14. A.M. Kempf, B.J. Geurts, J.C. Oefelein, Error analysis of large-eddy simulation of the turbulent non-premixed sydney bluff-body flame. Combust. Flame **158**(12), 2408–2419 (2011)
15. M. Klein, A. Sadiki, J. Janicka, A digital filter based generation of inflow data for spatially developing direct numerical or large eddy simulations. J. Comput. Phys. **186**(2), 652–665 (2003)
16. T. Lu, C.K. Law, A criterion based on computational singular perturbation for the identification of quasi steady state species: a reduced mechanism for methane oxidation with no chemistry. Combust. Flame **154**(4), 761–774 (2008)
17. C.N. Markides, E. Mastorakos, An experimental study of hydrogen autoignition in a turbulent co-flow of heated air. Proc. Combust. Inst. **30**(1), 883–891 (2005)
18. E. Mastorakos, Ignition of turbulent non-premixed flames. Prog. Energy Combust. Sci. **35**(1), 57–97 (2009)
19. E. Mastorakos, T.A. Baritaud, T.J. Poinsot, Numerical simulations of autoignition in turbulent mixing flows. Combust. Flame **109**(1), 198–223 (1997)
20. F. Nicoud, H.B. Toda, O. Cabrit, S. Bose, J. Lee, Using singular values to build a subgrid-scale model for large eddy simulations. Phys. Fluids **23**(8), 085106 (2011)
21. M.J. Papageorge, C.M. Arndt, F. Fuest, W. Meier, J.A. Sutton, High-speed mixture fraction and temperature imaging of pulsed, turbulent fuel jets auto-igniting in high-temperature, vitiated co-flows. Exp. Fluids **55**(7), 1763–1783 (2014)
22. M.J. Papageorge, C.M. Arndt, F. Fuest, W. Meier, J.A. Sutton, Erratum to: high-speed mixture fraction and temperature imaging of pulsed, turbulent fuel jets auto-igniting in high-temperature, vitiated co-flows. Exp. Fluids **57**(1), 14–20 (2016)
23. F. Proch, P. Domingo, L. Vervisch, A.M. Kempf, Flame resolved simulation of a turbulent premixed bluff-body burner experiment. Part I: analysis of the reaction zone dynamics with tabulated chemistry. Combust. Flame **180**, 321–339 (2017)
24. M. van Reeuwijk, P. Salizzoni, G. Hunt, J. Craske, Phys. Rev. Fluids **1**(7), 074301 (2016)
25. G.P. Smith et al., Gri-mech 3.0 (2000), http://combustion.berkeley.edu/gri-mech/
26. L.J. Spadaccini, M.B. Colket, Ignition delay characteristics of methane fuels. Prog. Energy Combust. Sci. **20**(5), 431–460 (1994)
27. J.G. Verwer, B.P. Sommeijer, W. Hundsdorfer, RKC time-stepping for advection-diffusion-reaction problems. J. Comput. Phys. **201**(1), 61–79 (2004)
28. G. Zhou, Numerical simulations of physical discontinuities in single and multi-fluid flows for arbitrary Mach numbers. Ph.D. thesis, Chalmers University of Technology, Goteborg, Sweden (1995)

Direct Numerical Simulation of Supersonic Film Cooling by Tangential Blowing

Johannes M. F. Peter and Markus J. Kloker

Abstract Film cooling is an effective method to thermally protect the nozzle extension of rocket engines from the hot exhaust gases. A cool secondary gas is blown
into the supersonic hot-gas turbulent boundary layer through a backward-facing step
to generate a cooling film that reduces the heat load of the structure. In this work the
complex interaction between the hot supersonic main-flow and the coolant stream is
investigated using high-order direct numerical simulations (DNS) to gain fundamental understanding of the mixing physics. The cooling gas is injected at a Mach number
of 1.8 into the turbulent Mach-3.3 flat-plate boundary-layer at zero pressure gradient. The main gas is steam (gaseous H_2O), the cooling gas is helium, and adiabatic
wall conditions are used. Results for various blowing ratios F at kept cooling-gas
temperature and Mach number are presented. The interaction of the main stream turbulence and the initially laminar cooling film is investigated in detail as well as the
evolution of the cooling effectiveness. The common Goldstein correlation formula
for the effectiveness is applied, but no satisfying scaling is achieved.

1 Introduction

The structure of modern, high-thrust rocket engines with their high chamber pressure and temperatures is subject to extreme thermal loads. Without active cooling,
the wall temperatures would far exceed the limits of today's available materials.
Hence, innovative and efficient cooling strategies have to be developed. An effective
method also for the nozzle-extension is film cooling, where a cool secondary gas is
blown into the hot supersonic main-flow boundary layer near the wall to generate a
cooling film that reduces the heat load of the structure near and possibly downstream
of the injection location. Beneficial coolant properties are a high heat capacity and

J. M. F. Peter (✉) · M. J. Kloker
Institute of Aerodynamics and Gas Dynamics, University of Stuttgart,
Pfaffenwaldring 21, 70569 Stuttgart, Germany
e-mail: peter@iag.uni-stuttgart.de

M. J. Kloker
e-mail: kloker@iag.uni-stuttgart.de

© Springer Nature Switzerland AG 2021   263
W. E. Nagel et al. (eds.), *High Performance Computing in Science and Engineering '19*,
https://doi.org/10.1007/978-3-030-66792-4_18

low thermal conductivity. The cooling gas can be injected either by wall-normal blowing through holes or slits (see, e.g., [12, 18]) or in wall-parallel fashion by blowing through a backward-facing step (see, e.g., [9, 11, 16]). The cooling efficiency depends on various parameters such as cooling-gas type and temperature, mass flux, and the mixing-flow characteristics. In this work, the complex interaction between the turbulent main and the coolant flow is investigated using high-order direct numerical simulation (DNS). The aim is to identify the fundamental parameters and physical phenomena governing the unavoidable mixing process and the subsequent decrease of the cooling effect downstream of the injection. Further research goals are to examine existing film-cooling correlations, to provide design-guidelines for film-cooling applications and reference cases for turbulence modelling used in faster simulations tools like RANS or LES. The DNS are performed for the introduction of a laminar, cold coolant stream into a turbulent, hot boundary-layer flow at formally zero streamwise pressure gradient. The principal flow setup is based on experimental investigations performed by sub-project A2 [19] of the German Collaborative Research Center SFB-TRR40 (in a nozzle extension). The main features of the flow have been approximately analyzed using RANS, and the DNS are now performed for a subdomain near the injection imposing a zero pressure gradient and an adiabatic wall in a first investigation step, see Sect. 3. The cooling gas is helium, injected through a backward-facing step at a Mach number of 1.8 into the turbulent main flow of hot steam (gaseous H_2O) at Mach-3.3.

The paper is organized as follows: In Sect. 2 the governing equations are provided and the numerical method is described. Section 3 shows the simulation setup, and the DNS results are shown in Sect. 4. Computational aspects of the DNS are discussed in Sect. 5. Finally, Sect. 6 summarizes the main findings and gives some concluding remarks.

2   Numerical Method

For the DNS we use our in-house high-order code NS3D, which has been successfully used for the calculation of film and effusion cooling in boundary-layer flow for non-reacting calorically perfect gases [11–14, 18]. The code is written in Fortran and parallelized using the MPI and OpenMP libraries. Detailed information about the fundamentals of NS3D can be found in, e.g., [1, 12, 13] and parallelization aspects are discussed in [5, 10, 11, 26].

2.1   Governing Equations

The governing equations for a gas-mixture flow of two non-reacting calorically perfect gases are the continuity equation, the three momentum equations, the energy equation, and the equation of state, all for the *mixture* values. Additionally, a second continuity equation for one of the gas species has to be considered and the energy equation has to be modified to include the effects of ordinary and thermal diffusion, caused by concentration and temperature gradients, respectively. The non-dimensionalized equations in vector notation can be found in [14].

For the non-dimensionalization a reference length $L^{\star} = \left(\mu_{\infty}^{\star} \cdot Re_{\infty} \right) / \left(\rho_{\infty}^{\star} \cdot u_{\infty}^{\star} \right)$ and the free-stream values of velocity, density, temperature, viscosity, and heat conductivity are used. Note that the pressure is made dimensionless by $\left(\rho_{\infty}^{\star} u_{\infty}^{\star 2} \right)$. The non-dimensional parameters are the Mach number Ma_{∞}, the Prandtl number Pr and the Reynolds number Re_{∞}. Dimensional quantities are marked by the asterisk *, the subscript $_{\infty}$ refers to free-stream values. Throughout this paper, species 1 is the main flow gas and species 2 is the cooling gas. Both species have constant Prandtl number Pr_i and constant ratio of specific heats $\kappa_i = c_{p,i}/c_{v,i}$, where the species number is indicated by the subscript i. The mass fraction is denoted by c_i. Sutherland's law is used to calculate the dynamic viscosity μ_i of the pure gases as a function of temperature and the mixing rule of Wilke [2] is then used to derive the mixture viscosity. For the diffusion coefficients D and D_T, see, e.g., [2].

2.2 Spatial Discretization and Time Integration

NS3D solves the governing equations in conservative formulation on a block-structured, curvilinear grid. High-order spatial accuracy is achieved by employing (alternatingly biased) compact finite differences of 6th-order. To enable a computationally parallel solution of the resulting equation system a sub-domain compact approach is used. Here, the sub-domains are decoupled by using explicit finite differences of 8th-order at overlapping grid points, thus breaking down the global tridiagonal equation system from the compact finite difference scheme to independent systems for each subdomain [11]. Time integration is performed by an explicit 4th-order 4-step Runge–Kutta scheme. To stabilize the simulation and to ensure de-aliasing, a 10th-order compact low-pass filter can be applied to the conservative variables at a chosen timestep interval [6]. Another effect of the filter is to strongly damp fluctuations in highly stretched grid regions, i.e. in buffer zones ahead of boundaries to minimize reflections. Additionally, sponge regions can be defined at boundaries to prevent undue reflections. If necessary, strong gradients due to shocks or at gas species interfaces can be treated by a shock-capturing procedure based on low-order filtering of the conservative variables [3]. A shock-sensor σ is applied based on the pressure and density gradients, and a 2nd-order filtering is locally performed. Then a blending is done between the original and the filtered flow field based on the value of the shock-sensor. The sensor, and therefore the influence of the low-order filter, is zero if the gradients are below a prescribed threshold.

Table 1 Free-stream conditions for the DNS and thermophysical parameters of superheated steam and helium

Free stream			Steam	Helium
Ma_∞	3.3	Pr	0.8	0.7
$U_\infty^\star$	3383 (m/s)	κ	1.15	1.66
$T_\infty^\star$	1980 (K)	$R^\star$	461.5	2077.3 (J/(kgK))
$p_\infty^\star$	0.28 (bar)	Sutherland $\mu_{ref}^\star$	$1.12 \cdot 10^{-5}$	$1.85 \cdot 10^{-5}$ (kg/(m s))
$\rho_\infty^\star$	0.0306 (kg/m³)	Sutherland $C^\star$	1064.0	79.44 (K)
		Sutherland $T_{ref}^\star$	350.0	273.1 (K)

3 Film-Cooling Setup

3.1 Flow Configuration

In the experimental facility, a hydrogen-oxygen mixture is burnt in a detonation tube to provide rocket-engine-like stagnation conditions for a short testing time in the order of 7−10 ms. The burnt gas—superheated steam/gaseous H_2O/GH_2O—is expanded in an axisymmetric conical nozzle. For film-cooling experiments a cooling gas is injected tangentially to the nozzle wall through a backward facing step downstream of the throat. Only cases with supersonic cooling gas injection are investigated. Due to the short testing time, the nozzle walls virtually remain at their initial temperature. For more details on the experimental setup see [19]. The experimental flow has been analyzed using steady-state RANS of a one-species gas to yield the necessary free-stream conditions for the DNS of the film cooling in a near-wall domain, see [21]. The resulting free-stream parameters, here used for the inflow boundary, are listed in Table 1, along with the used thermophysical properties of hot GH_2O and helium. In contrast to the experiment, where the nozzle wall is effectively strongly cooled due to the short-time experiment, all presented DNS use adiabatic wall conditions, and the free-stream pressure gradient is not considered.

3.2 Film Cooling

The first step in the DNS is a validated turbulent simulation setup for the given flow conditions. Therefore the turbulent boundary-layer alone, i.e. flat-plate without coolant injection, has been simulated and compared to reference data, see [21]. A backward-facing step is now placed at a main-stream Reynolds number based on momentum thickness of $Re_\theta \approx 1000$, the boundary-layer thickness being $\delta_{99}^\star \approx 7.0$ mm. The step has a height of $\delta_{step}^\star = 1.2$ mm and the lower part contains the cooling-slot opening with a height of $s^\star = 0.7$ mm.

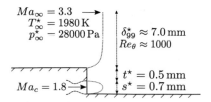

Fig. 1 Detailed view of the geometry in the step region

Table 2 Investigated blowing ratios and cooling stream condition

F	$p_c^\star$ [Pa]	p_c/p_∞	Cooling-channel exit condition
0.33	16 350	0.584	Overexpanded
0.66	32 700	1.168	Underexpanded
1.00	49 050	1.752	Underexpanded

The setup is depicted in Fig. 1. Helium is used as cooling gas and injected through the slot opening. The channel flow is not simulated but a modelled approach is taken, where a parabolic velocity profile is prescribed at the slot opening according to a laminar flow in the cooling-gas channel. The centerline Mach number of the helium stream is fixed to $Ma_c = 1.8$ and the temperature profile is then derived from the assumption of a total-temperature profile that varies linearly from the full value of $T_{0,c}^\star = 330$ K in the center to the recovery value at the walls; the subscript c denotes cooling gas values and the temperature on the centerline is about 200 K. The pressure p_c is taken constant over the slot height and the density ρ_c is derived from the equation of state. All coolant inlet values are kept constant due to the supersonic condition in the channel. Three blowing ratios, $F = 0.33, 0.66, 1.00$ are investigated in this work, were $F = (\rho^\star u^\star)_c / (\rho^\star u^\star)_\infty$. The blowing ratio is varied by varying p_c (and thus ρ_c with $\rho_c \propto p_c$ at constant T_c), leading to different ratios of cooling gas to free-stream pressure p_c/p_∞. The three cases are listed in Table 2. Note that the reported coolant inlet condition, i.e. the cooling-channel exit condition, is based on the free-stream pressure, not on the pressure behind the step without blowing.

3.3 Computational Setup, Initial Conditions, and Boundary Conditions

The full computational setup is depicted in Fig. 2. The length scales for the DNS are non-dimensionalized by the boundary-layer thickness $\delta_{99,i}^\star$ at the inlet. The origin of the coordinate system is placed at the upper edge of the backward-facing step, the regular domain extends from $-80 \le x/\delta_{99,i} \le 87$ in the streamwise direction and has a height of $y/\delta_{99,i} = 20$. This corresponds to approximately eight boundary-

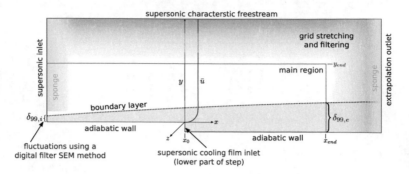

Fig. 2 Setup for the film cooling DNS. The flow is assumed periodic in the z-direction

layer thicknesses at the injection location. In the spanwise direction the domain has a width of $9\,\delta_{99,i}$ or approximately 3.7 boundary-layer thicknesses at the injection location. In both wall-normal and streamwise direction additional buffer regions with grid stretching and compact filtering are added to the regular domain to prevent numerical reflections at the boundaries. At the walls, the no-slip, no-penetration boundary conditions are imposed on the velocity components. The pressure and temperature are obtained by a 5th-order polynomial according to $(\partial p/\partial y)_w = 0$ and $(\partial T/\partial y)_w = 0$, respectively; the density is calculated from the equation of state. At the free-stream boundary, a spatial supersonic characteristic condition is used where all flow variables are computed such that the gradient along spatial characteristics is zero, except for the pressure, which is computed from the equation of state [8]. At the outflow, all flow quantities are extrapolated from the field using a 2nd-order parabola. Additionally, a sponge region is defined ahead of the outflow boundary to help absorb fluctuations before they reach the outlet. At the supersonic main flow inlet, all values are fixed to the profile used as initial condition; additionally, unsteady artificial turbulent fluctuations using a digital filtering synthetic-eddy method (SEM) are superimposed within the boundary layer, see [15, 25]. Although a SEM-boundary provides a pseudo-turbulent flow field at the inlet of the domain, the flow needs about $10\,\delta_{99,i}$ in streamwise direction to fully satisfy equilibrium turbulent-flow statistics. A sponge zone above the boundary layer in the inlet region prevents the far-field flow from being distorted by this transition process and also damps all shocks arising close to the inlet due to the supersonic condition.

4 Results

4.1 Main Flow Features and Cooling Effectiveness

For the three investigated cooling cases Figs. 3 and 4 show the time-averaged stream-wise and wall-normal velocity, respectively. The flow field is averaged over a period of

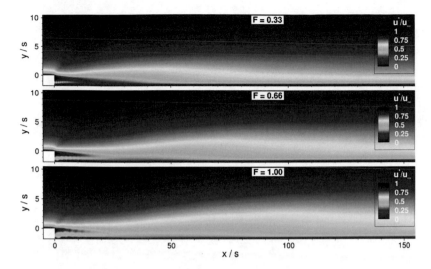

Fig. 3 Contours of streamwise velocity u

at least $u_\infty^\star t_{avg}^\star / s^\star = 250$ and additionally spanwise averaged. (Note that averaging is only started after the initial transient phase has passed; additionally, the turbulent time scales in the cooling region are much smaller than in a regular flat-plate boundary-layer.) Due to the exit pressure of the coolant not being matched to the free stream, the existence of a nozzle shock-train is clearly visible in the velocities, with the wave-length of the shock-structures significantly increasing with higher blowing ratios. The higher cooling-gas pressure for the underexpanded cases ($F = 0.66$ and $F = 1.00$) also leads to a distinct upwards deflection of the hot oncoming stream. In the temperature plot in Fig. 5 the hot oncoming boundary layer is visible in the lower left, with the main-flow recovery temperature being $T_{rec,\infty} = \left(1 + r\frac{\kappa-1}{2}Ma^2\right) \approx 1.76$, where $r = Pr^{1/3}$ is the turbulent recovery factor, and the *total* temperature of the helium being $T_{0,c} \approx 0.17$. The ratio of boundary-layer thickness to step-height is $\delta_{99}/\delta_{step} \approx 6$. Seban [23], Cary and Hefner [4], and Konopka et al. [17] investigated the influence of this ratio and found very little influence on the adiabatic cooling effectiveness. As can be expected, a higher blowing ratio leads to a longer sustaining cooling effect.

The adiabatic cooling effectiveness is defined as

$$\eta_{ad} = \frac{T_{rec,\infty} - T_w}{T_{rec,\infty} - T_{rec,c}}, \tag{1}$$

where $T_{rec,c}$ is the coolant recovery temperature and T_w is the wall temperature with cooling. Figure 6 shows the distribution of η_{ad} along the cooled wall. All three blowing ratios show an initial region with perfect cooling ($\eta_{ad} \approx 1$). Following Stollery [24] this is the "potential-core region" (where the mixing with the main-flow gas has not yet pierced through), which is terminated by the "film-breaking

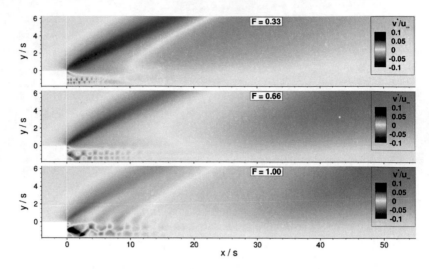

Fig. 4 Contours of wall-normal velocity v. Compared to Figs. 3 and 5, a smaller domain is shown

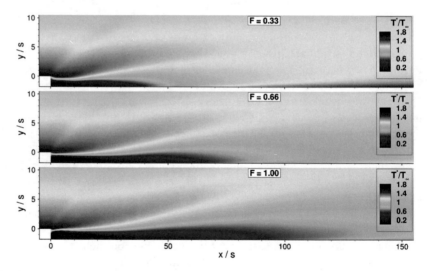

Fig. 5 Contours of temperature T

point". The following "boundary-layer region" shows a decay of the cooling effectiveness due to the transition of the laminar cooling-gas boundary layer to turbulence and the mixing with the hot gas. In this region, the cooling effectiveness can be approximated by

$$\eta = \left(\frac{x}{r}\right)^m, \tag{2}$$

Table 3 Length r of potential-core region and exponent m in $\eta = \left(\frac{x}{r}\right)^m$

F	r/s	m
0.33	26.55	−0.899
0.66	45.48	−0.741
1.00	57.34	−0.575

where r approximately describes the length of the potential-core region. This suggests that the obtained flow fields are self-similar in the respective boundary-layer region. The values for r and m for the three blowing ratios are listed in Table 3. This further highlights the better cooling performance of the higher blowing ratios; a higher cooling-gas injection rate leads to a thicker film that takes longer to be heated up by the main gas. The region with perfect cooling is increased and the following effectiveness decay is weaker, but both effects scale sub-linearly with the blowing ratio.

Another important quantity is the fluctuation of the wall temperature, as those fluctuations may cause problems with thermal fatigue. To that end, the fluctuation of the cooling effectiveness,

$$\eta'_{ad,rms} = \sqrt{\overline{\eta'_{ad}}^2} = \frac{\sqrt{\overline{T'^2_w}}}{T_{rec,\infty} - T_{rec,c}} = \frac{T'_{w,rms}}{T_{rec,\infty} - T_{rec,c}}, \tag{3}$$

is looked at, where the overbar denotes time-averaged data, the subscript rms refers to root-mean-square, and $'$ indicates a fluctuation. The distribution is also shown in Fig. 6. All cases show the same qualitative behavior: the fluctuation is close to zero in the potential-core region with perfect cooling, followed by a rise to a peak and a decay that approaches values on the order of those expected in supersonic turbulent flat-plate boundary-layers [27]. The two blowing ratios with $p_c > p_\infty$ show a similar quantitative behavior. The rise starts somewhat earlier for $F = 0.66$, but similar peak values of $\eta'_{ad,rms}$ are attained. In contrast, the lowest blowing rate with $p_c < p_\infty$ shows a much higher peak value. The overexpansion of the coolant for $F = 0.33$ leads to a shock train behind the slot with a much shorter wavelength than for the underexpanded cases, giving rise to short-wavelength disturbances (cf. Sect. 4.2).

For the prediction of the film-cooling effectiveness an often employed approach is the description of the mixing process using a boundary-layer model, in the attempt to scale different setups using a self-similarity correlation variable. A very common mixing-layer approach is by Goldstein [7], with the scaling variable

$$\xi = \frac{x^\star}{F s^\star} \frac{\rho^*}{\rho^\star_\infty} \left(Re_c \frac{\mu^\star_c}{\mu^*}\right)^{-0.25}, \tag{4}$$

where Re_c is the cooling-channel Reynolds number and μ_c is the coolant viscosity, here both evaluated using averaged properties of the coolant. Values with the super-

Fig. 6 Comparison of mean (solid lines) and fluctuating (rms) (dashed lines) cooling effectiveness. Dash-dotted lines show $\eta = \left(\frac{x}{r}\right)^m$-approximation in boundary-layer region

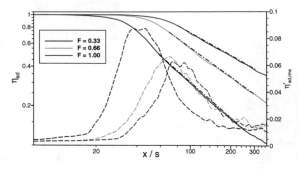

Fig. 7 Scaling of cooling effectiveness

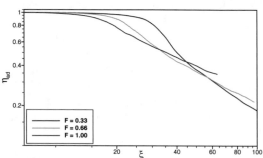

script * are evaluated using the reference-temperature method. A correlation for the reference temperature of a non-air gas is given by Rasmussen [22]:

$$T^* = 0.566 Pr^{1/3} T_\infty^* + \left(1 - 0.566 Pr^{1/3}\right) T_w^* +$$
$$\left(0.566 Pr^{5/6} - 0.421 Pr\right) \frac{\kappa_1 - 1}{2} Ma_\infty^2 T_\infty^* \tag{5}$$

In the derivation of this formula a laminar, self-similar flat-plate boundary-layer with zero pressure gradient and constant wall temperature is assumed. The application of this formula in the present case is therefore questionable. (We note that it was applied in [19].) The reference temperature is calculated using the main-flow Prandtl number and the local (therefore varying) wall temperature for each case. The resulting correlation $\eta(\xi)$ is shown in Fig. 7. As can be seen, no satisfying scaling collapse can be achieved. Neither the length of the potential-core region nor the slope of the effectiveness drop-off in the boundary-layer region collapse for the three blowing ratios. For a better match x must have an exponent depending on the blowing ratio in Eq. 4, which is so far effectively close to one. Note that for cooling in a laminar boundary-layer a scaling with ξ roughly proportional to x is successful [14].

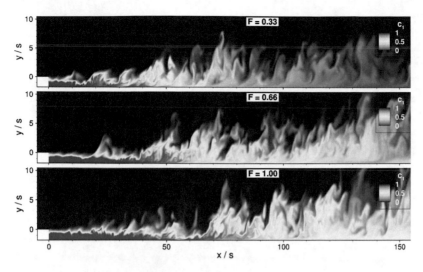

Fig. 8 Instantaneous snapshots of the main-gas mass fraction c_1

4.2 Turbulent Mixing

The strong impact of the main-flow turbulence on the initially laminar cooling stream can be seen in Fig. 8, where snapshots of the main-gas mass fraction c_1 are shown. Dynamical coherent structures (high-shear layers) known from laminar-turbulent boundary-layer transition scenarios quickly appear in the mixing layer, indicating a quick transition of the mixing shear-layer and break-up of the coolant stream.

The structures appear the farther downstream the larger the blowing ratio is. The higher-density and thus higher-momentum jets seem more resistant to distortion through the vortices present in the main-flow boundary-layer. Further evidence is provided by the turbulent kinetic energy (TKE)

$$k = \frac{1}{2\bar{\rho}} \cdot \left(\overline{\rho u'' u''} + \overline{\rho v'' v''} + \overline{\rho w'' w''} \right), \qquad (6)$$

where $''$ indicates a Favre-fluctuation. The time- and spanwise-averaged contours of k are shown in Fig. 9. High TKE values also indicate intense mixing due to turbulence. With an increase of the blowing ratio and overpressure the TKE in the mixing zone decreases and the mixing zones lift up, meaning less coolant is transported away from the wall. Generally, the turbulent kinetic energy levels in the mixing region are higher than in the oncoming boundary layer. Additionally, Fig. 9 shows wall-normal profiles of the streamwise velocity at downstream locations $x/s = 10$, $x/s = 30$, $x/s = 80$. Pink dots in the profiles mark locations of generalized inflection points (GIPs), determined from $\frac{\partial}{\partial y} \left(\rho \frac{\partial u}{\partial y} \right) = 0$ [20]. The existence of GIPs in the averaged

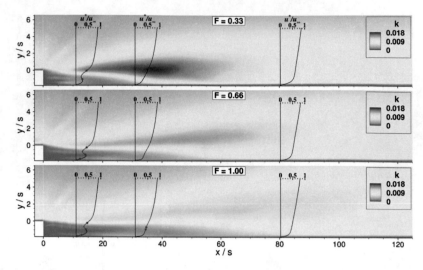

Fig. 9 Contours of turbulent kinetic energy k. The lines show wall-normal profiles of the velocity component u at $x/s = 10$, $x/s = 30$, $x/s = 80$, pink dots mark generalized inflection points

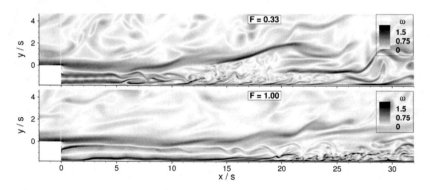

Fig. 10 Instantaneous snapshots of the absolute vorticity ω

velocity profile leads to the existence of an inviscid instability in the mean flow, feeding the generation of turbulence in the mixing zone.

Figure 10 shows details of the flow field in the mixing region by instantaneous total-vorticity contours. Three different disturbance scales can be identified: main gas turbulence (MGT, for $y/s > 0$ and x/s small), cooling gas turbulence (CGT, for $F = 1.00$ and $x/s > 20$), and the shock-train structures near the injection (CGS, for $F = 0.33$ and $0 \leq x/s \leq 5$). The CGT scales are much smaller than the MGT scales due to the higher density ($l \propto l^{+} \cdot \rho^{-\frac{1}{2}}$), and thus the direct infection of the laminar cooling stream by the MGT is impeded with higher F. Additionally, the CGS spatial wavenumber gets smaller. For $F = 0.33$ the CGT seems clearly triggered by an instability of the CGS due to its wavenumber scale matching the CGT scale.

5 Computational Aspects

The simulations for the presented film-cooling DNS are carried out on the Cray XC40 'Hazelhen' supercomputer at the federal high performance computing center Stuttgart (HLRS). Additional simulations, especially for the validation of the turbulent baseflow (see [21]), have been run on the NEC SX-ACE 'Kabuki' vector computer system at HLRS.

The simulation code NS3D is parallelized using a hybrid MPI/OpenMP approach. For the version used in this investigation, the computational grid can be split in the x- and y-direction using MPI for inter-domain communication and within a domain the z-direction is parallelized using the shared-memory OpenMP library. Recently, scaling aspects of this approach have been investigated and some shortcomings have been found [5]. The number of cores used for the z-direction is limited to the 24 available cores per node (then with one MPI process per node), independent of the number of grid points. Additionally, the code has been found to scale well up to 12 OpenMP threads with the parallel efficiency dropping to only around 55% for 24 threads due to socket-remote memory access in the ccNUMA architecture of the Cray XC40. This has been taken as incentive to invest time into improving the code performance and scalability to very high core counts by implementing a full MPI decomposition in all spatial directions. Figure 11 shows the speed-up and efficiency as a function of the CPUs used for the parallelization of the z-direction for the old code version (pure increase of OpenMP threads) and the improved variant (mixed MPI decomposition and increase of OpenMP-threads). The test case consists of $6912 \times 600 \times 256$ grid points in the x-, y-, and z-direction, respectively. For the baseline case the domain is split into 768 MPI sub-domains in the $x - y$-plane using one CPU per MPI process. The scaling is then done using parallelization of the z-direction up to 24 CPUs, giving a scaling from 768 to 18 432 CPUs total. Very good results have been achieved showing a large increase in efficiency for high core numbers and thus enabling even higher parallelization while keeping the already-good base performance of the code, see [26]. The scaling using parallelization of the z-direction is now on par with scaling (by pure domain decomposition) in the x-y plane, see [11].

These improvements were implemented in a from-ground revised version of the code, which was done for the single-species version of NS3D first. The computational work for the present results was started before the extension to two-component flows was implemented and thus the "old" NS3D was used. Further work for this project will be using the revised code version. The computational details of the presented film-cooling DNS are listed in Table 4. The grid is highly decomposed in the x- and y-direction with only 35×35 grid points per domain, while the z-direction consists of 1024 grid points (due to the limitations mentioned above).

Performance optimization workshop The HLRS biannually hosts the "Optimization of Scaling and Node-level Performance" workshop, where code developers jointly work with the cluster experts of the supercomputing center on their simulation software. The knowledge gained during these workshop has greatly helped

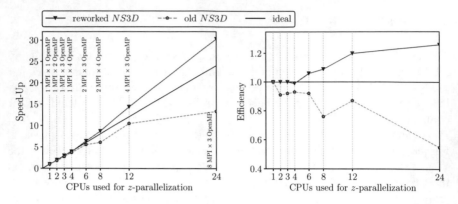

Fig. 11 (Left) Speed-up and (right) efficiency for a variation of the number of CPUs used for the z-parallelization. Scaling is done using pure OpenMP for the old NS3D and a mixture of MPI/OpenMP for the reworked code

Table 4 Computational details of the investigated film-cooling DNS

Parameter	Value
MPI ranks $\times$ OpenMP threads = total CPUs	$1248 \times 12 = 14\,976$
Grid points of the main domain $N_x \times N_y \times N_z$	$4130 \times 315 \times 1024$
Grid points of the step domain $N_x \times N_y \times N_z$	$2170 \times 105 \times 1024$
Grid points per MPI rank $N_x \times N_y \times N_z$	$35 \times 35 \times 1024 = 1\,254\,400$
Total grid points	1.57×10^9
Computed time steps	$300\,000$

in the performance optimization of the revised NS3D code and its scaling aspects. Due to the time investment from both the people working on NS3D and the team of HLRS, see the acknowledgments, the code is well suited for current and next-generation HPC systems at HLRS.

6 Conclusions and Outlook

High-order DNS of a supersonic film-cooling configuration have been performed for various blowing ratios $F = (\rho^\star u^\star)_c / (\rho^\star u^\star)_\infty$. Analysis of the adiabatic cooling effectiveness η_{ad} shows the expected better performance for higher blowing ratios. Here the laminar-cooling-gas density and pressure have been varied at kept temperature and Mach number. A high momentum by high density and pressure of the cooling stream makes it less vulnerable to turbulence infection and mixing, whereas blowing with an overexpanded cooling gas does the opposite. The results for η have been scaled using the common Goldstein formula, but no satisfying results could be achieved. The correlation formula does not include the turbulent mixing that depends strongly on the flow structure; the evaluation of the needed reference T^* is non-trivial by the mixing process. The DNS show that the turbulent-kinetic-energy maximum reduces with increasing F, thus the mixing is decreased and η does not decay as fast. A next step is the investigation of the slot height and the coolant Mach number.

Acknowledgements Financial support has been provided by the German Research Foundation (Deutsche Forschungsgemeinschaft—DFG) in the framework of the Sonderforschungsbereich Transregio 40 (SFB-TRR40, SP A4). The simulations were performed on the national supercomputers Cray XC40 'Hazelhen' and NEC SX-ACE 'Kabuki' at the High Performance Computing Center Stuttgart (HLRS) under grant GCS_Lamt, ID 44026. The authors and the people working on NS3D at IAG in the groups on "Transition and Turbulence" and "Boundary-Layer Flows" would like to thank the team of the HLRS, especially Björn Dick, Stefan Andersson (Cray), and Holger Berger (NEC), for their continued support in developing software for HPC applications.

References

1. A. Babucke, Direct numerical simulation of noise-generation mechanisms in the mixing layer of a jet. Ph.D. thesis, Universität Stuttgart (2009)
2. R.B. Bird, W.E. Stewart, E.N. Lightfoot, *Transport Phenomena* (John Wiley & Sons Inc., 1960)
3. C. Bogey, N. de Cacqueray, C. Bailly, A shock-capturing methodology based on adaptive spatial filtering for high-order non-linear computations. J. Comput. Phys. **228**(5), 1447–1465 (2009). https://doi.org/10.1016/j.jcp.2008.10.042
4. A.M. Cary, J.N. Hefner, Film-cooling effectiveness and skin friction in hypersonic turbulent flow. AIAA J. **10**(9), 1188–1193 (1972). https://doi.org/10.2514/3.50348
5. P.C. Dörr, Z. Guo, J.M.F. Peter, M.J. Kloker, Control of traveling crossflow vortices using plasma actuators, in *High Performance Computing in Science and Engineering '17*, ed. by W.E. Nagel, D.B. Kröner, M.M. Resch (Springer International Publishing, 2017). https://doi.org/10.1007/978-3-319-68394-2_15
6. D.V. Gaitonde, M.R. Visbal, Padé-type higher-order boundary filters for the Navier-Stokes equations. AIAA J. **38**(11), 2103–2112 (2000). https://doi.org/10.2514/2.872
7. R.J. Goldstein, Film cooling. Adv. Heat Transf. **7**, 321–379 (1971). https://doi.org/10.1016/S0065-2717(08)70020-0
8. P.J. Harris, HF Fasel, Numerical investigation of unsteady plane wakes at supersonic speeds, in *34th Aerospace Sciences Meeting and Exhibit* (1996). https://doi.org/10.2514/6.1996-686. aIAA-1996-686
9. K.A. Juhany, M.L. Hunt, Flowfield measurements in supersonic film cooling including the effect of shock-wave interaction. AIAA J. **32**(3), 578–585 (1994). https://doi.org/10.2514/3.12024
10. M. Keller, M.J. Kloker, Direct numerical simulations of film cooling in a supersonic boundary-layer flow on massively-parallel supercomputers, in *Sustained Simulation Performance* (Springer International Publishing, 2013), pp 107–128. https://doi.org/10.1007/978-3-319-01439-5_8. Or see ResearchGate.net
11. M. Keller, M.J. Kloker, DNS of effusion cooling in a supersonic boundary-layer flow: influence of turbulence, in *44th AIAA Thermophysics Conference* (2013). https://doi.org/10.2514/6.2013-2897. aIAA-2013-2897
12. M. Keller, M.J. Kloker, Effusion cooling and flow tripping and in laminar and supersonic boundary-layer and flow. AIAA J. **53**(4) (2015). https://doi.org/10.2514/1.J053251
13. M. Keller, M.J. Kloker, Direct numerical simulation of foreign-gas film cooling in supersonic boundary-layer flow. AIAA J. **55**(1), 99–111 (2016). https://doi.org/10.2514/1.J055115

14. M. Keller, M.J. Kloker, H. Olivier, Influence of cooling-gas properties on film-cooling effectiveness in supersonic flow. J. Spacecr. Rocket. **52**(5), 1443–1455 (2015). https://doi.org/10.2514/1.A33203

15. M. Klein, A. Sadiki, J. Janicka, A digital filter based generation of inflow data for spatially developing direct numerical or large eddy simulations. J. Comput. Phys. **186**(2), 652–665 (2003). https://doi.org/10.1016/s0021-9991(03)00090-1

16. M. Konopka, M. Meinke, W. Schröder, Large-eddy simulation of supersonic film cooling at finite pressure gradients, in *High Performance Computing in Science and Engineering* (Springer International Publishing, 2011), pp. 353–369. https://doi.org/10.1007/978-3-642-23869-7_26

17. M. Konopka, M. Meinke, W. Schröder, Large-eddy simulation of shock/cooling-film interaction. AIAA J. **50**(10), 2102–2114 (2012). https://doi.org/10.2514/1.J051405

18. J. Linn, M.J. Kloker, Effects of wall-temperature conditions on effusion cooling in a supersonic boundary layer. AIAA J. **49**(2), 299–307 (2011). https://doi.org/10.2514/1.J050383

19. S. Ludescher, H. Olivier, Experimental investigations of film cooling in a conical nozzle under rocket-engine-like flow conditions. AIAA J. **57**(3), 1172–1183 (2018). https://doi.org/10.2514/1.j057486

20. L.M. Mack, Boundary-layer linear stability theory, in *Special Course on Stability and Transition of Laminar Flow - AGARD-R-709*, AGARD (1984)

21. J.M.F. Peter, M.J. Kloker, Preliminary work for DNS of rocket-nozzle film-cooling, in *Deutscher Luft- und Raumfahrtkongress DLRK, DLRK-2017-450178* (2017), http://d-nb.info/1142014584. Or see ResearchGate.net

22. M.L. Rasmussen, *Hypersonic Flow* (Wiley, New York, 1994)

23. R.A. Seban, Effects of initial boundary-layer thickness on a tangential injection system. J. Heat Transf. **82**(4), 392–393 (1960). https://doi.org/10.1115/1.3679966

24. J.L. Stollery, A.A.M. El-Ehwany, A note on the use of a boundary-layer model for correlating film-cooling data. Int. J. Heat Mass Transf. **8**(1), 55–65 (1965). https://doi.org/10.1016/0017-9310(65)90097-9

25. E. Touber, Unsteadiness in shock-wave/boundary-layer interactions. Ph.D. thesis, University of Southampton (2010)

26. C. Wenzel, J.M.F. Peter, B. Selent, M.B. Weinschenk, U. Rist, M.J. Kloker, DNS of compressible turbulent boundary layers with adverse pressure gradients, in *High Performance Computing in Science and Engineering '17*, ed. by W.E. Nagel, D.B. Kröner, M.M. Resch (Springer International Publishing, 2018). https://doi.org/10.1007/978-3-030-13325-2

27. C. Wenzel, B. Selent, M.J. Kloker, U. Rist, DNS of compressible turbulent boundary layers and assessment of data-/scaling-law quality. J. Fluid Mech. **842**, 428–468 (2018). https://doi.org/10.1017/jfm.2018.179

On Parallelization Strategies for Multiple Representative Interactive Flamelets Combustion Models

Marco Davidovic, Mathis Bode, and Heinz Pitsch

Abstract A large fraction of today's greenhouse gas emission is produced from burning fossil fuels. Within the transportation energy sector, liquid fossil fuels are still the major energy source. Even though huge effort has been directed to alternative propulsion systems, such as electrified powertrains, it is expected that conventional combustion engines cannot be substituted within the next decades. Liquid fuels that are synthesized using renewable energy can improve the overall CO_2 balance on a short time scale, if they are usable in conventional combustion systems. Furthermore, it has been shown in engine experiments that some synthetic fuels might simultaneously reduce pollutant emissions. However, the exact reasons are not entirely understood and more fundamental knowledge is required in order to design synthetic fuels for technical applications. Therefore, the simulation framework *CIAO* is being developed for studying the entire process chain of fuel injection and combustion with highly accurate numerical methods and physical models. Due to the broad range of scales that occur in such combustion systems, the simulations can only be performed on modern supercomputers. In this study, the performance and scalability of the Multiple Representative Interactive Flamelets (MRIF) combustion model is analyzed and two different parallelization strategies are discussed.

1 Introduction

In order to reduce risks and effect of climate change, the increase in global average temperature should be limited to 1.5 °C according to the Paris agreement [25]. In order to achieve this goal, greenhouse gas (GHG) emissions need to be reduced drastically within the next two decades. Burning fossil fuels in propulsion systems contributes significantly to global warming, e.g. 14% of the global greenhouse gas emissions in 2010 can be attributed to the conversion of fossil fuels to CO_2 [8]. While substantial effort is being directed towards electrified propulsion concepts,

M. Davidovic (✉) · M. Bode · H. Pitsch
Institute for Combustion Technology, RWTH Aachen University, Templergraben 64, 52056 Aachen, Germany
e-mail: m.davidovic@itv.rwth-aachen.de

© Springer Nature Switzerland AG 2021
W. E. Nagel et al. (eds.), *High Performance Computing in Science and Engineering '19*,
https://doi.org/10.1007/978-3-030-66792-4_19

279

combustion engines are expected to provide the majority of transportation energy in the foreseeable future due to advantages in energy density, storage capabilities, ease of distribution, and existing infrastructure [8, 13]. Synthetic fuels that are produced from renewable energy and alternative carbon sources have a huge potential to reduce GHG emissions on a short time scale, since existing engines and infrastructure can be utilized. In addition to the closed CO_2 loop advantage, experiments have shown that renewable fuels might also reduce pollutant emissions or increase efficiency compared to conventional fuels [10]. However, the effect of individual fuel properties on the combustion and pollutant formation process is not entirely understood. Hence, the fuel and future engine design is extremely challenging due to the complexity of the system and large variety of fuel candidates.

There are several processes, which are affected by the fuel properties, e.g. the nozzle internal flow featuring cavitation, the fuel jet breakup, the evaporation process, and the chemical reactions during combustion and pollutant formation. Most of these processes have been individually studied and models have been developed. However, interactions between those processes are not entirely understood and predictive models do not yet exist. The broad range of scales prohibits numerical simulations of entire systems due to computational limitations. In the present study, the injection process is decomposed in three coupled simulations with different resolution and physics in order to reduce the computational effort. The nozzle internal flow is computed with a Large-Eddy Simulation (LES) accounting for the complex nozzle internal geometry and potential cavitation. The flow field at the nozzle outlet is applied as boundary condition to Direct Numerical Simulation (DNS) of the primary breakup using resolved interface methods. Recorded statistics of the droplet distribution are imposed to a Lagrangian particle based LES, which includes submodels for evaporation, secondary breakup, combustion, and pollutant formation. This approach, which has been successfully applied in previous studies [2, 4, 5], still requires the computational power of current supercomputers. Hence, the simulation code, *CIAO*, needs to scale up to high numbers of processors. While the generic flow solver of *CIAO* has shown very good scaling behavior [3], the effect of different parallelization strategies of the Multiple Interactive Representative Flamelets (MRIF) combustion model has not been studied in literature and will be investigated in the present study. The *Spray A* case, which has been defined by the Engine Combustion Network (ECN), is used in this study as target case, since comprehensive experimental data sets are available. All simulations have been performed on the national supercomputer Cray XC40 at the HLRS Stuttgart. The coupled nozzle-internal flow/primary breakup simulations (NIF/PBS) have significantly larger sizes compared to the Lagrangian spray simulations (LSS). Hence, the NIF/PBS scale up to the machine size [3], while the LSS are typically performed on 100–200 nodes.

2 Physical Models

The entire fuel injection and combustion process is decomposed into three coupled simulations:

1. LES of the nozzle internal flow
2. DNS of the primary breakup
3. LES of the secondary breakup, evaporation, and combustion

Since the computational performance of the combustion model is investigated in this paper, more details about the third simulation are given in the following. For more information on the first two simulations, the reader is referred to Bode et al. [1].

2.1 Lagrangian Spray Model

During the primary breakup process, the liquid fuel jet disintegrates into tiny and mostly spherical structures, which may further breakup into even smaller droplets and evaporate. Since resolving all structures is computationally unreasonable, Lagrangian particle models are typically applied for such kind of problems. In the Lagrangian approach, the shape of the droplets is presumed and the droplet quantities are integrated along their trajectory in time. Interactions with the surrounding gas need to be modeled. In the present study, standard drag and evaporation models [17, 19, 21] are applied to solve the position, $x_{d,i}$, the velocity, $u_{d,i}$, the mass m_d, and the temperature T_d of a droplet:

$$\frac{\mathrm{d}x_{d,i}}{\mathrm{d}t} = u_{d,i}, \tag{1}$$

$$\frac{\mathrm{d}u_{d,i}}{\mathrm{d}t} = \frac{f_1}{\tau_d}\left(\tilde{u}_i - u_{d,i}\right) = \dot{u}_{d,i}, \tag{2}$$

$$\frac{\mathrm{d}m_d}{\mathrm{d}t} = -\frac{\mathrm{Sh}}{3\mathrm{Sc}}\frac{m_d}{\tau_d}\ln\left(1 + B_M\right) = \dot{m}_d, \tag{3}$$

$$\frac{\mathrm{d}T_d}{\mathrm{d}t} = \frac{\mathrm{Nu}}{3\mathrm{Pr}}\frac{c_p}{c_l}\frac{f_2}{\tau_d}\left(\tilde{T} - T_d\right) + \frac{\dot{m}_d}{m_d}\frac{L_v}{c_l}$$

$$= \frac{\dot{Q}_d}{c_l m_d} + \frac{\dot{m}_d}{m_d}\frac{L_v}{c_l}. \tag{4}$$

i corresponds to the Cartesian direction, while $\tilde{u}_i$ and $\tilde{T}$ denote the Favre filtered gas velocity and temperature, respectively. The heat capacity of the liquid droplet is denoted by c_l, while c_p is the heat capacity of the surrounding gas at constant pressure. L_v is the latent heat of vaporization of the droplet. The Prandtl number, Pr, and Schmidt number, Sc, are evaluated from local transport properties. For more information on modeling the Nusselt number, Nu, Sherwood number, Sh, the particle time constant, τ_d, and the two correction coefficients f_1 and f_2 the reader is referred to Knudsen et al. [14]. The mass transfer number is evaluated based on the fuel vapor mass fraction at the particle surface $Y_{d,s}$ and the local filtered fuel mass fraction $\widetilde{Y}_F$ according to:

$$B_M = \frac{Y_{d,s} - \widetilde{Y}_F}{1 - Y_{d,s}}. \tag{5}$$

$Y_{d,s}$ is computed by vapor-liquid equilibrium at the interface.

The Kelvin–Helmholtz Rayleigh–Taylor model is applied for secondary breakup [20], while droplet collision is neglected. The initial droplet size distribution and spray cone angle are determined from droplet statistics from the DNS [5].

2.2 Turbulent Flow Model

The flow of a continuous fluid is governed by the balance of mass, momentum, and energy, which is expressed by the Navier–Stokes equations. Even though this set of equations is valid for turbulent flows, obtaining a direct numerical solution of high Reynolds number flows is typically not feasible due to computational limitations, since the broad range of scales cannot be resolved. In LES, the governing equations are filtered and thus, only the large scale turbulent structures need to be resolved. While the computational cost is drastically reduced, the filter operation introduces unclosed terms, which require modeling. The filtered balance equation for compressible flows read:

$$\frac{\partial \bar{\rho}}{\partial t} + \frac{\partial \bar{\rho} \widetilde{u}_i}{\partial x_i} = \bar{\bar{S}}_\rho, \tag{6}$$

$$\frac{\partial \bar{\rho} \widetilde{u}_i}{\partial t} + \frac{\partial \bar{\rho} \widetilde{u}_i \widetilde{u}_j}{\partial x_i} = -\frac{\partial \bar{P}}{\partial x_j} + \frac{\partial}{\partial x_i} \left[\bar{\tau}_{ij} - \bar{\rho} \left(\widetilde{u_i u_j} - \widetilde{u}_i \widetilde{u}_j \right) \right] + \bar{\bar{S}}_{\rho u_i}, \tag{7}$$

$$\frac{\partial \bar{\rho} \widetilde{e}}{\partial t} + \frac{\partial \bar{\rho} \widetilde{u}_i \widetilde{e}}{\partial x_i} = \overline{P \frac{\partial u_i}{\partial x_i}} + \overline{\tau_{ij} \frac{\partial u_i}{\partial x_j}}$$

$$\frac{\partial}{\partial x_i} \left[\overline{\lambda \frac{\partial T}{\partial x_i}} + \overline{\rho \sum_k^{n_s} h_k D_k \frac{\partial Y_k}{\partial x_i}} - \bar{\rho} \left(\widetilde{u_i e} - \widetilde{u}_i \widetilde{e} \right) \right]$$

$$+ \bar{\bar{S}}_e \tag{8}$$

In these equations, t denotes the time, x_i the spatial coordinate in direction of i, ρ the density, u_i the velocity in direction of i, P the pressure, τ_{ij} the viscous stress tensor, e the energy, and T the temperature. h_k, D_k, and Y_k denote the specific enthalpy, diffusivity, and mass fraction of species k, respectively, while n_s denotes the number of species. Contributions from the liquid phase are included in the volumetric source terms $\dot{S}_\rho$, $\dot{S}_{\rho u_i}$, and $\dot{S}_e$, which are computed according to:

$$\widetilde{\dot{S}}_\rho = -\frac{1}{\Delta V} \sum_p^{n_p} w_{p,xyz} n_{d,p} \dot{m}_{d,p}, \tag{9}$$

$$\widetilde{\dot{S}}_{\rho u_i} = -\frac{1}{\Delta V} \sum_p^{n_p} w_{p,xyz} n_{d,p} (u_{d,p} \dot{m}_{d,p} + \dot{u}_{d,p} m_{d,p}), \tag{10}$$

$$\widetilde{\dot{S}}_{\rho e} = -\frac{1}{\Delta V} \sum_p^{n_p} w_{p,xyz} n_{d,p} (\dot{Q}_{d,p} + \dot{m}_{d,p} e_{d,p}). \tag{11}$$

ΔV is the grid cell volume, n_p is the number of parcels in the surrounding cells, $w_{p,xyz}$ are weighting factors for distributing the source term of a parcel to its surrounding cells, and $n_{d,p}$ is the number of droplets in the parcel.

The unresolved Reynolds stresses $(\widetilde{u_i u_j} - \widetilde{u}_i \widetilde{u}_j)$ and energy fluxes $(\widetilde{u_i e} - \widetilde{u}_i \widetilde{e})$ require subgrid scale (SGS) turbulence models. In this study, a dynamic Smagorinsky model with averaging along Lagrangian trajectories [9] is applied. The filtered laminar diffusion fluxes are modeled using the simple gradient assumption. The ideal gas law

$$\bar{P} = \bar{\rho} \frac{\mathcal{R}}{W} \widetilde{T} \tag{12}$$

is employed as equation of state, where $\mathcal{R}$ is the universal gas constant and W the molecular weight of the gas mixture. The temperature of the mixture is calculated from the transported energy value mixture composition according to $\widetilde{e} = \sum_n^{n_k} [\widetilde{Y}_k h_k(\widetilde{T})] - R/W \, \widetilde{T}$. The local mixture composition is provided by the combustion model, which is briefly described in the following section.

2.3 Combustion Model

In LES, only the large turbulence scales are resolved on the numerical grid. However, combustion typically takes place on very small scales, especially at elevated pressure. Hence, chemical source terms cannot be evaluated using the filtered quantities. Several LES models exist for non-premixed combustion, such as flamelet based models. The flamelet concept views a turbulent flame as an aggregation of thin laminar flamelets. Based on this assumption, the chemistry and flow can be decoupled. In most flamelet models, the chemistry is represented in a one-dimensional space. For non-premixed combustion, the mixture fraction, which is a measure for mixing of fuel and oxidizer stream, is typically chosen as flamelet coordinate. The mixture fraction, Z, can be solved in the flow domain according to:

$$\frac{\partial \bar{\rho}\widetilde{Z}}{\partial t} + \frac{\partial \bar{\rho}\widetilde{u}_i\widetilde{Z}}{\partial x_i} = \frac{\partial}{\partial x_i}\left[\bar{\rho}D_Z\frac{\partial \widetilde{Z}}{\partial x_i} - \left(\widetilde{u_i Z} - \widetilde{u}_i\widetilde{Z}\right)\right] + \hat{\widetilde{S}}_\rho. \tag{13}$$

The species transport and energy equation can be transformed from physical into mixture fraction space. The unsteady non-premixed laminar flamelet equations read:

$$\frac{\partial Y_k}{\partial t} = \frac{\chi}{2}\frac{\partial^2 Y_k}{\partial Z^2} + \dot{\omega}_k, \tag{14}$$

$$c_p\frac{\partial T}{\partial t} = c_p\frac{\chi}{2}\frac{\partial^2 T}{\partial Z^2} + \frac{\chi}{2}\left(\frac{\partial c_p}{\partial Z} + \sum_k^{n_s} c_{p,k}\frac{\partial Y_k}{\partial Z}\right)\frac{\partial T}{\partial Z} + \frac{1}{\rho}\frac{\partial P}{\partial t} + \sum_k^{n_s} h_k\dot{\omega}_k. \tag{15}$$

In those equations, $\dot{\omega}_k$ is the chemical source term of species k and χ is the scalar dissipation rate, which is defined as $\chi = 2D_Z\left(\frac{\partial Z}{\partial x_j}\right)^2$. The scalar dissipation rate is a crucial parameter, since it enables diffusive transport in mixture fraction space, and needs to be provided by the flow solver. In the Representative Interactive Flamelet (RIF) model [22], the flamelet equations are advanced interactively with the flow in a two-way coupling sense, which is advantageous since the temporal evolution of temperature, species concentrations, and scalar dissipation rate are considered. However, the model does not account for spatial variations in scalar dissipation rate field as well as different fuel residence times. The latter becomes especially important, if the injection event is longer than the ignition delay. In this situations, both burnt and unburnt fuel are simultaneously present in the simulation domain, which cannot be represented by a single mixture fraction flamelet. In order to overcome these restrictions, multiple flamelets are solved, which are attributed to different fuel residence times and also represent different locations in physical space. A Multiple Representative Interactive (MRIF) LES model has been developed by the authors, which already showed good agreement with experimental data for the targeted simulation case [5]. In order to account for different fuel residence time, fuel that has been injected in a discrete time interval is tracked within the flow domain using $n_f - 1$ additional marking scalars M_f

$$\frac{\partial \bar{\rho}\widetilde{M}_f}{\partial t} + \frac{\partial \bar{\rho}\widetilde{u}_i\widetilde{M}_f}{\partial x_i} = \frac{\partial}{\partial x_i}\left[\bar{\rho}D_Z\frac{\partial \widetilde{M}_f}{\partial x_i} - \left(\widetilde{u_i M_f} - \widetilde{u}_i\widetilde{M}_f\right)\right] + \hat{\widetilde{S}}_{\rho,f}. \tag{16}$$

where n_f denotes the number of flamelets and the source term $\hat{\widetilde{S}}_{\rho,f}$ is equal to $\hat{\widetilde{S}}_\rho$ in a certain time interval and zero at all other times. Given the local values of the marking scalars, weighting factors $\widetilde{W}_f = \widetilde{M}_f/\widetilde{Z}$ are defined that are used within the scalar dissipation rate conditional averaging process. Thus, every flamelet has its unique scalar dissipation rate profile. After the flamelet equations have been advanced in time, a presumed probability density function (PDF) closure is used for evaluating the filtered species from the laminar solution. The weighting factors are then used for linearly combining all flamelet solutions in order to find the local species solution. For more details on the combustion model, the reader is referred to Davidovic et al.

[5]. The Hybrid Method of Moments (HMOM) soot model [18] is coupled to the MRIF combustion in order to predict formation of soot.

3 Numerical Methods

In this study, the Navier–Stokes equations are solved in a fully compressible formulation on structured staggered grids. The spatial discretization is based on finite differences with arbitrary order [6], while the time integration is performed using an explicit low-storage five-stage Runge–Kutta (RK) method [12, 23]. The momentum equations are discretized using central differences in order to achieve low numerical dissipation. WENO schemes are used for the convective terms of the scalar transport equations in order to ensure bounded solutions [15]. The Lagrangian particles are advanced on an adaptive particle specific time step using a 2nd order RK method. Information from gas flow fields are evaluated using 3D linear interpolation, while the particle source terms are distributed using the same weights. In order to reduce the computational effort, the stochastic parcel approach is applied, which pools droplets with similar properties into a so-called parcels [7]. The flamelet equations are solved using central differences. The flamelet grids are coarsened in the very fuel rich zones. Time integration is performed using CVODE from the SUNDIALS package [11]. Different Krylov methods are provided in SUNDIALS that can be used with the Newton iteration of CVODE. In this study, the Generalized Minimal Residual (GMRES) method is applied.

4 Parallelization

CIAO is parallelized for distributed memory computing clusters using Message Passing Interface (MPI). The structured grid of the flow solver is decomposed into Cartesian partitions. At the partition borders, overlap (buffer) cell values need to be communicated in all three spatial directions, if spatial gradients need to be evaluated. For those communications, the MPI Cartesian Communicator is utilized. The number of buffer cells depends on the spatial discretization scheme.

Since the Lagrangian particles interact with the flow field in a two-way manner, the MPI process that owns the flow partition, where the particle is physically located on, also solves the particle's equations. This approach reduces the communication overhead, but simultaneously increases the load-imbalance, if the particles are not homogeneously distributed in space. A particle is communicated when it has reached the overlap cells of its current partition. In addition, the overlap cell values of the source term fields need to be summarized in a communication step.

The chemistry is solved on a 1D mixture grid. CVODE offers a MPI parallelized version, which is applied in this study. The solution vector of the ODE system has the size $n_s \cdot n_z$, where n_s is the number of species and n_z is the number of flamelet grid points. The solution vector is decomposed in mixture fraction space and neighbor-to-neighbor communication is implemented in the user-specified routines for evaluating the gradients in the diffusive terms. CVODE also applies all-to-all communication for evaluating the residuals. Since the number of flamelet grid points is typically much smaller than the number of MPI ranks, sub communicators are created for executing the chemistry solver. However, since multiple flamelets have to be advanced at the same time, $n_{tasks} = n_z \cdot n_f$ are available for distribution. The distribution of those tasks is trivial, if the number of MPI ranks, n_{MPI}, exceeds the number of tasks $n_{MPI} \geq n_{tasks}$. However, if $n_{MPI} < n_{tasks}$ and presuming that all MPI tasks should participate in these operations, two parallelization strategies exist:

- Strategy I: Each process participates in the computation of one flamelet by computing more than one grid point.
- Strategy II: Each process participates at multiple flamelet computations but just owns fewer grid points on each flamelet compared to strategy I.

The simplified program flow of a simulation time step is visualized in Fig. 1. Each MPI process is depicted by a black arrow, while horizontal lines illustrate a communication on the global MPI communicator. The parallelized tasks are visualized by boxes, while the name of the task is given for the first MPI rank. First, each process advances the Lagrangian particles that are present on its flow domain in a frozen flow field. Particles that have reached the overlap cells during integration are communicated afterwards. The source terms in the buffer cells are also communicated but neglected in Fig. 1 for the sake of simplicity. Secondly, the flow field is advanced in multiple RK steps. The overlap cells of density, momentum, scalars, species, temperature and pressure need to be communicated in each RK-substep, which is not shown in Fig. 1. In the following step, the chemistry solution is advanced. Parallelization strategy I is applied and hence, each MPI rank participates in the calculation of just one flamelet. After the flamelets have been integrated to the new time, the solutions need to be broadcasted in order to update the species fields in the flow solver. After the species have been updated in the flow domain, a Newton iteration is performed in each cell in order to compute the temperature from the transported energy given the updated gas composition. Finally, the pressure is updated using the equation of state.

The second parallelization strategy is visualized in Fig. 2. All MPI ranks own partitions in multiple flamelets. The partitions are smaller compared to those of strategy I, such that each flamelet is computed by more MPI processes. Note that strategy I is computationally advantageous in terms of communication cost, while strategy II yields smaller data sizes in CVODE for each MPI rank, which might improve the cache utilization efficiency. In fact, in [3], it was already shown that the chemistry solver shows superideal scaling up to one grid point per MPI rank.

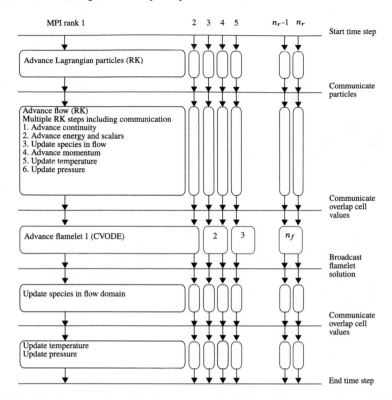

Fig. 1 Simplified program flow of the Lagrangian spray LES applying strategy I (marked in red)

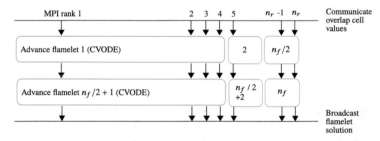

Fig. 2 Flamelet advancement section of the Lagrangian spray LES applying strategy II (marked in green)

5 Simulation Case and Numerical Setup

The *Spray A* target case, which has been defined by the Engine Combustion Network [24] is targeted in this study. *Spray A* is a single-hole injection into a constant volume spray chamber. The injector has been well characterized and the case has be measured extensively by multiple research groups, who applied advanced exper-

Table 1 Summary of the *Spray A* case

Nozzle outlet diameter	90 µm
Nozzle K factor	1.5
Discharge coefficient	0.86
Area contraction coefficient	0.98
Ambient gas temperature	900 K
Ambient gas pressure	60 bar
Ambient gas oxygen	15%
Ambient gas velocity	Quiescent
Injection pressure	1500 bar
Fuel	n-dodecane
Fuel temperature	363 K
Injection duration	1.5 ms
Chamber dimensions	$108 \times 108 \times 108$ mm^3

Table 2 Numerical setup of the *Spray A* case

Chamber dimensions	$20 \times 20 \times 90$ mm^3
Minimum grid spacing	80 μm
Time step size	2 ns
Number of cfd cells	$50 \cdot 10^6$
Numerical schemes	CD4/WENO5
Number of liquid parcels	$300 \cdot 10^3$
Number of flamelets	40
Number of flamelet grid points	96
Number of species	822
Number of reactions	4597

imental methods like laser-optical diagnostics. The conditions of the *Spray A* case are summarized in Table 1.

In the numerical setup, only a subdomain of $20 \times 20 \times 90$ mm^3 is computed. The nozzle geometry is considered in the simulation using a stair-step approach. The minimum grid spacing at the nozzle orifice is 80 μm yielding approximately $50 \cdot 10^6$ cells. Applying the Courant–Friedrichs–Lewy <1 restriction results in a time step of roughly 20 ns. 4th-order central differences (CD4) are applied in the simulation for the momentum equations and diffusive terms, while the convection terms of the scalar equations are approximated by a WENO5 scheme.

The number of droplets per parcels has been set such that the number of parcels does not exceed $300 \cdot 10^3$.

The utilized chemical mechanism has been developed at the ITV and consists of 822 species and 4597 reactions including NO$_x$ and PAH chemistry. A sensitivity

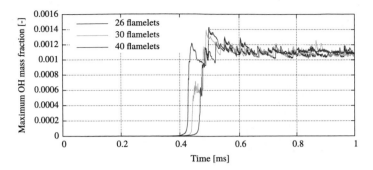

Fig. 3 Temporal evolution of the maximum OH mass fraction inside the flow domain for different numbers of flamelets

study has been performed in order to estimate the required number of flamelets. Figure 3 shows the temporal evolution of the maximum OH mass fraction, which is commonly used for determining the ignition delay, for different flamelet counts. Even though multiple realizations are required for a quantitative comparison, the results indicate that ignition delay increases with larger number of flamelets. Given those results, the following performance measurements are performed using 40 flamelets. Each flamelet has been discretized using 96 grid points using central differences of 2nd-order. A summary of the numerical setup is given in Table 2.

6 Results

6.1 Scaling

In this section, the scaling behavior of the different flamelet parallelization strategies is compared. For this test, a production job with the setup described in Sect. 5 has been restarted at 1.4 ms with different numbers of computing nodes and has been run for 20 time steps. I/O and initialization has not been measured. Figure 4 shows the measured timings for the entire code (a) and for the code modules that have the highest contribution to the total time—the scalar solver (b), the combustion module (c), the Lagrangian spray module (d). Flamelet parallelization strategy I and II are shown in red and green, respectively. Ideal scaling is depicted by dotted lines.

It can be seen that the entire code shows much better performance using parallelization strategy II at fewer numbers of nodes. Strategy II shows improved performance at moderate number of nodes, which is caused by more efficient cache utilization of the CVODE solver. The same behavior has already been found in cache performance measurements of the flamelet solver by Bode et al. [3]. Additionally, it can be observed that performance of the scalar and spray module is not affected by the flamelet parallelization strategy. This can easily be explained by the program

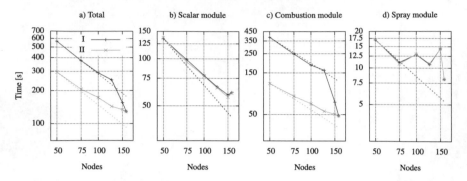

Fig. 4 Scaling of the Lagrangian spray LES and the most expensive code modules. Strategy I is shown in red and Strategy II in green. Ideal scaling is depicted by dotted lines

flow (Figs. 1 and 2). All MPI processes are synchronized when entering and leaving the flamelet solver in the MPI_ALLREDUCE communication of the scalar dissipation profiles χ_k and the MPI_BCAST communication of the flamelet solutions. At 160 nodes, the difference in computing time vanishes, since $n_{MPI} = n_{tasks}$ and hence, both strategies yield the same results. In general, it can be seen that the spray module already exceeded its scaling limit at 75 nodes, while the scalar module scales fairly well up to 100 nodes for present problem size and WENO5 schemes.

6.2 Comparison with Experiments

In this section, the recent simulation results are compared with experimental data that has been provided by Sandia National Laboratories through the ECN [24]. In the plot on the left-hand side of Fig. 5, a comparison of the experimentally measured and numerically predicted temporal evolution of the spray vapor penetration length is shown. Although the penetration length is slightly over predicted towards the end of injection (1.5 ms), an overall good agreement has been achieved. In the plot on the right-hand side of Fig. 5, the temporal evolution of the total soot mass is shown. Note that the HMOM model is only one-way coupled to the combustion model and hence does not contribute to PAH consumption in the gas phase. Considering that soot formation is affected by all subprocesses within the fuel spray combustion processes and hence all modeling errors are integrated into this quantity, the result is satisfying. More realizations are required however, in order to allow a quantitative comparison of statistical measures. Figure 6 shows a visualization of the simulation results performed on the rendering nodes of Cray XC40.

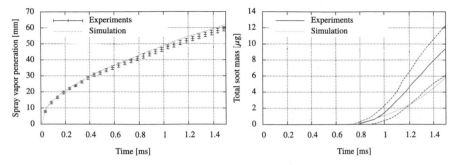

Fig. 5 Comparison of simulation results with experimental data provided by Sandia National Laboratories throughout the ECN. Left: Spray vapor penetration length over time—Experiments in red error bars and simulation results from a single realization in green. Right: Total soot mass over time—Experiments [16] (mean +/- sigma) in red and simulation results from a single realization in green

Fig. 6 Visualization of the simulation results. 3D volume rendering of the soot radiation (black-body radiation colormap) and 3D iso-surface of OH mass fraction (blue) and liquid particle in grey

7 Conclusions

This work focuses on the development of predictive models, which are well suited for studying the fuel spray injection process and its effect on combustion and pollutant formation for novel synthetic fuels. A Lagrangian spray LES model, which employs the MRIF combustion model, has been presented. The results of the presented simulation framework showed good agreement with experimental data in terms of spray penetration and soot formation. However, multiple realizations are required for a more quantitative comparison with experiments. Furthermore, two parallelization strategies for the MRIF combustion have been implemented into the code and the computational performance of those strategies has been compared. It was shown that it is beneficial to reduce the partition size in each flamelet domain of the MPI ranks to a minimum. Even though the communication overhead increases, the more efficient cache utilization overcompensates this penalty resulting in much better computational performance, especially at moderate number of MPI ranks.

Acknowledgements The authors gratefully acknowledge support by Stefan Andersson (Cray) and Björn Dick (HLRS, University of Stuttgart). Additionally, funding by the Cluster of Excellence 'The Fuel Science Center' and computing time on the national supercomputer Cray XC40 at the HLRS under the grant 'GCS-MRES' are acknowledged.

References

1. M. Bode, M. Davidovic, H. Pitsch, Multi-scale coupling for predictive injector simulations, in *Jülich Aachen Research Alliance (JARA) High-Performance Computing Symposium* (Springer, Berlin, 2016), pp. 96–108
2. M. Bode, T. Falkenstein, M. Davidovic, H. Pitsch, H. Taniguchi, K. Murayama, T. Arima, S. Moon, J. Wang, A. Arioka, Effects of cavitation and hydraulic flip in 3-hole GDI injectors. SAE Int. J. Fuels Lubr. **10**(2), 380–393 (2017)
3. M. Bode, M. Davidovic, H. Pitsch, Towards clean propulsion with synthetic fuels: computational aspects and analysis. High Perform. Comput. Sci. Eng. **18** (2019)
4. M. Bode, T. Falkenstein, H. Pitsch, T. Kimijima, H. Taniguchi, T. Arima, Numerical study of the impact of cavitation on the spray processes during gasoline direct injection. ICLASS
5. M. Davidovic, T. Falkenstein, M. Bode, L. Cai, S. Kang, J. Hinrichs, H. Pitsch, LES of n-dodecane spray combustion using a multiple representative interactive flamelets model. Oil & Gas Sci. Technol.-Revue d'IFP Energ. nouvelles **72**(5), 29 (2017)
6. O. Desjardins, G. Blanquart, G. Balarac, H. Pitsch, High order conservative finite difference scheme for variable density low Mach number turbulent flows. J. Comput. Phys. **227**(15), 7125–7159 (2008)
7. J.K. Dukowicz, A particle-fluid numerical model for liquid sprays. J. Comput. Phys. **35**(2), 229–253 (1980)
8. O. Edenhofer, R. Pichs-Madruga, Y. Sokona, S. Kadner, J. Minx, S. Brunner, Climate Change 2014: Mitigation. Technical summary (2014)
9. M. Germano, U. Piomelli, P. Moin, W.H. Cabot, A dynamic subgrid-scale Eddy viscosity model. Phys. Fluids A: Fluid Dyn. **3**(7), 1760–1765 (1991)
10. B. Heuser, F. Kremer, S. Pischinger, J. Klankermayer, Optimization of diesel combustion and emissions with tailor-made fuels from biomass. SAE Int. J. Fuels Lubr. **6**(3), 922–934 (2013)
11. A.C. Hindmarsh, P.N. Brown, K.E. Grant, S.L. Lee, R. Serban, D.E. Shumaker, C.S. Woodward, SUNDIALS: suite of nonlinear and differential/algebraic equation solvers. ACM Trans. Math. Softw. (TOMS) **31**(3), 363–396 (2005)
12. F. Hu, M.Y. Hussaini, J. Manthey, Low-dissipation and low-dispersion Runge–Kutta schemes for computational acoustics. J. Comput. Phys. **124**(1), 177–191 (1996)
13. G. Kalghatgi, Developments in internal combustion engines and implications for combustion science and future transport fuels. Proc. Combust. Inst. **35**(1), 101–115 (2015)
14. E. Knudsen, H. Pitsch et al., Modeling partially premixed combustion behavior in multiphase les. Combust. Flame **162**(1), 159–180 (2015)
15. X.-D. Liu, S. Osher, T. Chan, Weighted essentially non-oscillatory schemes. J. Comput. Phys. **115**(1), 200–212 (1994)
16. N. Maes, S. Skeen, *ECN5 workshop - topic 6 - soot*
17. R.S. Miller, J. Bellan, Direct numerical simulation of a confined three-dimensional gas mixing layer with one evaporating hydrocarbon-droplet-laden stream. J. Fluid Mech. **384**, 293–338 (1999)

18. M. Mueller, G. Blanquart, H. Pitsch, Hybrid method of moments for modeling soot formation and growth. Combust. Flame **156**(6), 1143–1155 (2009)
19. N.A. Okong'o, J. Bellan, Consistent large-Eddy simulation of a temporal mixing layer laden with evaporating drops. Part 1. Direct numerical simulation, formulation and a priori analysis. J. Fluid Mech. **499**, 1–47 (2004)
20. M.A. Patterson, R.D. Reitz, Modeling the effects of fuel spray characteristics on diesel engine combustion and emission. SAE Trans. 27–43 (1998)
21. H. Pitsch, Large-Eddy simulation of turbulent combustion. Annu. Rev. Fluid Mech. **38**, 453–482 (2006)
22. H. Pitsch, Y. Wan, N. Peters, Numerical investigation of soot formation and oxidation under diesel engine conditions. SAE Trans. 938–949 (1995)
23. D. Stanescu, W. Habashi, 2n-storage low dissipation and dispersion Runge–Kutta schemes for computational acoustics. J. Comput. Phys. **143**(2), 674–681 (1998)
24. The Engine Combustion Network (ECN)
25. The Paris Agreement (UNFCCC) (2016)

Drag Reduction by Surface Actuation

Marian Albers, Matthias Meinke, and Wolfgang Schröder

Abstract The flow over an DRA2303 wing section is controlled by spanwise traveling transversal surface waves. The actuated flow field is investigated by large-eddy simulation. Approximately 74% of the solid surface is deflected by a sinusoidal space- and time-dependent function in the wall-normal direction. Viscous drag reduction by 8.6% with a strong decrease of skin-friction in the favorable pressure gradient region and an overall drag decrease by 7.5% are achieved. Furthermore, a slight increase in lift is obtained for the external flow over a realistic geometry.

1 Introduction

Despite more efficient aircraft engines, the interest in reducing friction drag, which makes up to 50% of the total drag for standard narrow- and wide-body aircraft, is still high. A large variety of approaches has already been considered in the literature.

In general, the drag reduction techniques can be categorized in active or passive methods. Among passive methods, longitudinal grooves aligned with the streamwise direction, so-called shark-skin surfaces or riblets are well known. Early works by Walsh and Weinstein [55] and Bechert et al. [6] proved the general applicability of ribbed surfaces to lower turbulent friction drag. The responsible mechanism was shown in [24] to be related to the imposition of the no-slip condition for the spanwise velocity fluctuations at greater height into the flow than for the streamwise and wall-normal velocity components, thereby driving quasi-streamwise vortices further off the wall. García-Mayoral and Jiménez [16] investigated the degradation of the drag reduction at increasing riblet sizes and showed the link to a two-dimensional Kelvin-Helmholtz-like instability of the mean streamwise flow. In-flight tests confirmed the drag reducing effect using an aircraft partially covered with riblets [49]. However,

M. Albers (✉) · M. Meinke · W. Schröder
Institute of Aerodynamics, RWTH Aachen University, Wüllnerstrasse 5a, 52062 Aachen, Germany
e-mail: m.albers@aia.rwth-aachen.de

W. Schröder
JARA – High-Performance Computing, Forschungszentrum Jülich, 52425 Jülich, Germany

© Springer Nature Switzerland AG 2021
W. E. Nagel et al. (eds.), *High Performance Computing in Science and Engineering '19*,
https://doi.org/10.1007/978-3-030-66792-4_20

the sensitivity of the drag reducing effect of riblets to surface deterioration due to long-term use and the challenge to ensure the optimal riblet size to varying flow conditions have prevented their application in a production environment [43]. Besides riblets, other passive approaches are available. The use of compliant surfaces seems to be promising [10, 28, 34, 56]. The injection of gas reduces drag [9, 40]. Moreover, superhydrophobic surfaces yield interesting findings in turbulent flows [18]. Extensive research on large-eddy breakup devices showed hardly any energy saving in practical applications [3]. Other methods achieve drag reduction by stabilizing the wake in the downstream region of the airfoil [7].

A solution to the small-parameter-range drawback of passive methods can be overcome by active methods. They offer the possibility to alter the actuation parameters based on the operating conditions. The maximum drag reduction rates of most methods are higher than those of passive techniques. Following the observation of suppressed turbulent production by temporal spanwise pressure gradients [38], Jung et al. [26] measured turbulent flow with spanwise wall oscillations. In turbulent channel flows, they achieved drag reduction rates up to 40%.

Ever since, a wide variety of actuation types has numerically and experimentally been developed and tested. An excellent review on the different approaches is given by Quadrio [41]. Streamwise traveling waves of spanwise wall velocity were applied to turbulent channel flow by Quadrio et al. [42], yielding up to 60% drag reduction. Nakanishi et al. [39] investigated waves of wall-normal motions traveling in the streamwise direction determined relaminarized flow. A wave-like wall-normal excitation of piezo-actuator array decreased the wall-shear stress locally and a quick recovery was observed in [5]. Another promising approach is a forcing by spatial waves in the spanwise direction investigated by Du and Karniadakis [11] and Du et al. [12]. They found considerable reduction of the turbulent streamwise intensities, resulting in a reduced overall skin friction. Zhao et al. [57] advanced the idea by deriving an equivalent in-plane velocity. A similar type of control in turbulent boundary layer flow was conducted by Itoh et al. [22] and Tamano and Itoh [50]. The skin-friction was reduced by up to 13%. Numerical investigations by Klumpp et al. [30] and Koh et al. [31] confirmed these findings for a long wavelength, while a shorter wavelength enlarged the drag. The extension of these investigations to non-zero pressure gradient turbulent boundary layers [36] revealed a decreased drag reduction rate. More recent results [1] indicate a higher drag reduction potential for smaller wave periods and larger wavelengths.

The majority of the former research focuses on canonical flow problems, whereas only some investigations consider engineering-like flow geometries. Active modifications of the attached turbulent boundary layer on wing surfaces are primarily based on blowing and/or suction [33]. This, however, can result in increased boundary-layer thicknesses and more energetic fluctuations [27, 52]. Recent results by Atzori et al. [4] show promising reductions of the skin friction.

Combined passive and active measures are often considered in laminar flow control (LFC) [43], either by modifying the airfoil shape, natural laminar flow (NLF), or by an additional active technique, hybrid laminar flow control (HLFC), to control the development of instabilities [15]. HLFC has been tested in flight [19] and is currently applied to the Boeing 787-9. Note that the skin-friction is increased in the remaining turbulent boundary layer due to the lower thickness Reynolds number compared to the non-controlled case [47].

In brief, there is an enormous amount of literature on drag reduction available. However, hardly any of the active turbulent control approaches have been analyzed for more realistic configurations. The techniques which have been used in realistic flow setups do not directly interact with the turbulent flow field or possess technical difficulties. For this reason, the active drag reduction technique based on spanwise traveling transversal surface waves [36], will be applied to the wing section of an DRA2303 airfoil. It is the objective to reduce the overall drag for such an engineering geometry without an increase of the pressure drag and without lowering the lift. The full analysis of the drag reduction potential is published in [2]. This contribution represents a concise excerpt of the aforementioned paper.

2 Numerical Method

The computational approach is based on a finite volume approximation of the Navier-Stokes equations for unsteady compressible flows. The large-eddy simulation (LES) concept is used to determine the turbulent flow field. A second-order accurate formulation of the convective fluxes using the Advection Upstream Splitting Method (AUSM) [32] is applied. The viscous fluxes are discretized by a modified cell-vertex scheme [35]. The time integration is performed by a second-order accurate 5-stage Runge–Kutta scheme. The smallest turbulent scales are modeled by the monotonically integrated large-eddy simulation (MILES) approach [8]. A detailed discussion of the method and the subgrid-scale model was presented in [35]. To capture the wall motion, the Arbitrary Lagrangian–Eulerian (ALE) formulation of the Navier-Stokes equations is used [20].

A parallelization with distributed memory and communication via the message passing interface (MPI) is employed. The structured mesh is partitioned using a tree-based splitting algorithm to generate sub-blocks of the global mesh [17]. Each sub-block is assigned to one parallel process and the flow variables at the sub-block interfaces are exchanged in every Runge–Kutta sub-step.

This numerical method has thoroughly been extensively validated, see e.g.. [29, 30, 37, 44].

3 Computational Setup

The flow is defined in a Cartesian domain around an infinite wing section with the x-axis being the airfoil chord, the z-axis defining the direction of the infinite wing span, and the y-axis being perpendicular to the other two axes. Positions are $\mathbf{x} = (x, y, z)$ and velocities are denoted by $\mathbf{u} = (u, v, w)$, the pressure is given by p and the density

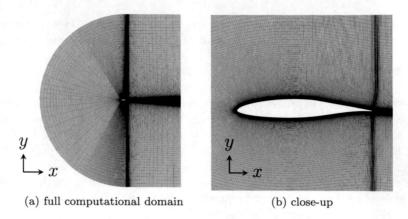

(a) full computational domain   (b) close-up

Fig. 1 C-type body-fitted mesh; the x-axis is aligned with the airfoil chord, only every 10th grid point in the wall-tangential direction is shown; **a** full computational domain; **b** close-up view

by ρ. The geometry of the wing section is defined by an DRA2303 airfoil with a maximum thickness of 14% chord. The flow field of this supercritical laminar-type airfoil has extensively been investigated for transonic flows [13, 14, 44, 48]. It can be considered a fundamental generic airfoil for this flow regime [31, 37]. The minimum of the pressure coefficient $c_{p,min} = \min((p - p_\infty)/(\rho_\infty u_\infty^2))$ on the suction side of the airfoil is reached almost at mid-chord, i.e., $(x/c)_{c_{p,min}} \approx 0.5 - 0.55$.

A block-structured three-dimensional body-fitted C-type mesh is used to discretize the physical domain with an extension of 15 chord lengths upstream and 10 chord lengths downstream of the wing. The size of the computational domain is substantially larger than in the DNS by Shan et al. [46]. In the spanwise direction, the wing section is 10% chord wide. This width suffices to capture at least three boundary layer thicknesses and is wide enough to consider multiple wave lengths of the actuation function. A two-dimensional representation of the numerical domain is shown in Fig. 1a with a close-up of the mesh in the vicinity of the airfoil shown in Fig. 1b.

A RANS simulation of the same setup with a larger domain size of $L_x \times L_y = 120c \times 120c$ yielded the same pressure distribution. The grid resolution in inner scaling on the wing surface near the leading and the trailing edge is $\Delta x^+ < 21.0$, $\Delta y^+ < 1.3$, and $\Delta z^+ < 8.0$. The final grid consists of 408 million cells.

No-slip solid wall boundary conditions are imposed on the surface of the airfoil. Characteristic inflow conditions are defined on the far-field boundaries on the left and lower boundary and characteristic outflow conditions including the ambient pressure p_∞ are used on the upper and right boundary. An additional sponge suppresses the reflection of outgoing pressure waves at the boundaries. In the spanwise direction, periodic boundary conditions are applied.

A numerical tripping at $x/c = 0.1$ on both sides of the airfoil triggers laminar-turbulent transition [21, 45].

The freestream Mach number is $M = 0.2$ and the angle of attack is $\alpha = 2.0$. The chord of the airfoil is aligned with the x-axis. The freestream Reynolds number based on the chord length of the airfoil is $Re = 400{,}000$, i.e., comparable to [21]. This problem can still be computed in an acceptable time frame. The traveling wave of the surface is given by

$$y_n^+|_{\text{wall}}(z^+, t^+) = A^+ \cos\left(\frac{2\pi}{\lambda^+}z^+ - \frac{2\pi}{T^+}t^+\right) . \tag{1}$$

The quantity $\lambda^+ = \lambda u_\tau / v$ is the wavelength, $A^+ = Au_\tau / v$ the amplitude, and $T^+ = Tu_\tau^2 / v$ the period in inner scaling. A smooth transition from the non-actuated to the actuated surface and vice versa is guaranteed by a $1 - \cos(x)$ function in the streamwise intervals $0.2 \leq x/c \leq 0.25$ and $0.9 \leq x/c \leq 0.95$. In terms of the chord-wise coordinate system 74.02% of the integrated wetted surface undergo an actuation. The limits of the actuated area ensure that a large percentage of the airfoil surface is influenced by the sinusoidal wall motion. Note that the temporal transition from the non-actuated to the actuated surface is also described by a $1 - \cos(t)$ function. The values of the parameters of the wave motion are based on results of previous analyses of zero-pressure gradient turbulent boundary layer flows [1, 30, 31, 37]. Using the previous findings, the amplitude is prescribed in the range $30 \leq A^+ \leq 50$. To keep the increase of the wetted surface due to the wall deflection small, a long wavelength is preferred. The most promising period of the actuation function is found to be in the range $30 \leq T^+ \leq 50$ [1, 30]. Due to the variation of the friction velocity u_τ in the streamwise coordinate x, the wave parameters are a function of the x-direction to approximate the optimal range in most parts of the wing. The amplitude varies in the streamwise direction. This is especially relevant in the adverse-pressure-gradient region, i.e., $x/c > 0.55$, to balance the reduced drag reduction potential due to the positive pressure gradient conditions [36].

The characteristic time scale of the large scales based on u_τ and δ_0 at $x/c = 0.8$ on the suction side is similar to that in [54] for a comparable setup. Note that no normalized eddy-turnover time [53] is given since no homogeneous direction exists for the actuated setup. The instantaneous solution of the non-actuated flow is used as initial distribution for the actuated flow. Again, when the steady state of the lift and drag coefficient of the actuated case is reached the computation is performed for another five flow-over times to sample enough data for the turbulence statistics. For the non-actuated flow, the solution fields are spanwise averaged and for the actuated case phase averaged.

4 Computing Resources

The computation of the actuated airfoil flow problem requires a high amount of computational resources. This is mainly due to the relatively high Reynolds number of $Re_c = 400{,}000$ and the requirements of the wall-resolved large-eddy simulation

(LES) approach. That is, most of the turbulent scales need to be resolved and a high mesh resolution is required, especially in the near-wall region and in the wake flow. The final mesh consists of 408 million cells and is furthermore time-dependent. Thus, in addition to the numerical flux discretization the geometrical properties, i.e., the metric terms, need to be recalculated in every Runge–Kutta sub-time step. Besides, to guarantee a stable and conservative solution the geometry conservation law (GCL) needs to be fulfilled, requiring the computation of the additional volume flux. The simulation was carried out on 200 compute nodes of the high-performance platform Hazelhen, where each node consists of two Intel® Xeon® E5-2680 v3 CPUs, i.e., a total of 4,800 cores were used. On average, each partition contains approximately 85,000 cells and one core is used for each partition. The total run time of one simulation was $2.5 \cdot 10^6$ iteration steps translating into 524 h of actual computation time and a total of $2.51 \cdot 10^6$ core hours required.

5 Results

In Fig. 2, the flow field over the actuated wing section is shown by contours of the λ_2-criterion [23] colored by the streamwise velocity component. The flow is tripped at $x/c = 0.1$ and the transition from the non-actuated to the actuated surface is imposed at $x/c = 0.2$. The turbulent structures in the outer boundary layer cover the changes in the near-wall structures that are responsible for the differences in the wall-shear stress of the actuated and the non-actuated flow.

The temporal evolutions of the combined viscous and pressure drag

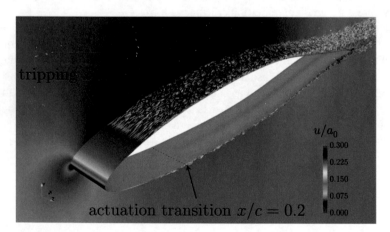

Fig. 2 Contours of the λ_2-criterion above the actuated airfoil, colored by the streamwise velocity component

$$c_d = \frac{2}{\rho_\infty u_\infty^2 A_{ref}} \left(\int_A \tau_w (\mathbf{n} \cdot \mathbf{e}_y) dA - \int_A p(\mathbf{n} \cdot \mathbf{e}_x) dA \right) ,$$

the viscous drag in the x-direction

$$c_{d,v} = \frac{2}{\rho_\infty u_\infty^2 A_{ref}} \int_A \tau_w (\mathbf{n} \cdot \mathbf{e}_y) dA ,$$

and the lift in the y-direction

$$c_l = \frac{2}{\rho_\infty u_\infty^2 A_{ref}} \left(\int_A \tau_w (\mathbf{n} \cdot \mathbf{e}_x) dA - \int_A p(\mathbf{n} \cdot \mathbf{e}_y) dA \right) ,$$

are presented in Fig. 3. The wall-shear stress and the pressure are integrated over the whole surface of the wing section. To emphasize the differences between the actuated and the non-actuated flow the distributions are scaled by the mean coefficients of the non-actuated reference case $\bar{c}_{d,ref}, \bar{c}_{d,v,ref}$, and $\bar{c}_{l,ref}$. During the temporal transition from the non-actuated to the actuated flow, the total drag of the actuated case starts to reduce and reaches a new quasi-steady state after about 0.25 flow-over times. A similar behavior is observed for the viscous drag coefficient in Fig. 3b. The temporal evolution of the lift coefficient in Fig. 3c shows a larger delay with a visible increase only after about 0.7 flow-over times. On average, a decrease of the total drag by 7.5% is achieved and the lift coefficient is even slightly increased by 1.4%. The large wavelength results in only a minor increase of the wetted surface by 1.6% such that the viscous drag coefficient is reduced by 8.6%.

The mean wall-tangential velocity distributions in the wall-normal direction at the streamwise positions $x/c = 0.5$ and $x/c = 0.7$ on the suction and the pressure side of the wing section are depicted in Fig. 4. The position $x/c = 0.5$ is located in the zero pressure gradient region and the location $x/c = 0.7$ in the adverse pressure gradient region in the aft part of the wing section. Note that the velocities are normalized by the friction velocity of the non-actuated reference case to enable a direct comparison. At all four locations, the velocity distribution in the trough region of the phase-averaged wave is significantly lowered, whereas only slight variations are observed in the near-wall region on the wave crest. This behavior is consistent with previous investigations of transversal surface waves on zero-pressure gradient turbulent boundary layers, where the essential drag reduction is achieved in the trough region of the wave [31]. In the following, we will focus on the two locations, i.e., $x/c = 0.5$ and $x/c = 0.7$, on the upper side.

A detailed look of the impact of the wave motion on the turbulent field is taken in Fig. 5, where the normal components of the Reynolds stress tensor in the wall-tangential, wall-normal, and spanwise direction are shown at the two positions on the suction side. The distributions of the shear-stress component are also illustrated. The fluctuations in the 4a direction in Fig. 4a, b are clearly reduced in the crest and the trough region. Due to the zero pressure gradient and the higher wave amplitude in inner units, the decrease of the fluctuations is stronger at the upstream position.

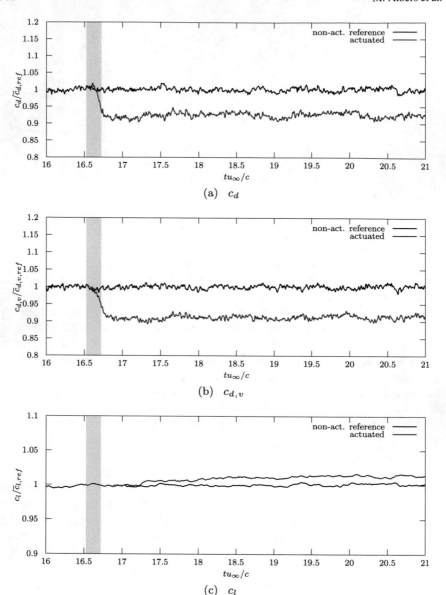

(a) c_d

(b) $c_{d,v}$

(c) c_l

Fig. 3 Temporal evolution of **a** the ratio of the instantaneous total drag coefficient of the actuated case to the averaged drag coefficient of the non-actuated reference case, **b** the ratio of the instantaneous viscous drag coefficient of the actuated case to the averaged viscous drag coefficient of the non-actuated reference case, and **c** the ratio of the instantaneous lift coefficient of the actuated case to the averaged lift coefficient of the non-actuated reference case; the grey column indicates the time of the temporal transition from non-actuated to the actuated case

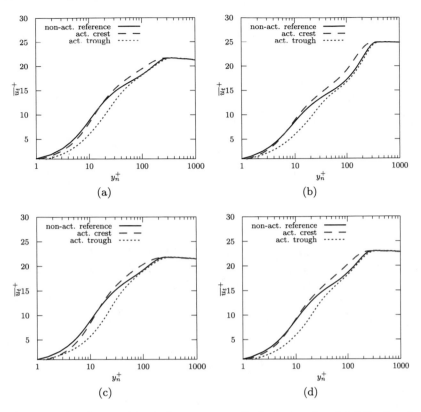

Fig. 4 Averaged wall-tangential velocity component distribution at $x/c = 0.5$ (left column) and $x/c = 0.7$ (right column) on the suction side (top) and on the pressure side (bottom) of the wing section for the non-actuated reference and the actuated case

In the wave trough, a shift of the peak of the fluctuations off the wall is apparent. This is caused by shielding the wall from quasi-streamwise vortices [31, 51]. While the fluctuations in the wall-normal direction in Fig. 4c, d are significantly decreased at the upstream position, they remain almost unchanged in the adverse pressure gradient region at $x/c = 0.7$. For the spanwise fluctuations in Fig. 4e, f the result is more diverse. An apparent decrease occurs in the zero pressure gradient region and an increase in the adverse pressure gradient region. In general, the zero pressure gradient conditions are beneficial for the wave motion, supporting the reduction of turbulent motion near the wall, whereas the adverse pressure gradient counteracts the desired effect of the actuation for the current set of wave parameters. The positive pressure gradient causes a lower wall-shear stress already in the non-actuated flow. This means that a reduced actuation, i.e., a smaller amplitude, is more efficient than the current surface motion [36].

Finally, the wall-tangential and the wall-normal vorticity fluctuations are considered. Jiménez and Pinelli [25] showed that a suppression of wall-normal vorticity

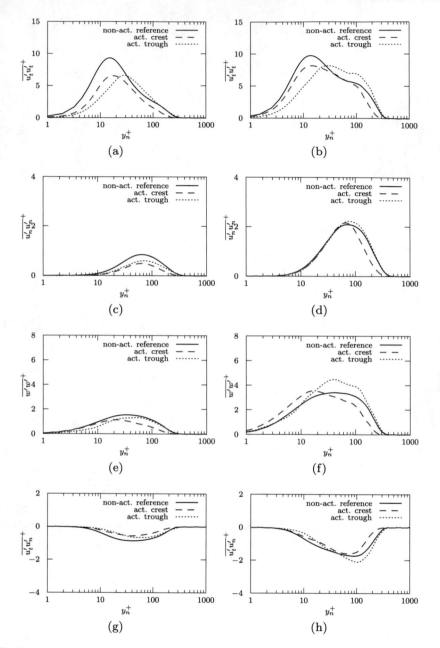

Fig. 5 Averaged distributions of the Reynolds stress tensor components at $x/c = 0.5$ (left column) and $x/c = 0.7$ (right column) on the suction side of the wing section

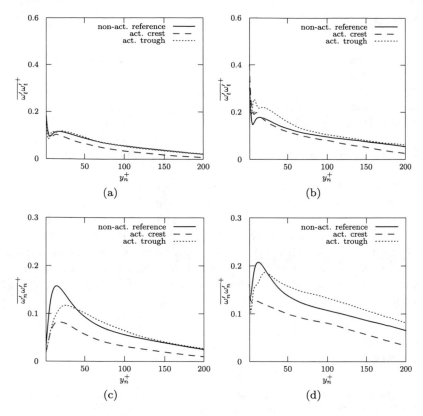

Fig. 6 Distribution of the root-mean-square of the vorticity fluctuations in inner units at $x/c = 0.5$ (left column) and $x/c = 0.7$ (right column) on the suction side of the wing section for the non-actuated reference and the actuated case; **a, b** wall-tangential vorticity component; **c, d** wall-normal vorticity component

fluctuations can be directly related to a decrease of the turbulent velocity fluctuations and thus to a decrease of the skin friction. This has been confirmed in [30] for an actuated flat plate flow and a similar conclusion can be drawn from Fig. 6. The decrease of the wall-normal component is observed in the near-wall region at both positions, with a stronger reduction in the zero pressure gradient region and generally on the wave crest. The distributions of the wall-tangential components show nearly unchanged distributions in the zero pressure gradient region and a slight increase in the wave trough in the adverse pressure gradient region.

6 Conclusion

Spanwise traveling transversal surface waves were applied to a wing section with a DRA2303 geometry in a turbulent flow. The purpose of this investigation was to prove that in engineering-like flow scenarios with favorable, zero, and adverse pressure gradients and non-zero surface curvature drag reduction can be achieved.

The turbulent flow was simulated by a high-resolution implicit LES. The boundary layer was tripped near the leading edge. About 74% of the surface underwent a transversal spanwise wave motion with a variation of the amplitude in the streamwise direction.

A reduction of the integrated total drag by 7.5% is obtained and the lift coefficient is enlarged by 1.4%. The simultaneous improvement of both coefficients is due to the reduction of the streamwise turbulence intensities and a decreased boundary-layer thickness. This is in contrast to findings of the technique of blowing into the turbulent boundary layer, where the turbulence production is primarily shifted off the wall, leading to a thickened boundary layer and an increased pressure drag [4]. The main reason for the drag reduction is the significantly lowered wall-normal vorticity fluctuations.

References

1. M. Albers, P.S. Meysonnat, W. Schröder, Drag reduction via transversal wave motions of structured surfaces, in *International Symposium on Turbulence & Shear Flow Phenomena (TSFP-10)* (2017)
2. M. Albers, P.S. Meysonnat, W. Schröder, Actively reduced airfoil drag by transversal surface waves. Flow Turbul. Combust. **102**(4), 865–886 (2019)
3. P.H. Alfredsson, R. Örlü, Large-eddy breakup devices - a 40 years perspective from a Stockholm horizon. Flow Turbul. Combust. **100**(4), 877–888 (2018)
4. M. Atzori, R. Vinuesa, A. Stroh, B. Frohnapfel, P. Schlatter, Assessment of skin-friction-reduction techniques on a turbulent wing section, in *12th ERCOFTAC Symposium on Engineering Turbulence Modeling and Measurements (ETMM12)* (2018)
5. H.L. Bai, Y. Zhou, W.G. Zhang, S.J. Xu, Y. Wang, R.A. Antonia, Active control of a turbulent boundary layer based on local surface perturbation. J. Fluid Mech. **750**, 316 (2014)
6. D.W. Bechert, G. Hoppe, W.E. Reif, On the drag reduction of the shark skin, in *AIAA Paper No. 85–0546* (1985)
7. D. Bechert, R. Meyer, W. Hage, Drag reduction of airfoils with miniflaps - can we learn from dragonflies? in *AIAA Paper No. 2000–2315* (2000)
8. J.P. Boris, F.F. Grinstein, E.S. Oran, R.L. Kolbe, New insights into large eddy simulation. Fluid Dyn. Res. **10**(4–6), 199–228 (1992)
9. S.L. Ceccio, Friction drag reduction of external flows with bubble and gas injection. Annu. Rev. Fluid Mech. **42**(1), 183–203 (2010)
10. K.S. Choi, X. Yang, B.R. Clayton, E.J. Glover, M. Atlar, B.N. Semenov, V.M. Kulik, Turbulent drag reduction using compliant surfaces. Proc. R. Soc. Lond. Ser. A **453**(1965), 2229–2240 (1997)
11. Y. Du, G.E. Karniadakis, Suppressing wall turbulence by means of a transverse traveling wave. Science **288**(5469), 1230–1234 (2000)

12. Y. Du, V. Symeonidis, G.E. Karniadakis, Drag reduction in wall-bounded turbulence via a transverse travelling wave. J. Fluid Mech. **457**, 1–34 (2002)
13. A. Feldhusen-Hoffmann, V. Statnikov, M. Klaas, W. Schröder, Investigation of shock-acoustic-wave interaction in transonic flow. Exp. Fluids **59**(1), 15 (2017)
14. J.L. Fulker, M.J. Simmons, *An Experimental Investigation of Passive Shock/Boundary-Layer Control on an Aerofoil* (Vieweg+Teubner Verlag, Wiesbaden, 1997), pp. 379–400
15. M. Gad-el-Hak, *Flow Control : Passive, Active, and Reactive Flow Management* (Cambridge University Press, Cambridge, 2000)
16. R. García-Mayoral, J. Jiménez, Hydrodynamic stability and breakdown of the viscous regime over riblets. J. Fluid Mech. **678**, 317–347 (2011)
17. G. Geiser, *Thermoacoustic Noise Sources in Premixed Combustion* (Verlag Dr. Hut, München, 2014)
18. J.W. Gose, K. Golovin, M. Boban, J.M. Mabry, A. Tuteja, M. Perlin, S.L. Ceccio, Characterization of superhydrophobic surfaces for drag reduction in turbulent flow. J. Fluid Mech. **845**, 560–580 (2018)
19. R. Henke, "A320 HLF Fin" flight tests completed. Air & Space Eur. **1**(2), 76–79 (1999)
20. C.W. Hirt, A.A. Amsden, J.L. Cook, An arbitrary Lagrangian-Eulerian computing method for all flow speeds. J. Comput. Phys. **14**, 227–253 (1974)
21. S.M. Hosseini, R. Vinuesa, P. Schlatter, A. Hanifi, D.S. Henningson, Direct numerical simulation of the flow around a wing section at moderate Reynolds number. Int. J. Heat Fluid Flow **61**, 117–128 (2016)
22. M. Itoh, S. Tamano, K. Yokota, S. Taniguchi, Drag reduction in a turbulent boundary layer on a flexible sheet undergoing a spanwise traveling wave motion. J. Turbul. **7**, N27 (2006)
23. J. Jeong, F. Hussain, On the identification of a vortex. J. Fluid Mech. **285**, 69–94 (1995)
24. J. Jiménez, Turbulent flows over rough walls. Annu. Rev. Fluid Mech. **36**(1), 173–196 (2004)
25. J. Jiménez, A. Pinelli, The autonomous cycle of near-wall turbulence. J. Fluid Mech. **389**, 335–359 (1999)
26. W.J. Jung, N. Mangiavacchi, R. Akhavan, Suppression of turbulence in wall bounded flows by high frequency spanwise oscillations. Phys. Fluids A **4**(8), 1605–1607 (1992)
27. Y. Kametani, K. Fukagata, R. Örlü, P. Schlatter, Effect of uniform blowing/suction in a turbulent boundary layer at moderate Reynolds number. Int. J. Heat Fluid Flow **55**, 132–142 (2015)
28. E. Kim, H. Choi, Space-time characteristics of a compliant wall in a turbulent channel flow. J. Fluid Mech. **756**, 30–53 (2014)
29. S. Klumpp, M. Meinke, W. Schröder, Numerical simulation of riblet controlled spatial transition in a zero-pressure-gradient boundary layer. Flow Turbul. Combust. **85**(1), 57–71 (2010)
30. S. Klumpp, M. Meinke, W. Schröder, Drag reduction by spanwise transversal surface waves. J. Turbul. **11** (2010)
31. S.R. Koh, P. Meysonnat, V. Statnikov, M. Meinke, W. Schröder, Dependence of turbulent wall-shear stress on the amplitude of spanwise transversal surface waves. Comput. Fluids **119**, 261–275 (2015)
32. M.S. Liou, C. Steffen, A new flux splitting scheme. J. Comput. Phys. **107**, 23–39 (1993)
33. P.Q. Liu, H.S. Duan, J.Z. Chen, Y.W. He, Numerical study of suction-blowing flow control technology for an airfoil. J. Aircr. **47**(1), 229–239 (2010)
34. M. Luhar, A. Sharma, B. McKeon, A framework for studying the effect of compliant surfaces on wall turbulence. J. Fluid Mech. **768**, 415–441 (2015)
35. M. Meinke, W. Schröder, E. Krause, T. Rister, A comparison of second- and sixth-order methods for large-eddy simulations. Comput. Fluids **31**(4), 695–718 (2002)
36. P.S. Meysonnat, S.R. Koh, B. Roidl, W. Schröder, Impact of transversal traveling surface waves in a non-zero pressure gradient turbulent boundary layer flow. Appl. Math. Comput. **272**, 498–507 (2016)
37. P.S. Meysonnat, D. Roggenkamp, W. Li, B. Roidl, W. Schröder, Experimental and numerical investigation of transversal traveling surface waves for drag reduction. Eur. J. Mech. B. Fluids **55**, 313–323 (2016)

38. P. Moin, T. Shih, D. Driver, N.N. Mansour, Direct numerical simulation of a three dimensional turbulent boundary layer. Phys. Fluids A **2**(10), 1846–1853 (1990)
39. R. Nakanishi, H. Mamori, K. Fukagata, Relaminarization of turbulent channel flow using traveling wave-like wall deformation. Int. J. Heat Fluid Flow **35**, 152–159 (2012)
40. M. Perlin, D.R. Dowling, S.L. Ceccio, Freeman scholar review: passive and active skin-friction drag reduction in turbulent boundary layers. J. Fluids Eng. **138**(9), 091104–091116 (2016)
41. M. Quadrio, Drag reduction in turbulent boundary layers by in-plane wall motion. Philos. Trans. R. Soc. Lond. Ser. A **369**(1940), 1428–1442 (2011)
42. M. Quadrio, P. Ricco, C. Viotti, Streamwise-travelling waves of spanwise wall velocity for turbulent drag reduction. J. Fluid Mech. **627**, 161 (2009)
43. J. Reneaux, Overview on drag reduction technologies for civil transport aircraft, in *European Congress on Computational Methods in Applied Sciences and Engineering (ECCOMAS)* (2004)
44. B. Roidl, M. Meinke, W. Schröder, Zonal RANS-LES computation of transonic airfoil flow, in *AIAA Paper No. 2011–3974* (2011)
45. P. Schlatter, R. Örlü, Turbulent boundary layers at moderate Reynolds numbers: inflow length and tripping effects. J. Fluid Mech. **710**, 5–34 (2012)
46. H. Shan, L. Jiang, C. Liu, Direct numerical simulation of flow separation around a NACA 0012 airfoil. Comput. Fluids **34**(9), 1096–1114 (2005)
47. P.R. Spalart, J.D. McLean, Drag reduction: enticing turbulence, and then an industry. Philos. Trans. R. Soc. Lond. Ser. A **369**(1940), 1556–1569 (2011)
48. E. Stanewsky, J. Délery, J. Fulker, P. de Matteis, Synopsis of the project EUROSHOCK II, in *Drag Reduction by Shock and Boundary Layer Control*, ed. by E. Stanewsky, J. Délery, J. Fulker, P. de Matteis (Springer, Berlin, 2002), pp. 1–124
49. J. Szodruch, Viscous drag reduction on transport aircraft, in *AIAA Paper No. 91–0685* (1991)
50. S. Tamano, M. Itoh, Drag reduction in turbulent boundary layers by spanwise traveling waves with wall deformation. J. Turbul. **13**, N9 (2012)
51. N. Tomiyama, K. Fukagata, Direct numerical simulation of drag reduction in a turbulent channel flow using spanwise traveling wave-like wall deformation. Phys. Fluids **25**(10), 105115 (2013)
52. R. Vinuesa, P. Schlatter, Skin-friction control of the flow around a wing section through uniform blowing, in *European Drag Reduction and Flow Control Meeting (EDRFCM 2017)* (2017)
53. R. Vinuesa, C. Prus, P. Schlatter, H.M. Nagib, Convergence of numerical simulations of turbulent wall-bounded flows and mean cross-flow structure of rectangular ducts. Meccanica **51**(12), 3025–3042 (2016)
54. R. Vinuesa, P.S. Negi, M. Atzori, A. Hanifi, D.S. Henningson, P. Schlatter, Turbulent boundary layers around wing sections up to $Re_c = 1,000,000$. Int. J. Heat Fluid Flow **72**, 86–99 (2018)
55. M. Walsh, L. Weinstein, Drag and heat transfer on surfaces with small longitudinal fins, in *11th Fluid and Plasma Dynamics Conference* (1978), p. 1161
56. C. Zhang, J. Wang, W. Blake, J. Katz, Deformation of a compliant wall in a turbulent channel flow. J. Fluid Mech. **823**, 345–390 (2017)
57. H. Zhao, J.Z. Wu, J.S. Luo, Turbulent drag reduction by traveling wave of flexible wall. Fluid Dyn. Res. **34**(3), 175–198 (2004)

Simulation of Particulate Flow Using HPC Systems

K. Fröhlich, M. Meinke, and W. Schröder

Abstract A standard strategy to predict the modulation of turbulence by the presence of particles is the two-way coupling approach, where the solid phase is approximated by point particles, which introduce sources in the momentum conservation equation. A validation of this approach is presented for isotropic decaying turbulence laden with prolate and oblate particles of Kolmogorov-length-scale size by generating highly accurate reference results via direct particle-fluid simulations, where all turbulent scales and the complete flow field in the vicinity of the particles are resolved. About 30,000 oblate and prolate particles with aspect ratios raging from 0.25 to 4 are released into the flow field. The simulation using the two-way coupled spherical and ellipsoidal Lagrangian model is compared against the reference results. The analysis of turbulent kinetic energy budgets reveals that the particles release kinetic energy into the flow field and simultaneously enhance the dissipation rate. This behavior is correctly predicted by both point-particle models. The kinetic energy of the particles, however, is significantly overestimated by the point-particle models. Moreover, the ellipsoidal Lagrangian model fails to predict the angular velocity of the particles due to the missing correlation terms for finite fluid inertia.

1 Introduction

Particle-laden turbulent flow is of importance in medical, natural, and technical environments such as blood flow, pollutant transport in the atmosphere and pulverized fuel combustion. A vast parameter space is introduced by the presence of the particles, which further enhances the complexity of turbulent flows. Two-way coupled Lagrangian point-particle models, where the feedback of the particles is included in the conservation equations of the flow field, have been frequently applied in direct numerical simulations (DNS) for spherical particles smaller than the Kolmogorov length η_k in the last decades. For particle diameters $d_p > \eta_k$, however, the validity of the spherical two-way coupled Lagrangian point-particle models (SLM) is still con-

K. Fröhlich (✉) · M. Meinke · W. Schröder
Institute of Aerodynamics, RWTH Aachen University, Wüllnerstraße 5a, 52062 Aachen, Germany
e-mail: k.froehlich@aia.rwth-aachen.de

© Springer Nature Switzerland AG 2021

W. E. Nagel et al. (eds.), *High Performance Computing in Science and Engineering '19*,
https://doi.org/10.1007/978-3-030-66792-4_21

troversially discussed [1]. Only few studies are available on the validity of ellipsoidal Lagrangian models (ELM) for LES. While one-way coupled ELMs are frequently applied in DNS with particles smaller than η_k [19], the accuracy of two-way coupled ELMs has not been assessed. This can be partially explained by a shortage of reference results, which can be possibly provided by fully resolved simulations. However, the resolution of thousands of Kolmogorov-length-scale size non-spherical particles requires extensive computational resources and advanced numerical algorithms.

Recently, a framework has been developed, which enables the efficient computation of sharply resolved freely moving boundaries interacting with the fluid, where the computational effort is considerably reduced by adaptive mesh refinement [13]. Subsequently, direct particle-fluid simulations (DPFS) have been generated for 45,000 spherical and ellipsoidal particles of Kolmogorov-length-scale size suspended in isotropic decaying turbulence [14], in which all particle and fluid scales are fully resolved. The identical setup has been considered in [6] with LES and DNS using SLMs, which showed a convincing accuracy for spherical particles.

In the current contribution, selected results of [5] are presented and details on computational aspects are provided. The study of [6] is extended towards non-spherical particles. Therefore, the DPFS performed in [14] will be supplemented by additional benchmark cases for oblate and prolate spheroids with varying aspect ratios. Like in [6], the quality of the point-particle models will be analyzed using DNS and LES.

2   Mathematical Models

2.1   Fluid Phase Equations

The conservation of mass, momentum, and energy in a control volume V may be expressed in non-dimensional integral form by

$$\int_V \frac{\partial Q}{\partial t} \, dV + \int_{\partial V} \bar{H} \cdot n \, dA = 0, \tag{1}$$

where $Q = [\, \rho_f, \; \rho_f u^T, \; \rho_f E \,]^T$ is the vector of conservative Eulerian variables and $\bar{H}$ is the flux hypertensor through the surface ∂V of V in outward normal direction n. The conservative variables are defined by the fluid density ρ_f, the vector of velocities u, and the total specific energy $E = e + |u|^2/2$ containing the specific internal energy e. The fluxes $\bar{H}$ can be divided into an inviscid part $\bar{H}_{inv}$ and a viscous diffusion part $\bar{H}_{visc}$, where

$$\bar{H} = \bar{H}_{inv} + \bar{H}_{visc} = \begin{pmatrix} \rho_f u \\ \rho_f uu + p \\ u\left(\rho_f E + p\right) \end{pmatrix} - \frac{1}{Re_0} \begin{pmatrix} 0 \\ \bar{\tau} \\ \bar{\tau} u - q \end{pmatrix}, \tag{2}$$

with the pressure p, the stress tensor $\bar{\tau}$, the vector of heat conduction q, and the Reynolds number Re_0. The latter is determined by $Re_0 = \rho'_{f,0} a'_0 L'_0 / \mu'_0$, given the reference quantities of the density $\rho'_{f,0}$, the speed of sound a'_0, the dynamic viscosity μ'_0, and the length L'_0. Using Stokes' hypothesis for a Newtonian fluid yields an equation for the stress tensor

$$\bar{\tau} = 2\mu\bar{S} - \frac{2}{3}\mu\left(\nabla \cdot u\right)\bar{I}, \tag{3}$$

in which $\bar{I}$ is the unit tensor and $\bar{S}$ holds the rate-of-strain tensor defined as $\bar{S} = \left(\nabla u + (\nabla u)^T\right)/2$. Fourier's law gives the heat conduction

$$q = -\frac{k_T}{Pr\,(\gamma - 1)}\nabla T, \tag{4}$$

using the temperature T, the constant capacity ratio $\gamma = c'_{p,0}/c'_{v,0}$, and the specific heat capacities $c'_{v,0}$ and $c'_{p,0}$ at constant volume and at constant pressure. The Prandtl number Pr is given by $Pr = \mu'_0 c'_{p,0}/k'_0$ containing the reference thermal conductivity k'_0. The dynamic viscosity μ and the thermal conductivity k_T are temperature dependent and can be approximated via Sutherland's law [20]. The system of equations can be closed by the caloric state equation $e = c_v T$, and the state equation of an ideal gas $p = \rho_f RT$, with R being the specific gas constant.

2.2 Rigid Particle Dynamics

The motion of a rigid particle can be described in the Lagrangian formulation by equations for the kinematics and dynamics, which have to be solved in point-particle models as well as in fully resolved simulations. The linear part of the kinematics in an inertial frame of reference is given by

$$\frac{d\boldsymbol{x}_p}{dt} = \boldsymbol{v}_p, \tag{5}$$

where $\boldsymbol{x}_p$ denotes the center of mass of the particle and the particle velocity $\boldsymbol{v}_p$. If the force $\boldsymbol{F}_p$ acting on the particle with mass m_p is known, the dynamic relation

$$m_p\frac{d\boldsymbol{v}_p}{dt} = \boldsymbol{F}_p \tag{6}$$

closes the linear motion.

The rotational dynamics are defined in a rotating frame of reference aligned with the particle fixed coordinate system $(\tilde{x}, \tilde{y}, \tilde{z})$. The torque $\widetilde{\boldsymbol{T}}_p$ acting on the particle causes an angular acceleration with

$$\widetilde{I}\frac{d\widetilde{\omega}_p}{dt} + \widetilde{\omega}_p \times \left(\widetilde{I}\widetilde{\omega}_p\right) = \widetilde{T}_p, \tag{7}$$

where $\widetilde{I}$ denotes the principal moments of inertia and $\widetilde{\omega}_p$ the angular velocity. The rotational kinematics can be described using quaternions, as described in e.g. [16].

In this contribution, volume forces such as gravity are omitted. Therefore, the coupling between the solid and the fluid frame is solely responsible for the force and the torque exerted to the particles. The coupling is established via the no-slip condition at the particle surface Γ_p

$$u = v_p + \omega_p \times \left(x_p - r_p\right), \tag{8}$$

with r_p the local distance between the surface the center of the particle. The momentum balance given by Eq. 1 at the material interface yields the net force F_p due to fluid pressure and fluid shear forces

$$F_p = \oint_{\Gamma_p} (-pn + \bar{\tau} \cdot n)\, dA. \tag{9}$$

Likewise, the torque T_p acting on the particle is given by

$$T_p = \oint_{\Gamma_p} (x - x_p) \times (-pn + \bar{\tau} \cdot n)\, dA. \tag{10}$$

The transfer of momentum at the material interface causes a direct transfer of kinetic energy into the fluid. Multiplying the momentum conservation in Eq. 1 and using Eqs. 8–10 yields the direct transfer of kinetic energy at the particle surface [14]

$$\psi_p = \oint_{\Gamma_p} (pn - \bar{\tau} \cdot n) \cdot u\, dA = - \left(F_p \cdot v_p + T_p \cdot \omega_p\right). \tag{11}$$

The determination of F_p and T_p as well as the coupling between the particles and the fluid in DPFS, SLMs and ELMs will be presented in the next Section.

3 Numerical Methods

In the following, the numerical methods for the solution of the system of equations given above will be presented. The solution scheme for Eq. 1 is based on a cell-centered finite-volume discretization on hierarchical Cartesian grids which has been described and validated in [13]. The conservative variables Q in Eq. 1 are integrated in time using an explicit five-step predictor-corrector Runge–Kutta method [13]. The inviscid fluxes are computed by a variant of the AUSM of second-order accuracy [11]. The primitive variables at the cell surfaces are extrapolated according to the MUSCL scheme [18] using cell-centered gradients of the primitive variables obtained by a second-order least-squares approach [13]. The viscous fluxes are computed using a low-dissipation variation of the central scheme proposed in [2].

The assessment of the point-particle models in LES requires a sufficiently accurate subgrid-scale model of the turbulent flow. Based on the monotone implicit LES approach (MILES) a residual stress model has been established in [6], where mixed central and upwind stencils provide the dissipative subgrid scale contribution.

In this Section, the discretization of the equations describing the particle dynamics is introduced. First, a brief description for DPFS is provided. Subsequently, the Lagrangian two-way coupled point-particle models are described.

3.1 Direct Particle-Fluid Simulations (DPFS)

Figure 1 illustrates the DPFS approach for an ellipsoidal particle of Kolmogorov-length-scale size. The Cartesian mesh is locally refined in the vicinity of the particles. More precisely, four refinement levels are introduced additionally to the DNS resolution. The geometry of the particles is analytically tracked via a level-set method developed in [8]. The interfaces of the particles are sharply resolved by reshaped Cartesian cut cells [9]. The no-slip condition is imposed on the particle surface segments and a conservative flux-redistribution technique stabilizes small cut cells [13]. The efficiency on high-performance computers is substantially improved using dynamic load balancing [12]. Isothermal conditions are considered and all cells within the solid frame can be discarded in the simulation. Due to the sharply resolved particle surface, mass, momentum, and energy are fully conserved in DPFS. The force F_p and the torque T_p can directly be obtained by summation over all discrete particle surface segments $A_i, i \in \Gamma_p$

$$F_p = \sum_{i \in \Gamma_p} [-pn + \bar{\tau} \cdot n]_i A_i, \tag{12}$$

$$T_p = \sum_{i \in \Gamma_p} [(x - x_p) \times (-pn + \bar{\tau} \cdot n)]_i A_i. \tag{13}$$

Equations 5–7 are solved by the predictor-corrector scheme of second-order accuracy [13] and coupled with Eq. 1. The collision model proposed in [7] is employed to avoid overlapping bodies. The DPFS approach has been validated in [13] for several laminar and turbulent flow problems.

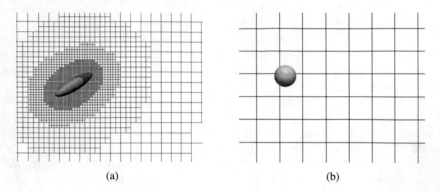

(a)   (b)

Fig. 1 Comparison of the DPFS approach and the point-particle model: **a** A snapshot of a single ellipsoidal particle of Kolmogorov-length-scale size using DPFS. All fluid scales of the background turbulence are fully resolved with the background DNS-mesh. Four additional refinement levels are introduced to fully resolve the particle scale, i.e., particle surface, wake, and boundary layer. The particle is freely moving such that the mesh has to be adaptively refined. To establish the no-slip condition, the Cartesian cut cells are reshaped at the material interface. **b** The illustration shows a DNS mesh with the grid cell length Δ_{DNS} and the spherical Lagrangian model with the same equivalent particle diameter $d_{p,eq}$

3.2 Point-Particle Models

The particle scale is not resolved in the point-particle model. To establish a feedback of the particles to the flow field, the forces and torques exerted on the particle are included in the momentum conservation in Eq. 1. A popular method is the two-way coupled point-particle model, where the force acting on the particle is projected onto the underlying grid in an actio-reactio sense similar as in, e.g., [17]. Following [6], self-induced disturbances are mostly avoided via a smooth distance-weighted feedback force. As shown in [5], the rotational contribution for turbulence modulation in isotropic turbulence is significantly smaller than the linear contribution. Therefore, the feedback of the rotational contribution of the point-particles is neglected. In this study, two different point-particle models are assessed, which will be described next.

3.2.1 Spherical Lagrange Model (SLM)

The SLM is not aware of the anisotropic particle shape. Hence, $\boldsymbol{F}_p$ is modeled for a sphere with the volume-equivalent particle diameter $d_{p,eq}$ neglecting the aspect ratio of the particles. That is, a simplification of the semi-empirical Maxey–Riley equation [10] for spherical particles can be used

$$\boldsymbol{F}_p = 3\pi \mu d_\mathrm{p}(\boldsymbol{u}_p - \boldsymbol{v}_p)\phi(Re_\mathrm{p}),\tag{14}$$

which represents the quasi-steady Stokes drag augmented by the Schiller-Naumann drag correlation $\phi(Re_p) = 24/Re_p(1 + 0.15Re_p^{0.687})$ with the undisturbed fluid velocity at the particle position u_p. The latter has to be interpolated using the fluid velocity at the centers of the neighboring cells. To mitigate filtering errors due to interpolations, u_p is approximated by the fluid velocity at the center of the nearest cell. Although the validity of Eq. 14 is restricted to particles which are smaller than the smallest scale of the flow field η_k, it has been shown that the SLM is capable to predict the results of the DPFS for spherical particles of Kolmogorov-length scale size [6].

3.2.2 Ellipsoidal Lagrange Model (ELM)

The ELM considers the hydrodynamic forces and torques acting on ellipsoidal particles in creeping flow conditions. That is, the ELM is not aware of fluid inertia and therefore restricted to vanishing particle Reynolds numbers. In the ELM, the force F_p is obtained by

$$F_p = \mu \bar{R}^T \tilde{\bar{K}} \bar{R} \left(u_p - v_p \right),$$

(15)

where $\bar{R}$ represents a rotational matrix. The hydrodynamic torque acting on the particle is given by

$$T_p = \mu \left(\tilde{\bar{K}}_s \tilde{S} + \tilde{\bar{K}}_\zeta \left(\tilde{\zeta} - \tilde{\omega}_p \right) \right),$$

(16)

with the fluid strain rate $\tilde{S}$ and $\tilde{\zeta}$ half times the fluid vorticity in the particle fixed coordinate system. The shape dependence of the hydrodynamic force and torque is established via the diagonal resistance matrices $\tilde{\bar{K}}$, $\tilde{\bar{K}}_s$, and $\tilde{\bar{K}}_\zeta$. A detailed definition of the resistance matrices is provided, e.g., in [16].

4 Results

The flow field of a fully periodic cube with an edge length of L is initialized randomly and divergence free. The initial microscale Reynolds number is set to $Re_{\lambda0} = u_0 \rho_f \lambda_0 / \mu = 79$, with the initial rms velocity u_0, the initial Taylor microscale λ_0, and μ the viscosity. The same initial flow field is used as in [14] for the DNS whereas in the case of LES, the initial energy spectrum is spectrally cut off at the highest resolvable wave number. Simulations with four resolutions are performed to assess the point-particle models. A mesh with 256^3 cells corresponds to a DNS and meshes with 128^3, 96^3, and 64^3 cells represent the LES cases. For all LES resolutions, the turbulent kinetic energy of the single phase DNS is accurately predicted, i.e., the subgrid turbulent kinetic energy is negligible. As in [6], the resolution has

Table 1 Parameters of the particle-laden simulations at injection time $t_i^* = 0.28$: Number of particles N_p, aspect ratio β, density ratio particle-to-fluid ρ_p/ρ, ratio of the minimum particle diameter $d_{p,min}$ to the Kolmogorov length scale η, ratio of the volume-equivalent particle diameter $d_{p,eq}$ to the grid cell length of the DNS Δ_{DNS}, the volume fraction ϕ_v, and the mass fraction ϕ_m of the particles. Case $P2$ has been analyzed in [15]

Case	N_p	β	ρ_p/ρ	$d_{p,min}/\eta$	$d_{p,eq}/\Delta_{DNS}$	$d_{p,eq}/\ell$	ϕ_v	ϕ_m
P2	45,000	2	1400	1.05	0.63	0.032	$3.5 \cdot 10^{-4}$	0.49
P4	25,000	4	1000	1.26	0.95	0.048	$6.7 \cdot 10^{-4}$	0.67
O4	20,000	0.25	1000	1.05	1.26	0.064	$12.5 \cdot 10^{-4}$	1.25

Table 2 Non-dimensional parameters of the DPFS and the single phase simulation (sP) at the time levels $t^* = 1.0$ and 2.0: Taylor-scale Reynolds number Re_λ; ratio of the volume-equivalent particle diameter $d_{p,eq}$ to the Kolmogorov length η and the Taylor length scale λ; ratio of the particle relaxation time and the Kolmogorov time scale τ_p/τ_η; mean particle Reynolds number $\langle Re_p \rangle$

Case	$t^* = 1.0$					$t^* = 2.0$				
	Re_λ	$d_{p,eq}/\eta$	$d_{p,eq}/\lambda$	τ_p/τ_η	$\langle Re_p \rangle$	Re_λ	$d_{p,eq}/\eta$	$d_{p,eq}/\lambda$	τ_p/τ_η	$\langle Re_p \rangle$
sP	36.6	–	–	–	–	31.8	–	–	–	–
P2	26.3	1.12	0.11	97.9	6.6	17.8	0.84	0.10	56.9	4.5
P4	24.7	1.73	0.18	167.6	10.1	15.6	1.32	0.17	98.0	7.25
O4	24.9	2.49	0.25	346.9	14.0	14.7	2.04	0.27	232.4	11.2

only minor impact on the statistics. Therefore, only the results of the LES cases with 64^3 cells are presented.

Three particle-laden cases are considered. The particles are introduced at the insertion time $t_i^* = 0.28$, where $t^* = t\epsilon_0/u_0^2$ is normalized by the initial eddy turnover time using the initial viscous dissipation rate ϵ_0. As in [14], the particles are initialized with the local fluid velocity and zero angular velocity. Table 1 lists the non-dimensional parameters related to the particles and Table 2 specifies non-dimensional parameters for $t^* = 1.0$ and 2.0. The prolate case $P2$ considered in [15] is supplemented by case $P4$ for prolate spheroids with an aspect ratio of $\beta = 4$ and by case $O4$ for oblate spheroids with an aspect ratio of $\beta = 0.25$. All particle-laden flows are dilute and assigned to the two-way coupling regime [3]. The single phase DNS will be denoted case sP.

4.1 Direct Particle-Fluid Simulation (DPFS)

Next, budgets of the turbulent kinetic energy defined by $E_k = \langle \rho u'^2/2 \rangle$, where the brackets $\langle \rangle$ denote the spatial ensemble average and u' the rms-velocity vector, are considered. For DPFS, the budget reads [14]

$$\frac{\partial E_k}{\partial t}(t) = \Psi(t) - \mathcal{E}(t) \tag{17}$$

where the total kinetic energy transfer $\Psi(t)$ on the particle surfaces can be computed by Eq. 11, i.e.,

$$\Psi(t) = \sum_{p=1}^{N_p} \psi_p = -\sum_{p=1}^{N_p} \left(\boldsymbol{F}_p \cdot \boldsymbol{v}_p + \boldsymbol{T}_p \cdot \boldsymbol{\omega}_p \right). \tag{18}$$

The integral viscous dissipation rate $\mathcal{E}(t)$ of incompressible flows can be determined by integrating the local dissipation rate ϵ over the fluid domain V_f, i.e.,

$$\mathcal{E}(t) = \int_{V_f} \epsilon \, dV = \int_{V_f} 2\mu \bar{\boldsymbol{S}} : \bar{\boldsymbol{S}} dV, \tag{19}$$

where $\bar{\boldsymbol{S}} : \bar{\boldsymbol{S}}$ denotes the inner product of the strain-rate tensor. It has been shown in [14], that the viscous dissipation rate can be decomposed in a background dissipation rate $\overline{\mathcal{E}}$ and a particle-induced dissipation rate $\mathcal{E}_p$ in the vicinity of the particles with

$$\mathcal{E}(t) = \overline{\mathcal{E}}(t) + \mathcal{E}_p(t) = \overline{\mathcal{E}}(t) + \sum_{p=1}^{N_p} \boldsymbol{F}_p \cdot \left(\boldsymbol{U}_p - \boldsymbol{v}_p \right) + \boldsymbol{T}_p \cdot \left(\boldsymbol{\Omega}_p - \boldsymbol{\omega}_p \right) + \mathcal{I}_f, \tag{20}$$

where $\boldsymbol{U}_p - \boldsymbol{v}_p$ represents the relative velocity vector and $\boldsymbol{\Omega}_p - \boldsymbol{\omega}_p$ the relative angular velocity vector of a particle and the surrounding fluid. The term $\mathcal{I}_f$ represents the contribution of the fluid inertia in the vicinity of the particles. The particle-induced dissipation $\mathcal{E}_p$ is an analytical expression, which has been derived using the momentum conservation equation in Eq. 1. The term $\boldsymbol{U}_p$ represents the ambient fluid velocity seen by the particles and is approximated as in [14].

Likewise, the global kinetic energy of the particles $K(t) = \sum(mv_p^2 + \tilde{\omega} \cdot (\tilde{I}\tilde{\omega}))/2$ is described by

$$\frac{dK}{dt} = -\Psi(t). \tag{21}$$

In the following, the different contributions of the turbulent kinetic energy budget generated via DPFS are presented. The contributions in Eq. 17 are normalized by the reference dissipation $\epsilon_{ref} = \rho u_0^3/L$ and the turbulent kinetic energy is normalized by its initial value $E_{k,0}$. Figure 2a shows the temporal development of the turbulent kinetic energy E_k. Moderate attenuation is observed in comparison to the single-phase DNS. Despite the substantially different setups of the particles for all cases, the modulation of the turbulent kinetic energy is very similar. Figure 2b shows the mean rate of total kinetic energy transfer $\langle \psi_p \rangle(t) = \Psi(t)/V_f$. All cases indicate a source of turbulent kinetic energy after an initial build-up. The case $O4$ is substantially larger than the cases $P2$ and $P4$. Figure 2c shows the mean viscous dissipation rate $\langle \epsilon \rangle(t) = \mathcal{E}(t)/V_f$. At insertion time, the viscous dissipation rate shows a sharp peak

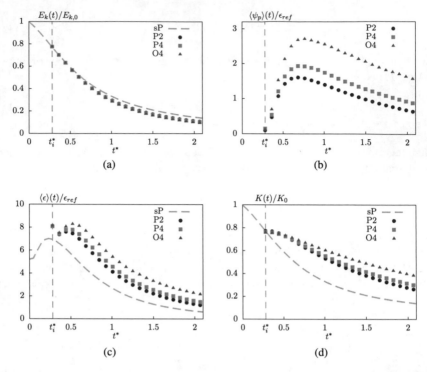

Fig. 2 Direct particle-fluid simulation of isotropic decaying turbulence laden with prolate ellipsoids with an aspect ratio of $\beta = 2$ ($P2$) and $\beta = 4$ ($P4$), and oblate particles ($O4$). The single phase results are shown as a reference (sP-DNS). Nondimensional temporal development of: **a** turbulent kinetic energy $E_k(t)$ normalized by its initial value $E_{k,0}$; **b** mean rate of total kinetic energy transfer $\langle\psi\rangle(t) = \Psi(t)/V_f$ normalized by the reference dissipation $\epsilon_{ref} = \rho u_0^3/L$; **c** mean viscous dissipation rate $\langle\epsilon\rangle(t) = \mathcal{E}(t)/V_f$ normalized by the reference dissipation ϵ_{ref}; and **d** total kinetic energy of the particles $K(t)$ normalized by $K_0 = \phi_m E_{k,0}$

due to the instantaneous build-up of the boundary layers. Similarly to the transfer of kinetic energy, the case $O4$ shows a larger dissipation rate. This explains the almost identical development of the turbulent kinetic energy for the three cases. The additional transfer of kinetic energy correlates with a higher viscous dissipation rate. The total kinetic energy of the particles $K(t)$ normalized by $K_0 = \phi_m E_{k,0}$ is presented in Fig. 2d. In all cases, the turbulent kinetic energy decays substantially faster than the kinetic energy of the particles. At insertion time, the particles have the same velocity as the surrounding fluid such that the total kinetic energy transfer vanishes. Immediately after insertion, the particles leave their initial position. The larger particles in case $P4$ and $O4$ maintain their kinetic energy for a longer time, which leads to a higher velocity difference vector and therefore a higher viscous dissipation rate and transfer of kinetic energy.

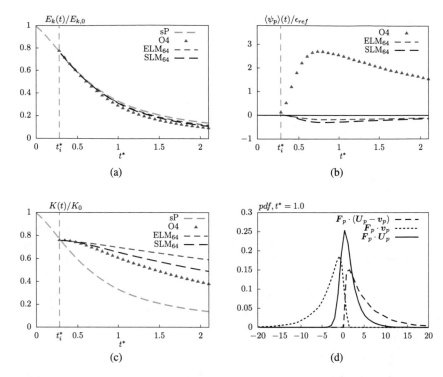

Fig. 3 Comparison of the LES using the spherical Langrangian model SLM$_{64}$ and the ellipsoidal Langrangian model ELM$_{64}$ against the reference results of DPFS for oblate particles ($O4$). The single phase results are shown as a reference (sP). Nondimensional temporal development of: **a** turbulent kinetic energy $E_k(t)$ normalized by its initial value $E_{k,0}$; **b** mean rate of total kinetic energy transfer $\langle \psi \rangle (t) = \Psi(t)/V_f$ normalized by the reference dissipation $\epsilon_{ref} = \rho u_0^3/L$; **c** total kinetic energy of the particles $K(t)$ normalized by $K_0 = \phi_m E_{k,0}$; **d** shows the probability density function (pdf) of the particle-induced contributions by the linear dynamics in the turbulent kinetic energy budgets normalized by the reference dissipation ϵ_{ref}. The probability density function (pdf) is generated using SLM$_{64}$ for case $O4$ at $t^* = 1.0$ and the pdf values are normalized by $u^3 d_p^2$. The dashed line represents the particle-induced dissipation rate, the dotted line is the direct transfer of kinetic energy, and the full line the total contribution of the particles in the budget

4.2 Assessment of LES Using Point-Particle Models

In this Section, the LES using an SLM (SLM$_{64}$) and an ELM (ELM$_{64}$) are validated against the reference results of the DPFS. Four different resolutions have been simulated for this study, where the mesh size ranges from 64^3 cells to 256^3 cells corresponding to a DNS. The statistics, which are presented hereafter, show only minor differences between the resolutions. Therefore, only the results of an LES with 64^3 cells are shown. For instance, case $O4$ will be presented in detail whereas the other cases are only outlined, if the results are significantly different.

Figure 3a shows the temporal development of the turbulent kinetic energy predicted by SLM$_{64}$ and ELM$_{64}$ compared to the DPFS. Consistently with the DPFS, the point-particle models predict a moderate attenuation. The ELM_{64} slightly over-estimates the turbulent kinetic energy, while the SLM_{64} is close to the DPFS. For point-particle models, the budget Eq. 17 reads [4]

$$\frac{\partial E_k}{\partial t}(t) = -\overline{\mathcal{E}}(t) - \boldsymbol{F}_p \cdot \boldsymbol{U}_p. \tag{22}$$

The point-particle model implicitly combines the particle-induced dissipation $\mathcal{E}_p$ with the direct transfer of kinetic energy $\Psi(t)$ [6, 14] neglecting the fluid inertia term $\mathcal{I}_f$. For the subsequent comparison, the mean rate of the turbulent kinetic energy transfer $\langle \psi_p \rangle(t)$ is redefined for point-particle models with $\langle \psi_p \rangle(t) = -\frac{1}{V_f} \sum_{p=1}^{N_p} \boldsymbol{F}_p \cdot \boldsymbol{U}_p$ to be consistent with other studies using two-way coupled point-particle models (e.g. [4]).

Figure 3b depicts the kinetic energy transfer of the point-particle models compared to the DPFS. Both models predict a sink of turbulent kinetic energy, which corresponds to the attenuated energy shown in Fig. 3a. Due to the different definitions of the kinetic energy transfer, the differences between the reference results of the DPFS and the point-particle models are substantial. The total kinetic energy of the particles $K(t)$ is shown in Fig. 3c. The point-particle models significantly overestimate the kinetic energy of the particles. Based on the reference results of the DPFS, the kinetic energy is 12.6% higher in the SLM_{64} and 24.8% higher in ELM_{64} at $t^* = 1.5$. The SLM_{64} is not aware of the anisotropy of the particles and uses empirical drag correlations for finite fluid inertia, which is valid for spherical particles. This leads to an underestimation of the forces acting on the particles and the initial kinetic energy of the particles remains longer during the decay of the turbulence. The ELM_{64} fails to predict the DPFS results. At this particle Reynolds number, the fluid inertia has to be taken into account to predict the linear dynamics of the particles. Figure 3d depicts the pdf of the linear contribution of the particles in the turbulent kinetic energy budget generated using SLM_{64}. The model predicts that the direct transfer of kinetic energy $\boldsymbol{F}_p \cdot \boldsymbol{v}_p$ is a source in the budget, which is largely eliminated by the particle-induced dissipation $\boldsymbol{F}_p \cdot (\boldsymbol{U}_p - \boldsymbol{v}_p)$. Note that both contributions are not resolved in the simulations using point-particle models but only recovered indirectly for the statistics. That is, the force is modeled via Eq. 14 and $\boldsymbol{U}_p$ is the velocity of the cell containing the point particle. The total contribution $\boldsymbol{F}_p \cdot \boldsymbol{U}_p$ is, however, different from the reference of DPFS. It can be expected that the differences are significantly larger for setups, in which the direct transfer of kinetic energy is not balanced by the particle-induced dissipation.

Figure 4a shows a comparison of the value of the particle angular velocity $|\boldsymbol{\omega}_p|$. The data of the ELM_{64} are compared against the DPFS. Since these statistics depend on the resolution of the background flow, the data of a DNS using the ELM (ELM_{256}) are included in the figure. A broad distribution of the angular velocity can be observed for the case O_4. The ELM fails to predict this distribution and shows a sharp peak at low angular velocities for both resolutions. The case P_4 is presented in Fig. 4b. The DPFS shows a narrower distribution than for case O_4, whereas the ELM predicts a wider distribution. The prediction of the ELM is better than for the case $O4$,

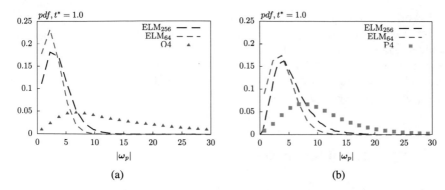

Fig. 4 Comparison of the results using the ellipsoidal Langrangian model ELM against the reference results of DPFS. A DNS using the ELM corresponds to ELM_{256} and the LES corresponds to ELM_{64}. The figures show the absolute value of the angular velocities of the particles $|\omega_p|$ normalized by u_0/L for ellipsoidal particles with aspect ratio of **a** $\beta = 0.25$ and **b** $\beta = 4$

but still shows significant deviations. Again a dependence on the resolution of the turbulence can be observed. As for the linear dynamics, the rotational dynamics are not accurately predicted, which can be explained by missing correlations for the fluid inertia in the formulation of the ELM.

5 Conclusion

The accuracy of LES using two-way coupled ellipsoidal and spherical Lagrangian models in isotropic decaying turbulence is analyzed. Reference results are generated via direct particle-fluid simulation, where an adaptively refined Cartesian cut-cell mesh is employed to completely resolve all fluid and particle scales. A dilute suspension laden with particles of Kolmogorov-length-scale size is considered, where turbulence modulation may appear but collisions are not statistically important, i.e., a two-way coupled suspension. The particle aspect ratio ranges from $\beta = 0.25$ to 4. Turbulent kinetic energy budgets are presented. The particles transfer kinetic energy into the flow field, which is mainly dissipated in the boundary layers and wakes of the particles. Moderate turbulence attenuation is observed in comparison to the single phase flow. This behavior is accurately predicted by the LES using point-particle models. The particle dynamics, however, are not accurately predicted by the Lagrangian models. The ellipsoidal model significantly overestimates the kinetic energy of the particles during the decay process, because it assumes a vanishing particle Reynolds number, i.e., it neglects the fluid inertia. The spherical Lagrangian model takes the fluid inertia into account but neglects the anisotropic particle shape. Although the results are closer to the reference, the spherical Lagrangian model still overestimates the kinetic energy of the particles. These observations hold for all resolutions of the LES, where the turbulent kinetic energy is not filtered significantly.

If an orientation dependent correlation for finite fluid inertia is used, the rotational dynamics define the orientation and thus, the linear dynamics of the particles. The rotation is, however, not correctly predicted by the ellipsoidal model. A full set of correlation terms for linear and rotational dynamics might be required, although the linear velocity difference between particles and surrounding fluid almost completely defines the turbulence modulation.

6 Computational Resources

The DPFS performed in this contribution contains up to 45,000 fully resolved particles to provide a sufficiently large number of samples for converged particle statistics. The parameter setup is chosen to study the effects of the particle shape on turbulence modulation and particle dynamics. A minimum resolution of 10 cells per particle diameter is required in the vicinity of each particle to provide sufficient accuracy of the near-particle hydrodynamics. In total, approx. 2 billion cells are required to resolve the near-particle flow field. Each of the simulation is performed on 2000 nodes using about 40 TB of RAM distributed over 48,000 CPU cores. Dynamic load balancing is applied using an automatic redistribution of the cells on a weighted Hilbert curve [12]. The post-processing of the results require the whole instantaneous flow field for a statistically sufficient amount of time steps. Therefore, the large amount of data has to be partially analyzed in-Situ, since the computation lasts $\mathcal{O}(10^5) - \mathcal{O}(10^6)$ time steps, where each time step would require 270 GB disk space.

Acknowledgements This work has been financed by the German Research Foundation (DFG) within the framework of the SFB/Transregio 'Oxyflame' (subproject B2). The support is gratefully acknowledged. Computing resources were provided by the High Performance Computing Center Stuttgart (HLRS) and by the Jülich Supercomputing Center (JSC) within a Large-Scale Project of the Gauss Center for Supercomputing (GCS).

References

1. S. Balachandar, J.K. Eaton, Turbulent dispersed multiphase flow. Annu. Rev. Fluid Mech. **42**, 111–133 (2010)
2. M. Berger, M. Aftosmis, Progress towards a Cartesian cut-cell method for viscous compressible flow, in *AIAA Paper 2012–1301* (2012)
3. S. Elghobashi, On predicting particle-laden turbulent flows. Appl. Sci. Res. **52**(4), 309–329 (1994)
4. A. Ferrante, S. Elghobashi, On the physical mechanisms of two-way coupling in particle-laden isotropic turbulence. Phys. Fluids **15**(2), 315–329 (2003)
5. K. Fröhlich, L. Schneiders, M. Meinke, W. Schröder, Assessment of non-spherical point-particle models in LES using direct particle-fluid simulations, in *AIAA Paper 2018–3714* (2018)
6. K. Fröhlich, L. Schneiders, M. Meinke, W. Schröder, Validation of Lagrangian two-way coupled point-particle models in large-Eddy simulations. Flow Turbul. Combust. **101**(2), 317–341 (2018)

7. R. Glowinski, T. Pan, T. Hesla, D. Joseph, J. Periaux, A fictitious domain approach to the direct numerical simulation of incompressible viscous flow past moving rigid bodies: application to particulate flow. J. Comput. Phys. **169**(2), 363–426 (2001)
8. C. Günther, M. Meinke, W. Schröder, A flexible level-set approach for tracking multiple interacting interfaces in embedded boundary methods. Comput. Fluids **102**, 182–202 (2014)
9. D. Hartmann, M. Meinke, W. Schröder, A strictly conservative Cartesian cut-cell method for compressible viscous flows on adaptive grids. Comput. Meth. Appl. Mech. Eng. **200**(9), 1038–1052 (2011)
10. M.R. Maxey, J.J. Riley, Equation of motion for a small rigid sphere in a nonuniform flow. Phys. Fluids **26**(4), 883–889 (1983)
11. M. Meinke, W. Schröder, E. Krause, T. Rister, A comparison of second-and sixth-order methods for large-Eddy simulations. Comput. Fluids **31**(4), 695–718 (2002)
12. L. Schneiders, J.H. Grimmen, M. Meinke, W. Schröder, An efficient numerical method for fully-resolved particle simulations on high-performance computers. PAMM **15**(1), 495–496 (2015)
13. L. Schneiders, C. Günther, M. Meinke, W. Schröder, An efficient conservative cut-cell method for rigid bodies interacting with viscous compressible flows. J. Comput. Phys. **311**, 62–86 (2016)
14. L. Schneiders, M. Meinke, W. Schröder, Direct particle-fluid simulation of Kolmogorov-length-scale size particles in decaying isotropic turbulence. J. Fluid Mech. **819**, 188–227 (2017)
15. L. Schneiders, M. Meinke, W. Schröder, On the accuracy of Lagrangian point-mass models for heavy non-spherical particles in isotropic turbulence. Fuel **201**, 2–14 (2017)
16. C. Siewert, R. Kunnen, M. Meinke, W. Schröder, Orientation statistics and settling velocity of ellipsoids in decaying turbulence. Atmos. Res. **142**, 45–56 (2014)
17. K.D. Squires, J.K. Eaton, Particle response and turbulence modification in isotropic turbulence. Phys. Fluids A **2**(7), 1191–1203 (1990)
18. B. van Leer, Towards the ultimate conservative difference scheme. V. A second-order sequel to Godunov's method. J. Comput. Phys. **32**(1), 101–136 (1979)
19. G.A. Voth, A. Soldati, Anisotropic particles in turbulence. Annu. Rev. Fluid Mech. **49**, 249–276 (2017)
20. F.M. White, *Viscous Fluid Flow* (McGraw-Hill, 1991)

Direct Numerical Simulation of a Wind-Generated Water Wave

Jonas Steigerwald, Jonathan Reutzsch, Matthias Ibach, Martina Baggio,
Adrian Seck, Brian K. Haus, and Bernhard Weigand

Abstract The interaction between an airflow and a water surface influences many environmental processes. For example in a rough ocean, entrained droplets from the water surface enhance transport processes above the ocean surface which can lead to the formation of hurricanes. In order to get a better understanding of the fundamental processes we perform direct numerical simulations (DNS) of a wind-generated water wave. To conduct these simulations we use our in-house code Free Surface 3D (FS3D) which is based on the Volume-of-Fluid (VOF) method and uses a Piecewise Linear Interface Calculation (PLIC) method. Two simulations with different grid resolutions are presented. In both cases a gravity-capillary wind-wave develops, starting with a quiescent water surface and solely driven by the turbulent air flow. We evaluate different aspects of the wind-wave characteristics such as topology, phase velocity, and temporal development of the wave energy. Furthermore, we compare the results with linear wave theory and experimental data obtained in the ASIST wave tank of the University of Miami. The comparison shows a very good agreement between experiments and numerical predictions, thus, FS3D is capable of numerically reproducing a gravity-capillary wind-wave with high accuracy. In addition, we continued our work to optimize the performance of FS3D by replacing large pointer arrays in our code with allocatable arrays. The analysis of strong and weak scaling shows an improvement in performance of up to 48%.

J. Steigerwald (✉) · J. Reutzsch · M. Ibach · M. Baggio · A. Seck · B. Weigand
Institute of Aerospace Thermodynamics (ITLR), University of Stuttgart, Pfaffenwaldring 31,
70569 Stuttgart, Germany
e-mail: jonas.steigerwald@itlr.uni-stuttgart.de

B. K. Haus
Department of Ocean Sciences, Rosenstiel School of Marine and Atmospheric Science,
University of Miami, 4600 Rickenbacker Causeway, Miama, FL, USA
e-mail: haus@rsmas.miami.edu

© Springer Nature Switzerland AG 2021
W. E. Nagel et al. (eds.), *High Performance Computing in Science and Engineering '19*,
https://doi.org/10.1007/978-3-030-66792-4_22

1 Introduction

The interaction between an airflow and a water surface is highly relevant for many environmental processes. When the airflow speed is high enough, droplets are entrained from the water surface, a scenario which is for example important for the formation and amplification of hurricanes. The intensity of such tropical cyclones depends on mass, momentum, and energy transfer between the air and the water and, thus, on size and number of entrained droplets. In order to get a reliable prediction of the intensity, an accurate estimation of the droplet size distribution in the sea spray environment is of great importance. However, the actual process of droplet entrainment is still not fully understood [31]. Precise experimental measurements of this small-scale process under such extreme conditions are hard to accomplish. Thus, our longterm objective is to uncover the mechanism of droplet entrainment from short wind-generated water waves by means of direct numerical simulation (DNS). In order to achieve this objective we use our in-house multiphase computational fluid dynamics (CFD) code Free Surface 3D (FS3D). FS3D is based on the Volume-of-Fluid (VOF) method and has already been used successfully to simulate complex hydrodynamic multiphase problems with droplet generation processes like primary jet breakup, droplet-wallfilm interactions, and droplet-droplet collisions [5, 6, 14, 19]. The first step towards our objective is the topic of this work: The generation of a pure wind-generated water wave, also called wind-wave, by using FS3D.

A DNS of a wind-wave is an ambitious task. Due to the high demands for the spatial and temporal resolution limits, the wind-speed, and the wavelength of wind-waves, the use of supercomputers is indispensable even today. One of the first DNS of a wind-wave in which both water and air layer were simulated was performed by Fulgosi et al. [8]. In their work they investigated the turbulent structures in the air above a deformable water surface. Both layers were driven by pressure gradients in opposite directions resulting in a countercurrent air-water flow. The surface elevation in their simulation was, however, only minimal due to the coarse mesh resolution. The same approach was used by Lakehal et al. [17] to evaluate the turbulent heat transfer across the deformable interface. Komori et al. [15] investigated a sheared gas-liquid interface due to wind-driven turbulence and the resulting turbulent structures in both water and air layer. In their simulations they initialized a fully developed wall turbulent airflow above a quiescent water layer. Furthermore, they used an arbitrary Lagrangian-Eulerian formulation (ALE) method with a moving grid.

The wind-wave generation process was investigated by Lin et al. [18]. Their air and water layer were separated by a deformable interface and they initialized a small constant velocity in the air layer to drive the waves. In addition, they superimposed the temperature field of both layers with small random perturbations to trigger turbulence caused by a distorted buoyancy force in normal direction to the interface. The very beginning of the wind-wave generation was investigated by Zonta et al. [32] with a countercurrent air-water flow.

The scalar transfer across an interface in a sheared wind-driven liquid flow and the effects of turbulent eddies, Langmuir circulation, and the influence of the Schmidt number was studied by Takagaki et al. in [27, 28], respectively. In both studies, the numerical procedure was the same as used by Komori et al. [15]. A very topical simulation was performed by Hafsi et al. [10] who focused on the scalar transfer across the interface during onset and growth of small-scale Langmuir circulations. They followed closely the numerical approach of Komori et al. [15] as well.

In this paper, a similar approach to Lin et al. [18] is used to generate a wind-wave. However, no perturbations are superimposed on any field to guarantee that the onset of wave generation starts naturally and only due to the transition from the initial laminar air flow to a turbulent one. In contrast to all above mentioned works, we also evaluate the temporal development of the wave energy to determine the actual wave state and its development. Consequently, we are able to link occurring phenomena to the wave state. In order to estimate the necessary resolution of our computational grid, we perform two simulations using a coarse grid and a fine grid with a doubled number of grid cells in each direction. The resolution of the coarse grid is already finer as the resolution used by Hafsi et al. [10]. Thus, to the best of our knowledge, our investigation is the highest resolved simulation of a wind-wave so far. Since a good parallel performance of FS3D is essential for performing DNS on thousands of cores, we also continued the work of Reutzsch et al. [24] to convert all data fields with a pointer attribute to allocatable arrays. The results are also presented in this paper.

2 Formulation and Numerical Method

In order to solve the incompressible Navier–Stokes equations for multiphase flows by means of DNS the CFD code FS3D is used. It was developed at the Institute of Aerospace Thermodynamics (ITLR) in Stuttgart for more than 20 years and is continuously enhanced. A wide variety of recent studies demonstrates the ability to simulate highly dynamic processes and complex topologies [5, 6, 14, 19, 24]. Since small-scale turbulent fluctuations are directly resolved without utilizing turbulence modeling, a high resolution in space and time needs to be achieved. Therefore, simulations in FS3D are parallelized using MPI and OpenMP and were executed at the High-Performance Computing Center Stuttgart (HLRS) satisfying the high demand for computational power. FS3D is well validated on the Cray XC40 Hazel Hen supercomputer and shows good performance even in test cases with up to eight billion computational cells.

2.1 Numerical Method

The in-house code FS3D solves the conservation equations for mass and momentum

$$\rho_t + \nabla \cdot (\rho \mathbf{u}) = 0, \tag{1}$$

$$(\rho \mathbf{u})_t + \nabla \cdot (\rho \mathbf{u} \otimes \mathbf{u}) = \nabla \cdot (\mathbf{S} - \mathbf{I}p) + \rho \mathbf{g} + \mathbf{f}_\gamma \tag{2}$$

on finite volumes (FV), where ρ denotes the density, $\mathbf{u}$ the velocity vector, p the static pressure; $\mathbf{g}$ denotes volume forces, such as gravity, and $\mathbf{f}_\gamma$ accounts for the body force which is used to model surface tension in the vicinity of the interface. The shear stress tensor $\mathbf{S}$ is defined by $\mathbf{S} = \mu \left[\nabla \mathbf{u} + (\nabla \mathbf{u})^T \right]$ for Newtonian fluids with μ representing the dynamic viscosity. The flow field is computed by solving equations (1) and (2) in a one-field formulation where the different phases are regarded as a single fluid with variable physical properties that discontinuously change across the separating interface. To identify the different phases, an additional indicator variable f according to the classical VOF-method by Hirt and Nichols [11] is introduced. The VOF-variable f determines the liquid volume fraction in each control volume and is defined as

$$f(\mathbf{x}, t) = \begin{cases} 0 & \text{in the gaseous phase,} \\ ]0; 1[& \text{at the interface,} \\ 1 & \text{in the liquid phase.} \end{cases}$$

The scalar f is advected across the computational domain by using the transport equation

$$f_t + \nabla \cdot (f\mathbf{u}) = 0. \tag{3}$$

Consistent with the one-field formulation, the local variables, e.g. the density, is defined by using the volume fraction f and the respective values of a pure gaseous (index g) and liquid (index f) phase in our present case, resulting in

$$\rho(\mathbf{x}, t) = \rho_f f(\mathbf{x}, t) + \rho_g (1 - f(\mathbf{x}, t)). \tag{4}$$

To prevent numerical diffusion of the transport variable f and to calculate the corresponding fluxes, FS3D makes use of a Piecewise Linear Interface Calculation algorithm (PLIC) proposed by Rider and Kothe [25]. The algorithm reconstructs a plane separating the fluid and the gas phase in an interface cell and is therefore capable of creating a sharp liquid-gas interface. For the advection method, three one-dimensional non-conservative transport equations are solved successively. With a permutation of the equation sequence and a specific divergence correction, second-order accuracy in space and time can be achieved [26]. Several models to compute the surface tension force are available. In the present study we use the conservative continuous surface stress (CSS) model by Lafaurie et al. [16]. A more detailed overview of the numerical implementation and methods as well as further capabilities and applications of FS3D are given in Eisenschmidt et al. [5].

2.2 Classification and Characterization of Wind-Waves

A classification of water waves can be achieved by considering the two acting restoring forces which return a displaced water surface to its equilibrium position: gravity and surface tension. The ratio of these forces is often used in context of water waves and is described by the non-dimensional Bond number

$$Bo = \frac{\Delta \rho g}{\sigma k^2}, \tag{5}$$

where $\Delta \rho = \rho_f - \rho_g$ and $k = 2\pi/\lambda$ is the wave number [30]. For $Bo \gg 1$ surface tension forces can be neglected and the main acting force is gravity. These waves are commonly referred to as gravity waves. For $Bo \ll 1$ surface tension forces become dominant and they are called capillary waves. Waves that lie in the transitional region where both forces have to be considered are referred to as capillary-gravity waves. This type of wave is simulated in this work.

Based on the wavelength λ, waves are governed by different physical processes and have different velocities of propagation c which is expressed by the linear dispersion relation

$$c^2 = \frac{g}{k} + \frac{\sigma}{\rho_f} k. \tag{6}$$

Another important parameter for characterizing the current state of a propagating wave is the total wave energy content per surface area E_{wave} as given by Tulin [29] consisting of the sum of three energy components, resulting in

$$E_{wave} = E_{kin} + E_{pot} + E_{\sigma}. \tag{7}$$

The energy components in Eq. (7) are the kinetic energy E_{kin} of the water, the gravitational potential energy E_{pot} with respect to the initially flat water surface, and the surface energy E_{σ}. A detailed description of the energy components can be found in Kaufmann [13].

3 Numerical Setup

The wind-waves are simulated using a three-dimensional rectangular computational domain with dimensions $\lambda \cdot 0.5\lambda \cdot 2\lambda$ in streamwise, spanwise, and normal direction, respectively, as shown in Fig. 1. The wavelength λ is set to $\lambda = 0.05$ m which ensures that the generated wind-waves lie within the range of gravity-capillary waves. The domain consists of an initially quiescent water layer and an air layer with an initial velocity of $U = 15$ m/s in streamwise direction. Both water and air layer have thickness λ. For the water layer this guarantees that a generated wave is not influenced by the bottom, meaning that the deep water assumption $d > 0.5\lambda$ is fulfilled [1]. The

Fig. 1 Computational
domain and coordinate
system. The arrays in the air
layer indicate the initial
direction of the wind

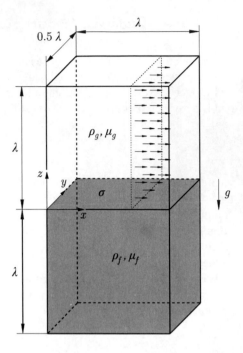

chosen thickness of the air layer ensures on the one hand that the air flow contains
sufficient kinetic energy to accelerate the water surface and to become turbulent,
leading to the generation of a wind-wave. It should be mentioned, that since the total
energy of the system consists only of kinetic energy at the beginning, the air velocity
is freely-decaying during the simulation due to the energy transfer into the water
layer. On the other hand we can ensure that the interaction between the turbulent air
flow and the water surface is not disturbed by the upper boundary of the computational domain. At the upper boundary as well as at the bottom of the domain free
slip boundary conditions are applied, whereas in streamwise and spanwise direction
periodic boundary conditions are used.

In this study, two different grid resolutions are used in order to analyze the necessary grid resolution for simulating accurately a gravity-capillary wind-wave. For the
simulation with the coarse resolution we discretize the domain with an equidistant
Cartesian grid using $256 \cdot 128 \cdot 512$ cells. For the fine resolution the number of cells
is doubled in each direction leading to $512 \cdot 256 \cdot 1024$ cells. Therefore, the number of used processors is 512 for the coarse and 2048 for the fine grid resolution,
respectively.

For both simulations the physical properties of air and fresh water are used at reference temperature $T = 293$ K and standard pressure $p = 1.013$ bar. The properties
are listed in Table 1. By using these physical properties and the selected wavelength
the Bond number results in $Bo = 8.52$ indicating that surface tension already plays
a significant role in our simulations and must not be ignored.

Table 1 Physical properties of air and water

Air density ρ_g (kg/m^3)	Air viscosity μ_g (Ns/m^2)	Water density ρ_f (kg/m^3)	Water viscosity μ_f (Ns/m^2)	Surface tension σ (N/m)
1.204	1.813×10^{-5}	998	1.002×10^{-3}	72.74×10^{-3}

In order to show that the grid resolution is sufficient for DNS, we calculated the ratio of grid spacing Δx to the smallest dissipative length scale, the Kolmogorov length ($\Delta x_{coarse}/l_{k,coarse} = 9.89$, $\Delta x_{fine}/l_{k,fine} = 5.11$). The chosen characteristic velocity is the highest velocity at the point in time when a single wave has formed. Since both grid spacing and Kolmogorov length are in the same order of magnitude for both grids, we are able to produce physically correct results.

4   Results

In the following the results of both simulations are compared regarding the overall shape, the phase velocity, and the wave energy. In addition, we compare the simulation on the fine grid with experimental data. For the evaluation, the non-dimensional time

$$\tilde{t} = \frac{tc}{\lambda} \tag{8}$$

was introduced, where c is the phase velocity, Eq. (6), for the corresponding wavelength. Both simulations were performed until $\tilde{t} = 7$ was reached meaning that the wave could run seven times through the computational domain. It is important to mention is that the growth of the waves at the early stages is influenced by the periodicity of the computational domain. For this reason, the results are not taken into account until a single wave has formed at $\tilde{t} \approx 4$.

In Fig. 2 both simulated gravity-capillary wind-waves are shown at three different times $\tilde{t} = 5$, $\tilde{t} = 6$, and $\tilde{t} = 7$. On the left side, Fig. 2a–c show the result using the coarse grid, whereas on the right side, Fig. 2d–e depict the wind-wave using the fine grid. For both cases the VOF-variable f is depicted as a contour plot, showing the iso-surface through $f = 0.5$ representing the water surface. Furthermore, the velocity magnitude is depicted on two slices in streamwise and spanwise direction, respectively. As one can see, for both resolutions a wind-wave has emerged showing the typical shape of a gravity-capillary wave with a pronounced bulge on the top of the wave's leeward side, caused by surface tension effects [4, 20]. Both waves are driven by a turbulent air flow above the water. In both cases a decrease of the velocity magnitude over time is visible due to dissipation effects and energy transfer from air to water. The turbulence structures as well as the recirculation region above the waves' leeward side is, however, more pronounced for the fine simulation, as expected. In addition, a clear difference on the leeward side between both simulations

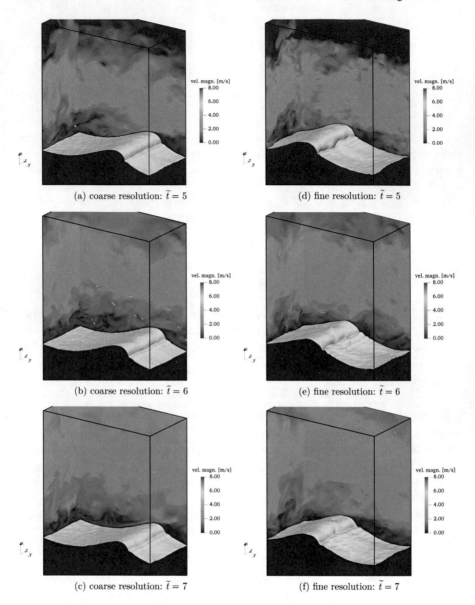

(a) coarse resolution: $\tilde{t} = 5$ (d) fine resolution: $\tilde{t} = 5$

(b) coarse resolution: $\tilde{t} = 6$ (e) fine resolution: $\tilde{t} = 6$

(c) coarse resolution: $\tilde{t} = 7$ (f) fine resolution: $\tilde{t} = 7$

Fig. 2 Comparison of the simulated gravity-capillary wind-wave using the coarse mesh resolution **a–c** (left) and the fine mesh resolution **d–f** (right) for $\tilde{t} = 5$, $\tilde{t} = 6$, and $\tilde{t} = 7$

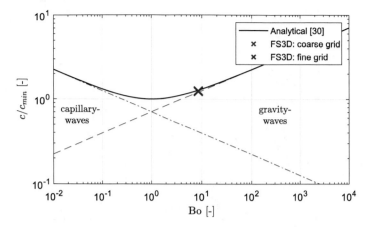

Fig. 3 Comparison of the evaluated phase velocity of both simulated wind-waves with analytical values from the linear wave theory, see Eq. (6) [30]. The phase velocity was evaluated between $\tilde{t} = 6$ and $\tilde{t} = 7$ and was made non-dimensional by means of $c_{min} = (4g\sigma/\rho_f)^{1/4}$

is visible. The fine resolved wind-wave shows a well-developed train of smaller waves on the forward face of the wave. In context of physical oceanography these small waves are called "parasitic[1] capillary waves". We will make use of this expression in the following. The parasitic capillary waves remain stationary in the frame of the moving wind-wave and are characteristic of gravity-capillary waves [20]. Down the forward side of the wave they finally disappear due to viscous damping [2]. The coarse resolved wave shows in contrast only one weak developed parasitic capillary wave indicating that the grid resolution is insufficient to accurately reproduce all characteristics of the gravity-capillary wind-wave.

To verify that the propagation of the wind-waves can be numerically reproduced we evaluated their phase velocity between $\tilde{t} = 6$ and $\tilde{t} = 7$ and compared the values with the analytical solution from the linear wave theory (Eq. (6)). This time interval was selected since the energy input by the wind is low during this interval and, therefore, the wave comes closest to a state of equilibrium. The point of reference for the evaluation was at both times the location of the maximum surface elevation averaged over the width of each wave. The resulting velocities are shown in Fig. 3 in non-dimensional form by scaling with the smallest possible phase velocity after the linear wave theory of $c_{min} = (4g\sigma/\rho_f)^{1/4}$ [30]. As can be seen, the phase velocity of the simulated wind-waves for both coarse and fine grid are nearly identical to the analytical solution. The deviation between the fine grid solution and the analytical solution is only around 0.8%, whereas the coarse grid solution overestimates the analytical solution by 2.6%. Even though these small deviations are present, this

---

[1]The term "parasitic" in context of physical oceanography must not be confused with the term "parasitic" in CFD simulation, where it describes artificial perturbations due to discretization errors leading to unphysical effects.

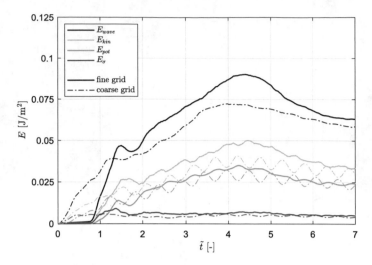

Fig. 4 Temporal evolution of the total wave energy and its three components for the coarse (dash-dotted lines) and fine (solid lines) computational grid

result clearly shows that our simulated wave exhibits a reasonable phase velocity and that already the coarse grid is sufficient to reproduce the velocity of wave propagation.

Besides the decreasing air flow velocity, also the bulge and the parasitic capillary waves seem to recede over time. This behavior can be explained by looking at the temporal evolution of the total wave energy which is shown together with its three energy components for both simulated wind-waves in Fig. 4. For both waves the temporal development of the wave energy is similar and differs only in few aspects. All energy components increase moderately and reach a maximum at $\tilde{t} = 4$ and $\tilde{t} \approx 4.4$ depending on the grid resolution. The maxima coincidence with the moment a single wind-wave has formed. After that, the energy decreases, however, differently strong for both grid resolutions. This energy decrease can be clearly linked now to the occurrence and recession of the parasitic capillary waves. Parasitic capillary waves extract energy from the main wave, enhance damping effects due to viscous dissipation, and, therefore, influence the dispersive and dissipative properties of the underlying main wave [7]. In case of the coarse grid resolution, the total wave energy E_{wave} decreases almost linear, as can be seen in Fig. 4. This is a consequence of the low resolved parasitic capillary wave whose shape remains nearly the same over time and whose dissipative effects cannot be numerically reproduced accurately. In contrast, for the simulation on the finer grid the decrease of the total wave energy E_{wave} shows strong nonlinear behavior caused by the dynamical behavior of parasitic capillary waves which was made possible due to fine grid resolution.

In order to show the physical soundness of our simulated gravity-capillary wind-wave, we compared our fine resolved simulation with our experimental data obtained by Ibach [12]. The experiments on short wind-waves were conducted in the Air-

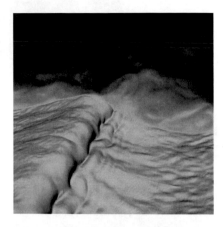 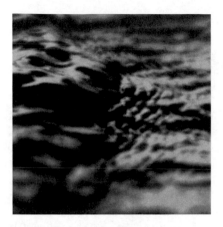

Fig. 5 Comparison of the simulated gravity-capillary wind-wave using the fine mesh (left) with experimental work by Ibach [12] (right). The train of parasitic capillary waves on the leeward side of the wind-wave can be numerically reproduced using FS3D

Sea Interaction Salt Water Tank (ASIST) of the University of Miami. A detailed description of the experimental setup can be found in [3, 12, 21, 22].

Figure 5 shows two close-up views of the leeward side of a wind-wave for both simulation on the lefthand side at $\tilde{t} = 6$ and experiment on the right hand side. As one can see, there is a good agreement between the simulated wind-wave with the fine structures on the wave's forward face and the observed wind-wave from the experiments. Both show a pronounced train of parasitic capillary waves in front of the bulge. From this result clearly follows, that only by using the fine grid resolution it is possible to simulate a gravity-capillary wind-wave and to reproduce important aspects of such a wave, as can be observed in wind-wave environments. It is also important to mention that the wavelengths of the parasitic capillary waves in simulation and experiment in Fig. 5 are not of the same size. The section visible in the experiment (see Fig. 5) is around 4 cm in the focal plane and shows only a part of the wind-wave. The wavelength of the observed complete wind-wave in the wave tank is, therefore, slightly longer than the one of our simulated wave. In contrast, the wavelength of the parasitic capillary waves in the experiment are shorter than the ones in our simulation. This observed relationship is as expected, since the wavelength of the parasitic capillary waves depends on the wavelength of the complete wind-wave due to the resonant condition and the linear dispersion relation [7]: The longer wavelength of the complete water wave, the shorter the wavelength of its parasitic capillary waves. This fact as well as the presence of the parasitic capillary waves indicate the good quality of our fine resolved gravity-capillary wind-wave. However, further simulations have to be performed in the future using finer grid resolutions in order to clarify if the train of capillary waves seen in the here presented simulation is sufficiently resolved.

5   Computational Performance

5.1   *Dynamic Memory Allocation: Pointer Versus Allocatable Arrays*

As reported by Reutzsch et al. [24], most of FS3D dynamic arrays have been declared with a pointer attribute for historical reasons. In contrast to allocatable arrays, objects with the pointer attribute can refer to different memory addresses during runtime. Because of this feature, pointer objects have more capabilities with respect to their allocatable counterpart, but they can also hinder optimization from the compiler so that a massive use of pointer arrays can result in a slower code. This was the reason that led us to switch from pointer to allocatable arrays wherever possible. This operation, already initiated in 2018 [24], has been finalized this year.

In the following, we present a performance analysis for this optimized version in terms of cycles per hour and code scalability. The benchmark case, analogous to the one described in [24], is an oscillating droplet representing a spume droplet entrained from a short wind-wave at high wind speeds. The droplet is initialized as an ellipsoid with the semi-principal axis $a = b = 1.357$ mm and $c = 0.543$ mm at the center of a cubic domain with an edge length of $x = y = z = 8$ mm. The fluid of the droplet is water at $T = 293$ K. We measured the performance of both weak and strong scaling, where we varied the number of processors from 2^3 up to 16^3. Since we did not use hyperthreading, the number of processors corresponds to the number of processes. The details of the analysis setups are shown in Table 2 for the strong scaling and in Table 4 for the weak scaling. Here, we used only spatial domain composition with MPI parallelization. For information on FS3D performance with hybrid OpenMP and MPI operation see [9, 23].

5.2   *Strong Scaling*

All simulated cases shown in Table 2 were carried out with a duration of 25 min and with a constant processor clock rate of 2.5 GHz. From the number of cycles performed in each case, we estimated the number of cycles per hour (CPH), which in turn was used to calculate the strong scaling efficiency (SSE). The latter is defined as:

$$SSE = \frac{CPH_N}{N \times CPH_1}. \tag{9}$$

As we don't have a case with just one processor, equation (9) changes to

$$SSE = \frac{CPH_N}{N/8 \times CPH_8}. \tag{10}$$

Table 2 Strong scaling setup

Problem size	512^3			
MPI-processes	2^3	4^3	8^3	16^3
Cells per process	256^3	128^3	64^3	32^3
Nodes	2	4	32	256
Processes per node	4	16	16	16

Table 3 Number of cycles per hour for the simulated cases

Processors	CPH allocatable	CPH pointer
8	79.2	64.2
64	403.2	364.8
512	2203.2	1994.4
4096	441.6	415.2

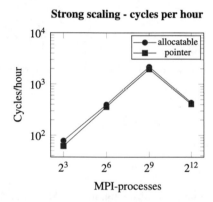
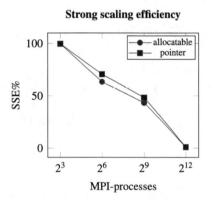

Fig. 6 Number of cycles per hour (on the left) and strong scaling efficiency (on the right) over the number of processes for the code versions with allocatable and pointer arrays

The results for the strong scaling test are shown in Table 3 and in Fig. 6. It can be seen that both code versions exhibit similar trends both in terms of cycles per hours and code scalability, whereas the code version with allocatable arrays is slightly faster for all simulated cases. The peak performance in terms of CPH is obtained in the 512 processors case, where the code version with allocatable arrays is about 10% faster (Table 3).

Table 4 Weak scaling setup

Cells per process	64^3			
Problem size	128^3	256^3	512^3	1024^3
MPI-processes	2^3	4^3	8^3	16^3
Nodes	2	4	32	256
Processes per node	4	16	16	16

Table 5 Number of cycles per hour for the simulated cases

Processors	CPH allocatable	CPH pointer
8	4428.794	3700.332
64	2893.369	2674.084
512	2142.857	1990.533
4096	300.592	203.283

5.3 Weak Scaling

The weak scaling was performed in analogy to the strong scaling. The simulations were carried out for 25 min with a constant processor clock rate (2.5 GHz). From that, the estimated number of cycles per hour (CPH) is used to calculate the weak scaling efficiency (WSE), defined by (11)

$$WSE = \frac{CPH_N}{CPH_1}. \tag{11}$$

Again, we don't have a case with just one processor, therefore equation (11) changes to

$$WSE = \frac{CPH_N}{CPH_8}. \tag{12}$$

Table 5 gives the number of cycles per hour for each simulation. Overall, the code version with the allocatable arrays is faster than the old version with pointers (see Fig. 7, left). For the small case with 8 processors (128^3 cells) a speed-up of 19.7% could be achieved. The difference is decreasing for larger cases up to a number of processes of 512 (512^3 cells), where the version with the pointers is 7.7% slower. Calculations with more processors, in this case 4096 (1024^3 cells), the total number of cycles per hour is decreasing drastically for both versions. This can also be seen when looking at the weak scaling efficiency in Fig. 7 on the right. However, the version with allocatable arrays is loosing less performance and, therefore, ends up with 47.9% more cycles per hour than the version with pointers.

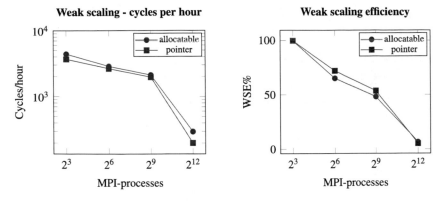

Fig. 7 Number of cycles per hour (on the left) and weak scaling efficiency (on the right) over the number of processes for the code versions with allocatable and pointer arrays

6 Conclusions

We presented our first DNS of a pure wind-generated water wave performed with the in-house code FS3D, which is based on the VOF- and PLIC-method. Whereas in most other studies dealing with this topic the initial wave structure or initial turbulence is imposed artificially, we set up a framework with a complete quiescent water layer and an air layer with a velocity profile. With this approach we ensure a natural generation of the wind-wave from scratch, only restricted to the applied domain length, hence, we chose only the final wavelength in advance. We performed two simulations of a wind-wave using a coarse and a fine grid resolution of the computational domain. We showed that in both cases a wind-wave has developed, solely driven by the turbulent air flow and with a typical shape of a gravity-capillary wave.

In both simulations the expected turbulent structures, the recirculation region on the leeward wave side and parasitic capillary waves could be identified. However, only with a fine grid resolution the dynamic behavior of the parasitic capillary waves could be captured. This was concluded by analyzing the temporal development of the wave energy. Furthermore, we verified the physical soundness of our simulation by making a comparison of the velocity of wave propagation to an analytical approach based on the linear wave theory. The evaluated phase velocity showed a good agreement for both simulations. As additional validation, we compared the simulation on the fine grid with experimental data obtained in the ASIST test rig of the University of Miami. A visual comparison of the topology and the wave type with a high speed camera image yielded promising results. For this reason, it is to planned increase the wind speed in future investigations in order to induce droplet entrainment from the water surface and to uncover the mechanism involved.

Furthermore, we finalized our work on replacing large pointer arrays in our Fortran code with allocatable arrays, which was initiated in our last report. Henceforth, with changing all possible pointers we conducted a full performance analysis, including strong and weak scaling. Both evaluations proved a much better performance with the conversion from pointers to arrays in our code.

Acknowledgements The authors kindly acknowledge the *High Performance Computing Center Stuttgart* (HLRS) for support and supply of computational time on the Cray XC40 platform under the Grant No. FS3D/11142. In addition, the authors kindly acknowledge the financial support of the Deutsche Forschungsgemeinschaft (DFG) through the projects GRK 2160/1, SFB-TRR75, WE2549/36-1, and WE2549/35-1.

References

1. R.G. Dean, R.A. Dalrymple, *Water Wave Mechanics for Engineers and Scientists*. Advanced Series in Ocean Engineering, Vol. 2 (World Scientific, 1991)
2. L. Deike, S. Popinet, W. Melville, Capillary effects on wave breaking. J. Fluid Mech. **769**, 541–569 (2015)
3. M.A. Donelan, B.K. Haus, W.J. Plant, O. Troianowski, Modulation of short wind waves by long waves. J. Geophys. Res.: Ocean. **115**(C10) (2010)
4. J.H. Duncan, Spilling breakers. Annu. Rev. Fluid Mech. **33**(1), 519–547 (2001)
5. K. Eisenschmidt, M. Ertl, H. Gomaa, C. Kieffer-Roth, C. Meister, P. Rauschenberger, M. Reitzle, K. Schlottke, B. Weigand, Direct numerical simulations for multiphase flows: an overview of the multiphase code FS3D. J. Appl. Math. Comput. **272**(2), 508–517 (2016)
6. M. Ertl, B. Weigand, Analysis methods for direct numerical simulations of primary breakup of shear-thinning liquid jets. At. Sprays **27**(4), 303–317 (2017)
7. A.V. Fedorov, W.K. Melville, Nonlinear gravity-capillary waves with forcing dissipation. J. Fluid Mech. **354**, 1–42 (1998)
8. M. Fulgosi, D. Lakehal, S. Banerjee, V. De Angelis, Direct numerical simulation of turbulence in a sheared air-water flow with a deformable interface. J. Fluid Mech. **482**, 319–345 (2003)
9. C. Galbiati, M. Ertl, S. Tonini, G.E. Cossali, B. Weigand, DNS investigation of the primary breakup in a conical swirled jet, in *High Performance Computing in Science and Engineering '15* (Transactions of the High Performance Computing Center, Stuttgart (HLRS), 2016), pp. 333–347
10. A. Hafsi, A.E. Tejada-Martínez, F. Veron, Dns of scalar transfer across an air-water interface during inception and growth of Langmuir circulation. Comput. Fluids **158**, 49–56 (2017)
11. C.W. Hirt, B.D. Nichols, Volume of fluid (VOF) method for the dynamics of free boundaries. J. Comput. Phys. **39**(1), 201–225 (1981). https://doi.org/10.1016/0021-9991(81)90145-5
12. M. Ibach, Investigations on sea spray generation and wave breaking processes on short wind waves. Master's thesis, University of Stuttgart, Institute of Aerospace Thermodynamics (2018)
13. J. Kaufmann, Direkte numerische Simulation der Instabilität einer überströmten Wasseroberfläche. Master's thesis, University of Stuttgart, Institute of Aerospace Thermodynamics (2016)
14. J. Kaufmann, A. Geppert, M. Ertl, R. Bernard, V. Vaikuntanathan, G. Lamanna, B. Weigand, Direct numerical simulations of one- and two-component droplet wall-film interactions within the crown-type splashing regime, in *ICLASS 2018, 14th triennial International Conference on Liquid Atomization and Spray Systems*, Chicago, USA (2018)
15. S. Komori, R. Kurose, K. Iwano, T. Ukai, N. Suzuki, Direct numerical simulation of wind-driven turbulence and scalar transfer at sheared gas-liquid interfaces. J. Turbul. **11**, N32 (2010)
16. B. Lafaurie, C. Nardone, R. Scardovelli, S. Zaleski, G. Zanetti, Modelling merging and fragmentation in multiphase flows with SURFER. J. Comput. Phys. **113**(1), 134–147 (1994)
17. D. Lakehal, M. Fulgosi, G. Yadigaroglu, S. Banerjee, Direct numerical simulation of turbulent heat transfer across a mobile, sheared gas-liquid interface. ASME J. Heat Transf. **125**(6), 1129–1139 (2003)

18. M.Y. Lin, C.H. Moeng, W.T. Tsai, P.P. Sullivan, S.E. Belcher, Direct numerical simulation of wind-wave generation processes. J. Fluid Mech. **616**, 1–30 (2008)
19. M. Liu, D. Bothe, Numerical study of head-on droplet collisions at high weber numbers. J. Fluid Mech. **789**, 785–805 (2016)
20. M.S. Longuet-Higgins, The generation of capillary waves by step gravity waves. J. Fluid Mech. **16**, 138–159 (1963)
21. L. Muser, Experimental and numerical investigations of sea spray generation processes on short wind waves. Master's thesis, University of Stuttgart, Institute of Aerospace Thermodynamics (2017)
22. D.G. Ortiz-Suslow, B.K. Haus, S. Mehta, N.J.M. Laxague, Sea spray generation in very high winds. J. Atmos. Sci. **73**(10), 3975–3995 (2016)
23. P. Rauschenberger, J. Schlottke, B. Weigand, A computation technique for rigid particle flows in an Eulerian framework using the multiphase DNS code FS3D, in *High Performance Computing in Science and Engineering '11*, Transactions of the High Performance Computing Center, Stuttgart (HLRS) (2011). https://doi.org/10.1007/978-3-642-23869-7_23
24. J. Reutzsch, M. Ertl, M. Baggio, A. Seck, B. Weigand, Towards a direct numerical simulation of primary jetbreakup with evaporation, in *High Performance Computing in Science and Engineering '18*, ed. by W.E. Nägel, D.H. Kröner, M.M. Resch (Springer International Publishing, 2019)
25. W.J. Rider, D.B. Kothe, Reconstructing volume tracking. J. Comput. Phys. **141**(2), 112–152 (1998). https://doi.org/10.1006/jcph.1998.5906
26. G. Strang, On the construction and comparison of difference schemes. SIAM J. Numer. Anal. **5**(3), 506–517 (1968)
27. N. Takagaki, R. Kurose, Y. Tsujimoto, S. Komori, K. Takahashi, Effects of turbulent eddies and Langmuir circulations on scalar transfer in a sheared wind-driven liquid flow. Phys. Fluids **27**(1), 016,603 (2015)
28. N. Takagaki, R. Kurose, A. Kimura, S. Komori, Effect of Schmidt number on mass transfer across a sheared gas-liquid interface in a wind-driven turbulence. Sci. Rep. **6**(37059) (2016)
29. M.P. Tulin, On the transport of energy in water waves. J. Eng. Math. **58**(1), 339–350 (2007)
30. J.M. Vanden-Broeck, *Gravity-Capillary Free-Surface Flows* (Cambridge University, 2010)
31. F. Veron, Ocean spray. Annu. Rev. Fluid Mech. **47**(1), 507–538 (2015)
32. F. Zonta, A. Soldati, M. Onorato, Growth and spectra of gravity-capillary waves in countercurrent air/water turbulent flow. J. Fluid Mech. **777**, 245–259 (2015)

Interface-Resolved Direct Numerical Simulation of Turbulent Flow over Porous Media

Xu Chu, Johannes Müller, and Bernhard Weigand

Abstract Direct numerical simulations (DNS) are conducted for turbulent flows over porous media. A high-order spectral/hp element method is adopted for solving the incompressible Navier-Stokes equations. Resolving flow details close to the interface relies on an adaptive polynomial refinement based on a conforming mesh. Four DNS cases up to bulk Reynolds number $Re = 15,000$ are conducted with a total mesh resolution up to 1 billion degrees of freedom. The highly-resolved DNS enables us to focus on two major physical phenomenon: (i) the turbulence modulation including drag reduction and flow control; (ii) the turbulent/non-turbulent interface close to the porous surface. The numerical solver exhibits an excellent scalability up to 96k cores on Hazel Hen. Strong scaling tests indicate an efficiency of 70% with around 5,000 mesh-nodes per core, which indicates a high potential for an adequate use of current and next-generation HPC platforms to investigate turbulent flows over porous media.

1 Introduction

Turbulent shear flows bounded by porous materials are encountered in various engineering applications such as transpiration cooling and selective laser melting. Understanding and manipulating the influence of the porous characteristics (e.g. morphology, topology) on turbulence owns strong significance in industry. Growing interest has been observed regarding the turbulent flow regime in porous structures, as the studies with direct numerical simulation (DNS) show [6, 7, 9, 11, 12, 24]. DNS allows microscopic visualization and analysis, which is hardly achievable within the experimental measurement in such confined and tortuous spaces.

Existing experiments provide information about the optically accessible areas. The modulation of turbulence by a permeable surface has been confirmed in early-years experiments with different configurations, e.g. turbulent open channel flows

X. Chu (✉) · J. Müller · B. Weigand
Institute of Aerospace Thermodynamics, University of Stuttgart, Pfaffenwaldring 31, 70569
Stuttgart, Germany
e-mail: xu.chu@itlr.uni-stuttgart.de

© Springer Nature Switzerland AG 2021   343
W. E. Nagel et al. (eds.), *High Performance Computing in Science and Engineering '19*,
https://doi.org/10.1007/978-3-030-66792-4_23

over porous media composed of spheres. A qualitative similar conclusion was drawn that the wall permeability is able to increase turbulent friction. Suga et al. [20] investigated a spanwise turbulence structure over permeable walls by using PIV measurements. Terzis et al. [21] examined experimentally the hydrodynamic interaction between a regular porous medium and an adjacent free-flow channel at low Reynolds numbers ($Re < 1$). In their study the porous medium consists of evenly spaced micro-structured rectangular pillars arranged in a uniform pattern, while the free-flow channel features a rectangular cross-sectional area. In a recent study [20], three different kinds of anisotropic porous media are constructed to form the permeable bottom wall of the channel. Their wall permeability tensor is designed to own a larger wall-normal diagonal component (wall-normal permeability) than the other components and the spanwise turbulent structures are investigated. However, because of the difficulty in performing measurements inside the porous media, it is not easy to discuss the turbulent flow physics inside the porous structures.

Direct numerical simulation exhibits an edge of observing and analyzing turbulent physics in a confined small space, not only for the canonical cases as channel flows and pipe flows [3, 5, 14–16, 18], but also for transitional and turbulent flow in a representative elementary volume (REV) of porous media [6, 7, 9]. However, an adequate resolving of the smallest length scales of the flow in the interface region requires enormous computational resources. To limit the computational cost, the size of the computational domain can be reduced. Therefore appropriate boundary conditions must be chosen. Jimenez et al. [10] performed DNS with a special boundary condition: They imposed no-slip conditions for the streamwise and spanwise velocities, and set the wall-normal velocity for the permeable wall to be proportional to the local pressure fluctuations. The friction is increased by up to 40% over the walls, which was associated with the presence of large spanwise rollers.

Another approach is to describe the flow inside the porous structure with the volume-averaged Navier–Stokes (VANS) equations [23] and couple them with the Navier–Stokes equations used for the free flow. Breugem and Boersma [1] belongs to one of the pioneers utilizing this method. They found a decrease in the peak value of the streamwise turbulence intensity normalized by the friction velocity at the permeable wall and an increase in the peak values of the spanwise and wall-normal ones. This can be explained by the absence of streaks and the associated streamwise vortices near a highly permeable wall. The existing Kelvin–Helmholtz instability is responsible for an exchange of momentum between the channel and the permeable wall. This process exhibits a significant contribution to the Reynolds-shear stress and leads therefore to a large increase in the skin friction. Rosti et al. [19] explored the potential of drag reduction with porous materials. They systematically adjusted the permeability tensor on the walls of turbulent channel flow via VANS-DNS coupling. The total drag could be either reduced or increased by more than 20% through adjusting the permeability directional properties. Configuring the permeability in the vertical direction lower than the one in the wall-parallel planes leaded to significant streaky turbulent structures (quasi 1-dimensional turbulence) and hence achieved a drag reduction. Recent studies achieved to resolve the porous media structures coupled with turbulent flows. Kuwata and Suga [13] used Lattice–Boltzmann method

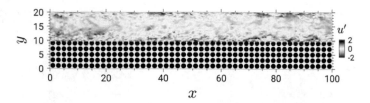

Fig. 1 Cross-Section of the computational domain in the $x - y$ plane. The spanwise direction is periodic. Depicted is a snapshot of the streamwise velocity fluctuation u'

(LBM) to resolve porous structures coupled with turbulent flows. The porous media is composed with interconnected staggered cube arrays. The difference between a rough wall and a permeable wall is elucidated.

The current study is intended to establish interface-resolved DNS research about turbulent flows over porous media. Through an adequate resolution of the flow field between the porous structure and the free-flow, both the turbulence modulation and the energy exchange across the porous surface is investigated. This physical knowledge can be used to support different levels of modeling like LES or RANS [22, 25–27]. Furthermore it will be possible to link the geometrical characteristics of the porous media with the turbulent structures of the flow field. This will enable the design of porous structures that generate specific flow properties.

2   Numerical Method

2.1   Simulation Method

Three-dimensional incompressible Navier–Stokes equations, given by Eqs. 1 and 2, are solved in non-dimensional form, where Π is the corresponding source term in the momentum equation to maintain a constant pressure gradient in x-direction.

$$\frac{\partial u_j}{\partial x_j} = 0 \tag{1}$$

$$\frac{\partial u_i}{\partial t} + \frac{\partial u_i u_j}{\partial x_j} = -\frac{\partial p}{\partial x_i} + \frac{1}{Re}\frac{\partial^2 u_i}{\partial x_i \partial x_j} + \Pi \delta_{i1} \tag{2}$$

A spectral/hp element solver Nektar++ [2] is used to perform the DNS and to resolve multiple-scales in the complex geometrical structures. The solver framework allows arbitrary-order spectral/hp element discretisations with hybrid shaped elements. Both modal and nodal polynomial functions are available for the high-order representation. In addition, the homogeneous flow direction can be represented with Fourier-spectral expansions and, therefore, enable the usage of an efficient parallel

Fig. 2 Mesh structure in the computational domain. The regions for the local polynomial refinement are $\Omega 1$, $\Omega 2$ and $\Omega 3$. The polynomial orders of the elements in this regions are $P_{\Omega 1} = 6$, $P_{\Omega 2} = 8$ and $P_{\Omega 3} = 4$

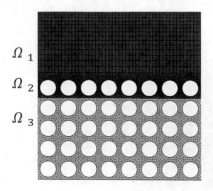

direct solver. The spectral-accurate discretisation combined with meshing flexibility is optimal to deal with complex porous structures and to resolve the interface region. The time-stepping is treated with a second-order mixed implicit-explicit (IMEX) scheme. The fixed time step is defined with $\Delta T/(h/\overline{u}) = 0.0002$. A complete run of 10 flow through time requires at least 5×10^5 time steps.

Figure 1 illustrates a two-dimensional sketch of the simulation domain, the spanwise direction is periodic. The domain size $L_x \times L_y \times L_z$ is $100 \times 20 \times 8\pi$, where the lower part $0 < y < 10$ is the porous media and the upper part $10 < y < 20$ is the free-flow channel. The porous media elements are circular cylinders arranged in-line. The porous layer consists of 20 porous media elements in streamwise direction and 5 elements in wall-normal direction, which indicates 100 porous elements in total. A no-slip boundary condition is defined for all surfaces of the porous structure (i.e. on the surfaces of the circular cylinders) as well as for the upper wall ($y = 20$) and for the lower wall ($y = 0$). Periodic boundary conditions are defined in x-direction between $x = 0$ and $x = 100$. The second pair of periodicity is defined in the spanwise direction $z = 0$ and $z = 8\pi$.

The geometry is discretizied with full quadrilateral elements on the $x - y$ plane with local refinement near the interface, as shown in Fig. 2. The third direction (z-direction) is extended with a Fourier-based spectral method. High-order Lagrange polynomials through the modified base are applied on the $x - y$ plane. The numerical solver enables a flexible non-identical polynomial order based on the conforming elements, which offers high meshing flexibility corresponding to laminar, turbulent or interfacial flows according to prior knowledge. For instance, the polynomial orders of the elements in different mesh regions are $P = 6$ in Ω_1, $P = 8$ in Ω_2 and $P = 4$ in Ω_3 (Fig. 2), corresponding to the free channel, interface region and laminar porous media flow. The 28.040 quadrilateral element discretization combined with 1.200 Fourier modes in the z-direction leads to a resolution up to 1 billion degrees of freedom (DOF).

Table 1 Summary of simulation cases, naming cases with the bulk Reynolds number and the porosity, e.g. Re7kP75 denotes Re=7371 and porosity $\varphi = 0.75$

Case	Re	Re_τ^p	Re_τ^t	φ	Mesh resolution	$(\frac{\Delta_x}{\eta})_{max,int}$	$(\frac{\Delta_y}{\eta})_{max,int}$	$(\frac{\Delta_z}{\eta})_{max,int}$
Re7kP75	7371	475	497	0.75	370×10^6	2.46	0.45	3.89
Re3kP56	3147	430	375	0.56	250×10^6	1.41	0.63	4.52
Re15kP75	15555	825	927	0.75	1.1×10^9	1.65	0.58	2.66
Re7kP56	7123	730	675	0.56	0.89×10^9	1.91	0.72	3.53

2.2 Simulation Conditions

The simulation cases are summarized in Table 1. Four cases are introduced here covering two porous topologies and two Reynolds numbers. The porous topology is characterised by the porosity φ which is defined by the ratio of the void volume V_V to the total volume V_T of the porous structure. The bulk Reynolds number is defined using the bulk averaged velocity $\bar{u}$, channel height h and kinematic viscosity v, i.e. $Re = \bar{u}h/v$. Additional flow characteristics are the shear Reynolds numbers $Re_\tau^t = u_\tau^t h/v$ and $Re_\tau^p = u_\tau^p h/v$. Wherein the superscript t refers to the evaluation at the top wall of the channel and p to the evaluation at the permeable interface of the friction velocity u_τ. The mesh size in each direction Δ_x, Δ_y and Δ_z is compared to the Kolgomorov length scale, i.e. $(\frac{\Delta_x}{\eta})_{max,int}$ as listed in Table 1. This ratio is evaluated at the permeable interface denoted with subscript int and the maximum is given denoted with the subscript max. The total mesh resolution ranges from 250×10^6 to 1.1×10^9 for the high Reynolds number condition.

3 Results

The distribution of the spatial and temporal averaged velocity $\bar{u}$ and the velocity fluctuations $\overline{u'u'}$ are depicted in Fig. 3 for case Re7kP75. The spatial averaging is applied on the streamwise and spanwise directions. Mean velocity profile $\bar{u}$ indicates an axis-asymmetry as a consequence of the porous media layer. This asymmetry can be seen since the velocity in the free-flow on the porous media side ($y < 15$) is clearly lower than on top wall side ($y > 15$) due to the suction effect from the porous media. The velocity magnitude between the porous layers in the porous structure (e.g. $y \approx 6$ and $y \approx 8$) is not comparable to the free-flow velocity. The averaged velocity fluctuation components $\overline{u'u'}$, $\overline{v'v'}$, $\overline{u'v'}$ and the turbulent kinetic energy k are illustrated in Fig. 3b). All the fluctuation components exhibit higher intensity on the porous media side.

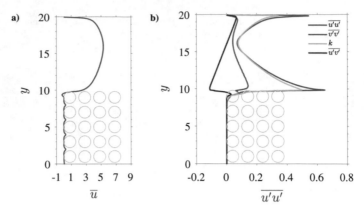

Fig. 3 Sections in the x-y plane: **a** Spatial and temporal averaged velocity $\bar{u}$; **b** velocity fluctuations $\overline{u'u'}$

The anisotropy of the Reynolds stress tensor can be represented by using the traditional anisotropy-invariant map (AIM). The Reynolds stress anisotropy tensor (b_{ij}) is written as:

$$b_{ij} = \frac{\overline{u'_i u'_j}}{\overline{u'_k u'_k}} - \frac{\delta_{ij}}{3} \tag{3}$$

Its invariants I, II and III are obtained to analyze the anisotropy of the Reynolds stress tensor. The first invariant I is characteristically zero whereas the other invariants are given by $II = \lambda_1^2 + \lambda_1\lambda_2 + \lambda_2^2$ and $III = -\lambda_1\lambda_2(\lambda_1 + \lambda_2)$. With the eigenvalues of the tensor λ_i. They can be directly used to depict the anisotropy in a two-dimensional visualization (Lumley triangle) or through a new $\eta - \xi$ coordinate system (turbulence triangle) with

$$\eta^2 = II/3, \ \xi^3 = III/2. \tag{4}$$

This turbulence triangle improves the visualization of the isotropic state aiming to evaluate trajectories of the return to isotropy of homogeneous turbulence. However, both approaches have difficulties to represent large amount of data in a complex geometry. A recent visualization technique is proposed based on the barycentric map construction as used in Chu et al. [3]. The barycentric map is based on a reconstruction of the eigenvalues λ_i. Placing the boundary states (one component x_{1C}, two component x_{2C} and isotropy x_{3C}) in a coordinate system at $x_{1C} = (1, 0)$, $x_{2C} = (0, 0)$ and $x_{3C} = (1/2, \sqrt{3}/2)$, the coordinate system (x_B, y_B) is defined as

$$x_B = C_{1c}x_{1c} + C_{2c}x_{2c} + C_{3c}x_{3c} = C_{1c} + C_{3c}\frac{1}{2}, \tag{5}$$

Fig. 4 Barycentric
anisotropy map

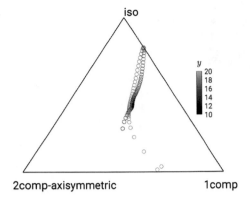

$$y_B = C_{1c}y_{1c} + C_{2c}y_{2c} + C_{3c}y_{3c} = C_{3c}\frac{\sqrt{3}}{2} \tag{6}$$

where the weight parameters C_{ic} are:

$$C_{1c} = \lambda_1 - \lambda_2, \tag{7}$$

$$C_{2c} = 2(\lambda_2 - \lambda_3), \tag{8}$$

$$C_{3c} = 3\lambda_3 + 1 \tag{9}$$

Figure 4 describes the turbulence anisotropy from the wall-normal range of $10 < y < 20$. A significant 1-component turbulence is observed close to the upper wall ($y = 20$), which corresponds to a typical fully developed turbulent channel flow. This character is however not observed close to the porous media side ($y = 10$). In the center of the channel ($y = 15$), the turbulence tends to be isotropic.

The averaged wall-normal velocity $\bar{v}$, which is relevant for the blowing/suction effects due to the permeability, is illustrated by Fig. 5. For the flow impinging onto the porous structure the wall normal velocity $\bar{v}$ shows suction, colored in blue, followed by ejection colored in red. The significant magnitude of the wall-normal velocity $\bar{v}$ is only observed in the first layer of porous media, indicating the interface region. This observation applies to all cases considered.

Further decomposition of the Reynolds shear stress leads to a quadrant analysis depicted in Fig. 6 for case Re7kP75. The signs of the velocity fluctuations are used to classify the products of these fluctuations into four categories. The Q2 motion ($u' > 0$, $v' < 0$) and the Q4 motion ($u' < 0$, $v' > 0$) are usually named as sweep and ejection respectively.

The one dimensional spectra of the kinetic energy E_{uu} as a function of the streamwise wave number k_x is given for two wall-normal positions in Fig. 7. For the position near to the permeable surface ($y = 11$) E_{uu} reveals a higher magnitude in the low wave number range. The inertial range indicates however a similar distribution for

Fig. 5 Exemplary
representation of the
averaged wall-normal
velocity $\bar{v}$ on the basis of
case Re7kP75

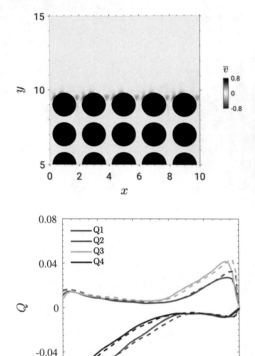

Fig. 6 Quadrant analysis of
the turbulent shear stress,
Re7kP75 with solid lines and
Re15kP75 in dashed lines

the position near the permeable surface as for the position near the impermeable to wall ($y = 19$).

Figure 8 shows the instantaneous streamwise velocity fluctuation u' on the porous media side $y = 10$ and for the upper wall side $y = 19$. Under the same color scale, the porous media side shows clearly higher intensity of fluctuations. Meanwhile, the spanwise size of fluctuation is significantly smaller than the upper wall side. It will be interesting to discover this modification of turbulence in detail with a systematic geometry variation in the future.

4 Computational Performance

The parallel computational performance will be discussed in this section. The hardware utilized for the computations was Hazel Hen located at the High-Performance Computer Center Stuttgart (HLRS). Hazel Hen is a Cray XC40 system that consists

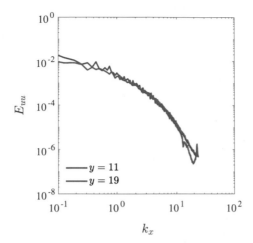

Fig. 7 One dimensional kinetic energy spectra: E_{uu} over the streamwise wavenumber k_x

a) porous media side: y = 10

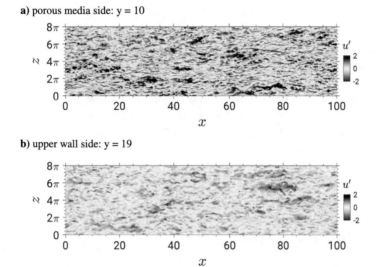

b) upper wall side: y = 19

Fig. 8 Instantaneous streamwise velocity fluctuation u' in the $x - z$ plane: in **a** for the porous media side $y = 10$ and in **b** for the upper wall side $y = 19$

of 7,712 compute nodes. Each node has two Intel Haswell processors (E5-2680 v3, 12 cores) and 128 GB memory. The compute nodes are interconnected by a Cray Aries network with a Dragonfly topology. This amounts to a total of 185,088 cores and a theoretical peak performance of 7.4 petaFLOPS.

Figure 9 shows the scalability tests based on two cases, one with 370×10^6 mesh element nodes and the other one with 1.1×10^9. Each core on Hazel Hen corresponds to one MPI rank in the simulations. A range from 2,400 to 96,000 MPI ranks is tested

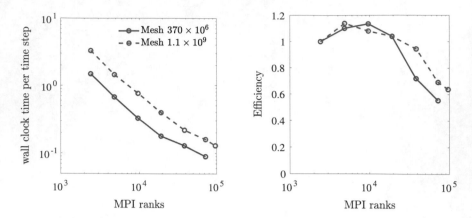

Fig. 9 Scaling behavior of the solver on Hazel Hen, wall clock time per time step on the left side. On the right side the efficiency for different numbers of MPI ranks

on the machine. A lower number of ranks is not possible due to the memory overload. An approximate linear relationship between the number of MPI ranks and the wall clock time per timestep can be seen on the left hand side. With a load of around 5,000 DOF per core, one time step takes about 0.08 s. The strong scaling efficiency on the right hand side shows a super linear speedup at low MPI ranks, followed by a slight drop for more than 10^4 MPI ranks. With the maximal number of MPI ranks, an minimum efficiency of 70% can be achieved. Compared to the previously employed open-source FVM code OpenFOAM [4, 5, 17] and the discontinuous-Galerkin spectral-element method (DGSEM) code [8], the open-source spectral/hp element code exhibits strong scalability on the HPC machine.

5 Conclusions and Future Works

Direct numerical simulations (DNS) are conducted for turbulent flows over porous media. A high-order spectral/hp element method is adopted for solving the incompressible Navier-Stokes equations. A fine resolution of the flow close to the porous surface is enabled by a local polynomial refinement based on conforming mesh elements. Four DNS cases up to a bulk Reynolds number of $Re = 15,000$ are ran with a total mesh resolution up to 1 billion. The highly-resolved DNS enables us to focus on two major physical interests: (i) the turbulence modulation including drag reduction and flow control; (ii) the turbulent/non-turbulent interface close to the porous surface. The numerical solver exhibits an excellent scalability up to 96k cores on Hazel Hen. The strong scaling tests show an efficiency of minimum 70% with only approximate 5,000 mesh-nodes per core. This indicates a high potential for an adequate use of current and next-generation HPC platforms to investigate turbulent flows over porous structures.

As the next step, the turbulent modulation and interfacial energy exchange will be studied in detail. The numerical framework enables us to alternate porous media of arbitrary shapes with a full unstructured mesh. Meanwhile, the fine scales can still be captured via the high-order numerical schemes.

Acknowledgements This work is funded by the Deutsche Forschungsgemeinschaft (DFG, German Research Foundation)—Project Number 327154368—SFB 1313. It is supported by MWK (Ministerium für Wissenschaft und Kunst) of Baden-Württemberg as a part of the project DISS (Data-integrated Simulation Science). The authors gratefully appreciate the access to the high performance computing facility 'Hazel Hen' at HLRS, Stuttgart and would like to thank the teams of HLRS and Cray for their kind support.

References

1. W. Breugem, B. Boersma, Direct numerical simulations of turbulent flow over a permeable wall using a direct and a continuum approach. Phys. Fluids **17**(2), 025,103 (2005)
2. C.D. Cantwell, D. Moxey, A. Comerford, A. Bolis, G. Rocco, G. Mengaldo, D. De Grazia, S. Yakovlev, J.E. Lombard, D. Ekelschot et al., Nektar++: an open-source spectral/hp element framework. Comput. Phys. Commun. **192**, 205–219 (2015)
3. X. Chu, W. Chang, S. Pandey, J. Luo, B. Weigand, E. Laurien, A computationally light data-driven approach for heat transfer and hydraulic characteristics modeling of supercritical fluids: From dns to dnn. Int. J. Heat Mass Transf. **123**, 629–636 (2018)
4. X. Chu, E. Laurien, Direct numerical simulation of heated turbulent pipe flow at supercritical pressure. J. Nucl. Eng. Radiat. Sci. **2**(3), 031,019 (2016)
5. X. Chu, E. Laurien, S. Pandey, Direct numerical simulation of heated pipe flow with strong property variation, in *High Performance Computing in Science and Engineering' 16* (Springer, 2016), pp. 473–486
6. X. Chu, B. Weigand, V. Vaikuntanathan, Flow turbulence topology in regular porous media: from macroscopic to microscopic scale with direct numerical simulation. Phys. Fluids **30**(6), 065,102 (2018)
7. X. Chu, G. Yang, S. Pandey, B. Weigand, Direct numerical simulation of convective heat transfer in porous media. Int. J. Heat Mass Transf. **133**, 11–20 (2019)
8. F. Föll, S. Pandey, X. Chu, C.D. Munz, E. Laurien, B. Weigand, High-fidelity direct numerical simulation of supercritical channel flow using discontinuous Galerkin spectral element method, in *High Performance Computing in Science and Engineering'18* (Springer, 2019), pp. 275–289
9. X. He, S. Apte, K. Schneider, B. Kadoch, Angular multiscale statistics of turbulence in a porous bed. Phys. Rev. Fluids **3**(8), 084,501 (2018)
10. J. Jimenez, M. Uhlmann, A. Pinelli, G. Kawahara, Turbulent shear flow over active and passive porous surfaces. J. Fluid Mech. **442**, 89–117 (2001)
11. Y. Jin, A.V. Kuznetsov, Turbulence modeling for flows in wall bounded porous media: an analysis based on direct numerical simulations. Phys. Fluids **29**(4), 045,102 (2017)
12. Y. Jin, M.F. Uth, A.V. Kuznetsov, H. Herwig, Numerical investigation of the possibility of macroscopic turbulence in porous media: a direct numerical simulation study. J. Fluid Mech. **766**, 76–103 (2015)
13. Y. Kuwata, K. Suga, Transport mechanism of interface turbulence over porous and rough walls. Flow, Turbul. Combust. **97**(4), 1071–1093 (2016)
14. D.M. McEligot, X. Chu, R.S. Skifton, E. Laurien, Internal convective heat transfer to gases in the low-Reynolds-number turbulent range. Int. J. Heat Mass Transf. **121**, 1118–1124 (2018)

15. S. Pandey, X. Chu, E. Laurien, Investigation of in-tube cooling of carbon dioxide at supercritical pressure by means of direct numerical simulation. Int. J. Heat Mass Transf. **114**, 944–957 (2017)
16. S. Pandey, E. Laurien, X. Chu, A modified convective heat transfer model for heated pipe flow of supercritical carbon dioxide. Int. J. Therm. Sci. **117**, 227–238 (2017)
17. S. Pandey, X. Chu, E. Laurien, Numerical analysis of heat transfer during cooling of supercritical fluid by means of direct numerical simulation, in *High Performance Computing in Science and Engineering'17* (2018), pp. 241–254
18. S. Pandey, X. Chu, E. Laurien, B. Weigand, Buoyancy induced turbulence modulation in pipe flow at supercritical pressure under cooling conditions. Phys. Fluids **30**(6), 065,105 (2018)
19. M. Rosti, L. Brandt, A. Pinelli, Turbulent channel flow over an anisotropic porous wall-drag increase and reduction. J. Fluid Mech. **842**, 381–394 (2018)
20. K. Suga, Y. Nakagawa, M. Kaneda, Spanwise turbulence structure over permeable walls. J. Fluid Mech. **822**, 186–201 (2017)
21. A. Terzis, I. Zarikos, K. Weishaupt, G. Yang, X. Chu, R. Helmig, B. Weigand, Microscopic velocity field measurements inside a regular porous medium adjacent to a low reynolds number channel flow. Physics of Fluids **31**(4), 042,001 (2019)
22. K. Weishaupt, A. Terzis, I. Zarikos, G. Yang, M. de Winter, R. Helmig, Model reduction for coupled free flow over porous media: a hybrid dimensional pore network model approach (2019)
23. S. Whitaker, *The Method of Volume Averaging*, vol. 13 (Springer Science & Business Media, 2013)
24. B.D. Wood, X. He, S.V. Apte, Modeling turbulent flows in porous media. Annu. Rev. Fluid Mech. **52**(1), null (2020)
25. G. Yang, B. Weigand, Investigation of the Klinkenberg effect in a micro/nanoporous medium by direct simulation Monte Carlo method. Phys. Rev. Fluids **3**(4), 044,201 (2018)
26. G. Yang, B. Weigand, A. Terzis, K. Weishaupt, R. Helmig, Numerical simulation of turbulent flow and heat transfer in a three-dimensional channel coupled with flow through porous structures. Transp. Porous Media **122**(1), 145–167 (2018)
27. G. Yang, E. Coltman, K. Weishaupt, A. Terzis, R. Helmig, B. Weigand, On the Beavers–Joseph interface condition for non-parallel coupled channel flow over a porous structure at high Reynolds numbers. Transp. Porous Media 1–27 (2019)

Uncertainty Quantification in High Performance Computational Fluid Dynamics

Andrea Beck, Jakob Dürrwächter, Thomas Kuhn, Fabian Meyer, Claus-Dieter Munz, and Christian Rohde

Abstract In this report we present advances in our research on direct aeroacoustics and uncertainty quantification, based on the high-order Discontinuous Galerkin solver FLEXI. Oscillation phenomena triggered by flow over cavities can lead to an unpleasant tonal (whistling) noise, which provides motivation for industry and academia to better understand the underlying mechanisms. We present a numerical setup capable of capturing these phenomena with high efficiency, as we show by comparison to experimental data and results from industry. Some of these phenomena are highly sensitive towards flow conditions, which makes an integrated approach regarding these conditions necessary. This is the goal of uncertainty quantification. We present software for both intrusive and non-intrusive uncertainty quantification methods. We investigate convergence and computational performance. The development of both codes was in parts carried out in cooperation with HLRS. Apart from validation results, we show a non-intrusive simulation of 3D turbulent cavity noise.

A. Beck (✉)
Laboratory of Fluid Dynamics and Technical Flows, University of Magdeburg"Otto von Guericke", Universitätsplatz 2, 39106 Magdeburg, Germany
e-mail: andrea.beck@ovgu.de

J. Dürrwächter · T. Kuhn · C.-D. Munz
Institute of Aerodynamics and Gasdynamics, Universität Stuttgart, Pfaffenwaldring 21, 70569 Stuttgart, Germany
e-mail: jd@iag.uni-stuttgart.de

T. Kuhn
e-mail: tkuhn@iag.uni-stuttgart.de

C.-D. Munz
e-mail: munz@iag.uni-stuttgart.de

F. Meyer · C. Rohde
Institute of Applied Analysis and Numerical Simulation, Universität Stuttgart, Pfaffenwaldring 57, 70569 Stuttgart, Germany
e-mail: fabian.meyer@mathematik.uni-stuttgart.de

C. Rohde
e-mail: christian.rohde@mathematik.uni-stuttgart.de

© Springer Nature Switzerland AG 2021
W. E. Nagel et al. (eds.), *High Performance Computing in Science and Engineering '19*,
https://doi.org/10.1007/978-3-030-66792-4_24

1 Introduction

The focus of our project is the development of a numerical framework which is capable of analyzing the influence of uncertain input and model parameters in large eddy simulations (LES). In particular, direct noise computations (DNC), which accurately resolve the interaction of hydrodynamics and acoustics, are within our research interest. The large bandwidth of scales between turbulent eddies and acoustic waves requires high-fidelity simulations leading to large computational requirements. In combination with uncertainty quantification (UQ), computational resources are stressed even further, which demands a highly efficient and scalable numerical framework.

Therefore, we aim at the development of a highly efficient UQ framework as well as a deterministic simulation setup that is capable to capture all physical mechanisms at minimal computational cost. With regard to UQ, we have developed both an intrusive as well as a non-intrusive forward uncertainty propagation environment. The intrusive solver is based on the Stochastic Galerkin method which was implemented into the high-order Discontinuous Galerkin solver FLEXI. FLEXI is developed within the Numerics Research Group of Prof. Munz at the IAG Stuttgart.

Our new non-intrusive UQ software POUNCE is a highly modular Python tool. Its design guarantees flexibility in terms of the chosen physics solver, the desired non-intrusive UQ method as well as the given system architecture. Here, a special focus was on queue-based HPC systems such as the Cray Hazel Hen in Stuttgart and HPC file system such as Lustre®.

In terms of deterministic simulations, we have developed a zonal LES approach that is suitable for aeroacoustic simulations of cavity flows. We have developed a new turbulent inflow method applicable for non-equilibrium wall bounded turbulent flow. Besides the zonal LES approach, we have investigated the suitability of a turbulence wall-model for aeroacoustic cavity simulations.

In this report, we will present first benchmark results of our Stochastic Galerkin solver as well as UQ simulations of an open cavity setup. For verification purposes, we compare to experimental results.

2 Numerical Methods

2.1 The Discontinuous Galerkin Spectral Element Method—FLEXI

The baseline CFD-solver we use in our project is FLEXI. FLEXI is based on a high-order Discontinuous Galerkin Spectral Element method. In this report we restrict ourselves to a short summary of the method. Further details are well-documented in literature and can be found for example in Hindenlang [5].

Starting with the three-dimensional compressible Navier–Stokes equations (NSE) written in conservation form

$$U_t + \nabla_x \cdot (\mathbf{G}(U) - \mathbf{H}(U, \nabla_x U)) = 0 \quad \forall \, \mathbf{x} \in D, \quad \forall \, t \in (0, T), \quad (1)$$

where U denotes the solution vector $U = (\rho, \rho u, \rho v, \rho w, \rho e)^T$, the subscript t represents the derivative with respect to time within the time interval $(0, T)$. ∇_x is the gradient operator in physical space $D \subset \mathbb{R}^3$. $\mathbf{G}(U)$ and $\mathbf{H}(U, \nabla_x U)$ describe the advective and viscous flux, respectively. To solve Eq. (1) numerically, the physical domain D is discretized with three-dimensional, non-overlapping elements. For reducing computational complexity, all elements are transformed from physical space onto the reference element $E(\boldsymbol{\eta}) = [-1; 1]^3$. Multiplication with a test-function ϕ^N followed by an integration over the reference space E, after integration by parts, finally leads to the weak form of Eq. (1):

$$\frac{\partial}{\partial t} \int_E J U \phi^N \, d\boldsymbol{\eta} + \oint_{\partial E} (G_n^* - \mathcal{H}_n^*) \phi^N \, ds - \int_E \mathcal{F}(U, \nabla_x U) \cdot \nabla_\eta \phi^N \, d\boldsymbol{\eta} = 0. \quad (2)$$

To get an efficient scheme, components of the solution vector are discretized by means of a tensor product of 1-D Lagrange polynomials l_i^N of degree N. To further reduce computational cost, we employ a collocation approach following Kopriva [9] using the $(N + 1)^3$ Gauss-Legendre quadrature points both for interpolation and integration.

To advance the solution in time we use a high-order explicit Runge–Kutta time integration scheme of Carpenter and Kennedy [2], which is implemented in a low storage version. The advective fluxes G_n^* normal to the cell interfaces require an evaluation of the current solution vector U at the surface in each adjacent cell. The numerical advective flux is then computed choosing an adequate Riemann solver. If shock waves or other discontinuities are present in the physical domain, the Discontinuous Galerkin scheme is prone to non-physical oscillations of the high-order polynomials. To address this issue, a shock capturing method is implemented in FLEXI based on Finite-Volume sub-cells. Here, the high-order cells are sub-divided into sub-cells, where the number of sub-cells corresponds to the degrees of freedom (DoFs) of the polynomial expansion within the high-order cell. Further details of the FV shock-capturing are given in Sonntag [14].

2.2 Uncertainty Quantification

In forward uncertainty propagation of the compressible Navier–Stokes equations (1), the solution $\mathbf{U}$ depends not only on space and time, but also on an N-dimensional real-valued random vector $\boldsymbol{\xi} = (\xi^1, \ldots, \xi^N)$ defined on a probability space $(\Omega, \mathcal{F}, \mathbb{P})$, i.e.

$$U = U(x, t, \xi) \ : \ D \times \mathbb{R}_+ \times \Omega \to \mathcal{U} \subset \mathbb{R}^5. \tag{3}$$

The goal is to propagate uncertainties in the simulation input through the PDE solver in order to obtain the response of the solution to the uncertain input, i.e. $U(\xi)$ at the point and time of interest. There are two fundamental approaches to this: In non-intrusive methods (called so because no changes have to be made to the baseline solver), a conventional code is run several times at different collocation points and the outputs are post-processed. In contrast, intrusive methods require changes in the code. However, the main advantage is that only one computation of the modified solver is needed to propagate the uncertainty. Both methods provide challenges in the context of high performance computing: while intrusive methods are coupled and therefore require an efficient parallelization strategy, non-intrusive methods necessitate a very large number of small computations and files, which most HPC file systems and schedulers are not designed for.

2.3   Stochastic Galerkin: An Intrusive Uncertainty Quantification Method

In stochastic Galerkin, at first only the stochastic dimension is discretized. To this end, a Galerkin projection is carried out. For the compressible Euler equations (i.e. omitting the parabolic flux in (1) for simplicity), this yields

$$\langle U, \phi \rangle + \langle \nabla_x \cdot G(U), \phi \rangle = 0 \tag{4}$$

with the test function ϕ and the inner product

$$\langle g(\xi), h(\xi) \rangle = \int_{\xi \in \Gamma} g(\xi) h(\xi) f_\Xi(\xi) \, d\xi, \tag{5}$$

where $\Gamma := \xi(\Omega) \subset \mathbb{R}^N$ is the image of the random vector and $f_\Xi(\xi) : \Gamma \to \mathbb{R}_+$ is its joint probability density function. The solution is approximated as a polynomial

$$U \approx \sum_i \widehat{U}_i(x, t) \, \Psi_i(\xi). \tag{6}$$

The basis is chosen orthonormal with respect to the joint probability density function of the uncertain input $\langle \Psi_i, \Psi_k \rangle = \delta_{i,k}$. For the Galerkin approach, the test functions are identical to the basis functions $\phi_i = \Psi_i$. The result of the stochastic Galerkin discretization

$$\widehat{U}_t + \nabla_x \cdot \widetilde{G}(\widehat{U}) = 0 \tag{7}$$

with kth component

$$\left\langle \widehat{\mathbf{U}}_{t,k} + \nabla_x \cdot \left(\mathbf{G} \left(\widehat{\mathbf{U}}^T \boldsymbol{\Psi} \right) \right), \Psi_k \right\rangle = 0 \tag{8}$$

is a new, strongly coupled, system of conservation equations. Its unknowns are the stochastic moments $\widehat{\mathbf{U}}_k$ of the solution $\mathbf{U}$. The SG discretized system is deterministic in the sense that the dependence on the random vector $\boldsymbol{\xi}$ is eliminated by integration. For highly nonlinear flux functions $\mathbf{G}(\mathbf{U})$, the second integral in (4) may not be evaluated analytically or only at very high computational cost. In the quasi-spectral approach [1], the nonlinear function $\mathbf{G}$ is split into elemental operations, which are individually projected onto the polynomial space with a Galerkin ansatz. A simple product thus becomes a double sum

$$\left\langle \widehat{\mathbf{g}}\widehat{\mathbf{h}}, \Psi_k \right\rangle = \sum_{i,j} \widehat{g}_i \widehat{h}_j \left\langle \Psi_i \Psi_i, \Psi_k \right\rangle = \sum_{i,j} \widehat{g}_i \widehat{h}_j C_{ijk}, \tag{9}$$

where $C_{ijk} = \langle \Psi_i \Psi_j, \Psi_k \rangle$ are the entries of a rank three tensor $\mathbf{C}$, which is called the polynomial chaos tensor. It is sparse, symmetrical in all directions and solution-independent, and its entries can be pre-computed and stored before computation.

In order to improve accuracy, especially in the presence of discontinuities, the stochastic domain can be subdivided into disjoint elements. Each of the elements can be treated separately with the SG method, and their solutions subsequently joined in a stochastic post-processing. The solvers for the different stochastic elements can be completely decoupled which yields a trivial parallelization strategy. This approach is called Multi-Element Stochastic Galerkin (ME-SG).

The SG discretized system (8) does in general not preserve hyperbolicity. A limiter was therefore developed in [13] and extended to the DG method in within the course of this project [3] to ensure hyperbolicity preservation. To this end, oscillations are limited towards the cell mean whenever negative densities or pressures occur.

2.3.1 Employed Software: SG FLEXI

Based on the DG flow solver FLEXI, a SG code for the 3D compressible Navier–Stokes equations has been developed. It can handle uniform and normal distributions and an arbitrary number of uncertain variables. Uncertainty can be introduced via initial and boundary conditions, uncertain source terms and uncertain viscosity. Adapted Lax-Friedrichs and HLLE Riemann solver are implemented. Finite volume sub-cell shock-capturing is possible and the limiter [3] can be employed. A multi-element approach is also implemented.

A visualization environment for FLEXI based on a plugin for ParaView was adapted to post-process results of the code. Four steps are necessary as the core of the visualization process:

1. interpolate in the physical space to a visualization grid. The DG nodes are non-equidistant and do not include element boundaries, so an interpolation with an arbitrary resolution is performed.

2. interpolate in the stochastic domain to

- A value $\boldsymbol{\xi}$ of the random input where solution is to be evaluated
- quadrature points in the stochastic domain for the calculation of stochastic moments such as expectation and standard deviation

3. calculate derived quantities such as pressure, if necessary
4. integrate in the stochastic domain to get expectation and/or standard deviation.

For the conservative variables, steps 2–4 are not necessary in order to compute expectation and standard deviation, as stochastic Galerkin already yields stochastic moments of the solution, e.g. $\mathbb{E}(\mathbf{U}) = \tilde{\mathbf{U}}_0$. But for derived quantities, an interpolation has to be carried out, because in general $\mathbb{E}(g(\mathbf{U})) \neq g(\mathbb{E}(\mathbf{U}))$.

Thread-level performance was considered. The calculation of double sums (9) with the sparse Polynomial Chaos Tensor $\mathbf{C}$ consumes the largest share of computational effort in the SG method. Its efficiency is therefore crucial to the code. The sparsity and symmetry of the tensor should be exploited for computational and memory efficiency, i.e. each non-zero value occurring in the tensor should only be stored once. In the case of one stochastic variable ($N = 1$), the entries and their indices can be addressed directly by nested loops. The number of stochastic variables and stochastic polynomial degree can be prescribed statically at the compiler pre-processing stage. They are the only input to all counter variables of these loops. Nonetheless, it was observed that these loops are not unrolled by the investigated compilers. As the employed loops are mostly very short, a considerable share of computational effort is spent for loop control, which is undesirable. Another approach is the use if index vectors for the entries of the tensor. This causes even more computational overhead. A third option was therefore investigated: A script was incorporated into the compile process, where the operations within the loops is explicitly converted to a list of statements, which then replaces the loops in the source code. This can be interpreted as explicit custom loop unrolling. Ordering of these statements was optimized in consideration of cache reuse. Vectorization is also promising in the context of these stochastic products, but was not looked into, as it would have necessitated considerable changes to the code structure.

There are two options for discretizing fractions $a = b/c$ in the flux function G with the quasi-spectral method, which are both implemented in the code:

- Rewrite the fraction to a multiplication, i.e. re-write $a = b/c$ to $ac = b$, discretize this equation and solve for the moments of a. This involves solving a linear equation system (for every point in time and space), which is very costly
- Integrate the fraction over the stochastic space numerically. In practice, this means evaluating enumerator and denominator at quadrature points, computing the fraction there, and projecting the solution back onto the orthonormal basis. This option is less expensive, but introduces an additional error due to the non-exact integration.

Parallel performance was also investigated, but the parallelization strategy of the baseline solver FLEXI [6] was used without considerable adaptions. In ME-

SG, several independent SG simulations are carried out. Nonetheless, in the current implementation they are incorporated into one parallel run of FLEXI. This facilitates organized parallel output to one common file based on the HDF5 library.

2.4 Non-intrusive Uncertainty Quantification Methods: POUNCE

In non-intrusive forward uncertainty propagation, a baseline solver is run at several points in the stochastic space ξ_i. In Non-intrusive spectral projection (NISP), a projection onto the orthonormal basis is carried out just as in the stochastic Galerkin method. However, the solution is not projected before the simulation. Instead, the simulation is carried out at quadrature points in the stochastic space. Subsequently, these quadrature points are used for the projection. Instead of the solution at end time T, a derived quantity of interest $Q(\mathbf{U})$ (such as an integral force, the maximum temperature at a wall boundary, or a sound pressure level) can be projected to gain stochastic properties of this quantity. The integral of the projection is approximated with Gaussian numerical quadrature

$$\widetilde{Q}_k = \langle Q, \Psi_k \rangle \approx \sum_i Q(\mathbf{U}(\boldsymbol{\xi}_i)) \, \Psi_k(\boldsymbol{\xi_i}) \, w_i, \tag{10}$$

where ξ_i are the quadrature points (at which the solver is run) and w_i are the according quadrature weights. Sparse grids can improve efficiency in the case of several random variables.

In Monte Carlo methods, samples are drawn randomly from the sample space. The expectation is the sample mean. Unbiased estimators for the variance exist as well. Monte Carlo are very robust and their convergence rate is independent of the stochastic dimension, but they feature only half-order convergence, so a very large number of samples is needed to achieve reasonable stochastic accuracy. Multilevel Monte Carlo (MLMC) methods tackle this issue. In MLMC, several spatial resolution levels are employed. A high stochastic accuracy is reached on the coarsest level via a very number of samples, where each sample is computationally cheap to compute. In order to remove the numerical bias, corrections (as differences to finer spatial resolution levels) are added to the solution on the coarsest level. For each of these differences, mean and variance are gained from Monte Carlo sampling. For a converging numerical method, the differences between two resolution levels become successively smaller with increasing resolution, so that fewer Monte Carlo samples are needed to reach the same absolute stochastic accuracy.

A framework for simulations with non-intrusive methods was introduced in the last report. During the last year, a new framework was developed under the name "Propagation of Uncertainties (POUNCE)". Similar to the previous framework, it involves changes to the baseline solver FLEXI (written in FORTRAN 95), stochastic post-processing routines (also written in FORTRAN 95) and a management environment written in Python3. So far, POUNCE can carry out MLMC and NISP simulations. It works on Hazel Hen and on local machines. Supported baseline solvers are FLEXI as well as a minimal example solver written in Python. POUNCE features several important advantages compared to its predecessor:

- In POUNCE, the baseline solver FLEXI is extended to handle several sample simulations with one call to the binary. A wrapper program calls several sample runs both sequentially and in parallel. All sample runs write to the same HDF5 file. This has considerable performance benefits. In Multilevel methods, sample runs can take as little as two seconds on a single core, but their numbers can be in the millions. On the employed HPC system, allocation of a command (i.e. a single executable call) takes several seconds, which necessitates grouping to one program execution to avoid large performance losses. Moreover, writing to HDF5 files simultaneously requires a coordinating MPI communicator. Writing separate files for every sample is very inefficient, since the file system is not designed to handle very large numbers of files efficiently.
- The Python part of POUNCE has a modular design, so that new non-intrusive UQ methods, new HPC machines and new baseline solvers can be added easily. These new possibilities are important for the sustainability of the framework and make it attractive for other groups in the field. We plan to make it open source in the near future.

POUNCE relies on external python libraries for parameter read-in, checkpointing/ restart capabilities, archiving and standard output. Sphinx has been used for source code documentation.

3 Numerical Results

3.1 Stochastic Galerkin: SG FLEXI

3.1.1 Validation

We validated our numerical solver against two benchmark problems. In a first step we consider a manufactured solution which depends on up to $N = 3$ random variables. The exact solution takes the following form

$$
\begin{pmatrix}
\rho(t, x, y, \xi_1, \xi_2, \xi_3) \\
m_1(t, x, y, \xi_1, \xi_2, \xi_3) \\
m_2(t, x, y, \xi_1, \xi_2, \xi_3) \\
E(t, x, y, \xi_1, \xi_2, \xi_3)
\end{pmatrix}
=
\begin{pmatrix}
\xi_3 + \xi_2 \cos(2\pi(x - \xi_1 t)) \\
\xi_3 + \xi_2 \cos(2\pi(x - \xi_1 t)) \\
0 \\
(\xi_3 + \xi_2 \cos(2\pi(x - \xi_1 t)))^2
\end{pmatrix},
\tag{11}
$$

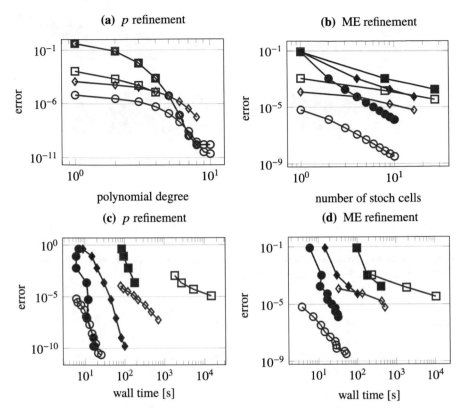

(a) p refinement

(b) ME refinement

(c) p refinement

(d) ME refinement

Fig. 1 Error plots for exact solution (11). (—●—) L_2-Mean, N=1, SC. (—◆—) $N = 2$, SC. (—■—) $N = 3$, SC. (—⊖—) $N = 1$, SG. (—◇—) $N = 2$, SG. (—⊟—) $N = 3$, SG

where $\xi_1 \sim \mathcal{U}(0.1, 1), \xi_2 \sim \mathcal{U}(0.1, 0.3), \xi_3 \sim \mathcal{U}(1.8, 2.5)$ are uniformly distributed random variables. We consider the absolute L^2-error in density in mean at time $T = 1$ in dependence of SG polynomial degree (p-refinement) and the number of stochastic elements (ME refinement). We also compared the SG method against a classical non-intrusive method, namely the Stochastic Collocation. In Fig. 1 we observe that the error rapidly decreases when increasing the number of ansatz polynomials or the number of stochastic elements. Moreover, we can see that for the same number of degrees of freedoms the SG methods yield a smaller absolute error, however if considering its efficiency, i.e. error versus wall time, it becomes clear that only for a one-dimensional random input the SG method is more efficient than SC. We would like to note that for SC we only used tensor-product collocation points. The efficiency of SC can be dramatically increased by using sparse-grids.

As second benchmark problem we consider an uncertain Double Mach reflection problem from [14]. In this example a Mach 10 shock wave hits a ramp with 30° inclination. We choose the angle of the ramp uncertain i.e. we let the angle be uniformly distributed between 28° and 32°. For this example we have to employ the

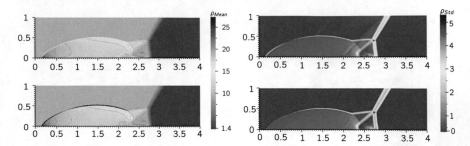

Fig. 2 Mean and standard deviation of density at $T = 0.2$ for uncertain Double Mach reflection problem

hyperbolicity preserving limiter otherwise we get negative pressure due to the high jump in pressure. We compare our numerical results to that obtained with SC. Figure 2 shows mean and standard deviation of density at time $T = 0.2$. Both methods show a very good agreement in mean and variance.

3.1.2 Performance

For the stochastic Galerkin code, both thread level performance (especially in regard of the optimization efforts) and weak scaling was tested. Core level performance is shown in Fig. 3. A viscous isentropic vortex with periodic boundary and a stochastic base flow velocity (one stochastic variable) is taken as a test example. The physical polynomial degree is 10. Performance is shown as a function of the number of stochastic degrees of freedom $(M + 1)$. Apart from a baseline version of the code, two optimized versions are plotted: Firstly, the polynomial degree M was prescribed during compiler pre-processing, allowing the compiler to vectorize and optimize the code. Secondly, the manual loop unrolling discussed in Sect. 2.3.1 was tested (label "optimized" in the figure). Furthermore shown are the computational effort of the baseline code FLEXI, a (theoretical) computational effort for a non-intrusive stochastic collocation approach consisting of several runs of the baseline solver, and slopes for cost $c \propto (M + 1)^p$, $p = 1, 2$.

In comparison with the baseline case, the fully optimized case yields a speed-up by a factor of about two for all investigated M. While at low polynomial degrees, choosing M as a constant during compiler pre-processing accounts for most of the speed-up, at higher polynomial degrees the manual loop unrolling accounts for most of the performance benefit. A possible explanation for this is that ta low polynomial degrees, the compiler can eliminate most of the unnecessary (not accessed) loops, while this is not possible at higher M. It also becomes obvious, that a (hypothetical) non-intrusive method with the same stochastic number of degrees of freedom needs less computation time for all investigated stochastic polynomial degrees. This is expected and due to the strong coupling in the SG flux function which is not present in non-intrusive methods, but potentially allows for a higher accuracy of SG compared to

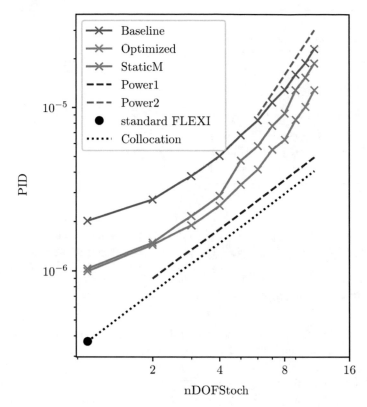

Fig. 3 Thread level performance of the stochastic Galerkin code; Computation time per physical degree of freedom and Runge Kutta step over number of stochastic degrees of freedom

non-intrusive methods at the same polynomial degree. The theoretical limit for large M is $p = 3$, since $M + 1$ double sums (with $M + 1$ elements) are to be computed. It is not reached, as other terms dominate for small M.

Results of weak scaling are shown in Fig. 4. The test case is the same as in Fig. 3. Starting from a computation on one node, Both the number of cores and the problem size (realized via a change of the number of physical elements) is increased and the relative efficiency (i.e. the inverse ratio of computation wall time). The stochastic polynomial degree as well as the number of physical elements per processor are varied. The excellent scaling properties of the baseline solver FLEXI are preserved here, with a parallel efficiency of close to 80 % up to more than 12.000 cores.

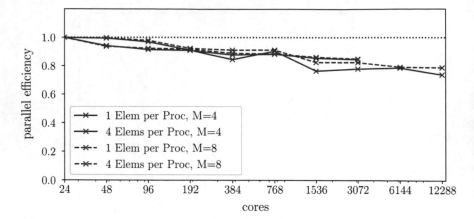

Fig. 4 Parallel efficiency of the stochastic Galerkin code (weak scaling) for different numbers of physical DG elements per core and for different stochastic polynomial degrees M

3.2 Cavity Aeroacoustics—Deterministic Results

Within this project, we aim at quantifying uncertainty in context of aeroacoustic cavity flows. Tonal noise emitted from open cavities is of industrial relevance in numerous applications. We investigate low Mach number flows regarding noise emission of the automobile exterior. Here, different noise generation mechanisms have been identified such as Rossiter feedback, Helmholtz resonance and standing waves [12]. Due to the large bandwidth of scales between acoustics and turbulence, the direct noise simulation of those effects results in a challenging multi-scale problem which requires high-order numerics to capture the interaction between acoustics and hydrodynamics.

Within in this project, we have extended our high-order discontinuous Galerkin solver FLEXI towards zonal LES simulation, in order to get a numerical framework that requires low computational effort while still being capable of capturing all relevant aeroacoustic effects. Therefore, we have developed a new turbulent inflow method that combines the anisotropic linear forcing method by de Meux [11] and a Recycling-Rescaling approach similar to the work by Lund et al. [10]. In contrast to the original method by de Meux weak forcing is now sufficient to control the inflow turbulence to follow desired boundary layer statistics, which helps the to reduce non-physical noise emission. Further, the method is now compatible with non-equilibrium turbulent flows.

To reduce computational cost even further, the wall model by Kawai and Larsson [8] has been investigated. In combination with an eddy viscosity closure, such as the Vreman model or the standard Smagorinsky model with van Driest damping, high Reynolds number LES are possible. Here, the explicit eddy-viscosity not only helps to give better results in terms of the boundary layer statistics, but also significantly reduces noise emission of artificial wall pressure fluctuations due to the wall model.

This modified high-order zonal LES setup has been applied to simulate an open rectangular cavity at various inflow velocities. The geometry of the cavity, which is incorporated into a flat plate, is $L = 25\,\text{mm}$ in length, $D = 50$ in depth and $W = 30$ in width. The rest of the whole experimental wind tunnel setup is described in [4, 7].

Simulations of the given cavity have been performed, using a 4th order polynomial approximation with 47,444 grid cells resulting in around 6 million DoFs. All simulations have been carried out on the HazelHen HPC system with 100 compute nodes (2400 processing cores). This leads to a computational load of 2500 DoF per processing core. The performance index (PID)

$$\text{PID} = \frac{\text{wall-clock-time} \cdot \#\text{cores}}{\#\text{DOF} \cdot \#\text{time steps} \cdot \#\text{RK-stages}} \tag{12}$$

was about $1.35\,\mu\text{s}$. The average physical time-step was $\Delta t = 0.34\,\mu\text{s}$. The time-interval considered is $\Delta = 0.3\,\text{s}$ in order to gain enough statistics for aeroacoustic post-processing. This results in around 880.000 time steps and a computational work of $29.000\ CPUh$.

All simulations have been post-processed using a DFT of the acoustic signal, recorded at the downstream wall of the cavity. Six Hanning windows with 50% overlap have been used resulting in a frequency resolution of $\Delta f = 12\,\text{Hz}$. As an example, the resulting sound pressure spectrum at a far field velocity of $u_\infty = 42.5\,\text{m/s}$ is given in Fig. 5. For verification, experimental results of the same setup are included in the spectrum. A comparison reveals excellent agreement especially for the lower frequencies. The dominant frequencies of the first and second Rossiter mode as well as the frequency of the first depth mode perfectly match with the experimental results. At this far field velocity, the second Rossiter mode approaches the first depth mode and starts to interact. This results in an increasing noise level at around $f = 1350\,\text{Hz}$.

3.3 Cavity Aeroacoustics—NISP Results

With an efficient numerical setup at hand, that is capable to perform direct noise computations of aeroacoustic cavity feedback at low computational cost, uncertainty quantification is now possible. Based on the new non-intrusive simulation framework POUNCE, 3D simulations with uncertain inflow conditions were performed and their effect on noise emission was analyzed. For now, only one uncertain parameter was considered per simulation. At a stochastic polynomial degree of $M = 9$ the total simulation time of the stochastic 1D and spatial 3D simulation increases to $(M + 1) \cdot CPUh_{det.} = 0.29\ Mio.\ CPUh$. In Fig. 6 the system response to an uncertain inflow Reynolds number Re_x is given. The inflow Reynolds number had a uniform distribution with $Re_x = [7.0Mio., 9.0Mio]$. The resulting stochastic noise spectrum provides both the average noise level and the corresponding standard deviation. An uncertain Reynolds number causes an uncertain boundary layer thickness which in turn influences the convection speed of the shear layer inside the cavity opening.

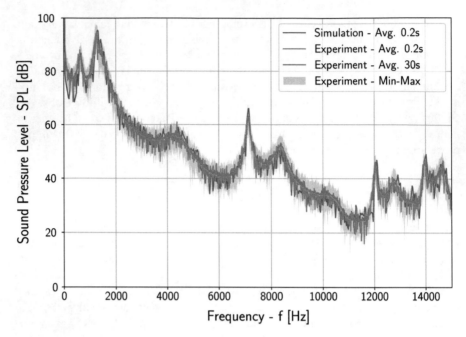

Fig. 5 Sound spectrum at the downstream cavity wall—comparison of numerical simulations and experimental data

Further, a change in Reynolds number also modifies the amount of turbulent kinetic energy.

The resulting stochastic spectrum reveals a change in broadband noise, resulting in a comparatively regular standard deviation over a large frequency range. Close to the geometric room modes the standard deviation is small demonstrating that an uncertain inflow Reynolds number has no influence on the frequency, fixed by the gap geometry. In turn, the Rossiter modes directly depend on the convection speed of the shear layer vortices. Here, standard deviation increases, predicting the influence of an uncertain inflow Reynolds number.

Besides an uncertain inflow Reynolds number, we have also quantified the influence of an uncertain far field velocity u_∞. The resulting stochastic noise spectrum is given in Fig. 7. Both mean and standard deviation reveal a similar behaviour as for an uncertain Reynolds number. As in the previous case, a modified far field velocity causes a significant change of the shear layer convection speed. In consequence both Rossiter modes and the first depth mode are directly influenced resulting in a higher standard deviation around the corresponding frequency peaks. The frequencies of the geometric resonance modes are again unaffected. Only the broadband noise level increase with higher far field velocities.

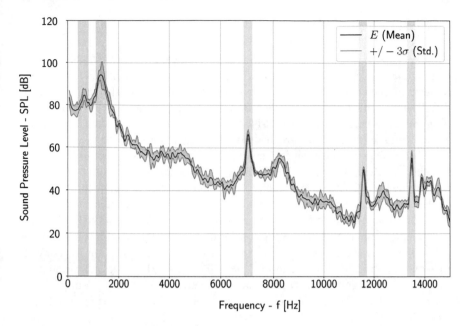

Fig. 6 Stochastic cavity simulation with an uncertain inflow Reynolds number. The inflow Reynolds number has a uniform distribution $Re_x \sim \mathcal{U}(7e6, 9e6)$. (Green) First Rossiter mode, (Purple) Second Rossiter mode, (Blue) Standing waves

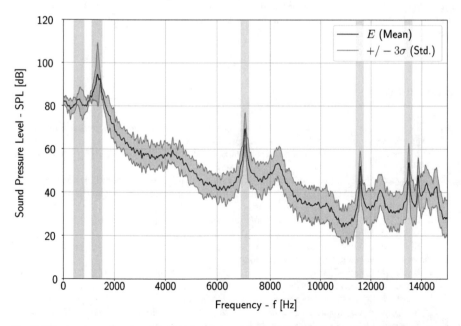

Fig. 7 Stochastic cavity simulation with an uncertain far field velocity u_∞. The far field velocity has a uniform distribution $u_\infty \sim \mathcal{U}(35.0 \, \text{m/s}, 50.0 \, \text{m/s})$. (Green) First Rossiter mode, (Purple) Second Rossiter mode, (Blue) Standing waves

4 Summary and Outlook

In this report we have presented the ongoing development of UQ methods and associated numerical schemes for the quantification of uncertainty in CFD, especially in context of open cavity flows. We have developed an intrusive stochastic Galerkin solver based on the discontinuous Galerkin solver FLEXI. We have presented validation test cases such as the double Mach reflection with an uncertain shock angle. Further, we have given details about its implementation as well as its parallel performance. Besides the intrusive framework, we have developed a non-intrusive UQ environment called POUNCE. POUNCE was designed highly modular in order to guarantee flexibility in terms of solver, UQ method and system architecture. In terms of application, we have developed a zonal LES framework to conduct aeroacoustic simulations at moderate computational cost. With both a computationally efficient, deterministic simulation setup and a non-intrusive uncertainty propagation framework at hand, we are now able to perform uncertainty quantification for complex aeroacoustic applications. In the future, we will conduct 3D zonal LES simulations of complex cavity geometries with multiple uncertain parameters. Using UQ, we hope to get a deeper insight into the non-linear interaction of acoustics and hydrodynamics.

Acknowledgements The research presented in this paper was supported via the HPCII-call of the Baden-Württemberg Stiftung and the Boysen-Stiftung. We thank Philipp Offenhäuser and Dr. Jing Zhang (both at HLRS) for their support in the optimization of our stochastic Galerkin code and Dr. Ralf Schneider (HLRS) for his advice on the efficient implementation of non-intrusive methods on Lustre® file systems.

References

1. H. Bijl, D. Lucor, S. Mishra, C. Schwab, *Uncertainty Quantification in Computational Fluid Dynamics* (Springer International Publishing, 2013)
2. M. Carpenter, C. Kennedy, Fourth-order 2n-storage Runge–Kutta schemes. NASA TM 109112 (1994)
3. J. Dürrwächter, T. Kuhn, F. Meyer, L. Schlachter, F. Schneider, A hyperbolicity-preserving discontinuous stochastic Galerkin scheme for uncertain hyperbolic systems of equations. J. Comput. Appl. Math. (2018). To appear
4. L. Erbig, N. Hu, S. Lardeau, Experimental and numerical study of passive gap noise, in *2018 AIAA/CEAS Aeroacoustics Conference* (2018), p. 3595
5. F. Hindenlang, Mesh curving techniques for high order parallel simulations on unstructured meshes. Ph.D. thesis, University of Stuttgart (2014)
6. F. Hindenlang, G.J. Gassner, C. Altmann, A. Beck, M. Staudenmaier, C.D. Munz, Explicit discontinuous Galerkin methods for unsteady problems. Comput. Fluids **61**, 86–93 (2012)
7. N. Hu, M. Herr, Characteristics of wall pressure fluctuations for a flat plate turbulent boundary layer with pressure gradients, in *22nd AIAA/CEAS Aeroacoustics Conference* (2016), p. 2749
8. S. Kawai, J. Larsson, Wall-modeling in large eddy simulation: length scales, grid resolution, and accuracy. Phys. Fluids **24**(1), 015105 (2012)
9. D. Kopriva, Implementing Spectral Methods for Partial Differential Equations: Algorithms for Scientists and Engineers (Springer Science & Business Media, 2009)

10. T. Lund, X. Wu, K. Squires, Generation of turbulent inflow data for spatially-developing boundary layer simulations. J. Comput. Phys. **140**(2), 233–258 (1998)
11. De Laage, B. de Meux, B. Audebert, R. Manceau, R. Perrin, Anisotropic linear forcing for synthetic turbulence generation in large eddy simulation and hybrid rans/les modeling. Phys. Fluids **27**(3), 035115 (2015)
12. S. Schimmelpfennig, Aeroakustik von Karosseriespalten. Ph.D. thesis, Friedrich-Alexander University Erlangen-Nürnberg (2015)
13. L. Schlachter, F. Schneider, A hyperbolicity-preserving stochastic Galerkin approximation for uncertain hyperbolic systems of equations. J. Comput. Phys. **375**, 80–98 (2018)
14. M. Sonntag, C.D. Munz, Efficient parallelization of a shock capturing for discontinuous Galerkin methods using finite volume sub-cells. J. Sci. Comput. **70**(3), 1262–1289 (2017). https://doi.org/10.1007/s10915-016-0287-5

Towards High-Fidelity Multiphase Simulations: On the Use of Modern Data Structures on High Performance Computers

Fabian Föll, Timon Hitz, Jens Keim, and Claus-Dieter Munz

Abstract Compressible multi-phase simulations in the homogeneous equilibrium limit are generally based on real equations of state (EOS). The direct evaluation of such EOS is typically too expensive. Look-up tables, based on modern data-structures significantly, reduce the computation time while simultaneously increasing the memory requirements during the simulation. In the context of binary mixtures and large scale simulations this trade off is even more important due to the limited memory resources available on high performance computers. Therefore, in this work we propose an extension of our tabulation approach to shared memory trees based on MPI 3.0. A detailed analysis of benefits and drawbacks concerning the shared memory and the non-shared memory data-structure is described. Another research topic investigates the diffuse interface model of the isothermal Navier–Stokes–Korteweg equations. A parabolic relaxation model is implemented in the open-source code *FLEXI* and 3D simulations of binary head on collisions at various model parameters are shown.

1 Introduction

Typical technical applications, in which multiphase processes can be found, are fuel injection systems such as rocket combustion chambers. The problems inherently contain multiple scales. First, the liquid fuel is injected as a jet with a liquid core. Over time, the jet breaks up and ligaments and droplets form. At the surface of the

F. Föll (✉) · T. Hitz · J. Keim · C.-D. Munz
Institute of Aerodynamics and Gas Dynamics, University of Stuttgart, Pfaffenwaldring 21, 70569 Stuttgart, Germany
e-mail: foell@iag.uni-stuttgart.de

T. Hitz
e-mail: hitz@iag.uni-stuttgart.de

J. Keim
e-mail: keim@iag.uni-stuttgart.de

C.-D. Munz
e-mail: munz@iag.uni-stuttgart.de

© Springer Nature Switzerland AG 2021
W. E. Nagel et al. (eds.), *High Performance Computing in Science and Engineering '19*,
https://doi.org/10.1007/978-3-030-66792-4_25

liquid interface, phase change occurs and the gaseous environment is mixed with evaporated fuel. This mixture is then ignited.

In this project, we aim to understand the mixing processes leading up to the burning of the fuel/oxidizer mixture. Due to the multiscale character, we split the investigation into large scale jet simulations and more detailed simulations of single droplets. These processes face extreme ambient conditions that often exceed the critical state of the fuel. In these regimes, the liquid phase cannot be described incompressible any more and we have to consider the full coupling of hydrodynamics and thermodynamics, requiring the fully compressible flow equations.

The macroscopic modelling for jet simulations is based on the Homogeneous Equilibrium Model (HEM) [22], which considers a mixture of saturated liquid and saturated vapor under full thermodynamic equilibrium. An extension of the intrinsic assumption of vapor-liquid equilibrium in the HEM approach, towards binary mixtures, is the nested procedure of tangent plane distance (TPD) function [15] analysis and classical TPn-flash calculation [16]. These methods are restricted to modifications of the underlying equations of state (EOS), only. Especially with more than one species, the evaluation of the EOS becomes very costly. Therefore, we use look-up tables which shifts the evaluation costs into a pre-processing step while during runtime, only the look-up in an octree data structure is required [9]. For binary mixtures, the look-up tables become huge in storage size which causes problems if the size exceeds the memory of the CPU. Therefore, in this paper we propose a shared memory parallelization of look-up tables based on the MPI 3.0 standard. We provide performance results on benchmark test cases and show its practical use with the simulation of a binary mixing layer.

This project also investigates modelling strategies of phase interfaces, e.g. for droplets, such as sharp and diffuse interface models. As an example of the latter, we use a parabolic relaxation model of the isothermal Navier–Stokes–Korteweg (NSK) equations to simulate the collision of two droplets at varying model parameters. Numerical experiments were conducted using an extension of the open source code *FLEXI*.[1] It is based on a high order nodal discontinuous Galerkin spectral element method (DGSEM) [12].

The outline of the paper is as follows. In the next section, the governing equations are presented. This is followed by the description of the numerical methods, thermodynamic modelling and the look-up table approach. We then present the results on the performance of the look-up tables. Numerical experiments are shown of a two component mixing layer at super-critical conditions using a Peng-Robinson EOS as well as two colliding droplets using NSK diffuse interface model.

---

[1] http://www.flexi-project.org.

2 Governing Equations

2.1 The Compressible Navier–Stokes System for Multi-components

The compressible Navier–Stokes equations with multiple components are given by

$$\frac{\partial \rho}{\partial t} + \nabla \cdot (\rho \boldsymbol{u}) = 0, \tag{1}$$

$$\frac{\partial \rho Y_k}{\partial t} + \nabla \cdot (\rho Y_k \boldsymbol{u}) = \nabla \cdot (-\boldsymbol{J}_k), \tag{2}$$

$$\frac{\partial \rho \boldsymbol{u}}{\partial t} + \nabla \cdot (\rho \boldsymbol{u} \otimes \boldsymbol{u} + p\underline{\boldsymbol{I}}) = \nabla \cdot (\underline{\boldsymbol{\tau}}), \tag{3}$$

$$\frac{\partial E}{\partial t} + \nabla \cdot [(E + p)\,\boldsymbol{u}] = \nabla \cdot (\underline{\boldsymbol{\tau}} \cdot \boldsymbol{u} - \boldsymbol{q}), \tag{4}$$

with

$$\underbrace{\underline{\boldsymbol{\tau}} = 2\mu \underline{\boldsymbol{S}} - 2/3\underline{\boldsymbol{I}} \cdot \nabla u}_{\text{Stokes law}}, \quad \underline{\boldsymbol{S}} = 1/2 \left(\nabla \boldsymbol{u} + (\nabla \boldsymbol{u})^{\mathrm{T}}\right), \tag{5}$$

where ρ is the density, $\boldsymbol{u} = (u, v, w)^{\mathrm{T}}$ is the velocity vector, p is the static pressure, E is the total energy per unit volume, $\underline{\boldsymbol{I}}$ is the unit tensor. By considering N_k species, the system is extended by $N_k - 1$ concentration equations where $\boldsymbol{Y} = (Y_1, Y_2, \ldots, Y_{N_k-1})^{\mathrm{T}}$ with $Y_k = \frac{\rho_k}{\rho}$ is defined as the mass fraction of each species. For multi-component simulations, the heat flux is usually comprised of $\boldsymbol{q} = \boldsymbol{q}^{\mathrm{f}} + \boldsymbol{q}^{\mathrm{d}} + \boldsymbol{q}^{\mathrm{c}}$, where

$$\boldsymbol{q}^{\mathrm{f}} = -\lambda \nabla T \tag{6}$$

is the specific heat flux according to Fourier law with thermal conductivity λ and temperature T. The second term is the inter-species energy flux due to diffusion

$$\boldsymbol{q}^{\mathrm{d}} = \sum_k h_k \boldsymbol{J}_k. \tag{7}$$

Here, $\boldsymbol{q}^{\mathrm{c}}$ are cross-effects, like the Dufour effect, which are not considered in this paper. The viscous stress tensor $\underline{\boldsymbol{\tau}}$ with the strain rate tensor $\underline{\boldsymbol{S}}$ is defined for a Newtonian fluid. The concentration diffusion flux is usually comprised of $\boldsymbol{J}_k = \boldsymbol{J}_k^{\mathrm{f}} + \boldsymbol{J}_k^{\mathrm{c}} + \boldsymbol{J}_k^{\mathrm{b}}$, where

$$\boldsymbol{J}_k^{\mathrm{f}} = -\rho D_k \nabla Y_k, \quad k = 1, \ldots, N_k - 1, \tag{8}$$

is the concentration diffusion flux according to Fickian law and D_k is the species diffusion coefficient. Here J_k^c are cross-effects, like the Soret effect, which are also neglected in this paper. The third term,

$$J_k^b = -\rho Y_k \sum_{j=1}^{N_k} \left(D_j \nabla Y_j \right), \quad k = 1, \dots, N_k - 1, \tag{9}$$

is a correction for the mass balance and recovers $\sum_{k=1}^{N_k} J_k = 0$ to guarantee conservation in cases where the species diffusion fluxes are significantly large [6]. Properties for the last species can be calculated via following relations

$$\sum_{k=1}^{N_k} Y_k = 1, \qquad \sum_{k=1}^{N_k} \rho_k = \rho. \tag{10}$$

Since there are $5 + (N_k - 1)$ unknown variables, a closure relation is required between the variables pressure, density, specific internal energy per mass, ϵ, and the species composition,

$$E = \rho\epsilon + \frac{1}{2}\rho u \cdot u, \quad \epsilon = \epsilon(\rho, p, Y), \quad p = p(\rho, \epsilon, Y). \tag{11}$$

Such a functional relation is called an equation of state, more precise caloric EOS, and defines the thermodynamic relations between the state variables. For the temperature a thermal EOS (12)

$$T = T(\rho, p, Y). \tag{12}$$

has also to be considered.

2.2 The Navier–Stokes–Korteweg Equations

The Navier–Stokes–Korteweg (NSK) equations are an extension of the Navier–Stokes equations where an interfacial stress is added that approximates capillary effects in phase interfaces of finite thickness. The NSK equations are given in the isothermal case for $T \equiv T_{\text{ref}}$ by

$$\rho_t + \nabla \cdot (\rho \mathbf{u}) = 0, \tag{13}$$

$$(\rho \mathbf{u})_t + \nabla \cdot \left(\rho \mathbf{u} \otimes \mathbf{u} + p\underline{I} \right) = \nabla \cdot \underline{\tau} + \nabla \cdot \underline{\tau}_K. \tag{14}$$

The NSK equations are non-dimensionalized such that the Stokes stress tensor, $\underline{\tau} \in \mathbb{R}^{d \times d}$, and the Korteweg stress tensor, $\underline{\tau}_K \in \mathbb{R}^{d \times d}$, are given by

$$\underline{\tau} = \frac{1}{\mathrm{Re}} \left(\nabla \mathbf{u} + (\nabla \mathbf{u})^{\mathrm{T}} - \frac{2}{3} \nabla \cdot \mathbf{u} \underline{I} \right), \tag{15}$$

$$\underline{\tau}_{\mathrm{K}} = \frac{1}{\mathrm{We}} \left(\rho \, \triangle \rho + \frac{1}{2} |\nabla \rho|^2 \right) \underline{I} - \frac{1}{\mathrm{We}} \nabla \rho \otimes \nabla \rho. \tag{16}$$

The Reynolds number, Re, and Weber number, We, are expressed in terms of the numbers $\epsilon_{\mathrm{K}} > 0$ and $\gamma_{\mathrm{K}} > 0$,

$$\frac{1}{\mathrm{Re}} = \epsilon_{\mathrm{K}}, \quad \frac{1}{\mathrm{We}} = \epsilon_{\mathrm{K}}^2 \gamma_{\mathrm{K}}. \tag{17}$$

Due to the capillary stress, Eq. (16), the momentum equation is a third order diffusion-dispersion equation. The system is closed by the pressure function of the Van-der-Waals law [24],

$$p = \frac{\rho R T_{\mathrm{ref}}}{1 - b\rho} - a\rho^2, \tag{18}$$

where a, b, R are material parameters. In reduced, non-dimensional, form, they are $a = 3, b = 1/3, R = 8/3$. For subcritical temperatures, Eq. (18) is non-convex and the eigenvalues of the hyperbolic flux Jacobian of the NSK equations may be imaginary numbers. The NSK system is therefore of hyperbolic-elliptic type and numerical methods that rely on the strict hyperbolicity of the conservation system cannot be used straight forward any more. To overcome these challenges, Corli et al. [7] proposed a parabolic relaxation scheme for diffusion-dispersion equations, which is extended to the isothermal NSK equations as

$$\rho_t^{\alpha} + \nabla \cdot (\rho^{\alpha} \mathbf{u}^{\alpha}) = 0, \tag{19}$$

$$(\rho^{\alpha} \mathbf{u}^{\alpha})_t + \nabla \cdot \left(\rho^{\alpha} \mathbf{u}^{\alpha} \otimes \mathbf{u}^{\alpha} + p^{\alpha} \underline{I} \right) = \nabla \cdot \underline{\tau}^{\alpha} + \alpha \rho^{\alpha} \nabla \left(c_{\mathrm{K}}^{\alpha} - \rho^{\alpha} \right), \tag{20}$$

$$\beta \left(c_{\mathrm{K}}^{\alpha} \right)_t - \epsilon_{\mathrm{K}}^2 \gamma_{\mathrm{K}} \, \triangle c_{\mathrm{K}}^{\alpha} = \alpha \left(\rho^{\alpha} - c_{\mathrm{K}}^{\alpha} \right). \tag{21}$$

An additional unknown, the relaxation variable c_{K}, satisfies a linear parabolic evolution equation with constant relaxation parameters $\alpha, \beta > 0$. The system is of second order and of mixed parabolic-hyperbolic type. For $\alpha \to \infty$, the solution of the parabolic relaxation model approaches the solution of the classical NSK equations, i.e. $(\rho^{\alpha}, \mathbf{u}^{\alpha}) \to (\rho, \mathbf{u})$. The total energy of the relaxation system is given by

$$\mathcal{E}^{\alpha}[\rho] = \int_{\Omega} \left(\frac{1}{2} \rho \, |\mathbf{u}|^2 + W(\rho) + \frac{\alpha}{2} (\rho - c_{\mathrm{K}})^2 + \frac{1}{2} \epsilon_{\mathrm{K}}^2 \gamma_{\mathrm{K}} \, |\nabla c_{\mathrm{K}}|^2 \right) \mathrm{d}\mathbf{x}. \tag{22}$$

Admissible solutions to Eqs. (19)–(21) are minimizers of Eq. (22).

3   Numerical Methods

The multiphase solver is comprised of several building blocks. The bulk solver is based on a high order discontinuous Galerkin spectral element method (DGSEM). We use an efficient look up table to incorporate real gas equations of state. For the modelling of the phase interface we apply diffuse interface methods. In the Homogeneous Equilibrium Model (HEM), we rely on the EOS to describe phase transition. In the NSK model, capillarity effects are resolved in a phase interface of finite thickness.

3.1   Discontinuous Galerkin Method

The compressible Navier–Stokes equations and the parabolic relaxation model for the NSK equations are discretized by a discontinuous Galerkin spectral element method as described by [11, 12, 14]. The approach is suitable for general systems of conservation equations. In this paper we restrict ourself to the conservation equations of the form

$$\mathbf{U}_t + \nabla_x \cdot \underline{F}(\mathbf{U}, \nabla_x \mathbf{U}) = Q, \tag{23}$$

where $\mathbf{U}$ is the vector of the solution unknowns, $\underline{F}$ is the corresponding flux containing the convective and the diffusive fluxes, and Q is the source term of the NSK relaxation model. The divergence operator in the physical space is defined as $\nabla_x = \left(\frac{\partial}{\partial x}, \frac{\partial}{\partial y}, \frac{\partial}{\partial z} \right)^T$.

In a three-dimensional domain we subdivide the computational space into non-overlapping hexahedral elements. Each element is mapped onto the reference cube element $E := [-1, 1]^3$ by a mapping $x(\boldsymbol{\xi})$, where $\boldsymbol{\xi} = (\xi, \eta, \zeta)^T$ is the coordinate vector of the reference element. The mapping onto the reference element E transforms Eq. (23) to the system

$$J\mathbf{U}_t + \nabla_\xi \cdot \underline{\mathcal{F}}(\mathbf{U}, \nabla_\xi \mathbf{U}) = J Q, \tag{24}$$

with the Jacobian J and the divergence operator in the reference space $\nabla_\xi = \left(\frac{\partial}{\partial \xi}, \frac{\partial}{\partial \eta}, \frac{\partial}{\partial \zeta} \right)^T$. In each element, the solution and the fluxes are then approximated as polynomials

$$\mathbf{U}_h = \sum_{i,j,k=0}^{N} \hat{\boldsymbol{U}}_{ijk} \psi_{ijk}(\boldsymbol{\xi}) \quad \text{and} \quad \mathcal{F}_h^m = \sum_{i,j,k=0}^{N} \hat{\boldsymbol{\mathcal{F}}}_{ijk}^m \psi_{ijk}(\boldsymbol{\xi}), \tag{25}$$

where the superscript $m = \{1, 2, 3\}$ denotes the flux in the direction of the Cartesian coordinates. The basis function $\psi_{ijk}(\boldsymbol{\xi}) = l_i(\xi)l_j(\eta)l_k(\zeta)$ is built by the tensor product of one-dimensional Lagrange polynomials l of degree N. As interpolation nodes we choose Gauss-Legendre points. Due to the nodal character of the Lagrange basis, the degrees of freedom $\hat{U}_{ijk}$ and $\hat{\mathcal{F}}_{ijk}^{m}$ are values of the approximations of the solution and the flux vectors at the interpolation nodes. To obtain the discontinuous Galerkin formulation, the approximations (25) are inserted into (24) which is then multiplied by a test function ϕ, identical to the basis function ψ, and then integrated in space. Integration by parts of the volume integral of the flux yields the weak formulation

$$\underbrace{\frac{\partial}{\partial t} \int_{\Omega} (J \mathbf{U}_h \phi) \, d\boldsymbol{\xi}}_{a} - \underbrace{\int_{\Omega} (\mathcal{F}_h \cdot \nabla_{\xi} \phi) \, d\boldsymbol{\xi}}_{b} + \underbrace{\int_{\partial\Omega} ([\mathcal{F}_h \cdot \boldsymbol{n}] \, \phi) \, dS}_{c} = \int_{\Omega} (J \boldsymbol{Q}_h \phi) \, d\boldsymbol{\xi} .$$

(26)

We identify three contributing parts: the volume integral of the time derivative of the solution (a), a volume integral (b) and a surface integral of the fluxes (c). The integrals are evaluated by Gauss-Legendre quadratures. To obtain an approximation of the flux $\mathcal{F}_h \cdot \boldsymbol{n}$ at the element surface, a numerical flux function $\boldsymbol{G} = \boldsymbol{G}(\mathbf{U}_L, \mathbf{U}_R)$ is introduced. It depends on the states left and right of the interface, $\mathbf{U}_L$ and $\mathbf{U}_R$, respectively. In case of the viscous and heat conduction fluxes, the gradients are needed in addition. For the numerical flux, we use standard approximative Riemann solvers of the HLL-type and Lax Friedrichs families [23]. The discrete formulation (26) is discretized in time using explicit third- or fourth-order Runge–Kutta schemes (RK) [13]. For the viscous fluxes, the approach of Bassi and Rebay [3, 4] is used.

The DG method with high order accuracy is favourable in smooth parts of the flow. At discontinuities or strong gradients we apply the shock capturing of Sonntag and Munz [20, 21]. We switch locally to a second order accurate finite volume (FV) scheme, where the interpolation nodes of the DG polynomials are reorganized as an equidistant sub-grid on which the solution is stored as integral mean values. A modal Persson indicator [19] is used to switch between DG and FV cells.

3.2 Equation of State and Thermodynamic Equilibrium

As thermodynamic coupling relation for the Navier–Stokes equations the cubic Peng-Robinson (PR) EOS [18] is used

$$p = \frac{R_m T}{\frac{M}{\rho} - b} - \frac{a}{\left(\frac{M}{\rho} + \delta_1 b\right) \left(\frac{M}{\rho} + \delta_2 b\right)},$$

(27)

with the universal gas constant R_m and the molar weight of the mixture M. The parameter a takes intermolecular attraction forces into account, b is the co-volume and the PR EOS specific parameters $(\delta_1, \delta_2) = (1 + \sqrt{2}, 1 - \sqrt{2})$. The transformation of the pressure explicit thermal EOS to a caloric one is provided by a residual function ansatz [17]. In case of two-phase phenomena a thermodynamic modelling by use of the HEM approach is performed. The underlying assumption of thermo-dynamic equilibrium is defined by

$$T_v = T_l, \tag{28}$$

$$p_v = p_l, \tag{29}$$

$$\mu_v^k = \mu_l^k, \tag{30}$$

where the symbols v and l represent the vapor and liquid side respectively, μ is the chemical potential and equation (30) has to hold for all N_k species. In case of single species systems the vapor-liquid calculation is performed by use of the algorithm presented by [1], for mixtures a combined approach of TPD analysis and multi-species VLE calculation is used. The TPD function is defined in mole fraction space $\mathbf{z}$ and given by

$$TPD(\mathbf{z}^{trial}) = \sum_{i=k}^{N_k} z_k^{trial} \left[\mu_k(\mathbf{z}^{trial}, T, p) - \mu_k(\mathbf{z}^{test}, T, p) \right]. \tag{31}$$

The superscript $(\cdot)^{test}$ indicates for the feed composition, which is provided from the flow solver and $(\cdot)^{trial}$ for all other possible molar compositions, which fulfill the mass balance condition $\sum_k^{N_k} z_k^{trial} = 1$. The TPD analysis is based on the idea of direct evaluation of the Gibbs free energy surface [2] and checks for a global minimum in Gibbs free energy at the present feed composition. Hereby TPD values greater zero correspond to a stable state, smaller ones to an unstable one. For the analysis of the TPD function the local minimization method with multiple initial guesses presented by [15] is used. The thermodynamic consistent modeling of the states in the two-phase region is provided by the HEM approach with

$$\epsilon^{EQ} = x_v \epsilon_v + (1 - x_v) \epsilon, \tag{32}$$

where the specific inner energy per mass works as a dummy value for any caloric state variable and x_v is the vapor mass fraction defined by

$$x_v = \frac{1/\rho - 1/\rho_l}{1/\rho_v - 1/\rho_l}. \tag{33}$$

Due to the loss of hyperbolicity inside the spinodale region with real gas EOS, the sound speed in the two-phase region in the HEM approach is modeled with the relation presented by [25]

$$\frac{1}{\rho a^2} = \frac{\alpha_v}{\rho a_v^2} + \frac{1 - \alpha_v}{\rho \alpha_l^2}, \tag{34}$$

where a is the sound speed and α the volumetric vapor fraction given by

$$\alpha_v = \frac{x_v \rho}{\rho_v}.$$ (35)

3.3 Look up Tables and Extension to Shared Memory Trees

The current Cray machine Hazel Hen has about 185,088 cores in the current expansion stage. These are provided with 24 cores each at 7712 nodes. Each node is comprised of 128 GB memory. The next expansion stage, Hawk, which is planned for spring 2020, will be approximately 640,000 cores at 5000 nodes. The ratio of nodes to cores will accordingly increase more than quintupled from the present time of $N_{cores}/N_{nodes} = 24$ to $N_{cores}/N_{nodes} = 128$. It is important to consider that the available capacity of memory on a node is not increased and will therefore be 1 GB per core. Scalable and highly efficient CFD codes for high-performance (HPC) computers, which are perfectly adapted to old architectures, should keep pace with such new developments. To maintain efficiency, the algorithms have to be modified. Examples are memory-consuming algorithms, which can be found in multi-phase and multi-component simulations in combination with so-called look up tables approaches [9, 10]. Today, these tables are composed of modern data structures such as quadtree or octree data structures, see Figs. 1 and 2. Quadtrees and octrees make use of properties from so called space filling curves for fast data localization. Here the *Morton curve* is popular due to the inherent possibility to access the data via bit operations

$$\text{data position} = f(\text{bit number}),$$ (36)

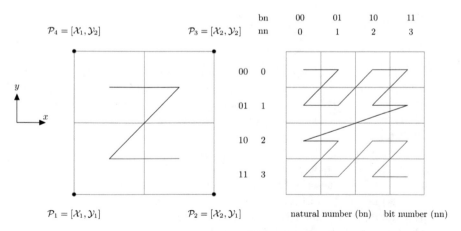

Fig. 1 Quadtree table: *left* first stage, *right* second stage, bit numbers (bn) are used for fast localization of quadtree elements in the definiton area given by $\mathcal{P}_i$

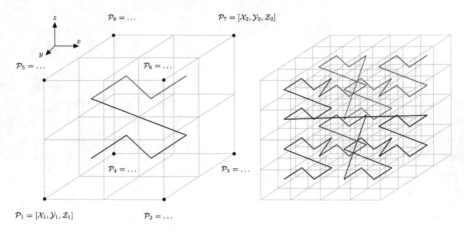

Fig. 2 Octree table—left: one stage, right: two stages

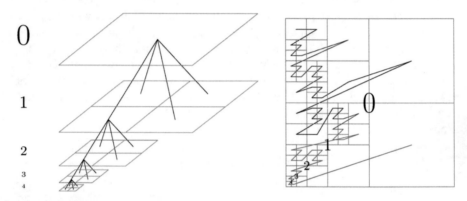

Fig. 3 Left: tree data structure, right: refined quad tree

see Fig. 1. Despite of the usage of such modern data structures, today's CFD simulations may reach the memory limits fast, if large scale high fidelity simulations are performed. In this context we want to discuss in this paper the implementation and application of tree structures on high-performance computers associated with MPI 3.0 and shared memory (Fig. 3).

In the last period, we have extended our tabulation framework, in order to use the look up tables as efficiently as possible on future high performance computers. Initially we will give some information about the parallelization strategy of the CFD solver FLEXI [8]. FLEXI is based on the so-called domain decomposition, which divides the computational grid into heterogeneously distributed MPI processes depending on the number of cores used, see Fig. 4. For the domain decomposition, again a space filling curve, more precisely the so-called *Hilbert curve*, is used. The curve has the special property to optimally distribute the different MPI regions with respect to the volume/surface ratio, even on unstructured grids. Figure 4 shows such

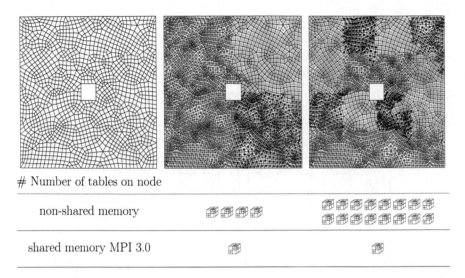

# Number of tables on node		
non-shared memory	🗄️🗄️🗄️🗄️	🗄️🗄️🗄️🗄️🗄️🗄️🗄️🗄️ 🗄️🗄️🗄️🗄️🗄️🗄️🗄️🗄️
shared memory MPI 3.0	🗄️	🗄️

Fig. 4 FLEXI [8] parallelization strategy with domain decomposition via space filling curve: *left* computational quad mesh, *middle* decomposition with 4 MPI processes, *right* decomposition with 16 MPI processes

a division. To ensure that each MPI process can access the data in the table, each MPI process has to initialize and allocate its own table when using standard MPI features. By considering MPI 3.0 features, like shared memory windows, the number of tables for each node can be reduced to one table for each node.

However, modern data structures generally consist of chained pointer lists, which are not directly applicable with the MPI 3.0 shared memory feature. This is due to the fact, that each MPI process is linked with its own virtual memory space, see Fig. 5. This has consequences for the way in which the tree structure has to be read in and accessed during the simulation on HPC systems. In Fig. 6, the standard approach to store and access the tree data is depicted. Here, each branch of the tree stores a small portion of the whole data. Furthermore, each MPI process reads and allocates the data during IO. In Fig. 7 the alternative approach to store and access the tree data with MPI 3.0 shared memory window is depicted. Here, unlike before, each branch of the tree only stores two integer IDs depicting a range in the global shared memory array. An important aspect is the fact, that during IO only one MPI process on the node is allowed to read and allocate the data. Nevertheless, each MPI process has to read and store the empty tree. This is necessary because each MPI process has still to know the relative path to the unique IDs in the last branch. With this approach it is possible to maintain the efficient tree data structure while simultaneously be capable to store and access several magnitudes of data.

virtual address   address mapping   physical address

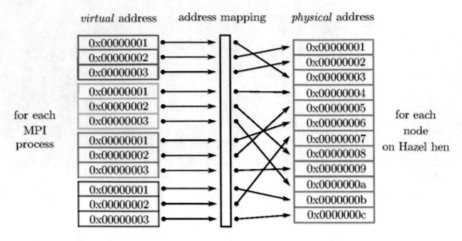

Fig. 5 Mapping from virtual address space for each MPI process to physical address space, here 4 MPI processes (the colors are chosen consistent to the domain decomposition in Fig. 4)

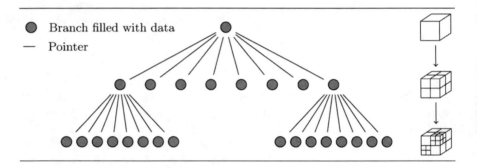

Fig. 6 Tree data structure without MPI 3.0 shared memory: each branch contains the data

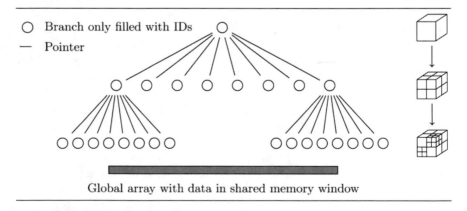

Fig. 7 Tree data structure with MPI 3.0 shared memory: each branch only contains an integer id for the global data array, the global data array is allocated in a shared memory window

4 Results

4.1 Performance Comparison of Tree Data Structures with and Without MPI 3.0 Shared Memory

In this section we investigate the different data structures in terms of performance and memory usage.

For the comparison we use the performance index

$$PID = \frac{\text{wall-clock-time} \cdot \text{\#processors}}{\text{\#DOF} \cdot \text{\#time steps} \cdot \text{\#RK-stages}}. \tag{37}$$

The results are obtained with the open source code FLEXI in combination with octree tables. Note that FLEXI is based on the HDF5 standard. To ensure a fair comparison, we have chosen a simple test case, the standard lid driven cavity in two dimensions, see [5]. We choose a binary mixture with two different ideal gases, instead of performing a one-component simulation as it is typically done in the literature. First, we look at the performance of both data structures that we defined in Sect. 3.3. We perform each simulation six times and average the measurement to cancel out hardware influences. The comparison was done on 8 nodes with 192 processes. The tree data was refined up to 7 levels resulting in about ≈ 2.8 GB memory size. Each octant represents the data in a three dimensional polynomial basis of degree 4. In the first two lines of Table 2, we have listed the results for the performance test. We notice a slightly higher PID for the MPI 3.0 implementation, which is most likely due to additional index mapping used to get the position in the global shared memory array. In the third line we compare the time which was used to read and allocate the data before the simulation starts. We note that the IO of the MPI 3.0 implementation is different in the way that we do not read in the whole tree from the HDF5 at once. By using the shared memory option, we read, allocate and deallocate each octant successively from the HDF5 file, to save as much memory as possible. Here, we notice a non negligible longer IO time for the MPI 3.0 implementation. The factor between the standard and the shared memory approach is about 6 (Table 1).

The next two Tables 3 and 4 contain memory comparisons.

Table 1 Numerical setup of the Lid driven cavity test problem

Species	Y_k	κ	N	# elements	Re
Helium	0.5	1.60	5	1024	100
Air	0.5	1.40			

Table 2 Performance comparison for the different data structures

	Standard (non-shared memory)	MPI 3.0 (shared memory)
PID (s)	2.70×10^{-5}	3.05×10^{-5}
Simulation time (s)	370.45	410.03
IO time (s)	116.05	742.34

Table 3 Memory usage depending on tree level, here we tabulated a binary mixture of Helium/Air

Tabulation type	Tree level	Memory usage in (GB)
Octree with N=4 in each direction	7	2.8
	8	10.0
	9	32.2

Table 4 Theoretical memory usage by using non-shared memory and shared memory data structures on different architectures

Tabulation of octrees	Standard (non-shared memory)	MPI 3.0 (shared memory)
Max memory on Hazel Hen (GB) 24 cores on each node	5.33	128.00
Max memory on Hawk (GB) 128 cores on each node	1.00	128.00

Here, we notice the huge improvement with the MPI 3.0 shared memory implementation. For the planned architecture Hawk we will (theoretically) be able to store and access about 128 times more memory than with the old algorithm.

4.2 Navier–Stokes Multi-component Simulations

The multi-component Navier–Stokes model was used for comparison simulations conducted with direct use of the EOS and tables with different refinement levels. As test case a two-dimensional shear layer of nitrogen and n-dodecane of the dimension $[0, 0.2] \times [-0.15, 0.15]$ m^2 was investigated. The initial states of the pure species are summarized in Table 5. As initial condition a base flow in x-direction superposed by a y-velocity disturbance was used, which are given by

Table 5 Specified initial conditions of the base flow for the pure species of the mixing layer test case

Species	Sum formula	u_0 (m/s)	p (MPa)	T (K)	ρ (kg/m^3)
Mitrogen	N_2	205.46	6	1000	19.88
n-heptane	C_7H_{16}	−85.59	6	600	255.63

$$u_{N_2} = 2M_{c,0}a_{N_2}\left[1 + \left(\frac{a_{N_2}}{a_{C_7H_{16}}}\right)\sqrt{\frac{\rho_{N_2}Z_{N_2}}{\rho_{C_7H_{16}}Z_{C_7H_{16}}}}\right]^{-1}, \tag{38}$$

$$u_{C_7H_{16}} = -\sqrt{\frac{\rho_{N_2}Z_{N_2}}{\rho_{C_7H_{16}}}}u_{N_2}, \tag{39}$$

$$u(x,t=0) = u_0\left|erf\left(\frac{\sqrt{\pi}y}{\delta_{\omega,0}}\right)\right|, \tag{40}$$

$$Y_{C_7H_{14}}(x,t=0) = 1 - y_{N_2}, \tag{41}$$

$$Y_{N_2}(x,t=0) = 0.5 + 0.5\, erf\left(\frac{\sqrt{\pi}y}{\delta_{\omega,0}}\right), \tag{42}$$

$$v(x,y,t=0) = 0.1\, max\,(u_0)\sin\left(\frac{8\pi x}{\delta_{\omega,0}}\right)\exp\left\{-\left(\frac{y}{\delta_{\omega,0}}\right)^2\right\} \tag{43}$$

and

$$\rho = \rho(T,p,\mathbf{Y}). \tag{44}$$

Here Z is the compressibility factor, $M_{c,0}$ is the Mach number which was chosen to 0.4 and $\delta_{\omega,0}$ is the initial blending thickness between the two species with $\delta_{\omega,0} = 6.859 \cdot 10^{-3}$ m.

The achieved results are visualized in Figs. 8 and 9. In both snapshots we can observe some differences in between the three computations. This is due to the fact, that the chosen Kelvin Helmholtz test problem is a highly sensitive initial value problem. The different thermodynamic approximations quickly lead to different results. In summary we can show that the tabulation approach is suitable for multi-component simulations in the super-critical regime, nevertheless future investigations are necessary.

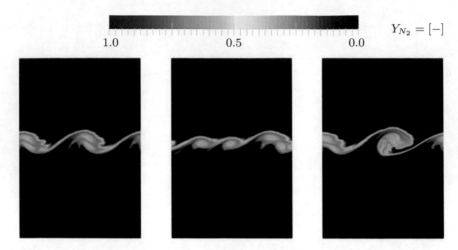

Fig. 8 Temporal snapshot of nitrogen mass fraction at $t = 4$ ms for simulations conducted by direct use of the EOS (left), coarse tables (middle) and refined tables (right)

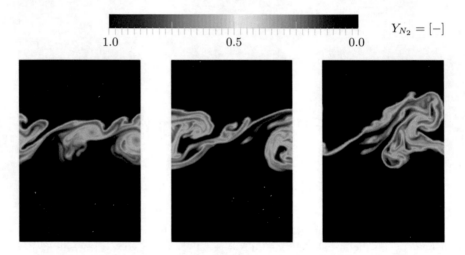

Fig. 9 Temporal snapshot of nitrogen mass fraction at $t = 6$ ms for simulations conducted by direct use of the EOS (left), coarse tables (middle) and refined tables (right)

4.3 Navier–Stokes–Korteweg

The parabolic relaxation model for the NSK equations was used to investigate head on collisions of two droplets.

4.3.1 Simulation Setup

The initial conditions were

$$\rho(\mathbf{x}, t = 0) = \rho_{\text{vap}} + \frac{\rho_{\text{vap}} - \rho_{\text{liq}}}{2} \sum_{i=1}^{2} \left(\tanh \left(\frac{d_i - r_i}{2\sqrt{\gamma_K \epsilon_K^2}} \right) \right) \tag{45}$$

$$u(\mathbf{x}, t = 0) = \begin{cases} \frac{v_{\text{ini}}}{2} + \left(1 - \tanh \left(\frac{d_1 - r_d}{2\sqrt{\gamma_K \epsilon_K^2}} \right) \right) & \text{if } x < 0.5, \\ \frac{-v_{\text{ini}}}{2} + \left(1 - \tanh \left(\frac{d_2 - r_d}{2\sqrt{\gamma_K \epsilon_K^2}} \right) \right) & \text{if } x \geq 0.5, \end{cases} \tag{46}$$

$$v(\mathbf{x}, t = 0) = 0, \tag{47}$$

$$w(\mathbf{x}, t = 0) = 0, \tag{48}$$

where $\rho_{\text{vap}} = 0.3197$, $\rho_{\text{liq}} = 1.8071$ are the Maxwellian densities at $T_{\text{ref}} = 0.85$. The droplet radii were $r_1 = r_2 = 0.5$ and the distance was given by

$$d_i = \| \mathbf{x} - \mathbf{x}_{0,i} \|, \tag{49}$$

where $\mathbf{x}_{0,1} = (0.3, 0.5, 0.5)^\top$ and $\mathbf{x}_{0,2} = (0.7, 0.5, 0.5)^\top$ are the initial positions of the droplets. Four cases were investigated where the droplet number, position, and size remained the same and the model parameters and initial velocities were changed. The parameters are summarized in Table 6. The computation domain was $\Omega = [0, 1]^3$ and it was discretized by 64 elements in each direction. The polynomial degree was $N = 3$ which yielded 256^3 degrees of freedom (DOF). Time integration was done implicit with CFL $= 100$ using a fourth order ESDIRK scheme with six stages. The simulations were performed on the Hazel Hen supercomputer at HLRS using 200 nodes.

Table 6 Parameters for head on droplet collision simulations

Parameter	ϵ_K [$-$]	γ_K [$-$]	α [$-$]	β [$-$]	v_{ini} [$-$]
Case A	1×10^{-3}	100	100	1×10^{-2}	3.0
Case B	1×10^{-2}	1.00	100	1×10^{-2}	3.0
Case C	1×10^{-2}	0.05	100	1×10^{-2}	3.0
Case D	1×10^{-2}	0.05	100	1×10^{-2}	4.0

4.3.2 Simulation Results

The isocontour of the mean density, $\rho_{mean} = 1.0634$, of the solution of case A is shown in Fig. 10 for different time instances. Two droplets were pushed towards each other and coalesce. A flat disc formed for $t > 0.12$ which broke up into a ring and a small droplet in the center at $t \approx 0.24$. Both the ring and the centered droplet evaporated and for $t \to \infty$ only vapour remained, since the average density was in the stable vapour phase.

In Case B ϵ_K was increased and γ_K was decreased such that different phenomena were observed. The isocontour of the mean density is shown in Fig. 11. Again, the two droplets merged and a disc formed. The disc flattened and break up occurred at its centre, however no droplet was formed and only a ring remained. Eventually, the ring evaporated and the domain was filled by a stable vapour phase.

Case C reduced γ_K further, which led to a thinner phase interface. The isocontour is shown in Fig. 12. After coalescence, the disc formed again but no break up occurred and the disc remained at that form until it evaporated completely.

Case D used the same parameters as Case C but increased the initial velocity of the droplets. The isocontour is shown in Fig. 13. The momentum of the droplets was increased and the impact was stronger such that the disc quickly broke up and a ring and centered droplet remained.

The total energy, Eq. (22), was calculated in each time step. As seen in Fig. 14, the total energy decreased monotonously until a minimum was reached. Hence, the solutions produced by the relaxation model were admissible.

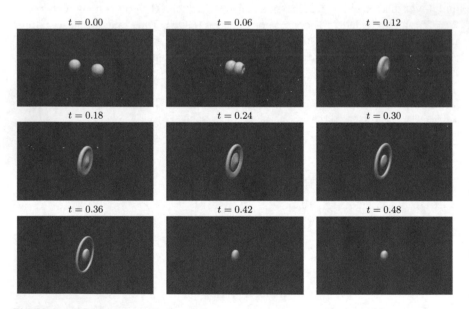

Fig. 10 Results for Case A with parameters $\epsilon_K = 1 \exp{-3}$, $\gamma_K = 100$, $v_{ini} = 3.0$

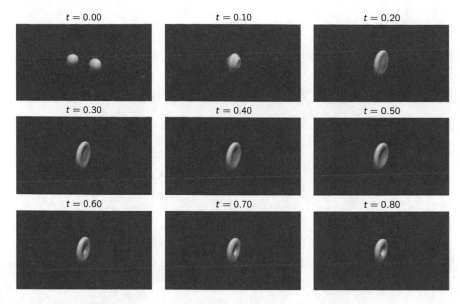

Fig. 11 Results for Case B with parameters $\epsilon_K = 1 \exp -2$, $\gamma_K = 1.00$, $v_{ini} = 3.0$

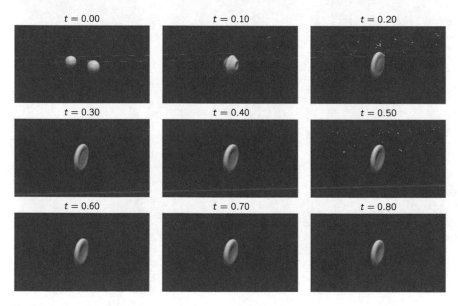

Fig. 12 Results for Case C with parameters $\epsilon_K = 1 \exp -2$, $\gamma_K = 0.05$, $v_{ini} = 3.0$

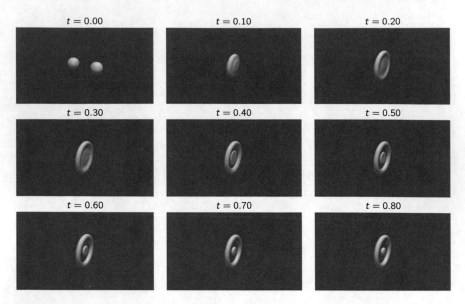

Fig. 13 Results for Case D with parameters $\epsilon_K = 1 \exp{-2}$, $\gamma_K = 0.05$, $v_{\text{ini}} = 4.0$

Fig. 14 Decay of total
energy for the head on
droplet collisions

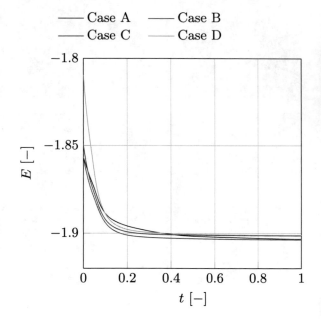

5 Summary and Conclusions

In this work we carried out investigations on the use of modern data structures on high performance computers. In this context, a new implementation strategy for shared memory look up tables for binary mixtures was introduced. We were able to show that a change in hardware architecture on high performance computers, e.g. from Hazel Hen to Hawk, has a great impact on the old algorithms. With the new implementation we are able to store and access about 128 times more memory than with the old algorithm. The simulation and comparison of a multi-component real gas shear layer with exact EOS and tabulation approach led to reasonable results, however further investigations are necessary.

In addition, 3D simulations of colliding droplets were carried out using a parabolic relaxation model of the Navier–Stokes–Korteweg diffuse interface model. A variation of model parameters produced a variation in the coalescence behaviour. Future research aims at validation with experimental results.

Acknowledgements We gratefully acknowledge the Deutsche Forschungsgemeinschaft (DFG) through SFB-TRR 40 "Fundamental Technologies for the Development of Future Space-Transport-System Components under High Thermal and Mechanical Loads" and SFB-TRR 75 "Droplet dynamics under extreme ambient conditions" Computational resources have been provided by the Bundes-Höchstleistungsrechenzentrum Stuttgart (HLRS).

References

1. R. Akasaka, A reliable and useful method to determine the saturation state from Helmholtz energy equations of state. J. Therm. Sci. Technol. **3**, 442–451 (2008)
2. L.E. Baker, A.C. Pierce, K.D. Luks, Gibbs energy analysis of phase equilibria. SPE J. **22**, 731–742 (1982)
3. F. Bassi, S. Rebay, A high-order accurate discontinuous finite element method for the numerical solution of the compressible Navier-Stokes equations. J. Comput. Phys. **131**(2), 267–279 (1997). https://doi.org/10.1006/jcph.1996.5572
4. F. Bassi, S. Rebay, Numerical evaluation of two discontinuous Galerkin methods for the compressible Navier-Stokes equations. Int. J. Numer. Methods Fluids **40**(1–2), 197–207 (2002). https://doi.org/10.1002/fld.338
5. O. Botella, R. Peyret, Benchmark spectral results on the lid-driven cavity flow. Comput. Fluids **27**(4), 421–433 (1998). https://doi.org/10.1016/S0045-7930(98)00002-4
6. T. Coffee, J. Heimerl, Transport algorithms for premixed, laminar steady-state flames. Combust. Flame **43**(Supplement C), 273–289 (1981). https://doi.org/10.1016/0010-2180(81)90027-4
7. A. Corli, C. Rohde, V. Schleper, Parabolic approximations of diffusive-dispersive equations. J. Math. Anal. Appl. **414**(2), 773–798 (2014). https://doi.org/10.1016/j.jmaa.2014.01.049
8. FLEXI, Description and source code (2018), https://www.flexi-project.org/. Accessed 02 Oct 2018
9. F. Föll, T. Hitz, C. Müller, C.D. Munz, M. Dumbser, On the use of tabulated equations of state for multi-phase simulations in the homogeneous equilibrium limit. Shock. Waves **1** (2019). https://doi.org/10.1007/s00193-019-00896-1
10. F. Föll, S. Pandey, X. Chu, C.D. Munz, E. Laurien, B. Weigand, High-fidelity direct numerical simulation of supercritical channel flow using discontinuous Galerkin spectral element method,

in *High Performance Computing in Science and Engineering ' 18*, ed. by W.E. Nagel, D.H. Kröner, M.M. Resch (Springer International Publishing, 2019), pp. 275–289

11. J. Hesthaven, T. Warburton, *Nodal Discontinuous Galerkin Methods: Algorithms, Analysis, and Applications*, 1st edn. (Springer Publishing Company, Incorporated, 2008). https://doi.org/10.1007/978-0-387-72067-8

12. F. Hindenlang, G. Gassner, C. Altmann, A. Beck, M. Staudenmaier, C. Munz, Explicit discontinuous Galerkin methods for unsteady problems. Comput. Fluids **61**, 86–93 (2012). https://doi.org/10.1016/j.compfluid.2012.03.006

13. C.A. Kennedy, M.H. Carpenter, R. Lewis, Low-storage, explicit Runge–Kutta schemes for the compressible Navier–Stokes equations. Appl. Numer. Math. **35**(3), 177–219 (2000). https://doi.org/10.1016/S0168-9274(99)00141-5

14. D. Kopriva, *Implementing Spectral Methods for Partial Differential Equations: Algorithms for Scientists and Engineers*, 1st edn. (Springer Publishing Company, Incorporated, 2009)

15. M.L. Michelsen, The isothermal flash problem. Part 1. Stability. Fluid Phase Equilib. **9**, 1–19 (1982a)

16. M.L. Michelsen, The isothermal flash problem. Part 2. Phase-split calculation. Fluid Phase Equilib. **9**, 21–40 (1982b)

17. M.L. Michelsen, J.M. Mollerup, *Thermodynamic Models: Fundamentals & Computational Aspects*, 2nd edn. (Tie-Line Publications, Holte, 2007)

18. D.Y. Peng, D.B. Robinson, A new two-constant equation of state. Ind. Eng. Chem. Fundam. (1976)

19. P. Persson, J. Peraire, Sub-cell shock capturing for discontinuous galerkin methods, in *44th AIAA Aerospace Sciences Meeting and Exhibit, Aerospace Sciences Meetings, American Institute of Aeronautics and Astronautics* (2006), https://doi.org/10.2514/6.2006-112

20. M. Sonntag, C.D. Munz, Shock capturing for discontinuous Galerkin methods using finite volume subcells, in *Finite Volumes for Complex Applications VII-Elliptic, Parabolic and Hyperbolic Problems*, ed. by J. Fuhrmann, M. Ohlberger, C. Rohde (Springer International Publishing, 2014), pp. 945–953

21. M. Sonntag, C.D. Munz, Efficient parallelization of a shock capturing for discontinuous Galerkin methods using finite volume sub-cells. J. Sci. Comput. **70**(3), 1262–1289 (2017). https://doi.org/10.1007/s10915-016-0287-5

22. H.B. Stewart, B. Wendroff, Two-phase flow: models and methods. J. Comput. Phys. **56**(3), 363–409 (1984). https://doi.org/10.1016/0021-9991(84)90103-7

23. E. Toro, *Riemann Solvers and Numerical Methods for Fluid Dynamics: A Practical Introduction* (Springer, Berlin, 2009). https://doi.org/10.1007/b79761

24. J. Van der Waals, Over de Continuiteit van den Gas-en Vloeistoftoestand. Ph.D. thesis, University of Leiden (1873)

25. A.B. Wood, *A Textbook of Sound*, 1st edn. (G. Bell and Sons, 1941)

Simulation of Flow Phenomena at a Rotor with a Discontinuous Galerkin CFD Solver

Fabian Genuit, Manuel Keßler, and Ewald Krämer

Abstract The Discontinuous Galerkin method is a high-order method in space reducing the amount of cells needed for calculations compared to standard computational fluid dynamics (CFD) solvers. At the Institue for Aerodynamics and Gas Dynamics the CFD code SUNWinT has been developed using a DG method with the aim to apply it to rotor flows. The present study concerns the progress in simulating the flow phenomena of an isolated rotor in hover. The results of the calculations are compared to experimental data and show good agreement. Furthermore, the first phenomenological results of the flow around an isolated rotor in forward flight are presented, which reveal promising results and should serve as a starting point for future investigations.

1 Introduction

Nowadays, computational fluid dynamics (CFD) has become a fundamental tool for the analysis of various kinds of flow problems in nature and for technical applications. The demand for precise results and short simulation times is growing as the available computer power has become an easily accessible and still steadily increasing resource. Therefore, High-Performance Computing (HPC) has become a standard tool in CFD. The simulations are performed on several computer processors (CPUs), which on the one hand shortens the time for the calculation. On the other hand, the parallelization of the simulation requires a communication between the CPUs which may become a bottleneck in terms of computation duration. Classical CFD methods like the finite volume (FV) method need to work on smooth and highly resolved meshes in order to produce reliable results. The creation of these computational meshes is usually a time-consuming process. Furthermore, the spatial order of accuracy of the FV method stagnates at second order due to an inefficiently high parallel communication effort in highly parallel calculations.

F. Genuit (✉) · M. Keßler · E. Krämer
Institute of Aerodynamics and Gas Dynamics, University of Stuttgart, Pfaffenwaldring 21, 70569
Stuttgart, Germany
e-mail: genuit@iag.uni-stuttgart.de

© Springer Nature Switzerland AG 2021
W. E. Nagel et al. (eds.), *High Performance Computing in Science and Engineering '19*,
https://doi.org/10.1007/978-3-030-66792-4_26

395

The Discontinuous Galerkin (DG) method, first used by Reed and Hill [14], has the potential to replace or at least augment the classical CFD methods. The advantage of DG is that the mesh resolution is less restrictive than for FV solvers because the spatial accuracy order may be chosen arbitrarily. Additionally, the parallelization of the DG method requires less communication effort. In fluid mechanics the DG method combines the ideas of the finite element (FE) and the FV methods. The approach of the method is based on the FE Galerkin discretization scheme in which the solution within a cell is given by a polynomial approximation of arbitrary order. In contrast to a classical FE discretization, the DG approach allows the solution to be discontinuous between the cells. Hence, a solver for that kind of Riemann problem is required, which is well known from FV methods.

The DG solver SUNWinT (Stuttgart University Numerical Wind Tunnel) has been developed at the Institute for Aerodynamics and Gas Dynamics (IAG) in the past 15 years. Landmann [9] introduced the first basic functionalities (up to the discretization of the RANS equations) in 1D and 2D space. In [10] Lübon implemented 3D functionality and a method for Detached Eddy Simulations (DES). Afterwards, Wurst [18] improved the solver and extended it for moving bodies in 2D space. This work will consistently pursue the global objective of the project: the turbulent simulation of a helicopter rotor considering of fluid-structure interaction.

Yang and Yang [20] presented results of the Euler flow around an isolated rotor using a DG solver with an overset grid method and third order. In this work first 3D simulations of a moving, isolated rotor in hover with the IAG DG solver SUNWinT have been performed. The results of fourth order are compared with the experimental data provided by Caradonna and Tung [4]. Furthermore, a first approach of the DG simulation of the flow around an isolated rotor in forward flight is presented.

2   Governing Equations

To determine the physics of a fluid in motion, conservation of mass, momentum and energy are required. If viscous, compressible flows the so-called Navier–Stokes equations are considered

$$\frac{\partial \mathbf{U}}{\partial t} + \nabla \left(\mathbf{F}_{inv}(\mathbf{U}) - \mathbf{F}_{vis}(\mathbf{U}, \nabla \mathbf{U}) \right) = \mathbf{S}(\mathbf{U}, \nabla \mathbf{U}) \tag{1}$$

describe their physical behavior. In Eq. (1) the state vector of conservative variables $\mathbf{U}$, the inviscid flux tensor $\mathbf{F}_{inv} = (\mathbf{F}_{inv}^x \ \mathbf{F}_{inv}^y \ \mathbf{F}_{inv}^z)^T$ and the viscous flux tensor $\mathbf{F}_{vis} = (\mathbf{F}_{vis}^x \ \mathbf{F}_{vis}^y \ \mathbf{F}_{vis}^z)^T$ read as

$$
\mathbf{U} = \begin{pmatrix} \rho \\ \rho u_i \\ \rho E \end{pmatrix}, \quad \mathbf{F}_{\text{inv}}^{j}(\mathbf{U}) = \begin{pmatrix} \rho u_j \\ \rho u_i u_j + p \delta_{ij} \\ (\rho E + p) u_j \end{pmatrix}, \quad \mathbf{F}_{\text{vis}}^{j}(\mathbf{U}, \nabla \mathbf{U}) = \begin{pmatrix} 0 \\ \tau_{ij} \\ u_i \tau_{ji} + q_j \end{pmatrix}.
$$

$$(2)$$

ρ represents the density, u_i the components of the velocity vector, p the pressure and E the specific total energy. To achieve a well-defined solution of the conservative variables, another equation is necessary describing the state of the fluid. The present work deals with air as the working fluid, which is assumed to be a perfect gas with a constant specific heat ratio $\kappa = 1.4$. Thus, the equation of state for ideal gases is used and can be written as a connection between the pressure p and the specific total energy E

$$
p = (\kappa - 1) \left(\rho E - \frac{1}{2} \rho \sum_{i=1}^{N_{\text{DIM}}} u_i^2 \right).
$$

$$(3)$$

Since air is considered a Newtonian fluid, the Stokes hypothesis is valid and the viscous stress tensor τ in (2) becomes symmetrical $\tau_{ij} = \tau_{ji}$. The entries are described as a function of the velocity gradients and the dynamic viscosity μ

$$
\tau_{ij} = \mu \left(\frac{\partial u_i}{\partial x_j} + \frac{\partial u_j}{\partial x_i} - \frac{2}{3} \frac{\partial u_k}{\partial x_k} \delta_{ij} \right).
$$

$$(4)$$

The energy flux vector $\mathbf{q}$ can be determined by Fourier's law

$$
q_j = -\lambda \frac{\partial T}{\partial x_j}
$$

$$(5)$$

where the thermal conductivity coefficient λ is assumed to be isotropic. λ can be expressed as a function of temperature, dynamic viscosity and the non-dimensional Prandtl number Pr

$$
\lambda = \frac{\mu c_p}{Pr} \quad \text{with } Pr = 0.72 \quad \text{(for air with } 200K \leq T \leq 700K).
$$

$$(6)$$

In Eq. (1), the source term $\mathbf{S}$ exists only for some specific applications, as it will be introduced in Sect. 4.

In [19], Wurst has already shown that the governing equations can easily be extended to the RANS equations with appropriate closure models for turbulence modeling beyond the scope of this work, which only considers inviscid and laminar flows.

3   Numerical Formulation

3.1   Spatial Discretization

The DG method is used to discretize the governing Eq. (1) in space. Since the DG method is a combination of the FV and the FE method, the starting point is a classical FE approach. The unknown conservative variables $\mathbf{U}$ are approximated on the basis of piecewise polynomial ansatz functions

$$\mathbf{U}(\mathbf{x}, t) \approx \tilde{\mathbf{U}}(\mathbf{x}, t) = \sum_{i=1}^{N_b} \mathbf{u}_i(t)\phi_i(\mathbf{x}) \tag{7}$$

with N_b basis functions $\phi_i(\mathbf{x})$ of a polynomial order p only depending on space [7]. In contrast, the unknown degrees of freedom $\mathbf{u}_i(t)$ are time-dependent.

Next, Eq. (7) is substituted into the Navier–Stokes equation (1) and multiplied by an arbitrary test function

$$\mathbf{v}(\mathbf{x}) = \mathbf{v}_h(\mathbf{x}) = \sum_{j=1}^{N_b} \mathbf{a}_j\phi_j(\mathbf{x}). \tag{8}$$

In the case of the FE Galerkin approach, the basis functions for the approximate solution $\tilde{\mathbf{U}}$ and the test function are selected equally ($\phi_i = \phi_j$). The values a_j in equation are arbitrary. An integration over the whole domain Ω follows, which is approximated with a finite number of elements E. Finally, the integration by parts yields the so-called semi-discrete weak form of the governing equations on the element level

$$
\begin{aligned}
& \int_E \mathbf{v}_h \frac{\partial \tilde{\mathbf{U}}_h}{\partial t} \, \mathrm{d}\Omega \\
& + \oint_{\partial E} \mathbf{v}_h \left(\mathbf{F}_{\mathrm{inv}}(\tilde{\mathbf{U}}_h) - \mathbf{F}_{\mathrm{vis}}(\tilde{\mathbf{U}}_h, \nabla\tilde{\mathbf{U}}_h) \right) \cdot \mathbf{n} \, \mathrm{d}\sigma \\
& - \int_E \nabla\mathbf{v}_h \cdot \left(\mathbf{F}_{\mathrm{inv}}(\tilde{\mathbf{U}}_h) - \mathbf{F}_{\mathrm{vis}}(\tilde{\mathbf{U}}_h, \nabla\tilde{\mathbf{U}}_h) \right) \, \mathrm{d}\Omega \\
& = \int_E \mathbf{v}_h \, \mathbf{S}(\tilde{\mathbf{U}}_h, \nabla\tilde{\mathbf{U}}_h) \, \mathrm{d}\Omega.
\end{aligned}
\tag{9}
$$

In the equation the index h marks the discrete functions on the element level. Up to this point there is no difference to a regular FE Galerkin scheme. However, the DG method has no requirements for any kind of continuity properties of the solution at the element boundaries. Due to the possibility of discontinuities in the solution, the surface integral of the inviscid fluxes is determined with a Harten, Lax and van Leer (HLL) Riemann solver according to [17]. The viscid flux terms have to be treated differently since they are not only dependent on the conservative variables but also on their gradients. For the determination of the gradients the so-called BR2 method, which was developed by Bassi et al. [2], is applied. The integrals in Eq. (9) are approximated numerically by Gaussian integration.

3.2 Temporal Discretization

The spatially discretized Eq. (9) can be written in compact form describe the global equation system

$$\mathbf{M}\frac{\partial \hat{\mathbf{U}}}{\partial t} = \mathbf{R}(\tilde{\mathbf{U}}_h). \tag{10}$$

$\hat{\mathbf{U}}$ denotes the unknown global solution vector and $\mathbf{M}$ the global mass matrix. On the right hand side the residual $\mathbf{R}(\tilde{\mathbf{U}}_h)$ is composed of the inviscid and viscous fluxes and source terms from Eq. (9).

The temporal integration of Eq. (10) is done using an implicit scheme. The advantage of these schemes is that their stability in relation to the Courant-Friedrichs-Lewy (CFL) condition is theoretically not restricted. Hence, the number of iterations needed until the calculation converges (e.g. to a steady state) is expected to be less than for explicit schemes. However, the computational effort is higher and the implementation is more complex. By applying a classical backward Euler method to Eq. (10), it can be rewritten as

$$\mathbf{M}\frac{\hat{\mathbf{U}}(t_{n+1}) - \hat{\mathbf{U}}(t_n)}{\Delta t} = \mathbf{R}\left(\tilde{\mathbf{U}}(t_{n+1})\right). \tag{11}$$

For the solution of the given nonlinear system, the Newton-Raphson method is used, and leading to

$$\left(\frac{\mathbf{M}}{\Delta t} - \frac{\partial \mathbf{R}\left(\tilde{\mathbf{U}}(t_{k-1})\right)}{\partial \tilde{\mathbf{U}}}\right)\Delta\hat{\mathbf{U}} = \mathbf{R}\left(\tilde{\mathbf{U}}(t_{k-1})\right) - \mathbf{M}\frac{\hat{\mathbf{U}}(t_{k-1}) - \hat{\mathbf{U}}(t_n)}{\Delta t}. \tag{12}$$

In order to solve this equation, several Newton iterations for each time step are required. Starting with the solution from the prior time step $\hat{\mathbf{U}}(t_n) = \hat{\mathbf{U}}(t_{k=0})$, the equation is solved for $\Delta\hat{\mathbf{U}}$ until a required accuracy ϵ is achieved. Then the new solution for the next time step can be determined by

$$\tilde{\mathbf{U}}(t_{n+1}) = \tilde{\mathbf{U}}(t_n) + \Delta\tilde{\mathbf{U}}. \tag{13}$$

Accordingly, for every Newton iteration k a linear system has to be constructed and solved. While the composition of the left and right hand side is done analytically, the linear system is solved by a GMRES solver, preconditioned with an ILU(0) method.

The implicit scheme can be applied in the way it is described above to solve unsteady problems. To achieve a faster convergence for $t \to \infty$, it is modified for steady simulations. Since there is no requirement for time precision, only a single Newton iteration per time step is performed in this case. Furthermore, a local or a global adaption of the time step can be chosen.

4 Arbitrary Lagrangian Eulerian Formulation

The kinematic description of motion for continuum mechanical problems is categorized into the Lagrangian and the Eulerian approach. In the Lagrangian specification of the field each individual node follows the material particle associated while moving. In contrast, in the Eulerian framework, which is mainly used in fluid mechanics, the material moves with respect to a fixed grid.

A combination of both descriptions results in the Arbitrary Lagrangian Eulerian (ALE) formulation, where the nodes of a mesh can either move or be fixed. Thus, a more universal description is attained. Additionally, it can handle any arbitrary velocity of the nodes of the computational grid in order to deal with problems e.g., like fluid-structure interaction. The basic principle of the ALE description can be described with a fixed computational domain Ω_{ref}, in which the governing equations are solved. A mapping $\mathbf{J}(x_{\text{ref}}, t)$ transforms the solution into the actual domain Ω, moving and deforming arbitrarily (see Fig. 1).

Hence, the governing equations have to be reformulated in the ALE space. For a detailed derivation the reader is referred to Persson et al. [13], leading to

$$\frac{\partial (j\mathbf{U})}{\partial t} + \nabla \cdot \left(j\mathbf{U}\mathbf{J}^{-1}\mathbf{F} - j\mathbf{U}\mathbf{J}^{-1}\mathbf{u}_g \right) = j\mathbf{S}. \tag{14}$$

Equation (14) serves as the basis for the DG discretization. As the ALE reference space the local reference space of each element is defined, which is also quite common in the literature [11, 12]. $\mathbf{J}$ respectively its determinant $j = |\mathbf{J}|$ are the mapping quantities from the local reference space to the moving, deformed physical space. The difference compared to the classical Eulerian formulation of the DG method is that the mapping quantities $\mathbf{J}$ and j now have to be calculated in every time step. Furthermore, additional fluxes arising from the grid velocity $\mathbf{u}_g$ have to be considered (see Eq. (14)).

Furthermore, the rotation of the mesh and the rotor investigated is considered rigid. Moreover, if the coordinate system is rotated with the same velocity as the mesh, the mapping quantities $\mathbf{J}$ and j stay constant and have to be calculated only once. Since there is no deformation of the mesh, there is no need to account for geometric conservation aspects. However, an additional source term has to be considered, describing the pseudo forces due to the rotation of the coordinate system

$$\mathbf{S} = \begin{pmatrix} 0 \\ -\omega \times \rho u_i \\ 0 \end{pmatrix} \qquad (15)$$

where ω is the angular velocity of the rotating system. Furthermore, the Riemann solver must be adapted as well as the grid velocity needs to be taken into account (for further details, see [18]).

5 Curved Elements

In classical CFD methods such as the FV method, the discretized domain usually is constructed with straight elements. Consequently, the resolution of curved surfaces on flow bodies is not accurate. This downside is countered by using a high number of cells near the body surface in FV calculations.

However, the DG method requires a lower number of cells, as mentioned in Sect. 1. At the same time the approximation of the solution in each cell becomes more accurate due to the high-order approach. Thus, using curved elements for the representation of curved flow bodies becomes a crucial part when high-order methods are applied. Generally, the high-order solution of curved surfaces with straight elements may not converge at all or could even become unphysical [1].

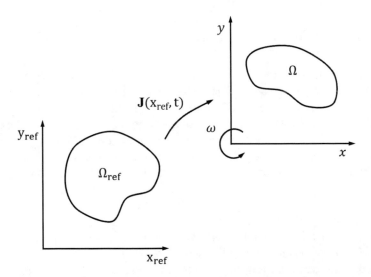

Fig. 1 Mapping between the ALE reference space and the arbitrary moving, physical space [13]

Since standard mesh generators are not able to create curved elements, a plethora of different approaches exists (see [1, 6, 9]) to handle the situation. In this work an agglomeration approach is chosen [6]. Each element of the mesh has a certain number of interpolation points $P_{ijk}^{\mathbf{x}} = (x_{ijk}^{Int}, y_{ijk}^{Int}, z_{ijk}^{Int})$ that contain information about the curvature of the flow body. To transfer this information to the high-order method, a mapping function with a third order Lagrangian polynomial is used. Considering 3D hexahedral elements this leads to 64 interpolation points, which are given as

$$P_{ijk}^{\mathbf{x}_{\text{ref}}} = \left(\frac{1}{3}i, \frac{1}{3}j, \frac{1}{3}k\right) \quad i, j, k \in \{0, 1, 2, 3\} \tag{16}$$

in the reference element. These points are mapped to the interpolations points of the high-order element by

$$\mathbf{x}(x_{\text{ref}}, y_{\text{ref}}, z_{\text{ref}}) = \sum_{i=0}^{3}\sum_{j=0}^{3}\sum_{k=0}^{3} P_{ijk}^{\mathbf{x}} l_i(x_{\text{ref}}) l_j(y_{\text{ref}}) l_k(z_{\text{ref}}) \tag{17}$$

with

$$l_i(x_{\text{ref}}) = \prod_{\substack{0 \le m \le 3 \\ m \ne n}} \frac{x_{\text{ref}} - x_{ref,i}}{x_{ref,i} - x_{ref,m}}. \tag{18}$$

$l_j(y_{\text{ref}})$ and $l_k(z_{\text{ref}})$ are analogous. These mapping quantities are also considered in the Jacobian $\mathbf{J}$ and its determinant j as introduced in Eq. (14).

6   Numerical Setup and Results

Since the validation of the above properties of our code was already provided in [15, 18], progress was made to perform calculations of a more advanced problem. In order to keep the complexity of the flow problem simple for the beginning, the simulation of a rigid rotor blade in hover was performed. By analyzing the results from these calculations, experience is gained with the difficulties of the case and later on the complexity of the problem can be gradually increased.

The numerical setup used for the simulations in this work is based on the experimental study of Caradonna and Tung [4]. The two bladed rotor uses a NACA0012 profile and is untapered and untwisted. In Fig. 2a the geometrical setup of one single blade is shown, where the aspect ratio of the outer blade radius R to the chord length c is $\frac{R}{c} = 6$. In the model the inner blade radius r_i was chosen to be $r_i = c$. In order to simplify the mesh generation of the two bladed rotor, a solid connection between both blades was constructed as it is illustrated in Fig. 2b with a view from the top onto the discretized domain. Obviously, the rotor hub was neglected. The collective pitch angle for each blade was set to $\alpha = 5°$.

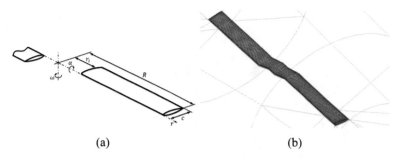

(a)   (b)

Fig. 2 Geometric (**a**) and numerical setup (**b**) of the rotor of Caradonna and Tung [4]

The numerical setup has about 560,000 hexahedral, curved elements (see Sect. 5). The distance of the outer boundary is twelve times the rotor radius and is chosen to be a farfield condition. For now, a slip wall boundary condition is chosen for the blade surfaces. Additionally, the Reynolds number is set to $Re = 10{,}000$ to expect a laminar flow since no appropriate turbulence model has been implemented yet. Although these conditions are not completely consistent with the experiment from Caradonna and Tung, their results were consulted as a reference.

6.1 Isolated Rotor in Hover

The simulation of the rigid rotor setup in hover was performed with a steady, implicit time integration scheme using the technique presented in Sect. 4 with a rotating coordinate system, which is a valid method as Krämer showed in [8]. The tip Mach number was set to $Ma_{tip} = 0.226$. With these settings simulations in ascending order up to four (p3) were performed.

The Q criterion was employed to visualize vortex structures, which are illustrated for isosurfaces of Q in Fig. 3a. The typical tip vortices convecting with the induced flow of the rotor are visible. Furthermore, the contraction of the rotor wake can clearly be seen. Figure 3b, c show the results for the pressure coefficient c_p over the chord length for different blade sections at $\frac{r}{R} = 0.68$ and $\frac{r}{R} = 0.96$ with the fourth order of accuracy (p3). Even though the blade surface is considered a slip wall, the comparison to the experimental data of [4] shows good agreement especially for the outer blade sections.

6.2 Isolated Rotor in Forward Flight

With respect to the work of Stangl [16], the numerical setup introduced above was used in a further study to perform the simulation of a rotor in forward flight.

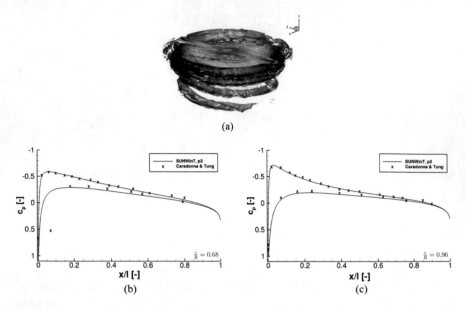

Fig. 3 Isosurfaces for the Q criterion (**a**) and pressure coefficient c_p at a blade section of $\frac{r}{R} = 0.68$ (**b**) and $\frac{r}{R} = 0.96$ (**c**) for the rotor in hover

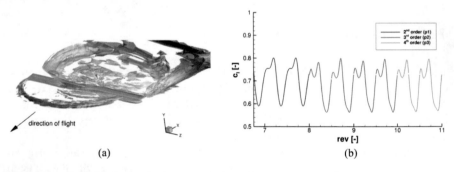

Fig. 4 Isosurfaces for the Q criterion (**a**) and lift coefficient c_l (**b**) for the rotor in forward flight

However, the collective pitch angle of $\alpha = 5°$ was not varied during the rotation of the rotor blade. Accordingly, an untrimmed rotor behavior with high roll moments was expected and observed. The simulations were executed with the unsteady, implicit time integration scheme for spacial orders up to four (p3). The advance ratio was chosen on $\mu = 0.2$, while the tip Mach number had a value of $Ma_{tip} = 0.25$.

In Fig. 4a, the isosurfaces for the Q criterion again depicts the vortices in the rotor wake. The blade-vortex interaction is visible, which is typical for a rotor in forward flight. This interaction is also observed in Fig. 4b, where the lift coefficient is plotted over the rotor revolution. Obviously, the interaction of the blade vortex with the following blade occurs twice each complete rotor revolution. With increasing accuracy order, the impact of the interaction on the lift coefficient is better resolved.

The purpose of these simulations was to verify the functionality of the solver with this kind of flow problem, which succeeded. Furthermore, the typical phenomenological effects of a rotor in forward flight were demonstrated.

7 Computational Performance

The promising parallel performance of the solver was reported previously for various cases with explicit calculations in [3] and even with implicit calculations in [19]. Hence, the simulation of rotating rigid bodies should essentially behave similarly as there is no additional effort in communication compared to a classical simulation without the ALE formulation and the curved element approach. As a test case a cycling sphere with radius $r = 0.5$ m was chosen, using 14,400 curved hexahedral elements.

Figure 5a shows the strong scaling of the case with $\omega = 0.01 1/s$, which causes a flow velocity of $Ma = 0.2$ at the sphere for simulations with an accuracy of second to forth order. Since the number of degrees of freedoms (DOFs) increases with the

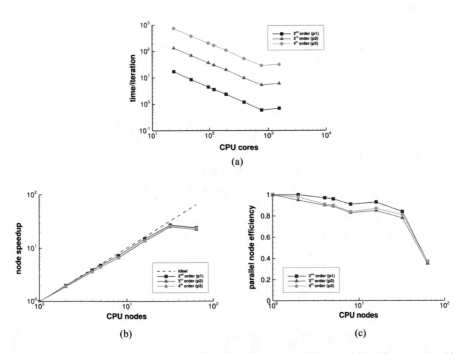

Fig. 5 Strong scaling (**a**), node speedup (**b**) and parallel node efficiency (**c**) for different orders of accuracy

accuracy order, it is obvious that the time per iteration increases. However, SUNWinT scales almost linearly for all orders. The calculation at 1536 cores indicates the limit of an efficient parallel execution. At this point the communication effort exceeds the local workload and thus the computational time increases.

Due to the memory required for the composition of the global system matrix, in this case a single core calculation was only possible with second order. Hence, a pseudo-speedup with computational nodes (1 node consists of 24 cores) was applied. In Fig. 5b the node speedup shows almost ideal behaviour for all accuracy orders that were investigated. The drop of the speedup for 64 nodes once again indicates the limit of the parallel work distribution. The parallel node efficiency in Fig. 5c reveals a slightly better performance for the second order calculations, whereas the behavior for third and fourth order is almost identical. Generally, the parallel efficiency stays above a value of 80% for a wide node range.

Finally, the focus is on the dependence of these performance results of the linear system solver. As introduced in Sect. 3.2, a GMRES solver with an ILU(0) precondi-tioner is employed, which is provided by the trilinos package [5] preinstalled at the CRAY XC40 system at HLRS. Thus, the performance of this iterative solver strongly influences the performance of the code and is identified as a crucial component.

8 Conclusion

The Navier–Stokes equations have been discretized with a DG method and extended with an ALE formulation to allow moving and deforming domains. Furthermore, the importance of an accurate representation of the flow body surface on a high-order method was discussed. The experiment of Caradonna and Tung [4] served as a test case for the first series of simulations of an isolated rotor in hover. Even though a few assumptions were made, it was shown that SUNWinT works in principle for these types of flow problems. The results were in good agreement with the experimental data and the typical flow phenomena at a rotor in hover have been detected. Moreover, the simulation of an isolated rotor with a constant pitch angle in forward flight confirms the phenomenologically expected results.

However, the aim of current and future work is to extend the code to fully tur-bulent simulations of multi-bladed rotors. Furthermore, the Chimera method will be employed, as it was introduced in [19] for 2D problems.

References

1. F. Bassi, S. Rebay, High-order accurate discontinuous finite element solution of the 2D Euler equations. J. Comput. Phys. **138**, 251–285 (1997)
2. F. Bassi, A. Crivellini, S. Rebay, M. Savini, Discontinuous Galerkin solution of the Reynolds-averaged-Navier–Stokes and k-w turbulence model equations. Comput. Fluids **34** (2005)

3. E.R. Busch, M. Wurst, M. Keßler, E. Krämer, Computational aeroacoustics with higher order methods, in *High Performance Computing in Science and Engineering '12*, ed. by W.E. Nagel, D.H. Kröner, M.M. Resch (Springer, Berlin, 2013)
4. F. Caradonna, C. Tung, Experimental and analytical studies of a model helicopter rotor in hover. Technical report TM 81232, NASA (1981)
5. M.A. Heroux, R.A. Bartlett, V.E. Howle, R.J. Hoekstra, J.J. Hu, T.G. Kolda, R.B. Lehoucq, K.R. Long, R.P. Pawlowski, E.T. Phipps et al., An overview of the Trilinos project. ACM Trans. Math. Softw. (TOMS) **31**(3), 397–423 (2005)
6. F. Hindenlang, Mesh curving techniques for high order parallel simulations on unstructured meshes. Ph.D. thesis, Institut für Aerodynamik und Gasdynamik, Universität Stuttgart (2014)
7. G.E. Karniadakis, S.J. Sherwin (eds.), *Spectral/hp Element Methods in CFD* (Oxford University Press, Oxford, 1999)
8. E. Krämer, Theoretische Untersuchungen der stationären Rotorblattumströmung mit Hilfe eines Euler–Verfahrens. Dissertation, Institut für Luftfahrttechnik und Leichtbau, Universität der Bundeswehr München (1991)
9. B. Landmann, A parallel discontinuous Galerkin code for the Navier–Stokes and Reynolds-averaged Navier–Stokes equations. Ph.D. thesis, Insitut für Aerodynamik und Gasdynamik, Universität Stuttgart (2008)
10. C. Lübon, Turbulenzmodellierung und Detached Eddy Simulation mit einem Discontinuous Galerkin Verfahren von hoher Ordnung. Ph.D. thesis, Institut für Aerodynamik und Gasdynamik, Universität Stuttgart (2009)
11. D.J. Mavriplis, C.R. Nastase, On the geometric conservation law for high-order discontinuous Galerkin discretizations on dynamically deforming meshes. J. Comput. Phys. **230**(11), 4285–4300 (2011). Special issue High Order Methods for CFD Problems
12. C.A.A. Minoli, D.A. Kopriva, Discontinuous Galerkin spectral element approximations on moving meshes. J. Comput. Phys. **230**(5), 1876–1902 (2011)
13. P.-O. Persson, J. Bonet, J. Peraire, Discontinuous Galerkin solution of the Navier–Stokes equations on deformable domains. Comput. Methods Appl. Mech. Eng. **198**(17–20), 1585–1595 (2009)
14. W.H. Reed, T.R. Hill, Triangular mesh methods for the neutron transport equation. Technical report, Los Alamos Scientific Laboratory (1973)
15. L. Schmitt, Erweiterung eines Discontinuous-Galerkin-Verfahrens auf rotierende Gitter (2012)
16. R. Stangl, Ein Euler–Verfahren zur Berechnung der Strömung um einen Hubschrauber im Vorwärtsflug. Dissertation, Institut für Aerodynamik und Gasdynamik, Universität Stuttgart (1996)
17. E.F. Toro, *Riemann Solvers and Numerical Methods for Fluid Dynamics: A Practical Introduction*, 2nd edn. (Springer, Berlin, 1999)
18. M. Wurst, Development of a high-order Discontinuous Galerkin CFD solver for moving bodies. Ph.D. thesis, Institut für Aerodynamik und Gasdynamik, Universität Stuttgart (2016)
19. M. Wurst, M. Kessler, E. Krämer, *A High-Order Discontinuous Galerkin Chimera Method for the Euler and Navier–Stokes Equations* (Chap. II.2) (2015), pp. 423–433
20. L. Yang, A. Yang, Implementation of spectral difference method on overset grids for compressible inviscid flows. Comput. Fluids **140**, 500–511 (2016)

Numerical Analysis of Relaminarization in Turbulent Channel Flow

Sandeep Pandey, Cenk Evrim, and Eckart Laurien

Abstract The turbulent channel flow subjected to a wall-normal, non-uniform body force is investigated here by means of direct numerical simulations (DNS). The DNS is performed with the spectral/hp element solver, Nektar++. Flow relaminarization and turbulence recovery were observed in the body-force-influenced flow when compared to the flow at the same Reynolds number. The interaction events assist the recovery at higher body forces and result in two peaks in the ratio of production to dissipation of turbulent kinetic energy. The spectra of velocity fluctuations are also analyzed which shows a drastic reduction in the energy and stretched tail in the dissipation range during the relaminarization which indicates the disappearance of streaky structures in the flow. A retrieval in the energy was observed as a result of turbulence recovery at low wave numbers. The strong scaling characteristics of the employed computational code shows a good scalability.

1 Introduction

Turbulent fluid flows are one of the most challenging problem in physics but are common in engineering applications. The turbulent flow subjected to any kind of body-force may suffer from the modulated turbulence. The body force has the potential to distort the mean velocity profile [3]. There are several investigations made to unveil the effects of the body force on the fluid flow. The effects of the Lorenz force, which is a result of the magnetic field, is widely examined in the literature due to possibilities of flow control [9, 15]. The Lorentz force can suppress the near-wall turbulence structures and it can also result in drag reduction in electrically conduct-

S. Pandey · C. Evrim (✉) · E. Laurien
Institute of Nuclear Technology and Energy Systems, University of Stuttgart, Stuttgart, Germany
e-mail: cenk.evrim@ike.uni-stuttgart.de

S. Pandey
e-mail: sandeep.pandey@ike.uni-stuttgart.de

E. Laurien
e-mail: eckart.laurien@ike.uni-stuttgart.de

© Springer Nature Switzerland AG 2021   409
W. E. Nagel et al. (eds.), *High Performance Computing in Science and Engineering '19*,
https://doi.org/10.1007/978-3-030-66792-4_27

ing fluids [15]. Kühen et al. [12] made a successful attempt to reduce the drag by modifying the streamwise velocity profile so that the flow fully relaminarizes. This is achieved by enhancing turbulence mixing which created a more uniform mean flow [13]. Flow subjected to streamwise acceleration also shows similar characteristics of turbulence attenuation [14, 18]. Sreenivasan [24] categorized turbulent flow subjected to the different phenomenon (e.g. acceleration, suction, blowing, magnetic fields, stratification, rotation, curvature, heating, etc.) into three regimes, namely, laminarescence, relaminarization, and retransition.

The flow of supercritical CO_2 is affected by the body force due to the gravity, which influences the flow drastically [6, 20] and it is the main motivation for this work. Both forced and natural convection play here a significant role in the flow and heat transfer. The characteristics of body force on the mixed convection can catastrophically affect the turbulence in certain circumstances. In heated vertical flow, the direction of fluid flow dictates the nature of heat transfer [10]. In buoyancy aided flow (heated upward flow and cooled downward flow), body force can significantly enhance the skin friction coefficient and decrease the convective heat transfer coefficient as a result of the suppression in turbulence [23]. Flow can even become laminar under a strong buoyancy force. The velocity profile flattens out during the relaminarization [25]. Flow recovers with a different kind of turbulence with a further increase in the buoyancy force and the velocity profile acquires an 'M' shape. The transformation of velocity profile is due to the external effects of buoyancy resulting from the local flow acceleration close to the wall which compensates the decrease in velocity in the core [3, 22].

Pandey et al. [23] showed that the reduction in sweep and ejection events results in turbulence attenuation in downward flow under cooling condition, while the inward and outward interaction events are responsible for a recovery. Chu et al. [5] performed a DNS study to examine the role of buoyancy in a strongly heated air flow through a pipe subjected to a buoyancy force. They reported that buoyancy production is suppressed as compared to turbulent kinetic energy production during relaminarization. They also observed longer streaks near the wall which separated the pipe flow into two layers with an increased anisotropy in the near-wall layer. Chu and Laurien [4] conducted a DNS investigation for horizontal flow of CO_2 with flow heating to find out the effects of gravity. In this type of configuration, thermal stratification was noticed in which low-density fluid accumulates at the top of the pipe. Due to the body force at supercritical pressure, most of the simple models do not perform well [19, 21]. However, the new machine learning based models proved to be a good alternative [2, 7], where data from DNS can be used.

This study aims to enhance our knowledge in understanding the role of a body force on the flow turbulence. As a high-fidelity approach, direct numerical simulations are used for a canonical geometry of a channel with periodic boundary conditions. Special attention is given to the flow relaminarization induced by the body force. The present work also aims to investigate the spectra of turbulent energy during the flow relaminarization and flow recovery.

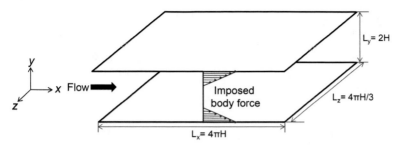

Fig. 1 Geometry for turbulent channel flow

2 Simulation Details

In the course of this project, different open source CFD codes have been used earlier. During this phase, a new CFD code is extended to analyze the effects of non-uniform body force. This section describes the governing equation, numerical method, computational details along with the supporting theory for the analysis.

2.1 Computational Domain and Boundary Conditions

An incompressible channel flow of a Newtonian fluid is considered here. The channel has the dimensions, $L_x \times L_y \times L_z = 4\pi H \times 2H \times 4\pi/3H$, where H is the half channel height as shown in Fig. 1. Periodic boundary conditions are used in the homogeneous streamwise (x) and spanwise directions (z) and no-slip boundary conditions are used at the two walls. The flow is driven by a temporally and spatially constant streamwise pressure gradient. The body force profile can take different shapes depending on the specific physical problem. Typically, the near-wall distribution of the body force results in a peculiar phenomenon. The primary aim of this work is to investigate the body-force-influenced flow at supercritical pressure. The density shows a steep variation due to the fact that temperature and pressure lie in the near-critical region. This density variation results in body force due to buoyancy (ρg). Therefore, a similar body force profile was employed in this work to resemble a buoyancy-aided-flow. Figure 2 illustrates the body force profile used in this work. Only the strength of the force was varied without considering its range of coverage. The body force here does not have exact quantitative variation similar to the buoyancy force in flow of sCO_2, rather its vary qualitatively. It also decouples the effects of variation in thermophysical properties which ultimately allow us to analyze the sole effect of body force.

Fig. 2 Wall normal profile
of body force employed in
this work

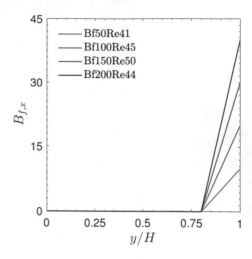

2.2   Governing Equations

The flow governing equations are given in Eqs. 1–2. In these equations, $\mathbf{u}$ denotes the velocity vector, and p is the pressure. The viscous stress is denoted by τ and defined in Eq. 3.

$$\nabla \cdot (\mathbf{u}) = 0 \tag{1}$$

$$\frac{\partial \mathbf{u}}{\partial t} + \nabla \cdot [\mathbf{uu} + p\mathbf{I} - \tau] = \mathbf{B_f} \tag{2}$$

$$\tau = \mu \left[[\nabla \mathbf{u} + (\nabla \mathbf{u})^T] - \frac{2}{3}(\nabla \cdot \mathbf{u})\mathbf{I} \right] \tag{3}$$

A forcing term ($\mathbf{B_f} = [B_{f,x}, 0, 0]^T$) was added to the streamwise momentum equation. The forcing term remains constant throughout the domain and it is governed by Eq. 4.

$$B_f, x = \begin{cases} Ay, & if \ |y/H| > 0.8 \\ 0, & if \ |y/H| \le 0.8 \end{cases} \tag{4}$$

Direct numerical simulations are performed with the spectral/hp element solver Nektar++ [1]. A velocity correction scheme was employed where the velocity system and the pressure are typically decoupled. Spatial discretization is based on the spectral element method in the y, z—plane combined with a Fourier decomposition in the x-direction. The time integration is treated using a second-order accurate mixed implicit-explicit (IMEX) scheme [11]. The simulation domain is discretized with a regular, structured quadrilateral mesh combined with eighth-order polynomials. This solver package has been used in many studies pertaining to fluid mechanics

Table 1 A summary of simulation conditions

Sr. Nr.	Case	Re_b	Re_τ	B_f	Δx^+	Δy^+_{min}-Δy^+_{max}	Δz^+
1	Bf50Re41	4190	256	B_f, x	8.0	0.15–2.5	5.8
2	Bf0Re41	4190	259	0	8.1	0.15–2.5	5.9
3	Bf100Re45	4565	266	B_f, x	8.3	0.15–2.7	6.1
4	Bf0Re45	4565	279	0	8.7	0.15–2.7	6.4
5	Bf150Re50	5000	310	B_f, x	9.7	0.15–3.1	7.1
6	Bf0Re50	5000	302	0	9.4	0.20–3.0	6.9
7	Bf200Re44	4490	337	B_f, x	10.5	0.20–3.3	7.7
8	Bf0Re44	4490	275	0	8.6	0.20–2.7	6.3

[8, 27]. The continuous Galerkin was selected for projection and advection was handled explicitly. The y,z- plane is discretized with 21×27 quadrilateral elements. The modified Legendre basis was used with eight modes (maximum polynomial order is 7). The streamwise direction contains 384 homogeneous modes. The streamwise mesh resolution normalized by the friction velocity ($\Delta x^+ = \Delta x u_\tau / \nu$) is kept below 10.5, and the dimensionless spanwise mesh resolution was below 7.7 in every case. The mesh was refined near to the wall and it has a maximum dimensionless spanwise mesh resolution 6.7 at the center and 0.4 at the wall. Table 1 shows the used resolutions for individual case. Spectral accuracy is one of the added advantage of using Nektar++ combined with high order nature of the code. The earlier work with OpenFOAM was having only second order of accuracy and finite volume implementation.

In the present study, eight distinct cases have been simulated, four of which include a varying body force and the remaining four are reference cases simulated without body forces, but with the same Reynolds numbers. These cases are selected to resemble a buoyancy-aided flow. Table 1 shows the parameters for all simulated cases. The bulk Reynolds number is defined as $Re_b = u_b H / \nu$, where ν is the kinematic viscosity and u_b is the bulk velocity. The body force, B_f follows a profile given by Eq. 4 and Fig. 2.

3 Results and Discussion

Direct numerical simulations were conducted for all eight cases mentioned in Table 1. Few important results are discussed here along with a code verification study. In this section, any generic quantity (say ϕ) is averaged in time as well as in streamwise and spanwise, and $\overline{\phi}$ shows the same while fluctuating part is shown as ϕ'.

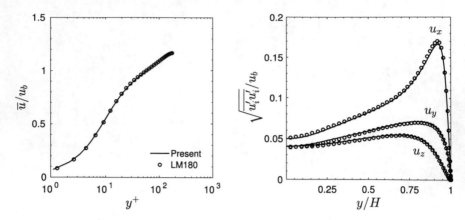

Fig. 3 Verification of the present DNS code with Lee and Moser [16]; **a**: mean streamwise velocity profile, **b**: root mean square fluctuations of velocity

3.1 Verification of the Code

The incompressible channel flow at a friction Reynolds number of 182 and a bulk Reynolds number of 2857 is verified with the well known incompressible channel flow DNS data of Lee and Moser [16] (acronym as LM180). The corresponding simulations were performed with the Nektar++ code. Figure 3 shows the wall-normal profiles for the mean velocity and the root mean square velocity fluctuations normalized with the bulk velocity (u_b). A very good agreement can be observed from Fig. 3 for the incompressible flow case.

The present study takes the advantage of the spectral analysis. Therefore, a verification study was also performed for the spectra as shown in Fig. 4 for $y^+ = 30$. The spectra is time-averaged and obtained by fast Fourier transform of the velocity fluctuations. The comparison is made between the classical DNS data of Moser et al. [17] (acronym as MKM) and a recent data from Vreman and Kuerten [26]. A good agreement can be seen for all three component of the velocity fluctuations.

3.2 Instantaneous and Mean Flow

The qualitative change of the flow field with increasing body forces is shown in Fig. 5, where the instantaneous streamwise velocity, normalized by bulk velocity, is depicted at one of the axial x-positions. With an increase in the body force, the contours change their shape. Relaminarization can clearly be visualized in Fig. 5b, corresponding to case Bf100Re45. With a stronger body force, a flow recovery can be seen in Fig. 5c, d. Also, it can be noticed from Fig. 5d that the fluid close to the

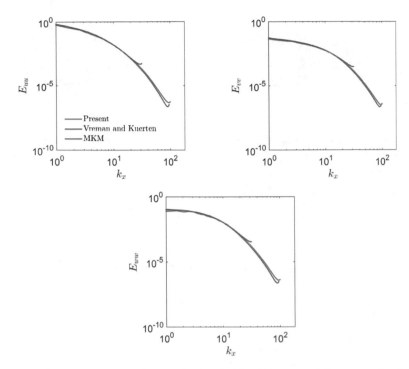

Fig. 4 Verification of the present DNS code with MKM and Vreman and Kuerten; **a**: E_{uu}, **b**: E_{vv}, and **c**: E_{ww} at $y^+ = 30$

wall, shows a higher velocity compared to the center in the recovered case which is contrary to Fig. 5a.

The temporal and spatial averaged streamwise velocity profile ($\bar{u}_x$) is shown in Fig. 6a in physical coordinates as well as in semi-logarithmic wall coordinates. As expected, the body force distorts the velocity profile. With a relatively low value of the body force (case Bf50Re41), the core velocity decreases to compensate the increased velocity near to the wall, as compared to the reference case. With further increasing body force (case Bf100Re45), the velocity profile flattens out and flow relaminarization was observed. The flow recovers as soon as the body force increases further and the velocity profile shows its typical 'M'-shape. The 'M'-shape profile in the channel flow is the effect of (strong) wall jets near each wall. Long and strong wall jets lead to an inflection point in the respective boundary layer and to a strong inviscid instability that holds for laminar as well as turbulent flow. It results in a second strong peak in the production of turbulent kinetic energy, which is discussed latter. The wall normal gradient of the streamwise velocity also varies with an increase in the body force as shown in Fig. 6b. The velocity gradient adjacent to the wall increases with

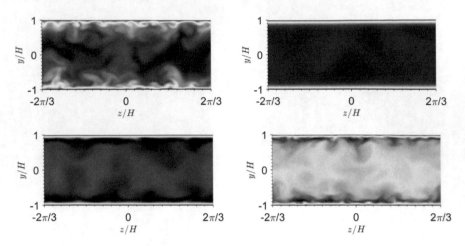

Fig. 5 Pseudo-colour visualization of the instantaneous streamwise velocity u_x, normalized by the bulk velocity u_b. **a**, **b**, **c**, **d**: Case 1,3,5, and 7 respectively

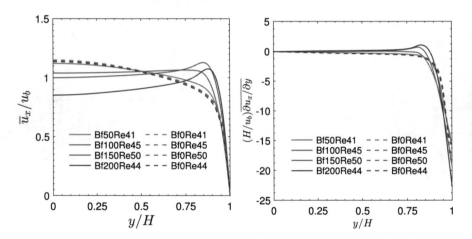

Fig. 6 Distribution of **a**: mean velocity profile, **b**: wall normal streamwise-velocity gradient

growing in body forces. Thus, positively affects the wall shear stress and thereby, the friction Reynolds number (as can be seen in Table 1). Due to the 'M'-shape profile, the mean shear rate increases close to the wall as a result of a local flow acceleration close to the wall. This assists a turbulence recovery by positively affecting the turbulence production. It can also be observe that the body force reduces the fluctuation until flow laminarize. The strength of streaks is significantly reduced in case Bf100Re45 (laminarized) and case Bf150Re50 (partially recovered). This certainly breaks the self-sustaining process and will result in elongated streaks in the near-wall region.

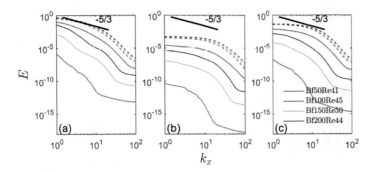

Fig. 7 One-dimensional spectra of **a**: E_{uu}, **b**: E_{vv}, and **c**: E_{ww} at $y/H = 0.99$. Dashed lines shows the reference case

3.3 Spectral Analysis

The streamwise turbulence energy spectra are analyzed in this section to investigate an inconsistent behavior due to the body force. Figure 7 illustrates the one-dimensional spectra for all three velocity components at $y/H = 0.99$ for both reference and body-force-influenced case. The reference case shows an expected behavior with fluctuating energies (E) of all velocity components, which decreases monotonically with an increase in streamwise wave number (k_x). An increase in the body force attributes to a reduction in energy. Due to the turbulence recovery, a retrieval in energy can also be seen at the low wave numbers. Interestingly, the energy containing range vanished out in the spectra of the streamwise velocity for the relaminarized case (case Bf100Re45) as shown Fig. 7a. Also, the tail of the spectra at high wave numbers is stretched out. Figure 8 shows the spectra at $y/H = 0.90$ for both reference and body-force-influenced case. It also shows a similar behavior with a drastic reduction in energy and a stretched tail in the dissipation range. Even after the recovery, this tail remains in the spectra, however, in a shorter range. As mentioned earlier, the recovery was contributed by the 'M' shape velocity profile which is associated with the positive turbulent production.

The drastic reduction in the energy at low wave numbers associated with the relaminarization can be explained by the ratio of production $\left(P_k = -\overline{u_i' u_j'} \frac{\partial u_i}{\partial x_j} \right)$ to dissipation $\left(\varepsilon_k = -\nu \overline{\frac{\partial u_i}{\partial x_j} \frac{\partial u_i}{\partial x_j}} \right)$ of turbulent kinetic energy and it is shown in Fig. 9. The low body force case (case Bf50Re41) shows a typical behavior observed at low Reynolds numbers with a peak in the buffer layer. This peak vanishes during the relaminarization due to negligible turbulence production. It leads to a reduced energy density along with a vanished energy containing range observed in Figs. 7 and 8. When the recovery starts in case Bf150Re50, two peaks appear in the ratio (see Fig. 9). A strong peak is located between the center and the wall of the channel while a weak peak is close to the wall. The strong peak is the result of the interaction events of the Reynolds shear stress. These events shift towards the wall

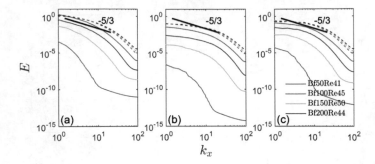

Fig. 8 One-dimensional spectra of **a**: E_{uu}, **b**: E_{vv}, and **c**: E_{ww} at $y/H = 0.90$. Dashed lines shows the reference case

Fig. 9 Distribution of the ratio of production to dissipation of turbulent kinetic energy

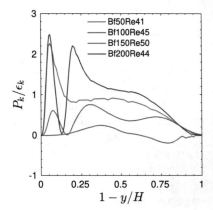

in the recovery (case Bf200Re44) which results in the second peak much closer to the wall. In this case, another peak with the same strength also appears at the usual location (buffer layer) which is an outcome of sweep and ejection events which take place in this region.

3.4 Computational Performance

In this section, we present strong scaling characteristics of the code without multi-threading. The computational code used in this study, Nektar++, is based on the spectral/ hp element framework. It is a tensor product based finite element package programmed in C++. The code is parallelized by OpenMP and for this study, HDF5 format was employed for IO. Fourier modes was use in the homogeneous stream-wise direction, which allow a pseudo-3D simulations and reduces the computational effort compared to pure-3D simulation without loosing any generality. The simulations were performed on *Hazel Hen* located at the High-Performance Computing

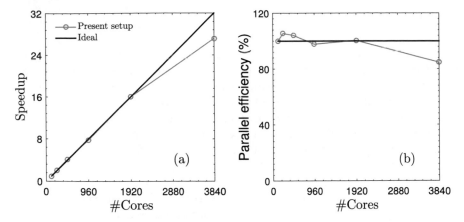

Fig. 10 Comparison of strong scaling of the present solver without multi-threading for a given ratio of number of cores to the number of partitions in the streamwise direction; **a** speedup **b** parallel-efficiency

Center (HLRS) Stuttgart for a range of 120–7680 physical cores. To demonstrate the computational performance, a strong scaling test was performed on *Hazel Hen*. For this purpose, the testcase has a maximum polynomial order of 4 and 768 modes in the streamwise direction. The degree of freedom for this testcase is approximately 60 Million. Typically, in our DNS, we have large number of Fourier modes in the homogeneous direction and relatively few elements in the y, z-plane. Therefore, it is more efficient to parallelize in the streamwise direction and this code allow us to do so by specifying a separate flag by which a single core handles particular modes.

Figure 10 illustrates the result from the strong scaling test. The curve shows the scaling when the ratio of number of cores to the number of partitions in the streamwise direction is constant (10 in this case). A very good scalability of the code can be seen until 1920 computational cores with a parallel efficiency up to 100%. Interestingly, there is a super-linear speed-up between 350 and 1920 cores. If we further double the number of cores then our performance degrades (not shown here). It is due the fact that for a given number of partitions, with the increase in the cores, the number of elements per cores decreases which creates an imbalance between computational time and communication time. In the future, we intend to seek the answers for influence of body force at higher Reynolds number, where the mesh resolution would be much higher due to the requirement for resolving Kolmogorov scale. The future work will require a mesh size of approximately 800 Million where it is expected to have a linear scaling even at higher nodes.

3.5   *Conclusion*

In this work, a systematic study of body-force-influenced flow in a channel has been performed by means of direct numerical simulations. The DNS are performed with a spectral/hp element method. A wall-normal varying body force was added to the streamwise momentum equation to elucidate the effects on turbulence. The body force influences the Reynolds numbers, therefore, additional simulations were also performed at the same Reynolds number without body force. The body force profile was motivated from the buoyancy force in supercritical fluids and have different strengths which allowed to observed all three possible states in such flows, viz., laminarescent (i.e. precursor of relaminarization), relaminarizing and recovery. The body force influences the mean velocity profile, thereby, affecting the gradient of the mean velocity which ultimately modulates the turbulence production. Two peaks in the ratio of production to dissipation of turbulent kinetic energy were observed during the recovery as a result of interaction events. The streamwise spectrum of the streamwise velocity fluctuations depicted a drastic reduction in energy and a stretched tail in the dissipation range along with a vanished energy containing range. This occurred due to a negligible production which itself was the result of flow relaminarization. The strong scaling of the computational code shows a good scalability even with a low degree of freedom. In the future, we intend to use much higher degree of freedom to analyze the flow at high Reynolds number.

Acknowledgements The authors are sincerely thankful to the High Performance Computing Center (HLRS) Stuttgart for providing access to *Hazel Hen* under project DNSTHTSC. SP is grateful to the Forschungsinstitut für Kerntechnik und Energiewandlung (KE) e.V., Stuttgart, for the fellowship. We are also thankful to the anonymous reviewers for his/her valuable suggestion on some of the technical aspects of this report.

References

1. C. Cantwell, S. Sherwin, R. Kirby, P. Kelly, From h to p efficiently: strategy selection for operator evaluation on hexahedral and tetrahedral elements. Comput. Fluids **43**(1), 23–28 (2011)
2. W. Chang, X. Chu, A.F.B.S. Fareed, S. Pandey, J. Luo, B. Weigand, E. Laurien, Heat transfer prediction of supercritical water with artificial neural networks. Appl. Therm. Eng. **131**, 815–824 (2018)
3. X. Chu, E. Laurien, Direct numerical simulation of heated turbulent pipe flow at supercritical pressure. J. Nucl. Eng. Radiat. Sci. (2) (2016)
4. X. Chu, E. Laurien, Flow stratification of supercritical CO2 in a heated horizontal pipe. J. Supercrit. Fluids **116**, 172–189 (2016)
5. X. Chu, E. Laurien, D.M. McEligot, Direct numerical simulation of strongly heated air flow in a vertical pipe. Int. J. Heat Mass Transf. **101**, 1163–1176 (2016)
6. X. Chu, E. Laurien, S. Pandey, Direct Numerical Simulation of Heated Pipe Flow with Strong Property Variation (Springer International Publishing, 2016), pp. 473–486

7. X. Chu, W. Chang, S. Pandey, J. Luo, B. Weigand, E. Laurien, A computationally light data-driven approach for heat transfer and hydraulic characteristics modeling of supercritical fluids: From DNS to DNN. Int. J. Heat Mass Transf. **123**, 629–636 (2018)
8. X. Chu, G. Yang, S. Pandey, B. Weigand, Direct numerical simulation of convective heat transfer in porous media. Int. J. Heat Mass Transf. **133**, 11–20 (2019)
9. C.H. Crawford, G.E. Karniadakis, Reynolds stress analysis of EMHD-controlled wall turbulence. Part I. Streamwise forcing. Phys. Fluids **9**(3), 788–806 (1997)
10. J. Jackson, M. Cotton, B. Axcell, Studies of mixed convection in vertical tubes. Int. J. Heat Fluid Flow **10**(1), 2–15 (1989)
11. G.E. Karniadakis, M. Israeli, S.A. Orszag, High-order splitting methods for the incompressible Navier–Stokes equations. J. Comput. Phys. **97**(2), 414–443 (1991)
12. J. Kühnen, D. Scarselli, M. Schaner, B. Hof, Relaminarization by steady modification of the streamwise velocity profile in a pipe. Flow Turbul. Combust. **100**(4), 919–943 (2018)
13. J. Kühnen, B. Song, D. Scarselli, N.B. Budanur, M. Riedl, A.P. Willis, M. Avila, B. Hof, Destabilizing turbulence in pipe flow. Nat. Phys. **14**(4), 386–390 (2018)
14. B. Launder, Laminarization of the turbulent boundary layer in a severe acceleration. J. Appl. Mech. **31**(4), 707–708 (1964)
15. D. Lee, H. Choi, Magnetohydrodynamic turbulent flow in a channel at low magnetic Reynolds number. J. Fluid Mech. **439**, 367–394 (2001)
16. M. Lee, R.D. Moser, Direct numerical simulation of turbulent channel flow up to $Re_\tau \approx 5200$. J. Fluid Mech. **774**, 395–415 (2015)
17. R.D. Moser, J. Kim, N.N. Mansour, Direct numerical simulation of turbulent channel flow up to $Re_\tau = 590$. Phys. Fluids **11**(4), 943–945 (1999)
18. R. Narasimha, K. Sreenivasan, Relaminarization in highly accelerated turbulent boundary layers. J. Fluid Mech. **61**(3), 417–447 (1973)
19. S. Pandey, E. Laurien, Heat transfer analysis at supercritical pressure using two layer theory. J. Supercrit. Fluids **109**, 80–86 (2016)
20. S. Pandey, X. Chu, E. Laurien, Investigation of in-tube cooling of carbon dioxide at supercritical pressure by means of direct numerical simulation. Int. J. Heat Mass Transf. **114**, 944–957 (2017)
21. S. Pandey, E. Laurien, X. Chu, A modified convective heat transfer model for heated pipe flow of supercritical carbon dioxide. Int. J. Therm. Sci. **117**, 227–238 (2017)
22. S. Pandey, X. Chu, E. Laurien, Numerical analysis of heat transfer during cooling of supercritical fluid by means of direct numerical simulation, in *High Performance Computing in Science and Engineering ' 17*, ed. by W.E. Nagel, D.H. Kröner, M.M. Resch (Springer International Publishing, Cham, 2018), pp. 241–254
23. S. Pandey, X. Chu, E. Laurien, B. Weigand, Buoyancy induced turbulence modulation in pipe flow at supercritical pressure under cooling conditions. Phys. Fluids **30**(6), 065,105 (2018)
24. K.R. Sreenivasan, Laminarescent, relaminarizing and retransitional flows. Acta Mech. **44**(1), 1–48 (1982)
25. A. Steiner, On the reverse transition of a turbulent flow under the action of buoyancy forces. J. Fluid Mech. **47**(3), 503–512 (1971)
26. A.W. Vreman, J.G.M. Kuerten, Comparison of direct numerical simulation databases of turbulent channel flow at $Re_\tau = 180$. Phys. Fluids **26**(1), 015,102 (2014)
27. H. Xu, S.M. Mughal, E.R. Gowree, C.J. Atkin, S.J. Sherwin, Destabilisation and modification of Tollmien–Schlichting disturbances by a three-dimensional surface indentation. J. Fluid Mech. **819**, 592–620 (2017)

DNS Study of the Turbulent Inflow Effects on the Fluid Dynamics and Heat Transfer of a Compressible Impinging Jet Flow

Gabriele Camerlengo and Jörn Sesterhenn

Abstract We present a DNS of a compressible impinging jet flow with Reynolds and Mach numbers of 8134 and 0.71, respectively. The jet is vertically confined between two isothermal walls and issues from a circular orifice of diameter D in the uppermost wall. The lowermost wall, placed at a distance of $5D$ from the other, serves as impingement plate. The temperature of the walls is constant and 85 K higher than the average total temperature of the jet at inlet. In order to resemble engineering configurations where the inflow will certainly not be laminar, we prescribe fully turbulent inlet conditions. To this end, the impinging jet simulation is coupled with an auxiliary fully developed turbulent pipe flow DNS. This approach circumvents the calibration issues that arise when a synthetic turbulence generator is used. Because of their relevance in cooling applications, the analysis focuses on the heat transfer at the impingement wall and its spatial distribution, whose peculiar shape is determined by the vortex dynamics in the proximity of the wall. Aiming at identifying the effects of the turbulent inflow, results are compared with those of previous computations performed with comparable configuration but laminar inflow.

1 Introduction

Impinging jets are widely used in various engineering configurations for the high heat flux they can generate at the impingement plate. They provide an indispensable cooling technique for gas turbine components, electronic parts and stock materials during material forming processes. Despite their importance and decades of research, the physical mechanism that governs the heat and mass transfer in proximity of the

G. Camerlengo (✉)
Institut für Strömungsmechanik und Technische Akustik, Technische Universität Berlin,
Müller-Breslau-Str. 15, 10623 Berlin, Germany
e-mail: gabriele.camerlengo@tu-berlin.de

J. Sesterhenn
Lehrstuhl für Technische Mechanik und Strömungsmechanik, Universität Bayreuth,
Universitätsstraße 30, 95447 Bayreuth, Germany
e-mail: joern.sesterhenn@uni-bayreuth.de

© Springer Nature Switzerland AG 2021
W. E. Nagel et al. (eds.), *High Performance Computing in Science and Engineering '19*,
https://doi.org/10.1007/978-3-030-66792-4_28

plate is yet not fully understood. This is due to the very small time and length scales involved in the phenomenon, which are often not detectable in experiments or are not at all resolved in Reynolds-averaged Navier–Stokes (RANS) or large eddy simulations (LES). Furthermore, it is still unclear how these small-scales feature of the flow are affected by parameters such as compressibility, nozzle-to-plate distance, Reynolds number, inflow conditions etc.

As summarized by Gauntner et al. [1], early experimental studies have been mainly devoted to identifying the characteristic flow regions and regimes of an impinging jet flow. These zones, each of which exhibits distinctive properties, are: the free jet region, the stagnation zone and the wall jet region. The free jet region is characterized by a shear layer that originates between the jet and the surrounding quiescent fluid. The shear layer gives rise to vortical structures, known as primary vortices, which are transported downstream while rolling up on themselves. In the stagnation zone, the flow is deflected radially, causing the primary vortex to break up. In the wall region, the flow evolves mainly radially and a boundary layer with a velocity profile similar to that observable in a wall jet flow originates. Near the wall, the shear layer between the wall jet flow and the quiescent fluid above produces new vortical structures, referred to as secondary vortices. By a LES, Hadžiabdić and Hanjalić [2] identified the vortex roll-up occurring above the impingement plate as the flow feature that influences most the fluid dynamics.

Gardon and Akfirat [3] observed that the mean heat flux distribution at the wall increases as the Reynolds number grows while exhibiting a similar qualitative shape if the nozzle-to-plate distance H is kept constant. Of particular interest has been the occurrence of a secondary peak in the heat flux distribution at a radial distance spanning between one and two jet diameters from the jet axis, in the proximity of which the primary peak is predictably observed. Among others, Jambunathan et al. [4] reported how several parameters affect the heat exchange at the impingement plate. The secondary peak is clearly visible for nozzle-to-plate distances lower than 3 diameters, whereas an inflection point is observed in the same region as H becomes larger. Several attempts of explaining the causes of such peak (or inflection point) have been made in the past decades. Wilke and Sesterhenn [5] showed that in the wall jet region primary and secondary vortices pair and produce concentric rings of alternatively high and low local heat flux travelling downstream on the wall. The area where these vortex rings originate corresponds to the high-heat-transfer area, suggesting that the second peak appears because cold fluid is transported towards the wall with the high velocity induced by the vortex pair in this particular region. Dairay et al. [6] indicate that the toroidal vortex ring structures undergo an azimuthal distortion that instantaneously drives cold fluid closer to the wall. This phenomenon occurs statistically more often in the high-heat-transfer region, giving a possible explanation to the existence of the secondary peak.

Despite the continued interest, most of the numerical studies existing in literature rely upon turbulence modelling for the closure of RANS (e.g. [7]) or LES (e.g. [8]). As previously mentioned, these approaches are inherently unable to explain the impingement heat transfer, because they model the process near the wall, which is exactly what one needs to understand prior to modelling. For instance, Dairay et al. [9] showed that even state-of-the-art large eddy simulations are still not able to satisfactorily predict the impingement heat flux. On the other hand, all previous direct numerical simulation (DNS) studies have specific shortcomings: they are either two-dimensional (e.g. [10]), consider low Reynolds numbers (e.g. [11]), exhibit an inadequate grid resolution (e.g. [12]) or do not implement turbulent inflow conditions (e.g. [5]). Certainly, the inflow in engineering configurations will not be laminar. Therefore, we perform a DNS in which we prescribe turbulent inflow conditions by coupling the impinging jet simulation with an auxiliary fully developed turbulent pipe flow DNS. When compared with synthetic turbulence generation methods as the one used by Dairay et al. [6], this procedure, which has never been used before for the simulation of an impinging jet flow, offers the advantage of not requiring any external calibration parameter and of giving a very accurate representation of all turbulence scales. To the best of the authors' knowledge, we report here on the first study addressing a compressible impinging jet flow with fully turbulent inflow conditions. We will consider, in particular, the effects of the turbulent inflow on the global flow features and on the heat transfer at the impingement plate. To this end, results will be compared with those obtained by Wilke and Sesterhenn [5], who simulated a comparable configuration with laminar inlet conditions.

2 Flow Configuration

We investigate the impinging jet flow between two horizontal flat plates. The jet issues from a straight pipe through an orifice in the uppermost wall and impinges on the lowermost one (target plate). The distance between the two plates is equal to $5D$, being D the diameter of the orifice. The plates are isothermal walls, the temperature of which is approximately 85 K higher than the average total temperature of the jet at the inlet. We consider, in particular, the flow of a compressible Newtonian fluid that obeys the ideal gas law. The characteristic Reynolds and Mach numbers are respectively defined as

$$Re = \frac{\rho_\infty U_\infty D}{\mu(T_\infty)}, \qquad M = \frac{U_\infty}{\sqrt{\gamma R T_\infty}}, \tag{1}$$

where ρ_∞, T_∞ and U_∞ are the density, temperature and velocity at the centreline of the jet inlet, R is the specific ideal gas constant and $\mu(T_\infty)$ the gas dynamic viscosity at the temperature T_∞. Prandtl number and heat capacity ratio of the gas are $Pr = \mu c_p/\lambda = 0.7$ and $\gamma = c_p/c_v = 1.4$, where c_p is the specific heat capacity at constant pressure, c_v the specific heat capacity at constant volume and λ the heat conductivity of the gas. The dynamics of the flow is described in Cartesian coordinates by the following formulation of the Navier–Stokes equations, introduced by Reiss and Sesterhenn [13]:

$$2\sqrt{\rho}\frac{\partial\sqrt{\rho}}{\partial t} + \nabla \cdot (\rho\mathbf{u}) = 0, \tag{2a}$$

$$\sqrt{\rho}\frac{\partial\sqrt{\rho}\mathbf{u}}{\partial t} + \frac{1}{2}[\nabla \cdot (\rho\mathbf{u} \otimes \mathbf{u}) + \rho\mathbf{u} \cdot \nabla\mathbf{u}] + \nabla p - \nabla \cdot \boldsymbol{\tau} = \mathbf{0}, \tag{2b}$$

$$\frac{1}{\gamma-1}\frac{\partial p}{\partial t} + \frac{\gamma}{\gamma-1}\nabla \cdot (p\mathbf{u}) - \mathbf{u} \cdot \nabla p + \mathbf{u} \cdot (\nabla \cdot \boldsymbol{\tau}) - \nabla \cdot (\boldsymbol{\tau}\mathbf{u}) - \nabla \cdot (\lambda\nabla T) = 0, \tag{2c}$$

$$\rho T - \frac{1}{R}p = 0, \tag{2d}$$

where $\mathbf{u}(\mathbf{x}, t)$ is the fluid velocity, $p(\mathbf{x}, t)$ the pressure, $\rho(\mathbf{x}, t)$ the density, $T(\mathbf{x}, t)$ the temperature,

$$\boldsymbol{\tau} = \mu\left(\nabla\mathbf{u} + \nabla\mathbf{u}^{\mathrm{T}} - \frac{2}{3}(\nabla \cdot \mathbf{u})I\right) \tag{3}$$

the viscous stress tensor and $R = 287.058\,\mathrm{J/(kg\,K)}$ the specific gas constant. The dependency of the viscosity on the temperature is taken into account by means of the Sutherland's law [14]. Accordingly, the dynamic viscosity of the fluid is given by:

$$\mu = \mu_0\left(\frac{T}{T_S}\right)^{3/2}\frac{T_S + C_S}{T + C_S}, \tag{4}$$

with coefficients $\mu_0 = 1.716 \cdot 10^{-5}\,\mathrm{kg/(m\,s)}$, $T_S = 273.15\,\mathrm{K}$ and $C_S = 110.4\,\mathrm{K}$.

Here we consider a configuration with a characteristic Reynolds number $Re = 8134$ and characteristic Mach number $M = 0.71$. This choice allows a direct comparison of the results with those of the study by Wilke and Sesterhenn [5], who addressed the same configuration, except the fact that laminar inflow conditions were used. Herein, Wilke and Sesterhenn's case is referred to as *reference case* or *laminar inflow case*, whereas the present as *turbulent inflow case*. More specifically, we consider the laminar inflow case with $Re = 8000$ and $M = 0.8$, so that Reynolds number, average total temperature at the inlet and average injected mass flux differ with the turbulent inflow case by less than 2%.

3   Numerical Methods

Direct numerical simulation (DNS) of the above-described flow configuration is performed by means of the in-house solver *NSF* developed over the years at the CFD group of the TU Berlin. The computational domain chosen for the solution of the fluid dynamics is shown in Fig. 1. It consists of two blocks: an uppermost block, which corresponds to the terminal part of the injection pipe, and the lower-most block, where the impinging jet flow is computed. The length of the uppermost block, equal to $3D$, is deemed sufficient to make the influence of the pipe outlet

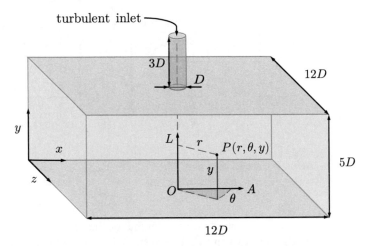

Fig. 1 Sketch of the computational domain, representing both the Cartesian and cylindrical reference systems adopted in this paper. Walls are colored in grey, whereas the injection pipe in orange. D indicates the orifice or pipe diameter

on its inlet not relevant. Both blocks are discretized by use of rectilinear grids with $1024 \times 1024 \times 1024$ points in the lower block and $144 \times 756 \times 144$ in the upper block. Equations (2a)–(2c) are discretized in space by using explicit 4th-order skew-symmetric finite differences, whereas a classical low-storage 4th-order Runge–Kutta method is adopted to advance in time. The use of a skew-symmetric scheme, in conjunction with the chosen formulation of the Navier–Stokes equations, results in a fully conservative finite-difference scheme [13].

The grid is refined in proximity of lowermost wall in order to ensure a $y^+ < 0.6$ at the first nodes above the wall, with y^+ the dimensionless wall coordinate. A grid refinement around the jet axis is also applied so that the grid spacing in x and z directions normalized with the orifice diameter D spans between 0.0099 and 0.0296. Figure 2 shows the turbulent energy spectra resulting from the present simulation. Given the axisymmetry of the geometry, it is sufficient to compute them on a xy-plane through the symmetry axis in both x and y direction. The monotonicity of the curve at large wave numbers indicates that no energy accumulation occurs at the smallest resolved scales, because the grid is sufficiently fine to dissipate it. Since identical grid refinement and resolution were used, the reader is referred to Wilke and Sesterhenn [5] for a validation of the grid in terms of Kolmogorov length scale.

As mentioned above, isothermal, non-slip boundary conditions are enforced at the walls. Inflow conditions for the upper block are discussed in Sect. 3.1. At the outlets, which laterally delimit the computing domain, non-reflecting characteristic outflow conditions are implemented. Having the uppermost block the shape a rectangular cuboid, a volume penalization method is used to obtain the cylindrical injection pipe. This method allows to apply non-slip boundary conditions at the wall of the pipe by modelling the solid as a porous media with a small permeability. In particular,

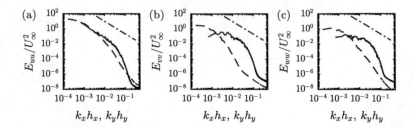

Fig. 2 Impinging jet flow simulation. Spectra of the x component (**a**), y component (**b**) and z component (**c**) of the turbulent kinetic energy normalized with U_∞ versus the dimensionless wave number in x and y direction: ———, x-direction (DNS data); — —, y-direction (DNS data); —· —, line with $-5/3$ slope

the additional forcing term acts on the entire volume of the solid an not just at its interface with the fluid. Proof of convergence to the solution of the Navier–Stokes equations has been given by Feireisl et al. [15].

3.1 Turbulent Inflow Conditions

Turbulent inflow conditions are prescribed at the inlet of the injection pipe by enforcing time-deponent flow data previously recorded from an auxiliary turbulent pipe flow DNS. The pipe flow is computed on a cylindrical structured grid with $192 \times 512 \times 4096$ points in the radial, azimuthal and axial direction. The auxiliary pipe length is equal to $18D$. Boundary conditions in axial direction are non-periodic, thus a recycling technique is implemented in order to maintain the flow in turbulent regime. This technique consists in copying velocity fluctuations from a *recycling station* downstream of the inlet of the auxiliary pipe. The recycling station is located in the auxiliary pipe at a distance of $15D$ from its inlet. This distance is deemed sufficient to avoid feedback-loop phenomena.

In order to validate the turbulent pipe DNS, the turbulent kinetic energy spectra of the turbulent kinetic energy components is calculated along the axial direction z. By looking at Fig. 3, it is possible to confirm the appropriateness of the grid resolution, because no energy accumulation at low scales can be observed. On the other side, the typical $-5/3$ slope of the inertial subrange is not wide enough to be detected because of the relatively low Reynolds number considered.

4 Computing Details

Investigating the fluid dynamics by means of direct numerical simulation requires significant computing resources, which can be solely made available by modern high performance computing centres. The direct solution of the Navier–Stokes equations

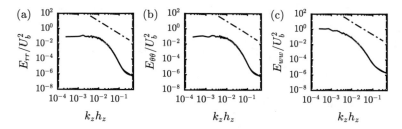

Fig. 3 Pipe flow simulation. Spectra of the radial component (**a**), azimuthal component (**b**) and axial component (**c**) of the turbulent kinetic energy normalized with the bulk velocity U_b versus the dimensionless wave number in axial direction: ——, DNS data; —·—, line with $-5/3$ slope

is inherently the only viable method to gain insights into the physical phenomena while resting assured that the observations do not stem from a flaw in the turbulence model adopted. The needful grid resolutions, apt to detect the smallest scales of turbulence, are such that millions of core-hours per computation are needed. Such simulations are therefore always parallelized on a large number of computing cores so that results can be obtained in reasonable times.

In the present case, the parallelization is achieved through a block decomposition. This technique consists in partitioning the total load between a number of computing processes, each of which operates on a fractional part of the computational domain (block). In particular, the partitioning is here realized at two different levels: the injection pipe and the domain within the two plates are main blocks, which are in turn subdivided into smaller sub-blocks. In order to compute derivatives, information from adjacent blocks (either main or sub-blocks) is needed. This information is made available with the aid of ghost points between the main blocks, or by changing the domain decomposition between the sub-blocks, so that each of them receives grid lines that span the entire (main) block in the direction along which the derivative is being computed. The required inter-process communication is implemented through Message Passing Interface (MPI). For further details regarding the domain decomposition and parallelization in NSF, refer to [16].

The NSF solver has been successfully ported and executed on the Cray XC40 (Hazelhen) supercomputer, on which scalability tests were performed for a computing configuration similar to the present one. As mentioned above, the uppermost block (injection pipe) has a grid resolution of $144 \times 756 \times 144$ points and is typically parallelized on 162 (or 324) cores. The lowermost block has a grid resolution of $1024 \times 1024 \times 1024$ points and is typically parallelized on 8192 (or 16,384) cores. In Fig. 4 strong and weak scaling plots for the lowermost block are displayed. The code performs well in both the weak and strong scaling tests, exhibiting in the latter case a nearly linear scaling. It should be noted that, although a setup with 64^3 grid points per core differs from the production setup, where 40^3 to 50^3 points per core are used, we consider the result of the weak scaling test conservative. Keeping constant the number of cores, an increment of points per core will indeed result in less

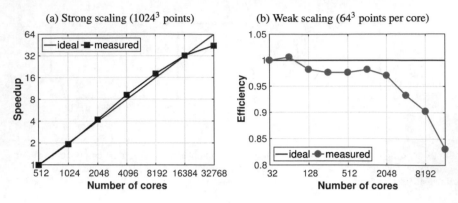

Fig. 4 Strong and weak scaling behavior of NSF on CRAY XC40 (Hazel Hen) at HLRS. The scalability tests were performed respectively on a grid with 1024^3 points and 64^3 points per core

Fig. 5 Contours on a xy-plane passing through the jet axis of the second invariant of the velocity gradient tensor Q normalized with U_∞^2/D^2. The x-axis spans from $r/D = 0$ to 4.4, whereas the y-axis form $y/D = 0$ to 5

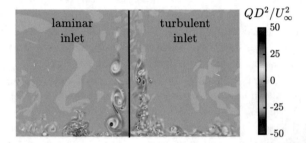

memory requirements and a lower amount of data transferred via MPI. Therefore, we expect in a production setup even more performant weak scaling results.

5 Results and Discussion

In this section, results are presented and discussed in comparison with those obtained by Wilke and Sesterhenn [5] (case with $Re = 8000$ and $M = 0.8$, see Sect. 3). As concerns the present case, statistics have been collected for a time equal to approximately $39\,t_r$, being $t_r = D/U_\infty$ the characteristic time of the simulation. Thanks to the homogeneity in θ (Fig. 1), averages are taken both in time and in the azimuthal direction. This makes it possible to compute statically converged averages in a relatively short simulated time.

   Figure 5 shows a snapshot of instantaneous Q contours on a xy-plane through the jet axis for both the laminar and turbulent inflow cases, being Q an indicator of the flow vorticity [17]. We observe that the typical Kelvin–Helmholtz structures in the shear layer region of the free jet are no longer easily discernible when a turbulent

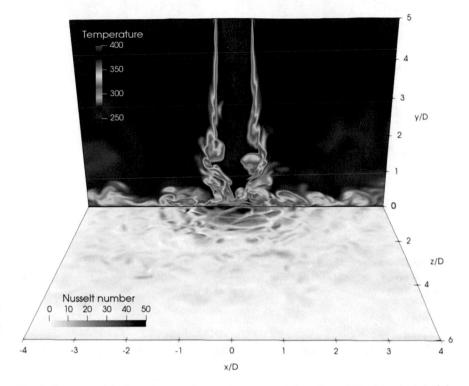

Fig. 6 Contours of the instantaneous temperature on a xy-plane through the jet axis and of the instantaneous Nusselt number on the impingement plate

inflow is used. In this case, also the wall-jet region appears more chaotic, making the vortex pairing mechanism not clearly observable.

In order to get more insight into the modified physics induced by the turbulent inflow, the instantaneous temperature contours on a xy-plane through the jet axis are plotted along with the Nusselt number at the impingement plate (Fig. 6), being the latter defined as the normalized local heat flux:

$$Nu = \frac{\dot{q}D}{\lambda \Delta T},$$ (5)

with $\dot{q}$ the heat flux, λ the thermal conductivity of the fluid and ΔT the difference between the total temperature of the jet at the inlet and the temperature of the isothermal plate.[1] It can be noted that high spots of Nu, which indicate strong heat removal from the plate, are distributed in a disorderly manner and high-heat-transfer annuli, generated by the vortex rings, are not visible.

In Fig. 7, the velocity and temperature wall boundary layers of the laminar and turbulent inflow case are shown by plotting the mean dimensionless radial velocity

---

[1] A positive Nusselt number indicates heat being transferred from the plate to the fluid.

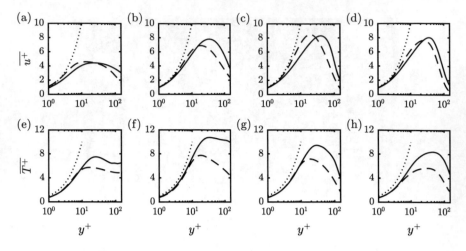

Fig. 7 Radial velocity (**a–d**) and temperature (**e–h**) boundary layer at $r/D = 0.3$ (**a, e**), $r/D = 0.8$ (**b, f**), $r/D = 1.4$ (**c, g**) and $r/D = 3.5$ (**d, h**): ——, laminar inflow; — —, turbulent inflow. As reference, the curves $u^+ = y^+$ and $T^+ = y^+$ are shown ($\cdots\cdots$)

$\overline{u^+}$ and the mean dimensionless temperature $\overline{T^+}$ as a function of the wall coordinate y^+ at different distances from the jet axis, with

$$y^+ = \frac{y\,u_\tau}{\nu}, \quad u_\tau = \sqrt{\frac{\tau_w}{\rho}}, \quad u^+ = \frac{\overline{u_r}}{u_\tau}, \tag{5a-c}$$

$$T^+ = \frac{T_w - T}{T_\tau} \quad \text{and} \quad T_\tau = \frac{\dot{q}}{\rho c_p u_\tau}, \tag{5d,e}$$

where τ_w is the wall shear stress, ρ the fluid density at the wall, ν the fluid kinematic viscosity at the wall and T_w the wall temperature. We note that in the near-wall region (for small y^+), the turbulent inflow $\overline{u^+}$ profile better approximates the law $\overline{u^+} = y^+$ at all radial distances. Moreover, the turbulent inflow velocity boundary layer shows a smaller thickness in the turbulent inflow case. The temperature boundary layers of the turbulent and laminar inflow cases match closely up to $y^+ \simeq 10$. Further away from the wall, $\overline{T^+}$ recovers quicker in the turbulent inflow case.

The wall shear stress can be analyzed by looking at the skin friction factor C_f, defined as:

$$C_f = \frac{2\mu}{\rho U_\infty^2} \frac{\partial u_r}{\partial y}. \tag{6}$$

Figure 8a shows the mean skin friction factor on the impingement plate. We observe that at $r/D = 0$, $\overline{C_f} \simeq 0$ in the turbulent inflow case, whereas $\overline{C_f} \simeq 0.01$ in the

Fig. 8 Mean skin friction coefficient $\overline{C_f}$ (**a**), mean Nusselt number $\overline{Nu}$ (**b**): ——, laminar inflow; – –, turbulent inflow

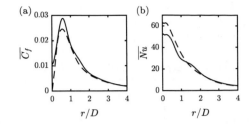

laminar inflow case. In both cases the maximum skin friction is attained at $r/D \simeq 0.6$, albeit the peak value in the laminar inflow case is approximately 20% higher than in the turbulent inflow case.

Figure 8b shows the mean Nusselt number $\overline{Nu}$ on the impingement plate. We note that the characteristic shoulder, which in the laminar inflow case appears at $r/D \simeq 1.4$, is no longer observable in the turbulent inflow case. In the latter case, the Nusselt number is approximately 20% higher in the region $r/D \lesssim 1.2$, whereas the laminar inlet case features a 5% larger heat flux at $r/D \simeq 1.8$. The heat flow rate in the region $r/D < 4$ expressed in terms of average Nusselt number is approximately equal to 12.55 in both laminar and turbulent inflow cases. This is due to the fact that the region where the turbulent inflow jet has a higher $\overline{Nu}$, being located at a small radial distance r, is about 2.5 times smaller than the area where the turbulent inflow $\overline{Nu}$ is lower than in the laminar inflow case.

7 Conclusions

Direct numerical simulation of a compressible impinging jet with fully turbulent inflow conditions has been performed at $Re = 8134$ and $M = 0.71$. Results were compared with those reported by Wilke and Sesterhenn [5] who analyzed the same configuration by using laminar inlet conditions.

We found that the use of a turbulent inflow hinders the formation of the Kelvin–Helmholtz instabilities in the free-jet region and hence prevents the formation of vortex rings at the wall, which were found responsible for producing high local heat flux at the wall [18]. As a consequence, the shoulder in the mean heat flux profile that was found for the laminar inflow case in the region where the vortex rings form is no longer observable. Even though the local heat flux is sensibly higher in the turbulent inflow case up to $r/D \simeq 1.2$, both jets surprisingly provide about the same heat transfer, because the region where the turbulent inflow jet has much better cooling rate is much smaller, too. Nonetheless, the local changes in $\overline{Nu}$ indicate that for the correct design of an impingement cooling device, attention shall be paid to the inflow conditions.

In order to further clarify these aspects, higher order statistics are object of current research. This will give a complete understanding of the physics involved, required to improve the current heat-transfer models.

Acknowledgements All simulations were performed on the Cray XC40 (Hazelhen) supercomputer at the High Performance Computing Center Stuttgart (HLRS) under the grant number Jet-Cool/44127. The authors gratefully acknowledge support by the Deutsche Forschungsgemeinschaft (DFG) as part of collaborative research center SFB 1029 "Substantial efficiency increase in gas turbines through direct use of coupled unsteady combustion and flow dynamics" on project B04.

References

1. J.W. Gauntner, P. Hrycak, J. Livingood, Survey of literature on flow characteristics of a single turbulent jet impinging on a flat plate. NASA Technical Report TN D-5652 (1970)
2. M. Hadžiabdić, K. Hanjalić, Vortical structures and heat transfer in a round impinging jet. J. Fluid Mech. **596**, 221–260 (2008)
3. R. Gardon, J.C. Akfirat, The role of turbulence in determining the heat-transfer characteristics of impinging jets. Int. J. Heat Mass Transf. **8**(10), 1261–1272 (1965)
4. K. Jambunathan, E. Lai, M. Moss, B. Button, A review of heat transfer data for single circular jet impingement. Int. J. Heat Fluid Flow **13**(2), 106–115 (1992)
5. R. Wilke, J. Sesterhenn, Statistics of fully turbulent impinging jets. J. Fluid Mech. **825**, 795–824 (2017)
6. T. Dairay, V. Fortuné, E. Lamballais, L.E. Brizzi, Direct numerical simulation of a turbulent jet impinging on a heated wall. J. Fluid Mech. **764**, 362–394 (2015)
7. N. Zuckerman, N. Lior, Impingement heat transfer: correlations and numerical modeling. J. Heat Transf. **127**(5), 544–552 (2005)
8. N. Uddin, S.O. Neumann, B. Weigand, Les simulations of an impinging jet: on the origin of the second peak in the Nusselt number distribution. Int. J. Heat Mass Transf. **57**(1), 356–368 (2013)
9. T. Dairay, V. Fortuné, E. Lamballais, L. Brizzi, LES of a turbulent jet impinging on a heated wall using high-order numerical schemes. Int. J. Heat Fluid Flow **50**, 177–187 (2014)
10. Y.M. Chung, K.H. Luo, Unsteady heat transfer analysis of an impinging jet. J. Heat Transf. **124**(6), 1039–1048 (2002)
11. M. Tsubokura, T. Kobayashi, N. Taniguchi, W. Jones, A numerical study on the eddy structures of impinging jets excited at the inlet. Int. J. Heat Fluid Flow **24**(4), 500–511 (2003)
12. H. Hattori, Y. Nagano, Direct numerical simulation of turbulent heat transfer in plane impinging jet. Int. J. Heat Fluid Flow **25**(5), 749–758 (2004)
13. J. Reiss, J. Sesterhenn, A conservative, skew-symmetric finite difference scheme for the compressible Navier–Stokes equations. Comput. Fluids **101**, 208–219 (2014)
14. W. Sutherland, The viscosity of gases and molecular force. Lond. Edinb. Dublin Philos. Mag. J. Sci. **36**(223), 507–531 (1893)
15. E. Feireisl, J. Neustupa, J. Stebel, Convergence of a Brinkman-type penalization for compressible fluid flows. J. Differ. Equ. **250**(1), 596–606 (2011)
16. L. Stein, J.: Sesterhenn, Direct numerical simulation of turbulent flow past an acoustic cavity resonator, in *High Performance Computing in Science and Engineering '18*, ed. by W.E. Nagel, D.H. Kröner, M.M. Resch (Springer, Berlin, 2019). https://doi.org/10.1007/978-3-030-13325-2_16
17. J. Jeong, F. Hussain, On the identification of a vortex. J. Fluid Mech. **285**, 69–94 (1995)
18. R. Wilke, J. Sesterhenn, Numerical simulation of subsonic and supersonic impinging jets II, in *High Performance Computing in Science and Engineering '16* (Springer, Berlin, 2016), pp. 425–441

Numerical Simulation of the FNG Wing Section in Turbulent Inflow

Jens Müller, Maximilian Ehrle, Thorsten Lutz, and Ewald Krämer

Abstract The influence of atmospheric turbulence on an extruded airfoil of the FNG wing in clean configuration is investigated using numerical simulation. Turbulence is injected into the flow field using a momentum source term. It is shown that the turbulence can be propagated accurately to the airfoil. Spectra of the pressure coefficient at different chordwise positions indicate a correlation between the inflow velocity spectrum and the local c_p spectra, especially for low to medium wave numbers. Furthermore, the applicability of the simplified Disturbance Velocity Approach (DVA) is evaluated, where the velocities of the atmospheric turbulence are added to the flux balance using superposition. The DVA shows satisfying results for the lift spectrum and the c_p spectrum at the leading edge over a broad wave number range. An overestimation of the amplitudes for the pitching moment and c_p spectra at $x/c = 0.2$ occurs at medium to high wave numbers. A scaling test of the TAU code in a development version with the implemented DVA is performed on this test case and shows satisfying scalability.

1 Introduction

Especially when flying at low altitudes, aircraft interact with atmospheric turbulence involving a broad wave number range. This interaction affects aerodynamic performance, passenger comfort, and leads to dynamic loads on the aircraft structure. According to the "Certification Specifications for Large Aeroplanes" CS-25 [2] gust as well as turbulence loads have to be taken into account in the aircraft design process. Hence, in the early phases of design it is necessary to understand and evaluate the effects of atmospheric turbulence on aircraft. Experimental investigations with grid generated turbulence were carried out, among others, by Hoffmann [6] and Hancock and Bradshaw [3]. These investigations show that free stream turbulence increases the maximum lift coefficient and also affects the level of turbulence kinetic

J. Müller (✉) · M. Ehrle · T. Lutz · E. Krämer
Institute of Aerodynamics and Gas Dynamics (IAG), University of Stuttgart, Pfaffenwaldring 21, 70569 Stuttgart, Germany
e-mail: jens.mueller@iag.uni-stuttgart.de

© Springer Nature Switzerland AG 2021   435
W. E. Nagel et al. (eds.), *High Performance Computing in Science and Engineering '19*,
https://doi.org/10.1007/978-3-030-66792-4_29

energy in the boundary layer. Numerical simulations of statically and dynamically disturbed inflow were carried out, for example, by Wawrzinek et al. [15]. Resolving atmospheric disturbances in numerical simulations requires high resolution of the computational grid and therefore high computational resources. However, an industrial application, for instance, in the aircraft certification process, requires a fast, cost-effective calculation. Therefore, Heinrich and Reimer [4] implemented a simplified method for simulating atmospheric disturbances, the so-called Disturbance Velocity Approach (DVA), into the flow solver TAU. Atmospheric disturbances are not propagated inside the flow field but added to the flux balance by superposition. This method has been successfully used for interactions of wake vortices and $1 - \cos$ gusts for gust wavelengths larger than twice the chord length c, see [4] and [5]. In the present test case the unsteady aerodynamic response of an extruded airfoil encountering broadband atmospheric turbulence is investigated. A reference case where the turbulence is resolved in the flow field is analyzed and compared to the simplified DVA.

2   Numerical Methods

Two different methods for representing atmospheric disturbances in CFD simulations are used within this work. As a reference case the atmospheric turbulence is fed into the flow field and then resolved within the discretized flow field. Consistent with Heinrich and Reimer [5] this physically correct method is called the Resolved Atmosphere Approach (RAA). The RAA simulations are carried out using the block structured FLOWer code developed by the German Aerospace Center (DLR) [9]. In recent years the spatial accuracy of FLOWer was enhanced by IAG from second to fifth order by implementation of a WENO scheme [13]. The high spatial accuracy allows for an accurate propagation and conservation of resolved atmospheric turbulence inside the CFD simulation necessary for the RAA. The DVA simulation is carried out using the unstructured flow solver TAU [12] since this approach is currently not available in FLOWer.

2.1   Resolved Atmosphere Approach

Simulating gust interaction, Heinrich and Reimer [4] use an unsteady far field boundary to feed the gust velocities into the flow field. A moving Chimera gust transport grid is applied in order to avoid numerical losses during gust propagation. While this approach is suitable for limited discrete signals, it is not applicable to broadband atmospheric turbulence. For statistical evaluations of the wing section's response to atmospheric turbulence, a large turbulence field is required. Hence, the Chimera transport grid cannot be applied and the turbulence has to be propagated from the far field domain to the airfoil. Since this distance is usually between $50c$ and $100c$, significant numerical losses occur during the propagation. With the spatial second order of TAU this approach is not applicable to simulations with broadband atmospheric turbulence.

To solve this issue the turbulence is fed into the flow field using a momentum source term instead of prescribing unsteady velocities at the far field. An essential advantage of this method is that the turbulence can be fed in at any position within the CFD grid. Hence, the turbulence injection plane can be shifted close to the airfoil, significantly reducing the amount of grid cells as well as numerical losses. This method has been implemented in the FLOWer code and successfully used in wind energy simulations, such as wind turbines with atmospheric turbulence on a complex terrain [11]. The velocities of the atmospheric turbulence field $\mathbf{u}'$ are transformed into a volume force $\mathbf{f}_s$, which is needed to accelerate the mean velocity field $\bar{\mathbf{u}}$ to $\bar{\mathbf{u}} + \mathbf{u}'$. The source term is applied to a plane perpendicular to the inflow. In FLOWer the source term formulation

$$\mathbf{f}_s = \frac{\rho \mathbf{u}'}{\Delta x} \left(\bar{u}_n + \frac{1}{2} u'_n \right) \tag{1}$$

of Troldborg [14] is applied where the velocities with the index n and the grid spacing Δx are perpendicular to the turbulent inflow plane. Since the atmospheric turbulence is fed in and propagated in the flow field, both the interactions of turbulent fluctuations with the airfoil as well as the interactions of the flow around the airfoil with the turbulent inflow are considered.

2.2 Disturbance Velocity Approach

The DVA is a simplified approach for the representation of atmospheric disturbances in CFD simulations. Unlike the RAA the atmospheric turbulence is not propagated in the flow domain. The disturbance velocities from the atmospheric turbulence $\mathbf{u}'$ are added to the flux balance of each cell while the convection across the cell interfaces changes from $\mathbf{u} - \mathbf{u}_b$ to $\mathbf{u} - \mathbf{u}_b - \mathbf{u}'$, which deactivates the convection of the disturbance velocities in the flow field. According to Heinrich and Reimer [4] the continuity equation can be written as

$$\frac{d}{dt} \int_V \rho \, dV + \oint_S \rho (\mathbf{u} - \mathbf{u}_b - \mathbf{u}') \cdot \mathbf{n} \, dS = 0 \,. \tag{2}$$

For a detailed description of the DVA, see [4] and [5]. Since the turbulent fluctuations are not convected across the cell interfaces, no refinement of the grid compared to standard CFD simulations is needed, which leads to a significant reduction in computational costs. The main drawback is that the influence of the flow around the aircraft on the turbulence field is not considered since the turbulent fluctuations remain unchanged during the simulation. In gust simulations this leads to inaccuracies in the DVA results for length scales smaller than twice the chord length [4], which is also expected for simulations with atmospheric turbulence.

2.3 Inflow Turbulence Generation

The atmospheric turbulence is created prior to the numerical simulations based on the model proposed by Mann [8]. The second order statistics of the velocity tensor are modeled using Taylor's frozen turbulence hypothesis. The model is based on the von Kármán isotropic spectrum in combination with the rapid distortion theory. Using an inverse Fourier transform the modeled atmospheric turbulence spectrum is transformed into a three dimensional velocity field. This velocity field is then passed to the flow solvers FLOWer and TAU and fed into the simulation using RAA and DVA respectively.

3 Setup

The wing of the "Flugzeug nächster Generation" (FNG, in English "next generation aircraft") [1] in clean configuration is used as a reference configuration within this work as a representative for today's conventional transport aircraft. The aspect ratio of the wing is $\lambda = 9.1441$ with a mean aerodynamic chord of $l_\mu = 5.15m$. The reference inflow condition for the current investigation is subsonic flow at $M = 0.25$ and $Re = 11.6 \times 10^6$ with respect to the mean aerodynamic chord of the FNG wing. The angle of attack is $\alpha = 4°$. In order to reduce the aerodynamic complexity an airfoil section at $\eta = 0.7$ is chosen for the current investigations. The airfoil section in the line of flight is transformed based on the local quarter-chord sweep angle. The inflow data and angle of attack are transformed accordingly. In the spanwise direction the airfoil section is extruded by $4l_\mu$. The extruded section of the FNG wing has a chord length of $c = 2.7m$ and is shown in Fig. 1. Isotropic atmospheric inflow turbulence based on the von Kármán spectrum is generated using the Mann model described in Sect. 2.3. A length scale $L = 5.15m$ with 5% turbulence intensity relative to the inflow velocity is chosen, where L is the length scale of the von Kármán energy spectrum [8]. The length scale is selected to lie within the valid range of the DVA for gust interactions with the smaller scales of the turbulent spectrum covering the critical range of the DVA to be accurate. This enables a reliable statement with regard to the applicability of the DVA for simulations with broadband atmospheric turbulence. The atmospheric turbulence field completely covers the extruded airfoil in the spanwise direction and is extended $\pm 3L$ above and below the airfoil. In the RAA simulation the turbulence is fed in at a distance of $5L$ upstream from the leading edge. This distance corresponds to the starting point of the turbulence for the DVA simulation.

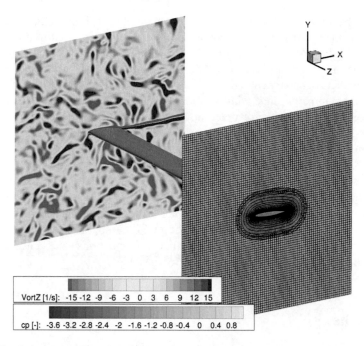

Fig. 1 Simulation setup: Extruded airfoil section with the corresponding grid used for RAA and DVA simulations. Every second grid point is shown in the figure. The contour plot shows the c_p distribution of the airfoil and the z-vorticity of the RAA in the flow field

A structured O-mesh was created around the airfoil with 144 cells along the upper and lower sides of the airfoil and 32 cells along the blunt trailing edge. The spanwise direction is resolved by 256 cells of constant grid spacing $\frac{L}{\Delta z} = 64$. $y^+ \leq 1$ is ensured in the first cell of the boundary layer. The mesh of the airfoil section is identical for FLOWer and TAU and integrated into a cartesian background mesh using the Chimera technique. The Chimera technique is also applied for the representation of the angle of attack as shown in Fig. 1. For the RAA simulations in FLOWer a fine grid resolution is needed in the turbulence propagation region to avoid significant numerical losses. Hence, a cartesian grid with $\frac{L}{\Delta x} = \frac{L}{\Delta y} = \frac{L}{\Delta z} = 64$ was created ranging from $5.25L$ upstream from the airfoil's leading edge to $5.75L$ downstream from the leading edge. In the y-direction the fine grid covers the complete field of turbulence. Further away from the airfoil the grid is coarsened using hanging grid nodes in order to save computational resources. To avoid any grid based numerical differences, the TAU background grid is equivalent to the FLOWer background grid in the vicinity of the airfoil section. Since the DVA does not require a highly resolved mesh in the propagation region, the high-resolution cartesian grid only covers the region between $2L$ upstream and $3L$ downstream from the leading edge in the TAU simulation. The grid is then slightly coarsened using unstructured tetrahedra to save mesh points. The far field distance for both FLOWer and TAU is $100c$. This results in about 150 million cells for the RAA and about 113 million cells for the DVA simulation. In FLOWer, periodic boundary conditions are applied in the spanwise direction. In the TAU release 2014.2.0 used within this work, the combination of periodic boundary conditions and the Chimera overset grid is not possible. Since the symmetry plane boundary condition led to singularities at the connection of the wing section and symmetry plane with the DVA, the Euler wall boundary conditions is applied in TAU in the spanwise direction. The error is expected to be negligible when evaluating the results at a sufficient distance from the spanwise boundaries. For both approaches unsteady RANS simulations are performed and the Spalart-Allmaras turbulence model is applied since the angle of attack is chosen to be out of the range in which separation occurs. A fully turbulent boundary layer is assumed. Apart from the deviations that cannot be eliminated like the different grid metrics, the numerical parameters of FLOWer and TAU are chosen to be consistent. The physical time step size of $2.37 \cdot 10^{-4} s$ corresponds to a local CFL number $CFL_l = \frac{u_\infty \Delta t}{\Delta x} \approx 0.22$, ensuring an accurate propagation of the turbulent fluctuations. The convective time cycle of the wing section corresponds to 150 time steps.

4 Results

Prior to the simulations of the FNG airfoil section in atmospheric turbulence the accurate propagation of the resolved turbulence from the turbulent inflow plane to the airfoil's position has to be analyzed for the RAA. The results of this preliminary investigation are presented in Sect. 4.1. The effect of the inflow turbulence on the pressure spectra at the airfoil surface is analyzed in Sect. 4.2. Finally, the results of the RAA and DVA are compared with regard to the applicability of the DVA.

4.1 Propagation of Inflow Turbulence Using the RAA

The accurate propagation of atmospheric turbulence in FLOWer using the Mann model in combination with a WENO scheme and the momentum source term described in Sect. 2.1 has been proven by Kim et al. [7] and Schulz [10] for low Mach number flows typical for wind turbine simulations. The inflow Mach number considered in this work is significantly higher than it is in the work of Kim et al. or Schulz. Therefore, the propagation of the inflow turbulence for the present test case is analyzed for the highly resolved region of the cartesian mesh presented in Sect. 3 with a grid resolution of $\frac{L}{\Delta x} = \frac{L}{\Delta y} = \frac{L}{\Delta z} = 64$ without the airfoil present. The statistics of the propagated turbulence are evaluated at a distance $5L$ downstream from the turbulent inflow plane, where the leading edge is located in the main simulation with the airfoil. Figure 2 shows the power spectral density (PSD) of u, v, and w velocities at a distance $5L$ downstream from the inflow plane compared to the u velocity PSD of the Mann box, which is the turbulent data fed into the simulation. The v and w

Fig. 2 PSD of u, v, and w velocities at a distance $5L$ downstream from the turbulent inflow plane compared to the u velocity PSD of the Mann box and the Kolmogorov $-5/3$ slope

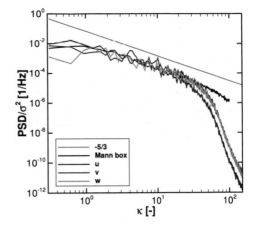

Table 1 Degree of isotropy, turbulence kinetic energy, and turbulence intensity for the turbulent input data (Mann box) and FLOWer (RAA) without the airfoil at a distance 5L downstream from the turbulent inflow plane

	σ_u/σ_v	σ_u/σ_w	$k[m^2/s^2]$	$TI[\%]$
Mann box	1.004	1.001	21.357	5.00
FLOWer, RAA 5L downstream	0.993	1.015	17.387	4.51

velocity spectra for the Mann box are not shown here since the difference to the u velocity spectrum is negligible for isotropic turbulence. The Kolmogorov $-5/3$ slope is given as a reference for the inertial sub-range. The wave number κ is normalized with the turbulent length scale L according to Kim et al. [7] with $\kappa = \frac{2\pi f L}{u_\infty}$.

Figure 2 shows that the energy spectrum of the Mann box fed into the FLOWer simulation is conserved during the propagation for a broad wave number range for all three velocity components. The Mann box and the propagated velocity spectra follow the Kolmogorov $-5/3$ slope. Hence, the inertial sub-range is captured in the inflow data as well as in the propagated turbulence. At higher wave numbers the propagated spectra drop since the chosen grid resolution is no longer able to conserve and resolve these small scales. The drop occurs at $\kappa = 32$ for v and w and at $\kappa = 28$ for the u velocity. In addition to the energy spectra, the propagated turbulence also conserves the isotropy of the flow as shown in Table 1. Consistent with Kim's observations for low Mach number flows [7], the turbulence kinetic energy (TKE) k decays with increasing propagation distances. Thus, the target turbulence intensity (TI) of 5% specified in Sect. 3, which corresponds to the TKE $k = 21.357 \frac{m^2}{s^2}$ of the Mann box, is not reached at the airfoil's leading edge. This effect is compensated for the RAA simulation with the airfoil section by scaling the input data for the RAA simulation with a scaling factor $SF = \sqrt{k_0/k} = 1.1$ as proposed by Kim et al. [7]. k_0 is the TKE of the Mann box and k the TKE of the propagated turbulence at a distance

$5L$ downstream from the turbulent inflow plane, where the airfoil is located in the subsequent simulations. With the scaling for the RAA it is ensured that the wing section interacts with turbulence of the same turbulence intensity for both the RAA and DVA simulations.

It can be concluded that the statistics of the atmospheric turbulence can be propagated correctly to the airfoil. The isotropy as well as the energy spectra are conserved and the loss in turbulence kinetic energy is offset by scaling the input data for the RAA simulation. This allows an accurate analysis of the wing section in turbulent inflow with the RAA and a reliable comparison to the DVA simulation.

4.2   Wing Section in Turbulent Inflow

Having ensured the accurate propagation of the turbulence to the airfoil for the RAA simulation, the wing section is included in the simulation. The evaluation of the results in the following sections start after about six turbulent length scales have passed over the airfoil. This ensures that the initial interaction of the wing section and turbulence is not taken into account for the statistical evaluations. The statistics are averaged over 10400 time steps for both the RAA and DVA simulations. All statistics are evaluated in the spanwise center of the wing section as the statistical results do not change in the spanwise direction. Additionally, it is ensured that the influence of the different boundary conditions can be neglected.

In the present simulations, the reference length scale of the inflow turbulence of $L = 5.15m$ is about twice as large as the chord length of the wing section. Accordingly, the main effect expected for the airfoil is a variation of the local angle of attack. However, since broadband turbulence is interacting with the airfoil, smaller scales can have a significant influence as well. The time averaged pressure coefficient with corresponding standard deviations is shown in Fig. 3. It is compared to a steady solution without turbulence to assess relevant chordwise positions for further investigations. The result of the DVA is shown because there is no damping of high frequencies of the inflow turbulence using this approach.

The mean c_p distribution does not differ significantly from the steady solution without inflow turbulence. There is a slight reduction of the suction peak on the upper side of the airfoil. The differences diminish downstream until there is no significant deviation at $x/c = 0.5$. On the lower side of the airfoil the differences between the mean DVA result and the steady solution are considerably smaller. Looking at the standard deviations of the DVA c_p distribution the main effect of the interaction with atmospheric turbulence occurs at the leading edge and the suction peak, respectively. On the lower side of the airfoil and downstream from $x/c \approx 0.3$ the standard deviations are small so that only small scales are likely to play a role here, which can affect local separation bubbles and the laminar turbulent transition. Since the chosen angle of attack is below values at which separation occurs and the wing section is assumed to be fully turbulent, this is not investigated further within

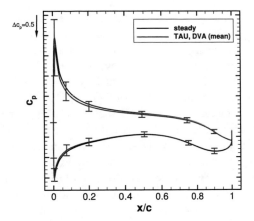

Fig. 3 Time averaged c_p distribution using DVA at $\eta = 0.5$ with corresponding standard deviations (blue) compared to a steady solution without atmospheric turbulence

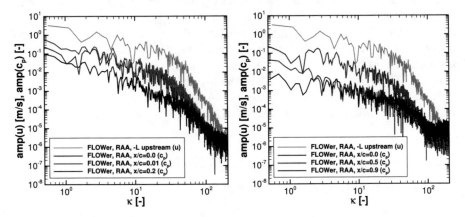

Fig. 4 FFT of the RAA c_p signal at different chordwise positions compared to the u velocity spectrum at a distance L upstream from the leading edge

this work. The impact on global loads at this part of the airfoil is expected to be small.

Based on the results presented in Fig. 3 the pressure coefficient of the RAA simulation at different chordwise positions is analyzed using a Fast Fourier Transform (FFT). Only the upper side of the airfoil is considered. The c_p spectra are compared to the u velocity spectrum at a distance L upstream from the leading edge. The RAA simulation represents the reference simulation for the later examination of the applicability of the DVA. On the left side of Fig. 4 the c_p spectra at $x/c = 0.0$, $x/c = 0.01$, and $x/c = 0.2$ are shown, where the first two positions correspond to the leading edge and the suction peak, respectively. The right side of Fig. 4 shows the c_p spectra at $x/c = 0.5$ and $x/c = 0.9$ compared to the spectrum at the leading edge and the inflow velocity spectrum.

In general, the slope of the spectra at the leading edge, suction peak, and 20% chord follow the slope of the u velocity spectrum upstream from the airfoil for wave numbers $\kappa \leq 12$. This indicates a correlation between the spectrum of the turbulent inflow and the local pressure at the region where the highest standard deviations from the mean pressure coefficient occur, which is the most relevant region for the evaluation of global loads acting on the configuration. For $\kappa > 12$ the gradient of the c_p amplitudes on the leading edge and suction peak is increased compared to the velocity amplitude, resulting in a higher reduction of the c_p amplitudes for these wave lengths. The smaller scales of the atmospheric turbulence are damped in this region due to the high velocity gradients of the flow around the leading edge. In the wave number range $12 \leq \kappa \leq 18$ the amplitude at $x/c = 0.2$ is reduced compared to the inflow velocity and the c_p at the leading edge and the suction peak. For higher wave numbers the slope again corresponds to that of the inflow up to $\kappa \approx 48$. This indicates a change of the medium to small scales corresponding with $12 \leq \kappa \leq 18$ at this position. There is almost no difference in the magnitude of the amplitudes at $x/c = 0.0$ and $x/c = 0.01$, whereas the magnitude of the amplitudes further downstream is reduced significantly. This corresponds with lower c_p fluctuations on the airfoil at the downstream positions. Looking at the c_p spectra at $x/c = 0.5$ and $x/c = 0.9$ on the right side of Fig. 4 the global amplitudes are further reduced, which results in lower c_p fluctuations in the pressure distribution. The amplitude reduction is more significant for the larger scales than for higher wave numbers. For wave numbers $\kappa > 32$ the differences in the c_p amplitudes are related to the boundary layer since the inflow turbulence is not resolved for higher wave numbers as shown in Sect. 4.1.

4.3 Applicability of the DVA

While the basic influence of broadband atmospheric inflow turbulence on the wing section was analyzed in Sect. 4.2 the question arises to what extent the simplified DVA can represent the physical processes. In contrast to the RAA, where the impact of the airfoil on the development of the incoming atmospheric turbulence is covered, the DVA only captures the influence of the atmospheric turbulence on the wing section. Based on the analysis of the c_p fluctuations with the RAA, only the chordwise positions with the most significant changes, $x/c = 0.0$ and $x/c = 0.2$ are analyzed within this section. The c_p spectra at both positions are shown in Fig. 5 for the RAA and the DVA. Since the atmospheric turbulence interpolated on the cells using the DVA is the velocity field from the Mann box, the u velocity spectrum of the Mann box is given as reference in addition to the u velocity spectrum at a distance L upstream from the leading edge for the RAA simulation. The maximum wave number $\kappa = 32$ where the velocity spectra of the atmospheric turbulence can be conserved in the RAA simulation is highlighted in the plots.

At $x/c = 0.0$ the DVA c_p spectrum corresponds to the c_p spectrum of the RAA simulation up to $\kappa \approx 32$. For higher wave numbers the amplitudes of the DVA spec-

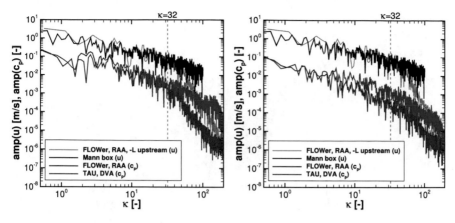

Fig. 5 FFT of the RAA and DVA c_p signal at different chordwise positions compared to the u velocity spectrum at a distance L upstream from the leading edge (RAA) and the one of the Mann box (DVA). left: $x/c = 0.0$, right: $x/c = 0.2$

trum are overestimated compared to the RAA. Since the velocity spectra of the atmospheric turbulence using the RAA can only be resolved up to $\kappa \approx 32$, as shown in Fig. 2, this is not a physical error of the DVA. For comparison at higher wave numbers the grid resolution in the RAA simulation has to be increased. The good agreement between RAA and DVA indicates that the feedback of the wing section's aerodynamics on the inflow turbulence is of minor importance at the leading edge. Hence, no significant change of atmospheric turbulence structures occurs at $x/c = 0$. Comparing the c_p spectrum of the DVA with the u velocity spectrum of the Mann box, it should be noted that the shapes correspond over the complete wave number range of the Mann box signal.

Further downstream the agreement between DVA and RAA decreases, as shown in the right part of Fig. 5. The c_p spectra at low wave numbers, i.e., the influence of the large scales, are well predicted by the DVA. However, for wave numbers $\kappa > 7$ the c_p amplitude is overestimated by the DVA. This indicates significant changes in the smaller scales of the inflow turbulence not covered by the DVA. The boundary layer thickness increases with increasing distance from the leading edge. Within the RAA simulation, the atmospheric turbulence interacts with the flow around the airfoil. This interaction influences the smaller scales of the inflow turbulence resulting in an amplitude reduction in the FFT. Since the inflow turbulence remains unchanged within the DVA, the effect of these scales on the pressure distribution is overestimated in comparison to the RAA.

The question arises how the overestimation of the c_p amplitudes with increasing x/c affects the global loads acting on the wing section. Therefore, the FFT of the lift and pitching moment history is plotted in Fig. 6. Taking the spectra of the lift signal, the DVA and RAA simulation match up to $\kappa \approx 25$. For higher wave numbers the amplitude of the lift spectrum is overestimated by the DVA compared to the RAA. As shown in Fig. 3 the major changes in the pressure coefficient, and thus also the

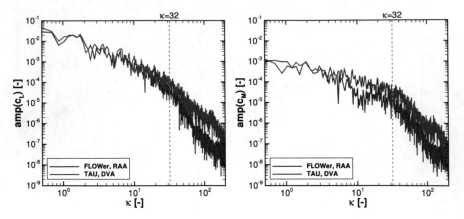

Fig. 6 FFT of the global loads for RAA and DVA. left: lift spectrum, right: pitching moment spectrum

main influence on the lift, occur at the area around the leading edge and the suction peak. At these positions, the DVA is capable of reproducing the RAA c_p spectrum up to $\kappa \approx 32$. Hence, the main influence on the lift is covered by the DVA. The c_p distribution of the positions further downstream, where the DVA lacks of agreement with the RAA, are of minor importance for the lift.

A different situation occurs when looking at the pitching moment around the quarter chord. Here, the amplitudes of RAA and DVA differ starting from $\kappa \approx 5$. While the c_p spectra of the DVA match the spectra of the RAA up to $\kappa \approx 32$ at the leading edge, the maximum wave number where the DVA c_p spectra correspond to the ones of the RAA decreases with increasing x/c as shown in Fig. 5. Due to the distance to the quarter chord the rear part of the airfoil is of significant influence when evaluating the pitching moment. The overestimation of the c_p amplitudes even for small wave numbers at this part of the airfoil directly influences the spectrum of the pitching moment. The larger distance to the quarter chord compensates the lower magnitude of the c_p amplitude compared to the leading edge. Hence, an accurate estimation of the RAA pitching moment with the DVA is only possible for small wave numbers.

5   Scaling Test

Improvements in runtime are mandatory in terms of effective use of valuable computational resources and enable more research simulations. This requires thorough testing of the scaling capabilities of the TAU code version 2014.2.0 with DVA implementation. It aims at investigating shortcomings of the implementation of the DVA routines, which are currently under development, as well as the overall performance of the code. In order to collect statistically relevant data for this test case, about

Table 2 Scaling test cases and according parameters

Case	Nodes	Cores	Points/Domain
1	9	216	163,460
2	18	432	81,730
3	37	888	39,760
4	74	1,776	19,880
5	148	3,552	9,940
6	296	7,104	4,970

8,800 node hours are required. This estimation is based on the assumption of 13,550 time steps on 37 nodes of the Cray XC-40 "Hazelhen". An effective use of computational resources requires a CFD code which scales well in terms of increasing number of nodes and thus an appropriate number of nodes for the present test case has to be chosen. For the investigation of the scaling behavior, the computational grid is decomposed in varying numbers of partitions. The simulations are performed accordingly with different node numbers resulting in a strong scaling test. Because the scaling test was conducted during the *Optimization of Scaling and Node-level Performance on Hazel Hen (Cray XC40)*, further performance data of the TAU code for this specific test case could be gathered and will be analyzed. This analysis is performed with *performance tools lite* by Cray and comprises a code sampling approach which only leads to minor influence on runtime. This enables a more in-depth analysis of code specific behavior. Thus, the scaling capabilities of each code routine can be analyzed separately.

The scaling test in the present work is performed on the DVA test case, which includes a hybrid grid with approx. 35 million grid points. Production runs presented in Sect. 4 are performed on 37 compute nodes with 888 cores which leads to approximately 40,000 grid points per domain. In order to investigate this code's scaling behavior, the node count is reduced to 18 and 9 nodes as well as extended to 74, 148 and 296 nodes respectively. Table 2 shows the scaling test cases with node numbers, core numbers and grid points per domain.

Figure 7 shows the influence of node number on the overall runtime of the test case. An increase from 9 nodes until 74 nodes leads to a nearly linear decrease of runtime. In comparison to previous TAU scaling tests, e.g., Wawrzinek et al. [15], which showed an almost linear scaling for the whole investigated node number range, a further increase of nodes decreases the advantage in runtime in the 148 nodes case and finally leads to an increase in overall runtime in case of 296 nodes. The difference of these results could be due to different versions of the TAU code as well as due to the not yet optimized implementation of the DVA, which is currently under development. This issue will be further investigated by means of a scaling analysis of the most time consuming TAU routines.

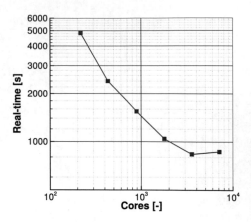

Fig. 7 Scaling of the present computational setup using TAU 2014.2.0 with implemented DVA. Scaling test cases according to Table 2. 15 physical time steps are used for the scaling test

5.1 Scaling of TAU Routines

Figure 8 shows the scaling behavior of the 9 most time-consuming routines of the TAU-Code by means of the percentage and absolute number of overall samples. The sampling approach of Cray's perftools lite samples the program counter at a time interval and delivers an absolute number of samples and the part of the overall runtime of each routine. The first two routines are attributed to MPI communication, routine 3 to 8 are functions implemented in the TAU solver, routine 9 is part of the DVA implementation and reads the turbulence data files. The first, most time consuming routine, shows a decrease in absolute samples and thus scales well in terms of an increasing number of nodes. The second routine does not scale well in terms of run time reduction when increasing the number of nodes. An increase of nodes leads to increasing percentage of samples on the overall computational time and an almost constant number of absolute samples between 18 and 296 nodes. The scaling behavior of these routines can be attributed to MPI communication, as an increase of nodes leads to an increase of data exchange between the different processes. Routines 3 to 8, which can be attributed to the TAU solver (e.g., calculation of fluxes and gradients) scale very well, which means less samples with increasing number of nodes for all test cases. The last routine shown in Fig. 8 is attributed to reading turbulence data for the DVA and therefore not part of the standard TAU implementation. It is clearly visible that the total number of samples stays constant with increasing node number, which is due to constant reading speed of the ASCII files containing turbulence data. This observation leads to a possible explanation of the overall scaling of this test case. While most of the standard routines of TAU show good scaling behavior, the part of reading gust data of the overall runtime increases with increasing number of nodes and therefore limits this code's scaling capabilities. As this part of the code is currently under development, the scaling analysis can be taken into account for optimization and parallel implementation of these routines e.g., by means of parallel reading of data.

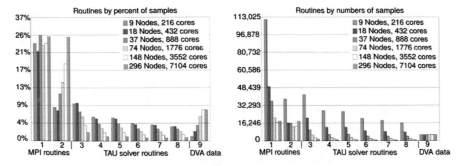

Fig. 8 Scaling behavior of the 9 most time consuming routines of this test case with the TAU solver, produced with Craypat pat view, relative and absolute number of samples

6 Conclusion

The influence of broadband atmospheric turbulence on an extruded airfoil was investigated in this study. The turbulence was fed in at a distance $5L$ upstream from the leading edge using a momentum source term. L represents the length scale of the atmospheric turbulence. It was shown that the turbulence can be propagated accurately from the turbulent inflow plane to the airfoil and isotropy is conserved within the flow. The main impact on the c_p distribution occurs at the leading edge and the suction peak. Up to a reduced wave number of $\kappa = 12$ the shape of the c_p spectra at this positions corresponds to that of the u velocity spectrum at a distance L upstream from the leading edge. Further downstream the amplitudes of the c_p spectra are reduced for small wave numbers with minor differences between $x/c = 0.5$ and $x/c = 0.9$. In addition to the effect of atmospheric turbulence on surface pressure and global loads, the applicability of the simplified DVA was investigated. At the leading edge the c_p spectrum of the DVA matches the one of the physically accurate RAA up to $\kappa = 32$, whereas the c_p spectrum of the DVA at $x/c = 0.2$ overestimates the c_p amplitudes for $\kappa > 7$. The lift spectrum of the DVA corresponds to the one of the RAA up to $\kappa = 25$, whereas the RAA pitching moment can only be estimated correctly by the DVA for $\kappa < 5$. Hence, the DVA is capable of covering the main effects of an airfoil in atmospheric turbulence in terms of lift spectrum and c_p spectra at leading edge and suction peak. It tends to overestimate the c_p spectra compared to the RAA at high wave numbers especially further downstream. Therefore, the RAA pitching moment can only be calculated correctly by the DVA for low wave numbers. Furthermore, the scaling capabilities of an extended version of the DLR TAU code that includes the DVA approach were investigated. This scaling test showed good scaling behavior of the TAU code with increasing node number, but revealed non-scaling routines of the DVA implementation. As this part of the code is currently under development, the results of this analysis could be taken into account for further implementation and optimization of the DVA routines.

Acknowledgements The authors gratefully acknowledge the Federal Ministry for Economic Affairs and Energy, which funded the work presented in this report as part of the LuFo project VitAM-Turbulence. Also, we acknowledge the High Performance Computing Center Stuttgart (HLRS) for the provision of computational resources and the continued support.

References

1. G. Dargel, H. Hansen, J. Wild, T. Streit, H. Rosemann, K. Richter, Aerodynamische Flüge-lauslegung mit multifunktionalen Steuerflächen. DGLR Jahrbuch p. 1605 (2002)
2. European Aviation Safety Agency, https://www.easa.europa.eu/sites/default/files/dfu/CS-25 %20Amendment%2022.pdf: Certification Specifications and Acceptable Means of Compliance for Large Aeroplanes CS-25, Amendment 22. Accessed: 2019-04-27
3. P. Hancock, P. Bradshaw, Turbulence structure of a boundary layer beneath a turbulent free stream. J. Fluid Mech. **205**, 45–76 (1989)
4. R. Heinrich, L. Reimer, Comparison of Different Approaches for Gust Modeling in the CFD code Tau. International Forum on Aeroelasticity & Structural Dynamics (2013)
5. R. Heinrich, L. Reimer, Comparison of Different Approaches for Modeling of Atmospheric Effects like Gusts and Wake-Vortices in the CFD Code Tau. International Forum on Aeroelasticity & Structural Dynamics (2017)
6. J.A. Hoffmann, Effects of freestream turbulence on the performance characteristics of an airfoil. AIAA J. **29**(9), 1353–1354 (1991)
7. Y. Kim, P. Weihing, C. Schulz, T. Lutz, Do turbulence models deteriorate solutions using a non-oscillatory scheme? J. Wind Eng. Industr. Aerodynam. **156**, 41–49 (2016)
8. J. Mann, The spatial structure of neutral atmospheric surface-layer turbulence. J. Fluid Mech. **273**, 141–168 (1994)
9. J. Raddatz, J.K. Fassbender, Block structured Navier-Stokes solver flower, in *MEGAFLOW - Numerical Flow Simulation for Aircraft Design*, ed. by N. Kroll, J.K. Fassbender (Springer, Berlin Heidelberg, Berlin, Heidelberg, 2005), pp. 27–44
10. C. Schulz, Numerische Untersuchung des Verhaltens von Windenergieanlagen in komplexem Gelände unter turbulenter atmosphärischer Zuströmung. Ph.D. thesis, Institute of Aerodynamics and Gas Dynamics, University of Stuttgart (2017)
11. C. Schulz, L. Klein, P. Weihing, T. Lutz, Investigations into the interaction of a wind turbine with atmospheric turbulence in complex terrain. J. Phys. Conference Series **753**(3), 032016 (2016)
12. D. Schwamborn, T. Gerhold, R. Heinrich, The DLR TAU-Code, Recent Applications in Research and Industry. European Conference on Computational Fluid Dynamics ECCOMAS CFD (2006)
13. C. Stanger, B. Kutz, U. Kowarsch, E.R. Busch, M. Keßler, E. Krämer, Enhancement and Applications of a Structural URANS Solver. In: High Performance Computing in Science and Engineering '14, pp. 433–446. Springer (2015)
14. N. Troldborg, J.N. Sørensen, R. Mikkelsen, N.N. Sørensen, A simple atmospheric boundary layer model applied to large eddy simulations of wind turbine wakes. Wind Energy **17**(4), 657–669 (2014)
15. K. Wawrzinek, T. Lutz, E. Krämer, Numerical Simulations of Artificial Disturbance Influence on a High Lift Airfoil. In: High Performance Computing in Science and Engineering '17, pp. 323–337. Springer (2018)

High Performance Computations of Rotorcraft Aerodynamics with the Flow Solver FLOWer

Constantin Öhrle, Johannes Letzgus, Manuel Keßler, and Ewald Krämer

Abstract Recent enhancements and applications of the flow solver FLOWer are presented. First, an Adaptive Mesh Refinement technique (AMR) is implemented and an AMR cycle is built around FLOWer. Depending on the application case, the usage of AMR significantly reduces the overall computational cost—and thus increases efficiency—of the flow solution. Exemplary, for a complete helicopter simulation a reduction in computational time of 45% is achieved without any observable loss of accuracy. Second, the implementation of a new shielding approach for detached eddy simulations is shown. This applies a sophisticated communication method to exchange flow variables in wall normal direction. As an application case, the computation of the highly resolved rotor wake in hover highlights the possibilities of high performance computing in combination with advanced numerical methods for rotorcraft flows.

1 Introduction

Thanks to the continuing increase in high performance computing (HPC) power in combination with advanced numerical methods and highly scalable tools, the simulation of complex rotorcraft aerodynamic phenomena becomes ever more accessible. On one hand, this enables the researchers to obtain deeper insight into the aerodynamic phenomena itself as well as the underlying physical mechanisms by using highly resolved simulations. As an example of current state of the art HPC simulations

C. Öhrle (✉) · J. Letzgus · M. Keßler · E. Krämer
Institute of Aerodynamics and Gas Dynamics (IAG), University of Stuttgart, Stuttgart, Germany
e-mail: oehrle@iag.uni-stuttgart.de

J. Letzgus
e-mail: letzgus@iag.uni-stuttgart.de

M. Keßler
e-mail: kessler@iag.uni-stuttgart.de

E. Krämer
e-mail: kraemer@iag.uni-stuttgart.de

© Springer Nature Switzerland AG 2021
W. E. Nagel et al. (eds.), *High Performance Computing in Science and Engineering '19*,
https://doi.org/10.1007/978-3-030-66792-4_30

451

in the field of rotorcraft, the high-fidelity simulation of the rotor's vortical wake in hover is shown in this paper. On the other hand, the increase in computational power paves the way for more application-oriented Computational Fluid Dynamics (CFD) simulations of rotorcraft during the development phase, as it is currently utilized for the de-risking of the RACER demonstrator [10].

Besides of these, the CFD code FLOWer is continuously enhanced at the Institute of Aerodynamics and Gas Dynamics (IAG) regarding both new functionality and computational performance. Therefore, within this paper, the scalability of FLOWer on the Hazel Hen system is shown. Furthermore, new functionalities which are an are Adaptive Mesh Refinement technique (AMR) and a shielding approach for detached eddy simulations (DES), are described and evaluated regarding their computational efficiency.

2 Flow Solver FLOWer

The block-structured finite volume flow solver FLOWer [13] builds the center of the internationally leading rotorcraft simulation framework at IAG. FLOWer was originally developed by the German Aerospace Center (DLR) and significantly extended by IAG in recent years. Besides of enhanced functionality, FLOWer is continuously optimized for high-performance computing. The most recent optimizations included a sophisticated communication method with shared memory communication for intra-node exchange. In case of node-to-node communication the data is gathered on each node and exchanged from only one process on each node [5]. Further improvements in code performance have been carried out by Stanger et al. [18] and Kranzinger et al. [3].

The additional functionalities include the implementation of several higher order spatial weighted essentially nonoscillatory (WENO) schemes of fifth [2] and sixth order [12]. Furthermore, a new coupling library for fluid-structure interaction, called IAGCOUPle [15], has been established and state of the art detached-eddy simulation (DES) methods [19] have been implemented. An Adaptive Mesh Refinement (AMR) functionality has been included [12], which promises higher computational efficiency without the loss of accuracy. Most recently, a method to conservatively compute the aerodynamic loads for overlapping surface meshes was implemented [11].

2.1 Scaling of FLOWer on Cray XC40

The scalability of the current version of the CFD code FLOWer on the Cray XC40 Hazel Hen system is shown in Fig. 1 in terms of weak (a) and strong (b) scaling. These results are comparable to the results shown by Letzgus et al. [5], where the last major parallel performance updates were presented.

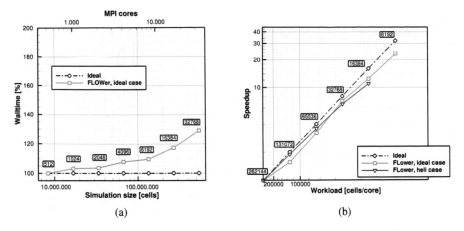

Fig. 1 Scaling studies of the flow solver FLOWer on the Hazel Hen system. **a** Weak scaling **b** strong scaling

Down to a workload of 16.384 or even 8.192 grid cells per physical core, the parallel scaling of FLOWer is good. Currently, most computations of complex rotorcraft flows at IAG are designed to meet a workload of around 16.000 cells per physical core as a trade-off between total simulation time and efficient use of computational resources. Regarding the weak scaling, simulation sizes of 500 Mio. grid cells on 32768 cores can still be performed efficiently.

For reasons of comparability and flexibility, the weak and strong scaling studies are performed using an ideal test case. This consist of blocks which all have the same size. Hence, no load imbalances exist. Furthermore, only one grid structure is considered, so no information transfer between different grid structures using the Chimera method is necessary. However, the strong scaling is additionally shown for a real application case of a rotorcraft simulation with 370 million grid cells. For this, it can be verified that the scaling does not significantly change and an efficient computation of the real application is asserted. It is remarkable that for the loadings of 131072 and 65536 cells per core, the rotorcraft case scales better than the ideal case. This might be related to the fact that the reference points with a loading of 262144 cells per core are computed with different core numbers.

It has to be noted that for the application case the simulation size is predefined and depends on the answers to be addressed. Thus, a weak scaling is not feasible in this context. Furthermore, the strong scaling is also limited by the maximal block size of the computational grids, which are defined during the generation of the grids in the preprocessing phase.

Fig. 2 Schematic of an
AMR cycle within the
simulation framework [12]
at IAG

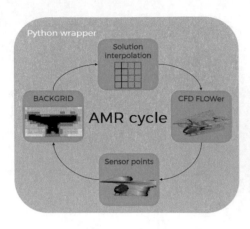

3 Implementation of Adaptive Mesh Refinement

3.1 Workflow

A key feature of AMR is the seamless continuation of the CFD computation after an
adaptation step is carried out. In order to achieve this, the implemented AMR cycle
within the simulation framework consists of four steps, as shown in Fig. 2. During the
CFD simulation, cells that meet a defined sensor criterion, such as a vortex criterion,
are tagged. After the CFD simulation, the tagged cells are gathered and coarsened
and provided to the grid generation tool Backgrid, which generates a new off-body
(OB) grid on the basis of the tagged cells. Finally, an additional tool performs the
transfer of the CFD solution from the old grid to the new grid by means of a trilinear
interpolation. Both the grid generation with Backgrid and the interpolation of the
solution are demanding in terms of computational power and disk I/O, so that the
time required for one AMR cycle depends on the performance and scaling of these
two tasks.

3.2 AMR Strategies

For the described AMR cycle, two different strategies are implemented, the periodic
and the time-dependent AMR. The periodic variant makes use of the inherent periodicity of the flow field in case of rotorcraft flows. Hence, the adaption is carried
out only once in a period (e.g. blade passage). Therefore, marked cells are gathered
during a complete period and subsequently used for the generation of the new OB
grid. This process is comparable to an averaging of the flow field. The periodic AMR
has two main advantages: First, the AMR cycle has to be performed only once in a

period and second, after several AMR cycles, the OB grid eventually converges and no additional AMR cycles are required.

The time-dependent AMR is a more advanced method for which a direct adaptation of the grid towards the solution for a specific adaptation frequency Δt_{AMR} can be performed. After the subsequent time integration of the time-step Δt_{AMR} of the CFD simulation, the AMR cycle is re-executed. The adaptation frequency Δt_{AMR} is normally in the range of 10–40 CFD time steps. However, it has to be ensured that resolved flow features of interest do not propagate into a coarse grid region during the time increment Δt_{AMR}. Therefore, an extrapolation technique based on the local flow velocity vector $\mathbf{u}^{\,t}$, according to

$$\Delta \mathbf{x}^{\,t+\Delta t_{AMR}} = \mathbf{u}^{\,t} \Delta t_{AMR}, \tag{1}$$

is applied. The advantage of the time dependent AMR is that the grid can be coarsened at regions where currently no relevant flow features are located. Hence, this saves overall grid cells. The disadvantage is the larger adaptation frequency, so that the AMR cycle has to be re-executed several times during one period. Thus, the performance of the AMR cycle is essential here for choosing an appropriate trade-off.

3.3 Application and Computational Performance of AMR

The implementation of the AMR aims at improving the performance of real application cases. Therefore, as a test case a state of the art complete helicopter simulation of the compound helicopter RACER in cruise flight with 220 knots is considered. One challenge is, that the simulation setup consists of 101 different grid structures and 115 million cells in the Near-Body (NB) meshes alone. Hence, the fraction of the OB grid is relatively small compared to other investigations applying AMR for rotorcraft simulations. However, even for this case the two AMR strategies implemented are able to sufficiently reduce the OB grid size compared to a manually created static grid. Slices through the resulting meshes are shown in Fig. 3. The sophisticated refinement in the regions where the main rotor tip vortices are located shows the robust application of the time-dependent AMR. Furthermore, large regions around the rotor are coarsened without influencing the flow solution.

With respect to the overall computational performance, when using AMR the total computational time includes both the computational time for the flow solution and the AMR. The aim is to reduce the overall computational time by reducing the time required for the flow solution (due to the smaller grids). Consequently, the additional expense for AMR should be smaller than the benefit gained during the flow solution in order to reduce the overall time. Table 1 shows the comparison of computational time usage for the time-dependent and the periodic AMR, as well as for a manually created grid. The most interesting number is in the last row and states how many core hours are required for one main rotor revolution. This is a direct measure of the computational efficiency as the quality of the flow solution is comparable between

Fig. 3 Slice through the grid in case of time-dependent (left) and periodic (right) AMR [12]

Table 1 Comparison of grid sizes and computation times for different OB grid generations using AMR

	Time-dependent AMR	Periodic AMR	Static OB grid
MPI processes	9812	12000	19200
OB grid cells [millions]	112	178	329
Overall cells [millions]	227	293	444
AMR time/cycle [core hours]	627	1800	–
CFD time/cycle [core hours]	3217	52210	–
AMR cycles/rev	60	5	–
Core hours/rev	230640	270050	423680

the three cases. Overall, the application of the AMR reduces the computational cost by approximately 45% for time-dependent and by approximately 36 % for periodic AMR.

Despite of this efficiency increase, it should be noted that the tools applied in the AMR cycle—Backgrid and solution interpolation—were initially not developed for HPC applications. Therefore, the scaling of these tools is investigated and is shown in Fig. 4. The scaling of the solution interpolation tool is acceptable up to 2400 MPI processes. Despite of using MPI I/O the disk I/O limits the performance. In contrast, the scaling of the tool Backgrid has room for improvements and usage of more than 600 MPI processes is currently not reasonable. In order to optimize the AMR performance, the parallelization of Backgrid will be addressed in the future work. Further speed-up potential of estimated 50% might be possible by utilizing in-memory data exchange between the different codes in the future. During one adaptation cycle, both tools require a comparable share of computation time. Thus they are equally important for the reduction of the overall AMR time.

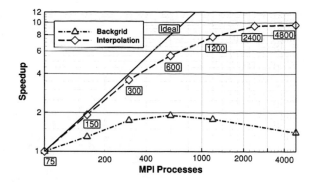

Fig. 4 Scaling of the tools (Backgrid and solution interpolation) used in the AMR cycle

4 Improvement of DES Boundary-Layer Shielding

It is known that a DES [16] is superior to URANS in simulating massively separated flow. This is due to the fact that, using DES, the artificial eddy viscosity introduced by the turbulence model is reduced outside the boundary layers, which allows to resolve energy-containing turbulent structures and physically realistic flow instabilities. However, it is crucial to treat the boundary layers in URANS mode, as an intrusion of the LES mode into the boundary layer leads to modeled-stress depletion (MSD) and eventually to premature, grid-induced flow separation (GIS). Although state of the art DES variants, as the here used DDES [1], contain so called boundary-layer shieldings that are designed to prevent MSD/GIS in theory, it showed that ever-increasing grid resolutions aggravate this key issue [6, 9]. Therefore, at IAG, research into alternative shielding functions, which are applicable not only to simple test cases but also to complex wind-turbine or helicopter-rotor flows, is carried out.

Fundamental to the boundary-layer shielding process is to define and determine the edge of the boundary layer, where the flow is supposed to reach free-stream conditions. Generally, the edge is defined to be reached moving in wall-normal direction, when the local flow velocity equals 99% of the free-stream value u_∞. However, in a block-structured flow solver run in parallel, there is a priori no way of locally, i.e. at every grid point, knowing the "real" free-stream value. Furthermore, e.g.. in case of accelerated flow as it occurs on an airfoil or a wing, the boundary-layer edge-velocity is even higher than the free-stream velocity u_∞, thus, the traditional 99% u_∞ criterion would fail. Alternatively, a pseudo-velocity u_{ps} can be computed by means of the piecewise wall-normal (here y-direction) integration of the vorticity Ω, as proposed by Spalart and Strelets [17]. This pseudo-velocity converges very fast outside the boundary layer, thus the 99% criterion can be applied to u_{ps} instead of u_∞ [7]:

$$u_{ps}(y) = \int_0^y |\Omega|\, d\tilde{y} \overset{?}{<} 0.99\, u_{ps_\infty}, \qquad u_{ps_\infty} = \int_0^\infty |\Omega|\, d\tilde{y}. \qquad (2)$$

Fig. 5 Schematic of the piecewise integration and block-exchange progress that is required due to grid decomposition

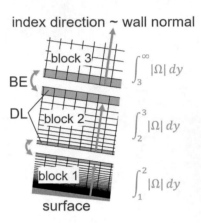

Given the decomposition of the grid, which is needed for HPC, the difficulty now is to carry out this wall-normal integration, since a single process does not have access to all the flow data beginning at the wall ($y = 0$) and ending sufficiently away from the wall ($y \rightarrow \infty$, for an airfoil typically a chord length). However, already existing ghost/dummy-layer (DL) and block-exchange (BE) code infrastructure can be exploited: At first, each process computes the integrals, which are approximated by Riemann sums, of its own block and saves the values at the boundary cells. Then, several block exchanges are carried out that distribute the single-block values throughout all relevant blocks, see Fig. 5. Given n blocks that cover the whole interval of integration (typically, $n = 2 - 5$), $n - 1$ exchanges are needed. Furthermore, the structured grid and the fact that it is always generated by wall-normal extrusion is utilized, in that the integration is not carried out in the real wall-normal direction but in the index direction of the grid that is only more or less wall-normal. This reduces the complexity and the computational effort of the approach significantly, without losing too much accuracy.

Although FLOWer's block exchange and node-to-node communication is well optimized [5] and the mentioned integration is carried out highly parallel, this shielding approach increases the computational cost. To keep the impact at a minimum, the shielding is only computed once a physical time step and not in every subiteration. The final shielding procedure, which is named vorticity-integrated algebraic DES (VIADES), also includes a sensor that distinguishes boundary layers from separated shear layers to allow the LES mode in case of flow separation. VIADES was tested for several cases of different complexity [7, 20] and it showed that it eliminated MSD leading to GIS and provides a robust, grid-independent and adaptive boundary-layer shielding.

To assess the impact of VIADES on the performance, a 240 million grid cells test case of a two-bladed model rotor was investigated, which was split into 15900 blocks and computed on 13200 cores of the Cray XC40 cluster. The shares of the dominant tasks within a FLOWer run are listed in Table 2. It shows that the influence of the shielding computation on the overall performance is very low.

Table 2 Shares of computation time of different tasks within a FLOWer run

Task	Share (%)	Time step
VIADES	0.6	Every
Surface output	1.1	Every 12th
Volume output	10.0	Every 12th
Chimera	13.7	Every
Sub-iterations	72.4	
Other	2.2	

5 Applications of HPC Computing: Highly Resolved Rotor Wake in Hover

The increased computational performance in combination with higher order numerical methods and improved code performance allows the computation of very highly resolved simulations that have not been possible several years ago. One investigation with FLOWer focusses on the development of the rotor wake in hover. The high resolution provides a deeper insight into the physics of such a vortex wake, where several phenomena are under investigations, such as vortex pairing, vortex instabilities, secondary vortices and wake breakdown. Figure 6 shows the wake of a five-bladed hovering rotor for three different setups: A second order and a sixth order computation on the baseline grid with 162 million grid cells (left and center, respectively) and a sixth order computation on a finer grid consisting of 816 million grid cells (right). The usage of the higher order WENO method shows significantly more resolved flow physics compared to the second order JST computation. For all three cases, the rotor wake looses its coherent structure within the first revolution and the wake disintegrates. However, in case of the higher-order computations, the wake breakdown is accompanied by the occurrence of small scale structures, such as secondary vortices around the tip vortices. These tend to additionally destabilize the wake and appear when the pitch of the helical wake (distance between successive vortex filaments) becomes small enough, as it is for this five-bladed rotor. The existence of these secondary structures is physically correct, but their highly destabilizing effect on the wake and the wholesale breakdown of the wake into a completely turbulent flow is not supported by measurements and is currently under investigation.

Furthermore, the size of the blade tip vortices, see Fig. 7, is significantly smaller for the higher order computation, and approaches the expected size that is extracted from several measurements (e.g. see McAlister [8], Ramasamy et al. [14] or Leishman et al. [4]). A further improvement in the computed vortex core size is achieved using the finer grid in combination with the higher order method. For this case, the computed vortex core sizes are within the measurement bandwidth, at least for the first 220° of wake age.

Fig. 6 Comparison of the rotor wake in hover for different grid sizes and numerical methods. Left: 162 M cells, second order, center 162 M cells 6th order, right: 816 M cells sixth order [12]

Fig. 7 Computed vortex core sizes with combined experimental data from literature from several measurements. Core size normalized with equivalent chord length

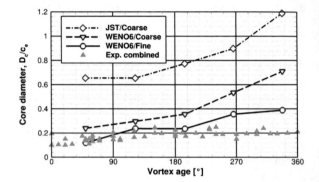

6   Conclusions

The scalability of the flow solver FLOWer was shown for an ideal test case and a rotorcraft application. It was ensured, that the real application scales equally good as the ideal case. Hence, the efficient use of computational resources is ensured.

Two new features of the tool chain at IAG have been introduced and assessed regarding their computational performance. The application of an AMR technique allows the clustering of grid cells in regions where they are required to resolve the relevant flow features (e.g.. vortices for rotorcraft applications). Depending on the application, this technique significantly reduces the overall computational time without loss of accuracy. The application for a complete helicopter simulation in high-speed flight showed a reduction of computational time—and thus an increased efficiency—of up to 45%. Despite of these efficiency gains, further possible performance improvements of the AMR tool chain were identified. The implementation of a new shielding approach for DES simulations uses a sophisticated communication method to exchange flow variables in wall normal direction. Furthermore, it is ensured that this exchange is performed efficiently as the overall computational time is only marginally increased when applying this method.

As an exemplary application of HPC in combination with advanced numerical methods, the high-fidelity simulation of the wake of a hovering helicopter rotor was presented. Highlights of this simulations are the improvements for the conservation of the tip vortices and the deeper insights into the wake breakdown mechanisms.

Acknowledgements The provided supercomputing time and technical support of the High Performance Computing Center Stuttgart (HLRS) of the University of Stuttgart within the HELISIM project is gratefully acknowledged. Parts of the research presented in this study were performed in cooperation of the Institute of Aerodynamics and Gas Dynamics of the University of Stuttgart and the Airbus Helicopters division of Airbus SE within the European research framework Clean Sky 2 under Grant 686530. We would like to express our thanks to the European Union for providing us the resources to realize this project. Also, parts of the simulations were carried out under the national project CHARME of the Luftfahrtforschungsprogramm, funded by the ministry of economics. Furthermore, this work is based on the long-standing cooperation with the German Aerospace Center (DLR) making us their CFD code FLOWer available for advancements and research purpose, which we would like to thank for.

References

1. M.S. Gritskevich, A.V. Garbaruk, F.R. Menter, Fine-tuning of DDES and IDDES formulations to the k-ω shear stress transport model. Prog. Flight Phys. **5**, 23–42 (2013)
2. U. Kowarsch, C. hrle, Hollands, M., Keßler, M., E. Krämer, Computation of Helicopter Phenomena Using a Higher Order Method, pp. 423–438. Springer International Publishing, Cham, Switzerland (2014)
3. P. Kranzinger, P., Kowarsch, U., Schuff, M., Keßler, M., Krämer, E, Advances in Parallelization and High-Fidelity Simulation of Helicopter Phenomena, pp. 479–494. Springer International Publishing, Cham, Switzerland (2016)
4. J. Leishman, A. Baker, A., Coyne, Measurements of rotor tip vortices using three-component laser doppler velocimetry. Journal of the American Helicopter Society **41**(4), 342–353(12) (1996)
5. J. Letzgus, J., Dürrwächter, L., Schäferlein, U., Keßler, M., Krämer, Optimization and HPC-Applications of the Flow Solver FLOWer, pp. 305–322. Springer International Publishing, Cham, Switzerland (2018). 10.1007/978-3-319-68394-2
6. J. Letzgus, A.D. Gardner, T. Schwermer, M. Keßler, E. Krämer, Numerical investigations of dynamic stall on a rotor with cyclic pitch control. J. Am. Helicopter Soc. **64**(1), 1–14 (2019). https://doi.org/10.4050/JAHS.64.012007
7. J. Letzgus, P. Weihing, P., Keßler, M., Krämer, E, Assessment of delayed detached-eddy simulation of dynamic stall on a rotor. In: Proceedings of the 7th Symposium on Hybrid RANS-LES Methods. Berlin, Germany (2018)
8. K.W. McAlister, Rotor wake development during the first revolution. In: Proceedings of the American Helicopter Society, 59th Annual Forum. Phoenix, Arizona (2003)
9. F. Menter, Stress-blended eddy simulation (SBES)–a new paradigm in hybrid RANS-LES modeling, in *Progress in Hybrid RANS-LES Modelling*, ed. by Y. Hoarau, S.H. Peng, D. Schwamborn, A. Revell (Springer International Publishing, Cham, 2018), pp. 27–37
10. C. Öhrle, C., Frey, F., Thiemeier, J., Keßler, M., Krämer, Coupled and trimmed aerodynamic and aeroacoustic simulations for Airbus Helicopters' compound helicopter RACER. In: American Helicopter Society Technical Conference on Aeromechanics Design for Transformative Vertical Flight. San Francisco, California (2018)
11. C. Öhrle, C., Frey, F., Thiemeier, J., Keßler, M., Krämer, E., Embacher, M., Cranga, P., Eglin, P., Compound helicopter X^3 in high-speed flight: Correlation of simulation and flight test. In: Vertical Flight Society, 75th Annual Forum. Philadelphia, Pennsylvania (2019)

12. C. Öhrle, U. Schäferlein, M. Keßler, E. Krämer, Higher-order simulations of a compound heli-
    copter using adaptive mesh refinement. In: American Helicopter Society 74th Annual Forum.
    Phoenix, Arizona (2018)
13. J. Raddatz, J., Fassbender, Block Structured Navier-Stokes Solver FLOWer, pp. 27–44. Springer
    Berlin Heidelberg, Berlin/Heidelberg, Germany (2005). 10.1007/3-540-32382-1_2
14. M. Ramasamy, M., B. Johnson, B., Leishman, Turbulent tip vortex measurements using dual-
    plane digital particle image velocimetry. In: Proceedings of the American Helicopter Society,
    64th Annual Forum. Montréal, Canada (2008)
15. M. Schuff, M., Kranzinger, P., Keßler, M., Krämer, Advanced CFD-CSD coupling: gener-
    alized, high performance, radial basis function based volume mesh deformation algorithm
    for structured, unstructured and overlapping meshes. In: 40th European Rotorcraft Forum.
    Southampton, United Kingdom (2014)
16. P.R. Spalart, Comments on the feasibility of LES for wings, and on a hybrid RANS/LES
    approach. In: Proceedings of first AFOSR international conference on DNS/LES. Greyden
    Press (1997)
17. P.R. Spalart, M.K. Strelets, Mechanisms of transition and heat transfer in a separation bubble.
    Journal of Fluid Mechanics **403**, 329–349 (2000)
18. C. Stanger, C., Kutz, B., Kowarsch, U., Busch, E.R., Keßler, M., Krämer, Enhancement and
    Applications of a Structural URANS Solver, pp. 433–446. Springer International Publishing,
    Cham, Switzerland (2015). 10.1007/978-3-319-10810-0_29
19. P. Weihing, P., Letzgus, J., Bangga, G., Lutz, T., Krämer, Hybrid RANS/LES capabilities of the
    flow solver FLOWer - application to flow around wind turbines. In: 6th Symposium on Hybrid
    RANS-LES Methods. Strasbourg, France (2016)
20. P. Weihing, Letzgus, J., Lutz, T., E. Krämer, Development of alternative shielding functions
    for detached-eddy simulations. In: Proceedings of the 7th Symposium on Hybrid RANS-LES
    Methods. Berlin, Germany (2018)

Influence of the Flow Physics on the Load Balancing During SPH Simulations

G. Chaussonnet, T. Dauch, M. Keller, M. Okraschevski, C. Ates,
C. Schwitzke, R. Koch, and H.-J. Bauer

Abstract This study identifies the main factors influencing the load balancing for the SPH method. The idle rate is introduced as an index to quantify the load balancing. We present the algorithmic strategies of our implementation of the SPH method, and their impact on the load balancing. Also, it is shown for the SPH method that due to the formulations of the physics and its Lagrangian aspect, the load balancing can significantly vary during the simulation. Especially, spurious pressure fluctuations arising at boundaries affect the load balancing.

1 Introduction

The increase of the number of processing units on supercomputers stresses the necessity to develop codes of high scalibility. Not only the pure scaling characteristics are important but also the capacity of the code to simulate a long physical time in the shortest execution time. We call that the restitution time. To achieve such specifications, the parallel and the serial optimization of the code must considered as a whole. The restitution time depends primarily on the performance of the code, but also the load balancing of the case, which depends on the geometry of the numerical domain.

In the numerical simulations of fluid mechanics, to solve the partial differential equations of motion, it is necessary to compute the differential vector operators (gradient, divergence, Laplacian) at each discretized elements. For a given element, this requires to have all the physical information of the neighborhood of this element, *i.e.* the stencil. Hence, an appropriate method of parallelisation is the domain decomposition. In this case, the numerical domain is decomposed into connected geometrical volumes, which guarantees that most of neighbors are located on the same CPU. In mesh-based methods for stationary grids, neither the size of the cells nor the connectivity change during the simulation, so that the load balancing is performed prior to the simulation, and remains constant during the simulation. In the case of particle

G. Chaussonnet (✉) · T. Dauch · M. Keller · M. Okraschevski · C. Ates · C. Schwitzke · R. Koch · H.-J. Bauer
Institut für Thermische Strömungsmaschinen KIT, Kaiserstraße 12, 76131 Karlsruhe, Germany
e-mail: geoffroy.chaussonnet@kit.edu

© Springer Nature Switzerland AG 2021
W. E. Nagel et al. (eds.), *High Performance Computing in Science and Engineering '19*,
https://doi.org/10.1007/978-3-030-66792-4_31

methods with the domain decomposition, the particles move at the fluid velocity through the domain, so that their number depends on the local state of the flow field. As seen latter, in some cases, their concentration is directly related to the thermodynamical state of the fluid. Hence, the number of elements inside a given physical volume, *i.e.* the load balancing, might changes during the simulation.

In this paper, the Smooth Particle Hydrodynamics (SPH) method is employed [6] to illustrate the issues of load balancing with particle methods. SPH is a mesh-free method that relies on a Lagrangian description, where the discretization points (called *particles*) carry the physical properties (mass, momentum and energy) and move at the fluid velocity. The advection is naturally handled by the particle motion and is not prone to numerical diffusion. This is also an advantage when capturing the phase interface in the case of multiphase flow: no surface reconstruction algorithm is needed. The SPH method will be applied to a case of air-assisted liquid atomization in a wall-bounded flow.

After this introduction, the principles of the SPH method are presented, followed by algorithmic aspects of the code used for this investigation which are relevant to the load-balancing. Then, the partitioning method is presented and an index to characterize load balancing is introduced. Finally, the evolution of the load balancing in the course of a simulation is shown on a practical case of liquid atomization.

2   Principle of the Smoothed Particle Method

In the SPH mehod, the physical quantities and their gradient at a particle location (a) are determined by the interpolation over the values of its neighbors (b) according to [6]:

$$f(\boldsymbol{r}_a) = \sum_{b \in \Omega_a} V_b \, f(\boldsymbol{r}_b) \, W(\boldsymbol{r}_b - \boldsymbol{r}_a, h) \tag{1}$$

where V_b is the volume of the adjacent particles. The term W is a weighting function (the *kernel*). It depends on the inter particle distance $\boldsymbol{r}_b - \boldsymbol{r}_a$ and a characteristic length scale h called the *smoothing length*. The kernel promotes the influence of closer neighbors as illustrated in 2D in Fig. 1 (*top*). When the neighbors are located outside the *sphere of influence* Ω of the central particle (Fig. 1 *bottom*), they are not taken into account. Hence, the kernel has a compact support, delimited by the sphere of influence of radius R_Ω.

In the present study, the motion of two isothermal weakly-compressible non-miscible fluids is solved by the Navier-Stokes equations. In the momentum equation the source terms are the pressure gradient, the viscous term proportional to the Laplacian of the velocity, the surface tension force and the gravity. No evaporation is taken into account. These equations are formulated in a Lagrangian frame of reference and solved using the SPH method. The interested reader is referred to [4] for further details about the modeling. The mass conservation and the equation of state are relevant for the present topic. They are expressed as:

Fig. 1 Top part: Surface of a 2-D kernel. Bottom part: Particle distribution superimposed with the kernel color map and illustration of the sphere of influence

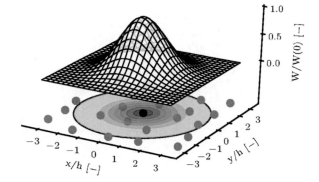

$$\rho_a = m_a/V_a \quad \text{with} \quad V_a = 1/\sum_{b \in \Omega} W_{ab} \tag{2a}$$

$$p_a = \frac{\rho_0 \, c^2}{\gamma} \left[\left(\frac{\rho_a}{\rho_0} \right)^{\gamma} - 1 \right] + p_{back} \tag{2b}$$

In Eq. 2a, m_a is the mass of the particle, and W_{ab} is the equivalent notation for $W(\mathbf{r}_b - \mathbf{r}_a, h)$. The density is therefore computed algebraically based on the virtual volume V_a, which is determined itself by the summation of the kernel over neighbor particles. This means that both the virtual volume and the density are directly dependent of the local inter particle distance. In Eq. 2b, the terms ρ_0, c, γ, and p_{back} are the nominal density, the fictive speed of sound, the isentropic ratio and background pressure, respectively. This equation of state is valid for weakly compressible flow, which is usually assumed in SPH simulations, because it allows to increase the time step and then decrease the restitution time of the method. The requirement is to ensure that the Mach number is always lower than 0.1, which leads to a pressure variation below 1%. However in our case of airblast atomization, the Mach number is typically of the order of 0.3, so that the density variation are above 1%. In this case, the speed of sound is set to its real value (*e.g.* 340 m/s for air at 20°C) and γ to one. It was shown in [4] that if the thermal effects due to compression do not influence the dynamics, Eq. 2b is still valid. Note that due to Eq. 2b, the pressure also directly depends of the local inter particle distance. This will have an impact on the load balancing, as it will be shown in the following.

The code used in this present study is super_sph, developed at ITS during the PhD thesis of Höfler [5] for physical models and Braun [1] for the HPC architecture. The parallel performances of super_sph in the configuration of air-assisted atomization were extensively investigated and compared to other methods by Braun *et al.* [2].

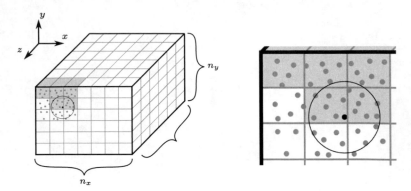

Fig. 2 Search cubes on one partition (left) and close up of the sphere of influence (right)

3 Strategies for Particle Search and Domain Decomposition

This part exhibits some of the algorithms used in the code super_sph. The frameworks for the particle search and the domain decomposition were originally developed by Braun [1], and the optimization of the load-balancing and the definition of the Idle Rate were developed in the present study.

Since the particles move at the velocity of the fluid inside the domain, the list of their neighbors must be updated at each iteration. Depending on the time integration scheme, the neighbor list is updated several time per time step. The selected method is the list search method. The complexity of list-search algorithm is in $\mathcal{O}(N_t)$ where N_t is the number of particles. It is depicted in Fig. 2. The numerical domain is decomposed into a Cartesian background grid of $N_e = n_x \cdot n_y \cdot n_z$ cubes of constant side length R_Ω, called search-cubes. The connectivity of the search-cubes, *i.e.* the list of neighboring search-cubes is determined for each search-cube at the beginning of the simulation. Each search-cube has at most 8 and 26 neighbors in 2D and 3D, respectively. In our formulation, all the source terms fulfill the third Newton's law of symmetrical interaction between two particles, *i.e.* $\mathbf{f}_{a \to b} = -\mathbf{f}_{b \to a}$. Hence, only half of the particle interaction is necessary and therefore the neighboring particles can be searched in only the half of the search-cubes. The search-cubes are attributed an ID-number from 0 to N_e. To locate the particles inside search-cubes, their coordinates are transformed into indices of the background Cartesian grid and converted to the corresponding search-cube ID. This conversion has a complexity of $\mathcal{O}(N_t)$ as it does not require any research algorithm.

Another aspect of SPH is the necessity to complete the sphere of influence for each particle. For particles located at an inter-processor interface, it is required to artificially extend the boundaries of the subdomain by adding a layer of particles of a thickness R_Ω at each interface. This extension is called the halo-zone and it is created by transferring the particles in the vicinity of the processor interface from the neighboring subdomain to the current subdomain. This constitutes most of the parallel communications. Even though the source terms in the equations of motion are not calculated for the halo-particles, their interaction with the inner particles is taken into account. This increases both the memory and computational overheads. Indeed, the proportion of halo-particles in percentage of the inner particles $H_p = N_{halo}/N_{inner}$ is estimated by considering the Cartesian boundaries of each subdomain increased by a distance of R_Ω. By considering a perfectly cubic domain of side L divided in N_{proc} cubic subdomains, the proportion of halo-to-inner particles H_p is given by:

$$H_p = 6\,CN_{proc}^{1/3} + 12\,C^2 N_{proc}^{2/3} + 8\,C^3 N_{proc} - H_{p,1} \tag{3}$$

where $C = R_\Omega/L$. The constant term $H_{p,1}$ is equal to $6C + 12C^2 + 8C^3$ because there are no halo zones on the boundaries of the overall domain. According to Eq. 3, H_p scales with $\sqrt[3]{N_{proc}}$ for low N_{proc} and have an asymptotic behavior linear with N_{proc}. Note that Eq. 3 can be written $H_p = 3\Gamma + 3\Gamma^2 + \Gamma^3 - H_{p,1}$ where $\Gamma = 2R_\Omega (N_{proc}/V_t)^{1/3}$, V_t being the total volume of the domain. To give an order of magnitude of Γ, the halo-to-inner particle ratio reaches 20, 50 and 80% for $\Gamma=0.068$, 0.149 and 0.220, respectively.

The parallelisation method is the domain decomposition. The numerical domain is decomposed according to its geometry. Given that in a typical simulation the particles are homogeneously distributed on a large scale, the density of particle is rather constant. The domain is decomposed in such a way that the subdomains have a structured Cartesian connectivity. It is illustrated in Fig. 3. The number of slices in the x-, y-, z-direction is prescribed as in input parameters of the decomposition. First, the domain is split along y in $N_y - 1$ slices. The algorithm moves the boundaries to reach the best balance, *i.e.* the number of particle per slice is the closest to $N_t/(N_y - 1)$ in each slice (Fig. 3b). The increment to move the boundary is the size of a search cube. At the end of this step, the position of N_y interfaces in y is set. Second, the same process is applied in the x-direction, individually to each y-slice (Fig. 3c), leading to the position of $N_y \times N_x$ interfaces in x. In the last step, the each $xy-$columns are decomposed along z, leading to $N_y \times N_x \times N_z$ interfaces in z.

The advantage of this partitioning method is that the inner subdomains are in the form of a parallelepiped, which is appropriate with the Cartesian background grid used for the list search. When the boundaries of the domain are round or tilted, the method may lead to subdomains of high skewness ratio. Such subdomains have a detrimental effect because their inter-processor interface is large, which leads to (i) a larger volume of inter-processor communication and (ii) an increase of the number of halo-particles.

Despite the optimizing loops to find the best load balance, a residual imbalance is observed in practice. We define the idle rate I_r, which corresponds to the time proportion of CPUs in idle state when they wait for the slowest CPU before synchronization. We assume here that the waiting time is proportional to difference of particle number between two CPUs. The idle rate is expressed as:

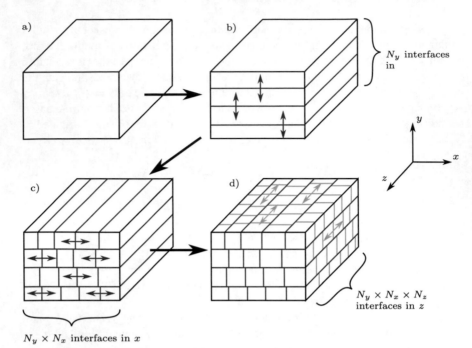

Fig. 3 Overall strategy to decompose the domain

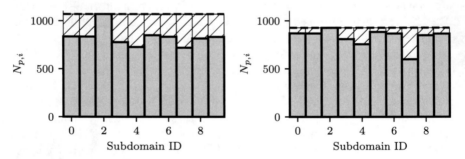

Fig. 4 Oversized domain (left) strongly increase the idle rate (29.0%) whereas undersized domain (right) does not impact the idle rate (11.8%)

$$I_r = \frac{\sum_i (\max(N_{p,i}) - N_{p,i})}{\sum_i N_{p,i}} = \frac{\max(N_{p,i})}{\overline{N}_p} - 1 \qquad (4)$$

where $N_{p,i}$ and $\overline{N}_p$ are the number of particle in the subdomain i and the mean number of particle per subdomain, respectively. As illustrated in Fig. 4, a single oversized subdomain has a great impact on the idle rate (left) whereas a single undersized subdomain has a minor effect (right).

As it was seen that a residual load imbalance subsists after decomposition, this means that for a given domain and an approximate number of partitions, there exist different triplets (n_x, n_y, n_z) leading to different idle rates. Therefore, before performing the production simulations, the triplets that minimize the idle rate must be determined. This will be illustrated later.

In our expression of the mass conservation (Eq. 2a), the density of the gas is directly related to the density of the particles, *i.e.* how dense they are packed, and from Eq. 2b, the pressure depends solely on the gas density. Hence, there is link between the pressure inside the fluid and the spatial density of the particles. Thus, in flows exhibiting a mean pressure gradient (*e.g.* a pipe flow, and more generally, in wall-bounded flows), the density of the particles varies accordingly to the local value of the mean pressure. This aspect has a consequence on the load balancing during the simulation. Usually, the partitioning is performed on an initial solution where the flow is not established and where all particles are regularly distributed. When a wall-bounded flows reaches the steady-state, the pressure gradient induces a gas density gradient, which in turns, leads to a gradient of particle density, and possible load imbalance. Therefore, for such flows, it is expected that the load balance evolves considerably during the simulation. In this case, it is recommended to perform a new partitioning of the numerical domain once the steady is reached. An illustration is provided in the following.

4 Application

In this section, the previous considerations are applied to a real case. First, the strong scaling of the code is assessed, then the idle rate is monitored during a production case. The case consists of an air-assisted atomizer operating at an ambient pressure of 11 bar. The domain is depicted in Fig. 5. It contains approximately 1.12 billion of particles in average. The gas and liquid are injected through permeable inlets and they exit the domain at the outlet where an outflow boundary condition is imposed. Note that the particle velocity at inlets and outlet are modified to reduce reflected pressure waves [3]. As a no-slip condition is imposed at the wall of the injector, a mean pressure gradient is expected in the inlet ducts. The speed of sound is set to 340 m/s for the two phase, and the isentropic ratio is classically set to 1 and 7 for the gas and liquid phase, respectively.

In order to produce strong scaling curves, the domain was partitioned in N_{proc} subdomains, where N_{proc} ranges from 750 to 15000. For each targeted N_{proc}, the best triplet (n_x, n_y, n_z) that minimizes the idle rate was set. The idle rate and the halo-particle proportion are shown in Fig. 6. Black dots are the triplet (n_x, n_y, n_z) minimizing the idle rate for a given $\approx N_{proc}$. Globally, the idle rate shows a linear increase up to 10000 cores, after which it reaches a saturation. Up to 15000 cores, it remains below four percent. The proportion of halo-particle is more regular and follows the trends predicted by Eq. 3. In this case, H_p is larger than the estimation from Eq. 3 (dashed lined) because the subdomains as well as the overall domain

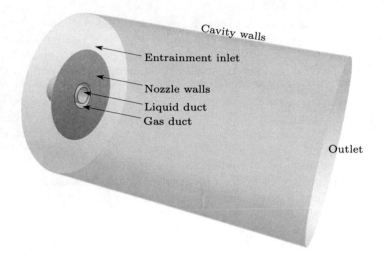

Fig. 5 Numerical domain. The fluids are flowing from the left to the right

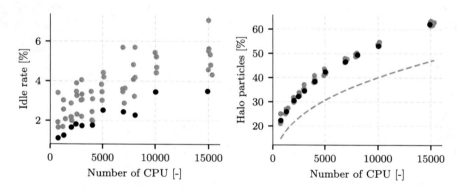

Fig. 6 Idle rate (left) and proportion of halo-to-inner particles (right) versus the total CPU number. Black dots are the triplet (n_x, n_y, n_z) minimizing the idle rate for a given $\approx N_{proc}$. Dashed line is the smallest H_p possible according to Eq. 3

are not perfectly cubic. Not only the growth but also the percentage of the number of halo-particle is significant and it suggests the existence of a limit beyond which computational overheads due to halo-particles would excess the costs of regular particles. Hence, there would be a limit to the total number of cores with the halo-particles technique. The trends for the development of new supercomputer shows that more and more cores sharing the same memory are mounted on the same node. In this context, the use of OpenMP would be greatly beneficial to circumvent the limit due to halo-particles. With the use of OpenMP, no halo-particles would be necessary for subdomains on the same node.

Table 1 Best partitioning parameters to minimize the idle rate, that were used to assess the strong scaling.

N_{proc} [−]	(n_x, n_y, n_z) [−]	Idle rate [%]	Halo particles [%]
2016	(12, 12, 14)	1.66	30.5
3024	(12, 12, 21)	1.74	34.4
4032	(12, 12, 28)	1.76	38.4
5040	(12, 12, 35)	2.52	42.4
6975	(15, 15, 31)	2.44	46.5
8100	(15, 15, 36)	2.28	49.4

To produce the scaling curve, the smallest number of cores on which the case could run was ≈ 2000 and the largest tested is 8000. The partitioning characteristics for the test cases are reported in Table 1.

An illustration of the domain decomposition is provided in Fig. 7 (top) for 2016 and 5040 subdomains and shows a variety of different aspect ratios. Except for (i) subdomains covering the inlet ducts and the cavity at the same time and for (ii) subdomains at the round boundaries, all subdomains are parallelepipedic. The corresponding repartition of particles among the subdomains is presented in Fig. 7 (bottom). They both show moderate idle rates of 1.37 and 2.62%, respectively. As mentioned earlier, the processors with <0.22 million particles for $N_{proc} = 5040$ has a very small influence on the idle rate.

For the determination of the speedup, the reference case is $N_{proc} = 2016$. For each processor number, the simulation was performed six times during 1000 iterations. Only the time loop was accounted for, *i.e.* the time for initialization and read of the input files was dismissed. In order to depict the pure scaling feature of the code, the simulation times were weighted by the idle rate. The strong scaling curve is shown in Fig. 8, compared to the ideal scaling. The deviation from the ideal scaling starts at 4000 CPU and further increase. A possible explanation is the requirement of halo-particles at the inter-processor interface that represent almost half of the inner particle for 8100 processors.

As mentioned earlier, the spatial repartition of the particles depends on the flow, especially on the pressure distribution. For the initial solution all the particles are prescribed a zero velocity. At the beginning of the simulation, the inlet imposes directly the nominal velocity of 60 and 1.4 m/s for the gas and the liquid respectively. This create a pressure wave that propagates in the cavity. Despite the modification of the outlet and inlet velocity to damp the pressures waves, there are some residual pressure fronts that are reflected at the inlet and outlet. The influence of this pressure waves on the idle rate is discussed below.

The simulation is run up to 5 ms and solutions are exported every 0.04 ms. For each exported solution, the idle rate is computed. It is plotted in Fig. 9 During the simulation, different triplets (n_x, n_y, n_z) were tested. The zones a, b, c, d correspond

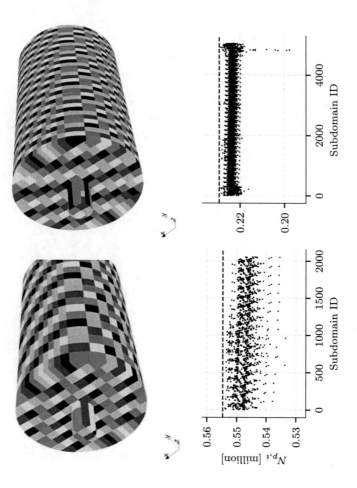

Fig. 7 Top: illustration of the domain decomposition for 2016 (left) and 5040 (right) processors. Bottom: corresponding repartition of particle per subdomain,

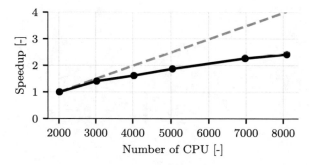

Fig. 8 Strong scaling for 1.4 billion particles, superimposed with the ideal curve in dashed line

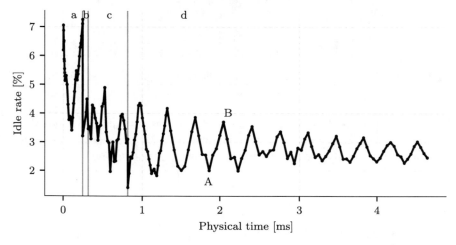

Fig. 9 Evolution of the idle rate during the transient regime. Zones a, b, c and d correspond to decomposition on 2000, 5040, 2000 and 2057 subdomains, respectively

to 2000, 5040, 2000 and 2057 cores, respectively. At the beginning of the simulation, the idle rate on 2000 cores is very large compared to the values given in Table 1 because the geometry is slightly different. In addition, no attention was paid to obtain the best idle rate because the focus was on the stability of the simulation at the very beginning. After the third decomposition (c), the idle rate is between 2 and 5% for the same number of cores as in (a). This highlights the usefulness to redecompose the domain after a few acoustics times. The fourth decomposition (d) was performed after one convective time, and the case was run for almost 4 ms. The idle rate shows decreasing oscillations of period 350 µs. This period is approximately four time the propagation time τ_{pw} of a pressure wave through the cavity. This is an evidence that the evolution of load balancing during the transient state is related to the pressure waves arising at the start of the calculation. To give a better insight of this phenomenon, the deviation of the particle volume to the reference volume dx^3 is depicted in Fig. 10 on the center-slice of the domain for a physical time located at

Fig. 10   Volume deviation at location A (top) and B (bottom) in Fig. 9

A and B in Fig. 9. These times correspond to a local minimum and maximum of the
idle rate. It is recalled that, in the present configuration, the relative variations of the
particle volume, density and pressure in the gas phase are related by:

$$\frac{\delta V}{V} = -\frac{\delta \rho}{\rho} = -\frac{\delta p}{p} \tag{5}$$

Therefore the relative variation of the volume is opposite to the one of pressure. For
a local minimal idle rate (A), the volume globally deviates by -1.5%, *i.e.* a positive
variation of the pressure. If the particle volume is smaller, the particles are closer
to each other, *i.e.* there are more particles inside the domain. When the idle rate is
maximal (B), the particle volume is 1.5% larger than dx^3 and the numerical domain
contains less particles. At the outlet the volume is always approximately constant to
dx^3, which is imposed by the pressure outlet condition.

   The pressure reflection occurs as follow. We start from the situation A where the
pressure (p^+) is larger in the cavity than at the outlet. The outlet impose the reference
pressure p_{ref}, thus generating a falling pressure wave that propagates upstream. After
a period of τ_{pw}, the wave has propagated through the whole cavity, and the mean
pressure inside the domain is the reference pressure. This leads to an intermediate
number of particles inside the domain and an intermediate idle rate (see Fig. 9). At
this time, the wave reaches the inlet which provides a constant velocity and thus it
generates another falling pressure wave (of level p^-) that propagates downstream. At
$t = 2\tau_{pw}$, the pressure inside the domain is p^-, equivalent to fewer particles and the
largest idle rate, *i.e.* point B in Fig. 9. At this time, the outlet boundary condition, since
it imposes p_{ref}, generates a rising pressure wave propagating upstream to balance
the pressure. When this wave reaches the inlet at $t = 3\tau_{pw}$, the pressure inside the

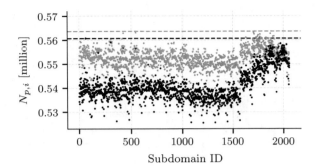

Fig. 11 Repartion of particles amongs the subdomains for point A (grey) and B (black) in Fig. 9

domain is p_{ref} and the idle rate is moderate. Finally, the inlet generates another rising pressure wave of level p^+, which reaches the outlet at $t = 4\tau_{pw}$. At this moment, the domain contain a larger number of particles and the idle rate is minimal. The cycle is then completed.

The repartition of particles inside the subdomains at point A and B is shown in Fig. 11. As expected the number of particle is larger at point A where the idle rate is the lowest. At point B, the total number is lower, but this diminution is not homogeneously distributed among the subdomains. Indeed, there are still subdomains with a number of particles similar to point A. This leads to a few subdomains with a number of particle significantly larger than the rest, thus degrading the idle rate.

To circumvent this phenomenon, one could develop more efficient non reflecting boundary conditions. However, with the present approach for the mass conservation, *i.e.* the density expressed algebraically with the particle distance, the density cannot be directly set to damp the pressure wave. The only way to modify the density is to impose a different velocity on neighboring particle, which is not straightforward.

In the present case, the pressure waves are generated at the beginning of the simulation due to a to steep velocity rising front imposed at the inlet. Therefore, a lighter case similar to the previous configuration is tested, where the initial inlet velocities are set to zero and gently increased to the prescribed value. This way, the initial pressure waves are weaker and the current damping method is sufficient to ensure a smooth pressure field at the outlet. The Error-Function is used for the time profile of the inlet velocities, so that it reaches 95% of the target velocity at $t = 3$ ms. To avoid a too steep velocity growth, it is verified that the time derivative of the inlet velocity is much lower than a characteristic acceleration due to compressibility, expressed here as c^2/h. One simulation (labeled 'front') is run with constant velocities imposed at the inlet, and the other simulation (labeled 'smooth') is run with the smooth increasing velocities.

The total number of particles versus the physical time is shown in Fig. 12 (left), superimposed with the shape of the time profile of the inlet velocities. With the 'smooth' case, the particle number does not sharply oscillate as compared to 'front'. However, when the velocity reaches ≈40% of its final value, oscillations are observed, with an amplitude smaller than with 'front'. Concerning the idle rate (Fig. 12 right),

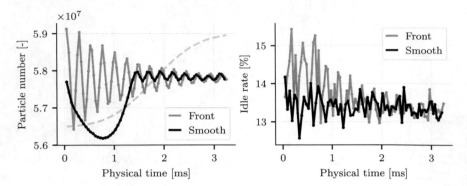

Fig. 12 Left: time evolution of the total of number of particles, superimposed with the shape of the imposed inlet velocities (dashed line). Right: time profile of the idle rate

there is a clear improvement at the beginning of the simulation. Hence, the presented strategy to decrease initial pressure waves is effective.

5 Conclusion

In this study, it was shown that decomposing the domain with a Cartesian connectivity fits well with the list-search algorithm but it leads to a slight residual load imbalance in practice. Also, it is recommended to search for the triplet (n_x, n_y, n_z) that minimizes the idle rate before running production cases.

With the SPH method, the computational overheads of halo-particles may limit the total number of cores. However, the increase of the number of cores per node together with the use of OpenMP is a promising solution to circumvent this limitation. The determining parameter is $\Gamma = 2 R_{\Omega} (N_{proc}/V_t)^{1/3}$. As a rule of thumb, the halo-to-inner particle ratio reaches 20, 50 and 80% for Γ=0.068, 0.149 and 0.220, respectively.

When the pressure in the fluid is related to the density of particles, significant variations of the load balancing can be observed during the simulation. This is especially visible in the presence of pressure waves together with when the gas density is computed algebraically from the particle spatial distribution. A smooth increase of the velocities from zero to their target values successfully decreases the amplitude of the pressure waves inside the domain at the beginning of the simulation.

Acknowledgements The authors like to thank the Helmholtz Association of German Research Centres (HGF) for funding (Grant No. 34.14.02). This work was performed on the computational resource ForHLR Phase II funded by the Ministry of Science, Research and the Arts Baden-Württemberg and DFG ("Deutsche Forschungsgemeinschaft").

References

1. S. Braun, Zur Simulation der Zerstäubung flüssigen Kraftstoffs mit der Smoothed Particle Hydrodynamics Methode. Logos Verlag Berlin GmbH (2018)
2. S. Braun, R. Koch, H.J. Bauer, Smoothed Particle Hydrodynamics for Numerical Predictions of Primary Atomization, pp. 321–336. Springer (2016). https://doi.org/10.1007/978-3-319-47066-5_22
3. S. Braun, L. Wieth, R. Koch, Bauer, A framework for permeable boundary conditions in SPH: Inlet, outlet, periodicity. In: Proc. of the 10th Internat. SPHERIC Workshop (2015)
4. G. Chaussonnet, S. Braun, T. Dauch, M. Keller, A. Sänger, T. Jakobs, R. Koch, T. Kolb, H.J. Bauer, Toward the development of a virtual spray test-rig using the smoothed particle hydrodynamics method. Comput. Fluids **180**, 68–81 (2019). https://doi.org/10.1016/j.compfluid.2019.01.010
5. C. Höfler, Entwicklung eines Smoothed Particle Hydrodynamics (SPH) Codes zur numerischen Vorhersage des Primärzerfalls an Brennstoffeinspritzdüsen. Logos Verlag Berlin GmbH (2013)
6. J.J. Monaghan, Smoothed particle hydrodynamics. Rep. Prog. Phys. **68**, 1703–1759 (2005)

Transport and Climate

Markus Uhlmann

The simulations in the category "Transport and Climate" are representative of the numerical investigations performed by the meteorological research groups at Karlsruhe Institute of Technology and at University Hohenheim. In the present granting period there have been five projects running on the system HazelHen (HLRS) and 12 on the twin systems ForHLR I/II (SCC), consuming approximately 53 million core-hours in total. The consumption of computational resources in this category was roughly evenly distributed between these two computing centers. The entire set of projects was based upon only four different numerical model families: COSMO, ICON, WRF, and EMAC.[1]

The present collection features contributions from three climate science studies, two of which are employing the "Weather Research and Forecasting" (WRF) model developed by NCAR, and one is based on the "COnsortium for Small scale MOdelling" (COSMO) model of DWD.

The authors of the first study ("HRCM" by Schädler et al.) report on their work with the climate version of the DWD community model COSMO-CLM. The focus is on the analysis of the impact of land use change upon climate change, and on the effect of using very high spatial resolution upon the quality of prediction of heat wave scenarios. For the first part of the study, a separate land surface model (VEG3D) has been coupled to the principal model, and it was concluded that the computational overhead could be kept small by making use of communication overlap. Concerning the prediction of heat waves, the authors could demonstrate a clear benefit of further increasing the spatial resolution down to 7km mesh width on the finest nesting level. The research performed in the project "HRCM" is also directly benefiting ongoing attempts to quantify the time of arrival of two-degree average global warming, and the spatial heterogeneity thereof.

The second project "W2W-WRFHYDRO", authored by Arnault and Kunstmann, focuses on the influence of terrestrial hydrological modeling on the prediction of convective events in the atmosphere. The project makes use of the WRF model with an enhanced soil–vegetation–atmospheric water tagging, and simulations are presented

---

[1]M. Uhlmann, Institute for Hydromechanics, Karlsruhe Institute of Technology, Kaiserstr. 12, 76131 Karlsruhe, e-mail: markus.uhlmann@kit.edu

for the upper Danube river basin over a period of 40 months with a spatial resolution of 500m on the finest nesting level. This setup enables the authors to perform a very fine-grained analysis of the partitioning in the terrestrial water balance.

The third project by Bauer et al. ("WRFSCALE") aims at land–atmosphere coupling processes with the aid of large-eddy turbulence modeling in the context of the WRF platform. The authors have investigated a variety of aspects in different scenarios: from high-impact weather in the American great plains area, land-surface modification effects and forecasting in the United Arab Emirates, assimilation of LIDAR data into their model, to water vapor content predictions over the Ammer catchment area. Overall, the project "WRFSCALE" impressively demonstrates how nested large-eddy simulations on different scales can help to advance our knowledge of atmospheric processes, in particular where a strong coupling to surface conditions prevails.

Regional Climate Simulations with COSMO-CLM:CORDEX FPS Convection and Heat Stress in Cities

G. Schädler, H.-J. Panitz, and B. Brecht

Abstract The IMK-TRO at the Karlsruhe Institute of Technology (KIT) presents in the HLRS annual report for 2018 selected projects in which the CRAY XE40 "Hazel Hen" was used. Different topics have been covered in 2018. In this report however, we focus on our very high resolution regional climate simulations within CORDEX FPS Convection and the simulation of present and future heat stress, especially in cities. The simulations are performed with the regional climate model COSMO-CLM (CCLM) and cover spatial resolutions from 50 to 2.8 km. The required Wall-Clock-Time (WCT) ranges from about 2000 to more than 30000 node-hours per simulated year.

1 Overview

The working group "Regional Climate and Water Cycle" of the Institute for Meteorology and Climate Research—Department Troposphere Research (IMK-TRO) at the Karlsruhe Institute of Technology (KIT) (www.imk-tro.kit.edu) uses the climate version of the COSMO model (CCLM) on the CRAY XC40 "Hazel Hen" at the HLRS high performance computing facilities to investigate past, present and future regional climate in subregions of Europe, the Arctic and Africa.

Topics of our studies include regional climate change with a focus on extremes, ensemble simulations, land surface processes, decadal predictability, very high resolution climate simulations at convection permitting scales and simulations of regional paleoclimates. We are also involved in the CORDEX FPS and CORDEX CORE activities.

G. Schädler (✉) · H.-J. Panitz · B. Brecht
Institute for Meteorology and Climate Research, Department Troposphere Research (IMK-TRO),
Karlsruhe Institute of Technology (KIT), Karlsruhe, Germany
e-mail: gerd.schaedler@kit.edu

© Springer Nature Switzerland AG 2021

481

W. E. Nagel et al. (eds.), *High Performance Computing in Science and Engineering '19*,
https://doi.org/10.1007/978-3-030-66792-4_32

Topics include

- Very high resolution regional climate prognoses and projections
- Analysis of climate extremes (heat waves, floods, draughts)
- Assessment of uncertainty via ensemble simulations
- Interaction between land surfaces and the atmosphere
- Simulation of regional paleoclimates.

In this report, we describe in some detail the current state of two of these topics which are interesting for both their practical relevance and their numerical/computational aspects: our activities within the CORDEX Flagship Project (FPS) Convection and the assessment of present day and future heat stress in cities with very high resolution regional climate simulations.

Section 2 describes CCLM and Sect. 3 describes the projects mentioned above in more detail. Section 4 gives an overview over the resources used.

2 The CCLM Model

The regional climate model (RCM) COSMO-CLM (CCLM) is the climate version of the operational weather forecast model COSMO (Consortium for Small-scale Modeling) of the German Weather Service (DWD). It is a three-dimensional, non-hydrostatic, fully compressible numerical model for the atmosphere and also includes a land surface model. The model solves prognostic equations for wind, pressure, air temperature, different phases of atmospheric water, soil temperature, and soil water content.

Further details on the COSMO model and its application as a RCM can be found in [1, 2] and on the web-pages of the COSMO consortium (www.cosmo-model.org) and of the CLM Community (www.clm-community.eu).

3 Regional Climate Simulations Using the HLRS Facilities

3.1 CORDEX FPS (Flagship Pilot Studies)

CORDEX is the acronym for the Coordinated Regional Climate Downscaling Experiment CORDEX (www.cordex.org), which has been initiated by the World Climate Research Programme (WCRP) in 2009 to produce standardised regional climate and climate change projections for the terrestrial regions of our planet; Fig. 1 shows the standard CORDEX Europe domain. CORDEX also contributes to the IPCC Assessment reports. The major aims of CORDEX are to provide a coordinated model evaluation framework, a climate projection framework, and an interface to the applicants of the climate simulations in climate change impact, adaptation, and mitigation

Fig. 1 The CORDEX
Europe domain

studies. Within CORDEX, Flagship Pilot Studies (FPS) have been created with the aim to provide a coordinated approach to specific research questions. Our group is contributing to two of these: one dealing with very high resolution (convection permitting) regional climate simulations (CORDEX FPS Convection) and the other with the impact of land use changes on climate (CORDEX FPS LUCAS). Below, we describe our activities within CORDEX FPS Convection.

3.1.1 CORDEX-FPS Convection

CORDEX-FPS Convection is one of the first endorsed Flagship Pilot Studies (FPS) in CORDEX and is a shared initiative by Med-CORDEX and Euro-CORDEX. Its aim is to model convective phenomena at high resolution over Europe and the Mediterranean and to assess the added value of such high resolution simulations, also in view of the considerably higher computational costs. The chosen resolution is about 3 km, which is considered a resolution where deep convection is explicitly—and therefore possibly better—simulated by the models. The scientific objectives are to investigate convective-scale events, the processes involved and their trends in key regions of Europe and the Mediterranean using a range of RCMs at convection-permitting (i.e. high) resolution in order to answer the questions

- How do convective events and associated damaging phenomena (heavy precipitation, wind storms, flash-floods) respond to changing climate conditions in different climatic regions of Europe?
- Does an improved representation of convective processes and precipitation at convection permitting scales lead to added value?

Fig. 2 The CORDEX FPS Convection Mandatory Domain (green frame)

To make the results of the different groups involved comparable, standard simulation domains (so-called mandatory domains) and test domains have been defined (see Fig. 2). In order to save CPU time and storage, these are currently subsets of the CORDEX Europe domain. During the last year, a series of sensitivity studies has been performed to prepare the long term—full domain simulations. These studies included the model setup, the sensitivity to domain choice and differences between runs in weather versus climate mode.

In a shared effort, a large series of sensitivity tests (we did more than 30) was performed in order to find the optimal configuration for the computationally very expensive multidecadal high resolution simulations which are to be performed in 2019. The reference model setup was DWD's COSMO-DE setup, and comparisons were made for near-surface (2m) temperature and precipitation. All simulations were driven by ERA40 Reanalyses and run for one year in the domain shown in Fig. 2. The following (not exhausting) list gives an overview over the configuration options which were considered:

- change to a different basic model setup (COSMO-DE vs. MeteoSwiss)
- nesting strategy (several steps vs. direct downscaling)
- number of vertical levels in the atmosphere
- update frequency of boundary data
- width of lateral sponge zone
- changing the background albedo
- parameterisation of subgrid scale orography

- type of aerosol climatology
- frequency of radiation routine call
- parameterisation of shallow convection
- different modeling of bare soil evaporation
- weighting of horizontal diffusion
- ...and several more.

As a result of these tests, we arrived at the following conclusions:

- COSMO-DE (reference) seems to be an adequate basic setup, and the MeteoSwiss setup does not improve the results sufficiently to invest more CPU time required with a higher number of vertical levels.
- slight improvements of 2 m temperature results can be achieved with alternative albedo data (monthly varying background albedo) and the resistance version of bare soil evaporation, together with increase of the evaporating fraction of grid cells over land.
- alternative aerosol distributions (Tegen, for example) have negative impact.
- precipitation results, especially over mountainous regions, can be improved by changing parameters that affect the strength of horizontal diffusion.
- updating of boundary (forcing) data every three hours seems to be a reasonable compromise between hourly and six-hourly updating.

These and the results of other groups were discussed at the Lisbon meeting in November 2018 and will be the basis for the production runs which are to be performed in 2019.

In addition to these sensitivity studies, we contributed CCLM simulations to three test cases which had been agreed upon during a 2017 meeting in order to assess the ensemble performance with a view to the representation of extreme precipitation events. We quote here from [3]: 'The test cases covered a summertime extreme precipitation event over Austria, a fall Foehn event over the Swiss Alps and an intensively documented fall event along the Mediterranean coast. The test cases were run in both "weather-like" (WL, initialized just before the event in question) and "climate" (CM, initialized 1 month before the event) modes. Ensembles of 18–21 members, representing six different modeling systems with different physics and modelling chain options, were generated for the test cases (27 modeling teams have committed to perform the longer climate simulations). Results indicate that, when run in WL mode, the ensemble captures all three events quite well with ensemble correlation skill scores of 0.67, 0.82 and 0.91. They suggest that the more the event is driven by large-scale conditions, the closer the agreement between the ensemble members. Even in climate mode the large-scale driven events over the Swiss Alps and the Mediterranean coasts are still captured (ensemble correlation skill scores of 0.90 and 0.62, respectively), but the inter-model spread increases as expected. In the Mediterranean case, the effects of local-scale interactions between flow and orography and land-ocean contrasts are readily apparent. However, there is a much larger, though not surprising, increase in the spread for the Austrian event, which was weakly forced by the large-scale flow. Though the ensemble correlation skill score

Table 1 Thermal comfort versus UTCI equivalent temperature

Thermal stress	UTCI emperature range ($°C$)
Extreme heat stress	Above 46
Very strong heat stress	From 38 to 46
Strong heat stress	From 32 to 38
Moderate heat stress	From 26 to 32
Thermal comfort	From 18 to 26
No thermal stress	From 9 to 18
Low cold stress	From 0 to 9
Moderate cold stress	From −13 to 0
Strong cold stress	From −27 to −13
Very strong cold stress	From −40 to −27
Extreme cold stress	Below −40

is still quite high (0.80). The preliminary results illustrate both the promise and the challenges that convection permitting modeling faces and make a strong argument for an ensemble-based approach to investigating high impact convective processes.'

3.2   Heat Stress in Cities

Extended extreme events like the long-lasting heat waves and droughts like the dry and hot summers of 2003 in Western Europe and 2018 in Central and Northern Europe are likely to become more frequent in the future. Apart from affecting vulnerable sectors like agriculture, forestry, water management and transport, such events are a threat to public health due to the heat stress involved. This affects especially urban areas due to the high population density there, reduced ventilation, reduced evapo-transpiration and anthropogenic heat input. To estimate the scale and impact of such events on thermal comfort, very high resolution simulations with CCLM for a present and a future period were performed within a doctoral thesis (B. Brecht).

Thermal comfort (or discomfort) is the result of several environmental factors like air temperature, radiation, humidity and wind, as well as human activity, metabolic rate and clothing. These factors are accounted for in the UTCI (Universal Thermal Climate Index, [4]), which was chosen in this study as a measure of heat stress.

The UTCI translates thermal comfort into an equivalent temperatures as indicated in Table 1:

As data for the calculation of present day and future heat stress, an ensemble of regional climate simulations with CCLM was used. The ensemble was created with driving data of four global CMIP5 models, namely ECHAM6, EC-Earth, CNRM-CM5 and HadGEM2. Periods considered were the control period 1981–2000 and the projection period 2031–2050. Future emission scenarios were RCP 4.5 and RCP

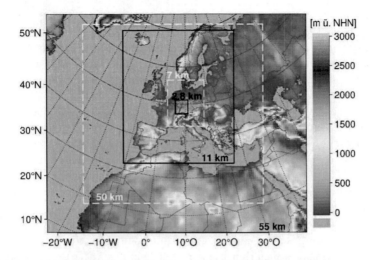

Fig. 3 Simulation domains used in the heat stress study

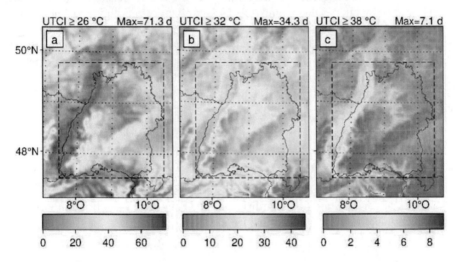

Fig. 4 Number of days per year with a given UTCI for the control period

8.5. Spatial resolution of the analysed grids was 7 and 2.8 km. For one detailed study of urban climate, a resolution of 925 m was used.

The various simulation domains (including the coarser nests) are shown in Fig. 3.

Although high resolution simulations tend to produce better results than coarser resolution ones [5], a bias correction turned out to be necessary to calculate the UTCI. For the first time, a bias correction using seasonal hourly values with the multivariate bias correction for climate model projections of several climate variables was performed.

Figure 4 shows the number of days per year with a given UTCI for the control period: moderate (left), strong (middle) and very strong (right) heat stress. One can see that extreme heat affects mainly the Rhine Valley with the Rhine-Neckar and Rhine-Main metropolitan regions. The highest number of days with extreme heat occurs in its southern parts, but also the Neckar (Stuttgart) and the Main (Frankfurt) valleys still have several days per year.

A comparison between the future and the control period showed that the average UTCI as well as the number, duration and intensity of heat stress events will increase considerably in the future. In some regions of Germany about 50 % more days with strong heat stress (UTCI $\geq 32\,°C$) and more than 100 % more days with very strong heat stress (UTCI $\geq 38\,°C$) are projected. Our results also indicate that in an average future summer of the years 2031–2050 more hours with heat stress occur than in 85 % of the summers in the years 1981–2015.

In a detailed study, the thermal climate in the cities of Karlsruhe and Freiburg was studied. For this purpose, simulations at 925 m resolution over the area of Baden-Württemberg were performed (innermost rectangle in Fig. 3). The results confirmed that heat stress is considerably more frequent in the urban areas than in their surroundings and that this is due mainly to a lack of nighttime cooling, thus increasing the heat load from day to day. At this resolution, local drainage wind systems are captured by the model and interact with the urban heat island. It could be shown that the nocturnal heat stress is reduced in those quarters of Freiburg which are reached by the local drainage wind system, as is also observed. This corroborates the requirement that existing cold air production areas and drainage flow corridors should be kept open.

4 Technical Aspects

The issues of parallelization, scalability and efficiency have been briefly discussed in earlier Annual Reports (e.g. Annual Report 2015).

4.1 Nesting Strategy

Weather and climate include processes over a wide range of spatial scales. The aim of nesting is to efficiently obtain detailed (= high resolution) information in the region of interest while also taking the large scale (synoptic, global) current weather situation at a coarser resolution into account. The nesting strategy is illustrated in Fig. 5: information (e.g. wind, pressure, temperature) flow at the respective domain boundaries is from the global (resolution about 1–2°) via the European (resolution about 0.5 °) to the German nest (resolution about 0.05 °); update frequency is about three to six hours. This means that nesting has to be done in a sequential instead of a parallel way, i.e. multiple nesting requires consecutive simulations, one for each

Fig. 5 From global
resolution to very high
regional resolution:
illustration of the multiple
nesting strategy (here two
nests) used in CCLM
applications

nest. Local grid refinement is not possible in COSMO-CLM. In most cases, the
nesting sequence is global to 0.44° to 0.0625° to 0.025°. Refining the horizontal grid
spacing and keeping the domain of the coarse grid would increase the computational
needs and the required storage capacity considerably: simulating the whole European
domain at 0.025° resolution would result in linear increase of CPU time by a factor
of about 1000. Although the real increase of computing time would be less than this
linear estimate because more horizontal grid points would allow the usage of more
computational nodes, the required CPU time would still be prohibitive and make such
simulations on climate time scales (several decades) impossible. The number of cores
and thus the number of nodes that can be used reasonably is limited by the number of
grid-points in each horizontal coordinate direction and by the numerical advection
scheme used; in our simulations, we used about 50 to 90 cores. Experiments with the
CCLM model shows that the ratio between the number of grid points and the number
of computational cores in each horizontal direction should be in the order of ten in
order to achieve a good balance between the times needed for the pure computations
and the communication between parallel cores.

Table 2 Summary of computing demands on CRAY XC40 "Hazel Hen" at HLRS

Project	Domain Size	Grid resolution (km)	Time-step (sec)	Computing time (node-h/year)	No. of sim. years	Storage Needs (Gbyte/year)	Total Computing Time (node-h)	Total Storage Needs (Tbyte)
FPS-Convection	522 * 490 * 50	3	25	3900	36	5500	141000	198
HiRes 2.8 km	322 * 328 * 40	2.8	25	2000	10	1000	21900	10
Urban Climate 1 km	480 * 480 * 76	0.925	10	32600	0.25	1600	8150	0.4

4.2 Coupling of Model Components

In the CORDEX LUCAS project, the atmospheric model CCLM was coupled to the land surface model VEG3D using the OASIS coupler. In these simulations , the extra costs of the OASIS3-MCT coupler were small. In comparison to a standard CCLM simulation (with built-in land surface model), the computational costs are increased by just about 5%. This is achieved by a fully parallelized exchange of the coupling fields between CCLM and VEG3D, provided by the new OASIS version (OASIS3-MCT), which significantly improves the coupling performance in comparison to older versions.

4.3 Ressources Used

Table 2 summarizes the resources used on the CRAY XC40 "Hazel Hen" at HLRS.

References

1. M. Baldauf, A. Seifert, J. Förstner, D. Majewski, M. Raschendorfer, T. Reinhardt, Operational convective-scale numerical weather prediction with the cosmo model: Description and sensitivities. Mon. Wea. Rev. **139**(12), 3887–3905 (April 2011)
2. B. Rockel, A. Will, and A. Hense. The regional climate model COSMO-CLM (CCLM). *Meteorologische Zeitschrift*, 17(4):347–348, 08 2008
3. E. Coppola, S. Sobolowski, E. Pichelli, F. Raffaele, B. Ahrens, I. Anders, N. Ban, S. Bastin, M. Belda, D. Belusic, A. Caldas-Alvarez, R. M. Cardoso, S. Davolio, A. Dobler, J. Fernandez, L. Fita, Q. Fumiere, F. Giorgi, K. Goergen, I. Güttler, T. Halenka, D. Heinzeller, Ø. Hodnebrog, D. Jacob, S. Kartsios, E. Katragkou, E. Kendon, S. Khodayar, H. Kunstmann, S. Knist, A. Lavín-Gullón, P. Lind, T. Lorenz, D. Maraun, L. Marelle, E. van Meijgaard, J. Milovac, G. Myhre, H.-J. Panitz, M. Piazza, M. Raffa, T. Raub, B. Rockel, C. Schär, K. Sieck, P. M. M. Soares, S. Somot, L. Srnec, P. Stocchi, M. H. Tölle, H. Truhetz, R. Vautard, H. de Vries, and K. Warrach-Sagi.

A first-of-its-kind multi-model convection permitting ensemble for investigating convective phenomena over europe and the mediterranean. *Climate Dynamics*, 2018

4. G. Jendritzky, R. de Dear, G. Havenith, UTCI - why another thermal index? International Journal of Biometeorology **56**(3), 421–428 (2012)
5. J. Hackenbruch, G. Schädler, J.W. Schipper, Added value of high-resolution regional climate simulations for regional impact studies. Meteorologische Zeitschrift **01**, (2015)

The Role of Soil Moisture and Surface and Subsurface Water Flows on Predictability of Convection

J. Arnault and H. Kunstmann

Abstract Since June 2018 we have further assessed the mechanism through which a more sophisticated treatment of terrestrial hydrological processes in a numerical weather prediction model potentially improves the predictability of convection. In order to achieve this, we have implemented the so-called soil-vegetation-atmospheric water tagging procedure in WRF and WRF-Hydro (Skamarock and Klemp in J Comp Phys 227:3465–3485, 2008 [3]; Gochis et al. in TheWRF-Hydro model technical description and user's guide, version 3.0, NCAR Technical Document :120, 2015 [2]). This tagging procedure is used to track a source of water through the terrestrial and atmospheric water compartments in the model. The tagging enhanced versions of WRF and WRF-Hydro are named WRF-tag and WRF-Hydro-tag. A publication detailing the implementation of WRF-tag and WRF-Hydro-tag with an application case-study has been recently published in Water Resources Research (Arnault et al. in Water Resour Res 55:6217–6243 (2019) [1]). In particular, WRF-tag and WRF-Hydro-tag are applied to the case of a precipitation event in the Upper Danube river basin. A comparison between WRF-tag and WRF-Hydro-tag results allows to deduce the role of lateral terrestrial water flow on land-atmospheric water pathways, including precipitation.

Implementation of WRF-tag and WRF-Hydro-tag at ForHLR2

The development of the WRF-tag and WRF-Hydro-tag models started at ForHLR2 in 2016. However, several issues have been encountered, which involved a lot of debugging and several re-run of the simulations presented in Arnault et al. [1]. The two main improvements which have been considered since June 2018 are:

This project is part of the DFG (German Research Foundation) Collaborative Research Center (CRC) 165/1 "Wave to Weather (W2W)", funded for the period 07/2015–06/2019.

J. Arnault (✉) · H. Kunstmann
Karlsruhe Institute of Technology (KIT), Institute for Meteorology and Climate Research (IMK-IFU),Garmisch-Partenkirchen, Germany
e-mail: joel.arnault@kit.edu

W. E. Nagel et al. (eds.), *High Performance Computing in Science and Engineering '19*,
https://doi.org/10.1007/978-3-030-66792-4_33

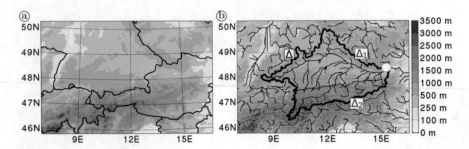

Fig. 1 **a** Terrain elevation [m a.s.l.] of the 5 km-resolution WRF domain. The solid black lines delineate the political boundaries. **b** Terrain elevation of the routing grid at 500 m resolution coupled with the WRF domain in the WRF-Hydro setup. The thin black lines show river channels with a Strahler stream order equal to or above 5. The bold black contour delineates the part of the upper Danube river basin referred as the study region Δ. Δ is divided in two subregions delineated by the bold magenta contour: Δ_1 characterized by moderate topography in the North, and Δ_2 characterized by steep topography in the South

- Implementation of the double precision for the computation of subrid lateral transport of tagged soil moisture and tagged ponded water, in order to improve the closure of the tagged terrestrial water balance.
- Introduction of a depth-weighting factor equal to the relative amount of tagged liquid soil moisture for the vertical distribution of the tagged lateral subsurface flow, in order to prevent tagged liquid soil moisture in the first soil layer to be directly moved to the bottom soil layer when the tagged liquid soil moisture has not infiltrated to that bottom layer yet.

It is worth to mention here that the access to the ForHLR2 facility contributed to accelerate the development and debugging of WRF-tag and WRF-Hydro-tag. Indeed, the supercomputing environment allowed to test and improve the tagging procedure for a real case, a relatively large simulation domain ($150 \times 100 \times 50$ grid points), a relatively long simulation period (5 years including 1 year of spinup time), and a relatively short simulation time (about two weeks for the 5 year-simulation). Furthermore, the possibility to process the model output directly at ForHLR2 allowed to quickly diagnose and correct potential bugs in the implementation of the tagging procedure.

Research highlights obtained with WRF-tag and WRF-Hydro-tag

The newly developed joint soil-vegetation-atmospheric water tagging procedure allows (1) tracking a source of water through the hydrological compartments resolved in WRF-tag and WRF-Hydro-tag, and (2) quantifying the contribution of the WRF-Hydro-resolved lateral terrestrial water flow to the hydrological cycle at regional scale. Arnault et al. ([1], in review) have applied WRF-tag and WRF-Hydro-tag to the case of the upper Danube river basin, using the domain setup displayed in Fig. 1 and a simulation period of 5 years from January 2007 to December 2011.

Arbitrarily, the precipitation event which occurred in the upper Danube river basin between 1200 UTC 14 August 2008 and 1800 UTC 16 August 2008 has been chosen as the source of water for the soil-vegetation-atmospheric water tagging procedure, tagged and tracked for a 40 month-period until December 2011. The tagged results from Arnault et al. ([1], in revision) are analysed in this section with the two following equations, which are the budget of tagged terrestrial water Eq. 1 and the budget of tagged terrestrial water Eq. 2.

$$P^{source} = E^{tagged} + R_{S^{tagged}} + R_G^{tagged} + (S_{tagged})_t \tag{1}$$

$$E^{tagged} = O_{NET^{tagged}} + (W_{tagged})_t + P^{tagged} \tag{2}$$

Equation 1 allows to quantify the partitioning of the source precipitation P^{source} among the tagged evaporation E^{tagged}, the tagged surface runoff $R_{S^{tagged}}$, the tagged underground runoff $R_{G^{tagged}}$, and the tagged terrestrial water change $(S^{tagged})_t$. Equation 2 further allows to quantify the partitioning of E^{tagged} among the tagged net atmospheric outflow $O_{NET_{tagged}}$, the tagged atmospheric water change $(W^{tagged})_t$ and the tagged precipitation P^{tagged}.

Figure 2 provides the spatial distribution of tagged water fluxes from Eqs. 1 and 2, accumulated for the whole study period. The left column of Fig. 2 shows the WRF-tag result, whereas the right column shows the differential result between WRF-tag and WRF-Hydro-tag. In WRF-tag, P^{source} covers the entire study region Δ , with lower amounts in the northern subregion Δ_1 where topography is moderate, and higher amounts in the southern subregion Δ_2 where topography is steep. $R_{S^{tagged}}$, $R_{G^{tagged}}$ and E^{tagged} display spatial patterns close to those in P^{source}. P^{tagged} is comparatively much smaller, with enhanced values in Δ_2 which suggests that the mountainous part of the basin has a blocking effect on the tagged atmospheric water.

In comparison to WRF-tag, WRF-Hydro-tag generates slightly different spatial patterns of P^{source}, which directly affect E^{tagged} and contribute to the small differences in P^{tagged} (see right column of Fig. 2). The WRF-Hydro-tag induced changes in $R_{S^{tagged}}$ and $R_{G^{tagged}}$ rather follow the differences in topography. In particular, WRF-Hydro-tag generates much more $R_{S^{tagged}}$ and much less $R_{G^{tagged}}$ in Δ_2, but slightly less $R_{S^{tagged}}$ and slightly more $R_{G^{tagged}}$ in Δ_1. This is related to the more realistic description of the surface runoff generation mechanism in WRF-Hydro, which is more efficient in steep topography gradient areas. Indeed, the subsurface lateral water flow accumulates soil moisture towards valley bottoms. In steep terrain this generates exfiltration, whereas in moderate terrain this accelerates the deep infiltration and increases the underground runoff.

The tagging results are further investigated with time series of accumulated tagged terrestrial water fluxes in Fig. 3 and tagged atmospheric water fluxes in Fig. 4, spatially averaged for the study region Δ and the subregions Δ_1 and Δ_2. In comparison to WRF-tag, WRF-Hydro-tag produces less $R_{S^{tagged}}$ during the tagging event in relation with an increase of the amount of tagged soil moisture storage in both subregions Δ_1 and Δ_2, as shown in Fig. 3. In Δ_1, the excess of tagged soil moisture in WRF-Hydro

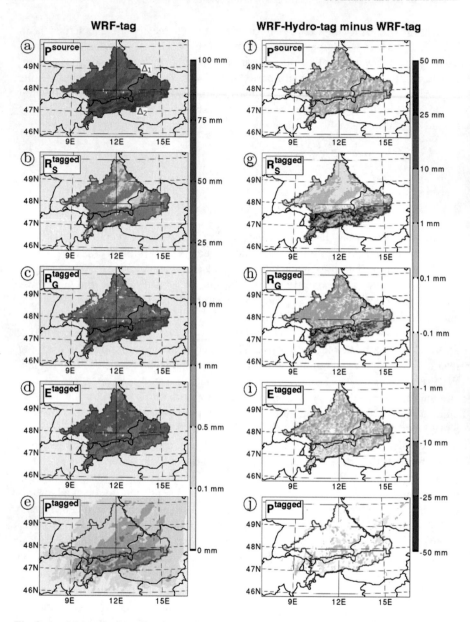

Fig. 2 a–e Maps of accumulated tagged water fluxes [mm] for the whole study period from 1 January 2008 to 31 December 2011, namely source precipitation P^{source}, tagged surface runoff R_S^{tagged}, tagged underground runoff R_G^{tagged}, tagged surface evaporation E^{tagged}, and tagged precipitation P^{tagged}, derived from WRF-tag. (**f–j**) As in (**a–e**), except for the difference between WRF-Hydro-tag and WRF-tag

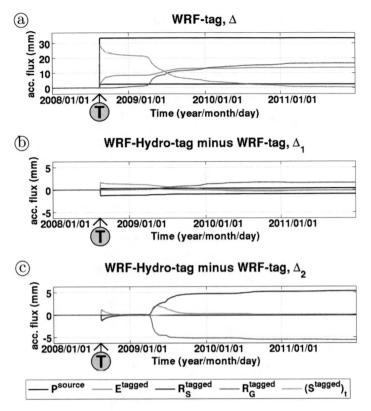

Fig. 3 **a** Daily time series of the tagged terrestrial water fluxes from the budget Eq. 1 derived from WRF-tag, spatially averaged in the study region Δ and displayed as daily accumulated sums [mm] from 1 January 2008 to 31 December 2011. **b** As in (**a**), except for the difference between WRF-Hydro-tag and WRF-tag spatially averaged in the subregion Δ_1. **c** As in (**a**), except for the difference between WRF-Hydro-tag and WRF-tag spatially averaged in the subregion Δ_2

(see Fig. 3b) is slightly depleted by E^{tagged} during 2008. During 2009 and 2010, this excess of tagged soil moisture is mostly depleted by $R_{Gtagged}$, and to a much lower extend by $R_{Stagged}$ as well. The fact that the tagged subsurface lateral water flow in WRF-Hydro-tag accelerates the deep infiltration of tagged soil moisture in Δ_1 induces a relatively smaller Δ_1-average tagged soil moisture amount from 2010. This can be related to Sprenger et al. [4] who showed that the travel time of precipitated water in the root zone is mainly driven by subsequent precipitation amounts. Our result further shows that the lateral subsurface flow contributes to accelerate this washing out of precipitated water in moderate topography areas. In Δ_2, the excess of tagged soil moisture in WRF-Hydro-tag (see Fig. 3c) is mostly depleted by $R_{Stagged}$ during the few weeks following the tagging event. From 2009, the delayed production of $R_{Stagged}$ in WRF-Hydro-tag is mostly counterbalanced by a reduction of $R_{Gtagged}$.

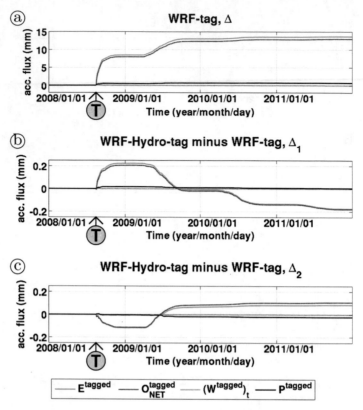

Fig. 4 **a** Daily time series of the tagged atmospheric water fluxes from the budget Eq. 2 derived from WRF-tag, spatially averaged in the study region Δ and displayed as daily accumulated sums [mm] from 1 January 2008 to 31 December 2011. **b** As in (**a**), except for the difference between WRF-Hydro-tag and WRF-tag spatially averaged in the subregion Δ_1. **c** As in (**a**), except for the difference between WRF-Hydro-tag and WRF-tag spatially averaged in the subregion Δ_2

The tagged ponded water infiltration in WRF-Hydro-tag enhances E^{tagged} during the tagging period and the following weeks, especially in Δ_1 as shown in Fig. 4b. In the case of Δ_2, E^{tagged} is rather reduced in association with the exfiltration of tagged soil moisture towards the surface, as shown in Fig. 4c. Accordingly, Δ_1-average P^{tagged} is slightly enhanced, whereas Δ_2-average P^{tagged} remains almost unchanged during the tagging period and the following weeks.

During the warm months of 2009, 2010 and 2011, E^{tagged} in Δ_1 is generally lower in WRF-Hydro-tag than in WRF-tag, in relation with a faster deep infiltration of the tagged soil moisture in WRF-Hydro-tag. In Δ_2, this enhanced tagged infiltration is overbalanced by the tagged exfiltration which maintains some tagged soil moisture in the upper soil layers and enhances E^{tagged}. As Δ_2 is much smaller than Δ_1, the increase of Δ_2-average E^{tagged} in 2009 underbalances the decrease of Δ_1-average

E^{tagged}, so that both Δ_1-average P^{tagged} and Δ_2-average P^{tagged} decrease from 2009 as shown in Fig. 4b and c.

In summary for the case of the upper Danube river basin, the consideration of the fate of ponded water in WRF-Hydro-tag primarily enhances the tagged infiltration, thereby increasing the transit time of the precipitated water during the few weeks following the tagging event, which slightly increases the tagged evaporation and very slightly increases the tagged precipitation. This primary effect is however reduced in steep topography gradient areas where the enhanced tagged exfiltration counterbalances the enhanced tagged infiltration. Secondarily, after a few months the subsurface lateral water flow in WRF-Hydro-tag induces a faster deep infiltration of the tagged soil moisture in low topography gradient areas, thereby decreasing the root zone transit time of the precipitated water, which slightly decreases the tagged surface evaporation and very slightly decreases the tagged precipitation.

References

1. J. Arnault, J. Wei, T. Rummler, B. Fersch, Z. Zhang, G. Jung, S. Wagner, H. Kunstmann, A joint soil-vegetation-atmospheric water tagging procedure with WRF-Hydro: Implementation and application to the case of precipitation partitioning in the upper Danube river basin. Water Resour. Res. **55**, 6217–6243 (2019)
2. D.J. Gochis, W. Yu, D.N. Yates, The WRF-Hydro model technical description and user's guide, version 3.0, NCAR Technical Document, 120 p (2015), http://www.ral.ucar.edu/projects/wrf_hydro/
3. W.C. Skamarock, J.B. Klemp, A time-split nonhydrostatic atmospheric model for weather research and forecasting applications. J. Comp. Phys. **227**, 3465–3485 (2008)
4. M. Sprenger, S. Seeger, T. Blume, M. Weiler, Travel times in the vadose zone: Variability in space and time. Water Resour. Res. **52**, 5727–5754 (2016)

WRF Simulations to Investigate Processes Across Scales (WRFSCALE)

Hans-Stefan Bauer, Thomas Schwitalla, Oliver Branch, Rohith Thundathil, Stephan Adam, and Volker Wulfmeyer

Abstract Several scientific aspects ranging from boundary layer research and land modification experiments to data assimilation applications were addressed with the Weather Research and Forecasting (WRF) model from the km-scale down to the turbulence-permitting scale. Case study simulations in as different regions as the central United States, the United Arab Emirates and southwestern Germany were performed to investigate the evolution of the convective boundary layer. The multi-nested WRF setup, driven by the operational analysis of the European Centre for Medium-range Weather Forecasts (ECMWF), high-resolution terrain, and land cover data sets simulated a realistic evolution of the internal turbulent structure of the boundary layer including the transitions between daytime and nighttime conditions. Land modification simulations in the United Arab Emirates demonstrated that plantations as small as $10 \times 10 \, \text{km}^2$ could modify the weather pattern in this area in a way that more precipitation reaches the desert. Data assimilation experiments demonstrated the beneficial influence of state-of-the-art lidar measurements on the forecast performance of WRF. A further improvement was found when the more sophisticated hybrid 3DVAR-ETKF method was applied, since this method includes a flow-dependent model error contribution and thus more realistically spreading the information of the observations.

H.-S. Bauer (✉) · T. Schwitalla · O. Branch · R. Thundathil · S. Adam · V. Wulfmeyer
Institute of Physics and Meteorology, Garbenstrasse 30, 70599 Stuttgart, Germany
e-mail: hans-stefan.bauer@uni-hohenheim.de

T. Schwitalla
e-mail: thomas.schwitalla@uni-hohenheim.de

O. Branch
e-mail: oliver_branch@uni-hohenheim.de

R. Thundathil
e-mail: rohith.thundathil@uni-hohenheim.de

S. Adam
e-mail: stephan_adam@uni-hohenheim.de

V. Wulfmeyer
e-mail: volker.wulfmeyer@uni-hohenheim.de

© Springer Nature Switzerland AG 2021
W. E. Nagel et al. (eds.), *High Performance Computing in Science and Engineering '19*,
https://doi.org/10.1007/978-3-030-66792-4_34

1 Introduction and Motivation

Numerical models have the potential to address the inherent weaknesses of observation-based approaches, since they can provide a full 4D representation of the atmosphere and produce a consistent state with respect to the 3D thermodynamic atmospheric fields, cloud water, cloud ice, and diagnostic variables such as precipitation. Therefore, numerical models are excellent tools to improve our understanding of atmospheric processes across scales.

Recent results have demonstrated that grid resolutions of less than around 4 km are necessary for a realistic representation of mesoscale processes, especially in terms of the spatial and temporal distribution of precipitation. Further improvements are expected if a chain of grid refinements is performed down to the Large-Eddy Simulation (LES) scale (100 m and below), as further details of land-surface atmosphere (LSA) interaction are resolved. Through application of extremely high resolution and low-pass filtering, larger eddies and the dominant spectra for turbulent transport of heat and moisture can be simulated explicitly.

The WRF model system, described with more detail in earlier reports, provides the opportunity to perform LES simulations under realistic conditions because of the wide range of scales it can be applied over. Using a nesting strategy that covers the synoptic weather situation and the mesoscale circulations, in combination with data assimilation, will ensure a forcing of the LES that is as close as possible to the real weather situation.

2 Work Done Since July 2018

2.1 LES Simulations to Better Understand Boundary Layer Evolution and High-Impact Weather (LES-PROC)

The four-dimensional results of high-resolution numerical weather prediction models provide excellent data sets for detailed investigations of the evolution of the boundary layer.

In August 2017, the Land Atmosphere Feedback Experiment (LAFE) took place at the Southern Great Plains site of the Atmospheric Radiation Measurement Program (ARM) in Oklahoma. Many different instruments were brought to the site and the measurement strategy was optimized in a way to derive as much information as possible. The goals of the campaign were:

1. Determine turbulence profiles and investigate new relationships among gradients, variances, and fluxes.
2. Map surface momentum, sensible heat, and latent heat fluxes using a synergy of scanning wind, humidity, and temperature lidar systems.
3. Characterize land-atmosphere feedback and the moisture budget.

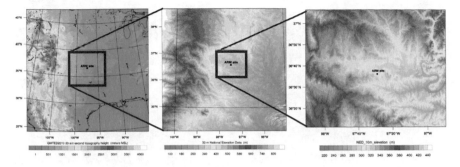

Fig. 1 Domain configuration of the first LAFE simulation. From left to right the domains with 2500 m, 500 m and 100 m are shown

4. Verify large-eddy simulation model runs and improve turbulence representations in mesoscale models.

More details about the campaign and the applied measurement strategy can be found in [4].

Within LES-PROC simulations of selected case studies during the campaign will be carried out to perform detailed comparisons with observations, contributing to goal four of LAFE. A first simulation down to turbulence-permitting resolution of 100 m was carried out for the 23rd of August 2017, one of the golden measurement days during the campaign. It was a clear sky day where the operation of the lidar systems was temporally extended to include the evening transition of the convective boundary layer.

The simulation was started at 06 UTC (01 local time) and run for more than 22 h within the 24 h walltime limit. With this, it was our first simulation containing both the morning and the evening transitions between the nighttime stable and the daytime convective boundary layer. WRF was set-up in a three-domain configuration with 2500 m, 500 m and 100 m resolution and the outer domain was driven by the operational analysis of the European Centre of Medium Range Weather Forecasting (ECMWF). The size of the domains is 1000×1000, 1001×1001 and 1201×1001 grid cells with 100 vertical levels in each domain. Figure 1 illustrates the domain setup. The location of the ARM site is marked in the different domains.

Figure 2 shows the development of the 10 m horizontal wind field for two time steps in the morning and the afternoon local time. In the morning local time (left panel), the turbulence is still weak, resulting in typical role-like circulation patterns. Later in the afternoon (right panel), with fully developed turbulence, the roles break apart into turbulent eddies, leading to the development of small-scale gust fronts. Interestingly, coherent structures caused by the underlying orography are seen, demonstrating the benefits of simulations with such high horizontal resolution.

Figure 3 illustrates the fully developed convective boundary layer 1000 m above sea level in the afternoon of August 23. The vertical velocity field 1000 m above sea level illustrates a typical convective boundary layer undisturbed by clouds. The cir-

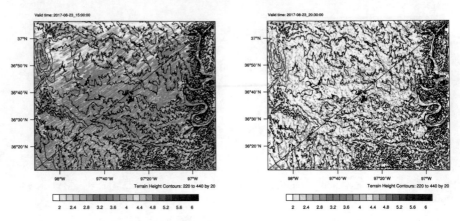

Fig. 2 10 m horizontal wind velocity $[ms^{-1}]$ for 10 a.m. (left) and 3:30 p.m. (right). Shown is the innermost domain with 100 m horizontal resolution

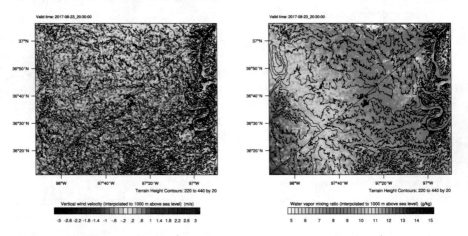

Fig. 3 Vertical wind velocity $[ms^{-1}]$ (left) and water vapor mixing ratio $[gkg^{-1}]$ (right) 1000 m above sea level in the well-developed boundary layer at 3:30 p.m., 23 August 2017 as simulated by WRF with 100 m horizontal resolution

culation broke apart into turbulent eddies of different size. Small regions of intense updrafts (blue) are surrounded by larger regions of weaker compensating downdrafts (red). The sizes of the developing structures correspond to observations. Furthermore, regions over flat terrain where the turbulence develops isotropically can be separated from regions where coherent structures develop influenced by the underlying orography.

The updraft regions can also be identified in the water vapor field (right panel) as moist bubbles. Furthermore, a large-scale southwest-to-northeast gradient in moisture with larger values in the elevated western and especially southwestern parts of the domain is seen. The large-scale meteorological forcing provided by the ECMWF analysis via the two coarser domains is the main cause for the observed behavior.

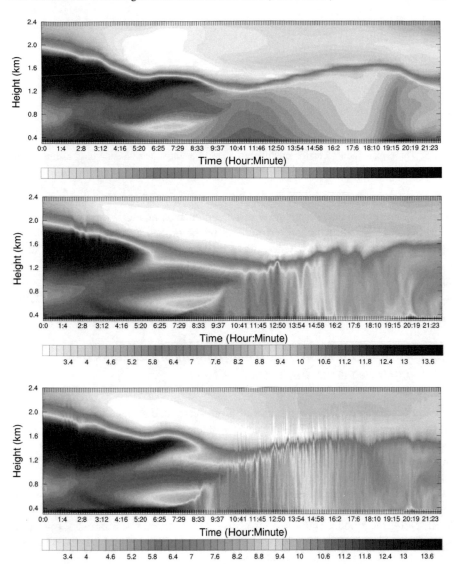

Fig. 4 Time-height cross section of the water vapor mixing ratio [gkg^{-1}] for the grid cell the Hohenheim lidar systems were located over the full forecast range. From top to bottom: 2500 m, 500 m and 100 m horizontal resolution

Figure 4 shows time-height cross sections of water vapor mixing ratio for the grid cell the Hohenheim lidar systems were located for the three different horizontal resolutions of the model chain.

The coarsest resolution, applying a turbulence parameterization, shown in the top panel, is only capable to reproduce the correct height of the boundary layer. The expected internal structure of the boundary layer and the undulating top of the

Fig. 5 Three plantation scenarios of different sizes showing square plantations of dimensions 10 km, 20 km and 30 km placed in the desert of the UAE

boundary layer cannot be simulated with a turbulence parameterization. Furthermore, spurious wave-like structures are simulated in the moisture field. In the simulation with 500 m resolution and switched-off turbulence parameterization, internal structure starts to develop, but the resolution is too coarse to capture the expected fine-scale turbulent eddies. The simulation with the highest 100 m resolution nicely captures the morning and evening transitions and the internal structure of the boundary layer.

2.2 Seasonal Land Surface Modification Simulations over the United Arab Emirates (UAE-1)

Previous studies have demonstrated that large-scale land surface modifications can affect the weather, and potentially in a controllable and predictable way [1–3]. However, these studies have all been based on very large scales of $100 \times 100\,km^2$. Furthermore, these simulations were in other regions and it is not known if similar impacts could occur in the UAE region.

The goal of sub-project UAE-1 is to assess the impact of smaller and more feasible land surface modification scenarios. This was done via single day simulations with the WRF-NOAHMP model under varying atmospheric conditions. Figure 5 shows three selected scenarios with three different sizes of plantation—10 km, 20 km and 30 km (square). These simulations were compared with control runs without plantations.

Figure 6 shows phenomena occurring in one of the case studies (27 July 2015). The left panel shows the change in net surface radiation at the surface caused by the reduced albedo of the plant canopy. This leads to a rise in 2m temperatures (vector colors) and a low pressure disturbance. This leads to a wind convergence and corresponding ascent of air. The right panel shows an increase in turbulence over the plantation caused by both heating and increased surface roughness.

With ascending air and increased turbulent transport, the level of free convection can be reached upon which a convective cell develops. Figure 7 shows the respective impacts on rainfall over the whole day from all three scenarios. It is clear that all sizes have a large impact on rainfall (In the control, not shown, there is little rainfall

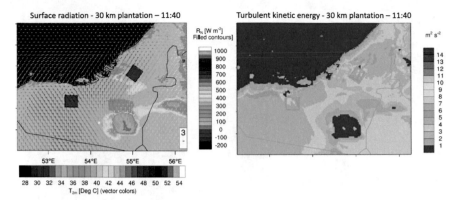

Fig. 6 Instantaneous impact on surface radiation, windflow and 2m temperatures (left) and on turbulent kinetic energy (right)

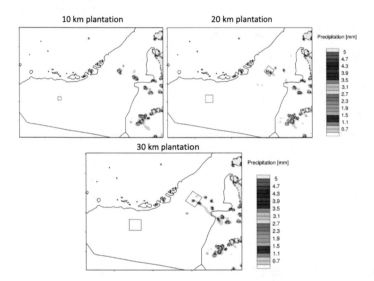

Fig. 7 Accumulated rainfall on 27 July 2015 for all three scenarios [*mm*] ending at 00:00 UTC

around the plantation areas). However, there is a difference in impact from the scale of the plantations both spatially and in the amounts.

Figure 8 shows the integrated rainfall amounts in a 50 km radius around the plantation areas in m^3. This larger integration zone was chosen because some rainfall occurs outside the plantation boundaries. We can see that all scenarios exhibit an increase in precipitation compared to the control and also that the amounts scale with size. However, the scaling is not that significant and a key finding is that plantations as small as 10 km have a very large impact. If we assume there are at least five cases during a summer period we could expect water amounts of at least 1,000,000

Fig. 8 Integrated rainfall amounts reaching the ground from each scenario [m^3]

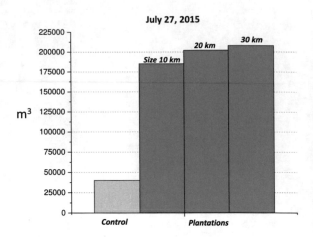

m^3. Taking into account the amount of irrigation water required for Jojoba this would be enough to offset 3 km^2 of plantation area.

In summary, these simulations indicate that conditions in summer in the UAE are conducive for impacts from small plantations down to 10 km size. It is also clear from rainfall amounts that there is a dependency on plantation size but it is not as significant as one might have estimated.

These findings have profound implications for all arid regions and in particular the UAE, which suffers greatly from water scarcity. Furthermore, there are exciting implications for using plantations as a weather modification method, a field of study, which until now, has focused primarily on cloud seeding. If the weather modification is predictable, it can form a synergy with the potential of plantations for significant carbon storage.

2.3   Data Assimilation and Forecasting Over the United Arab Emirates (UAE-2)

The initial set up with a nesting ratio of 1:5 ending at 111 m grid increment led to difficulties in simulating the cloud development over the Al Hajar Mountains because the steep gradient leads to an unstable behavior. The WRF model is based on a terrain following coordinate system and a true horizontal diffusion scheme is applied. According to the WRF developers, this pushes the applied numeric to its limits as the slope at this particular resolution exceeds 40°. A suggestion by the WRF developers is to heavily smooth terrain but this is not our intention on the turbulence permitting scale.

Therefore, it was necessary to revise our setup and we then applied a nesting ratio of 1:3 so that currently the finest grid increment is 309 m with an intermediate nest of 926 m grid increment. We also tested a fourth domain with 103 m grid increment

Table 1 Model domain configuration and resolutions

Domain	Grid Cells	Vertical levels	Grid increment	Time step (s)	Domain size
D01	900 * 700	100	0.025° (2.8 km)	10	2500 km × 2500 km
D02	790 * 790	100	0.0083° (926 m)	3	732 km × 732 km
D03	826 * 772	100	0.0027° (309 m)	1	255 km × 239 km

but we could not generate results due to numerical instability. Table 1 summarizes the grid dimensions and resolutions.

Domains D01 and D02 were run concurrently in a one-way-nested approach while D03 was forced by the 5 min output of D02 using the NDOWN approach. The NDOWN approach allows for running a domain not directly coupled to the outer domains. It makes use of high-frequency temporal output files of the parent domains available in few minute intervals. For D03, two simulations with and without a PBL scheme were conducted.

The upper row of Fig. 9 shows an example for the different results with and without a PBL parametrization of the 2 m temperatures together with the 10 m wind field at 309 m resolution. The lower row of Fig. 9 shows the simulated cloud liquid water path. From the upper row, it is clearly visible that a simulation without PBL parametrization simulates higher 2 m temperature with a difference of almost 4 K. Both simulations show similar cloud patterns, although it appears that the simulation with an applied PBL parameterization simulates less strong convection, which is partly reflected in the rainfall patterns (not shown here). The maximum rainfall amount is approx. 13 mm in case no PBL scheme is applied and 24 mm when a PBL scheme is applied. Apparently, the PBL scheme supports the development of convection in this particular case.

Figure 10 shows every minute model time series of vertical velocity at one grid point west of Al Ain where daytime turbulence is nicely visible as indicated by the changing colors.

The turbulence permitting simulation was conducted for 18 h forecast time and is initialized six hours after D01 and D02 at 06 UTC on 14 July, 2015. The computation required 7200 compute cores on the Cray XC40 system for about 12 h wall clock time resulting in an output data volume of 7.2 TB assuming 1 min output only in D03 for process studies.

In addition to the turbulence permitting simulation, we decided to set up a five-member physics ensemble with different combinations of PBL and cloud microphysics in order to achieve a sufficient spread among the simulations at the convection permitting resolution of 2.7 km.

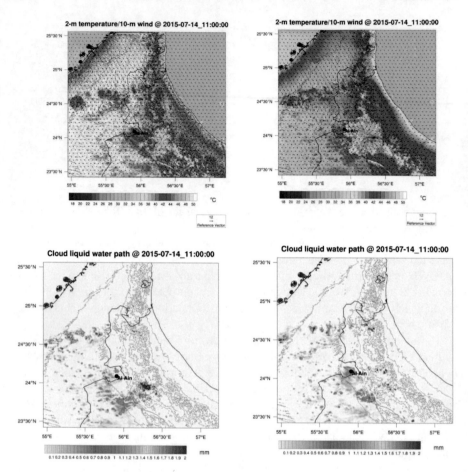

Fig. 9 Upper row: 2-m temperatures [^{o}C] of the turbulence-permitting domain with PBL (left) and without a PBL parametrization. Lower row: Same as the upper row but for the cloud water path [mm]

Downscaling simulations for the 14 July 2015 case study without data assimilation were performed and are currently being analyzed. The output data volume of the whole ensemble is around 7 TB with 15 min output frequency.

With the available station observations from the National Center for Meteorology (NCM) in Abu Dhabi, we validated the diurnal cycle of 2 m temperature, 2 m dew point as well as 10 m wind speeds. Figure 11 shows an example of the diurnal cycles of the 2-m temperature over the UAE for the different configurations compared to the NCM station observations.

All WRF configurations show a negative temperature bias before sunrise and after sunset. As the 2-m dewpoint and 10-m wind speeds do not show a significant deviation, it is assumed that the model has difficulties in predicting the strong nocturnal inversion layer. This is a well-known problem for almost all mesoscale models.

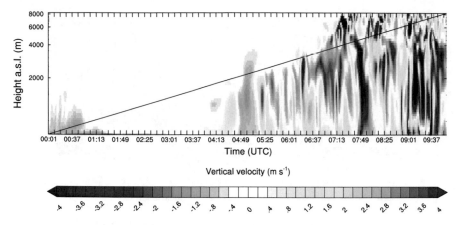

Fig. 10 Time series of the simulated vertical velocities west of Al Ain. The temporal resolution is 60 s

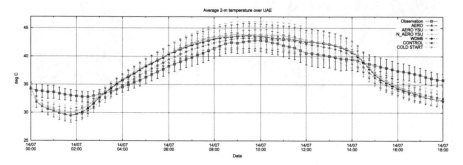

Fig. 11 Mean diurnal cycle of the 2-m temperatures over the UAE on July 14th, 2015

However, further details are very difficult to examine, as no vertical profiling data of temperature are available yet.

Regardless of these deviations, our initial results demonstrate a most realistic representation of the cloud pattern by the combination of the YSU PBL and Thompson Aerosol-aware cloud microphysics, when compared to the cloud liquid water (LWP) path obtained from the Climate Monitoring Satellite Application Facility (CMSAF) based on the SEVIRI imaginary from MSG-2. The product has a spatial resolution of 0.05° and is currently available every 15 min during daytime until the end of 2015.

Figure 12 shows an example of the observed and simulated LWP at 14 July 2015 12 UTC. It is seen that the simulation shows more dense water clouds in case the YSU PBL is applied (right panel). Compared to the simulations with the Thompson cloud microphysics schemes, the configuration with the WDM6 microphysics scheme (middle panel) simulates considerably less water clouds but instead produces too many ice clouds which does not happen to the other configurations.

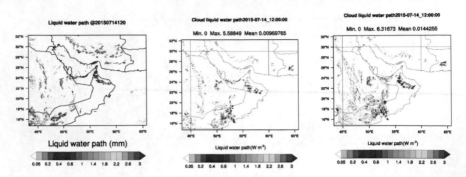

Fig. 12 Observed cloud liquid water (CWP) path [*mm*] from MSG SEVIRI (left). Simulated CWP applying different combinations of PBL and cloud microphysics schemes

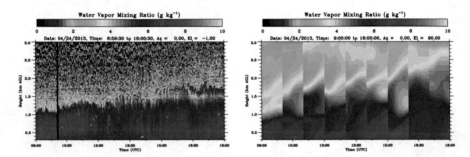

Fig. 13 Left panel: 1-min Water vapor mixing ratio from the DIAL from 09 UTC to 18 UTC. Right panel: 1-min Water vapor mixing ratio model output after assimilation with 3DVAR RUC DA in the same time period

2.4 Assimilation of Lidar Water Vapor Measurements (VAP-DA)

Water vapor plays a crucial role in the atmosphere although its fraction is comparatively less among the other constituents. Humidity content in the atmosphere has its effect on the overall dynamics, radiation and significantly in the development of clouds and precipitation as well. With the advent of lidar systems, promising research on the interaction of water vapor on various scales of the atmosphere had its commencement. With its robust and fast measurements with high temporal and spatial resolution, a huge amount of high-quality data can be measured. These measurements are crucial for process understanding in the atmosphere and especially in the boundary layer where rapid changes result in significant influence on the future state.

Assimilation of the high quality data obtained from lidar instruments is a good leap in numerical weather prediction. For NWP models, a better initial state determines the quality of the forecast. A lidar data network has a huge potential to close the voids in the data availability present in the lower troposphere where the only

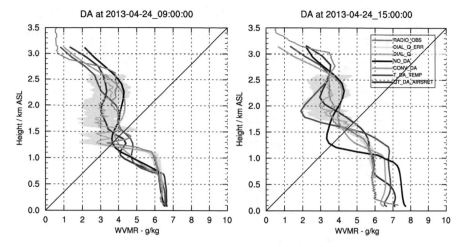

Fig. 14 Left panel: Water vapor mixing ratio [g/kg] profile during the first assimilation. Right panel: Water vapor mixing ratio profile at 15 UTC after 6 assimilation cycles

quality data is provided by radiosondes, which still has shortcomings. Instantaneous measurements like radiosondes cannot capture the changes in the atmosphere that can be registered by continuous measurements by the lidars. Assimilation of water vapor lidar measurements into the NWP model WRF was a limitation in the past few years since there hadn't been a suitable forward operator for the direct assimilation of water vapor mixing ratio which is one of the major prognostic variables in the WRF model. Within WRFSCALE, a forward operator for the direct assimilation of water vapor mixing ratio profiles has been accomplished with the modification of the atmospheric infrared sounding retrieval (AIRSRET) operator already available in the WRFDA system.

The assimilation of water vapor profiles into the model was successful and resulted in a positive impact. The assimilation was performed with the three-dimensional variational (3DVAR) data assimilation approach initially and now being tested for ensemble data assimilation methods. The test case for the DA technique was the data obtained from one intensive observation period (IOP) of the HOPE campaign which was held in western Germany during the month of April and May 2013. A 10-hour dataset was assimilated into the model with hourly data inputs. A rapid update cycle 3DVAR data assimilation was performed at hourly intervals for the entire European domain at a spatial resolution of 2.5 km and a vertical resolution of 100 levels up to 50 hPa. More vertical levels were defined below two km of height where assimilation of more data points are required due to the strong variability of the mixing ratio in the boundary layer.

The assimilation of the water vapor mixing ratio profiles corrects the model at hourly intervals where the DIAL data is fed into the model. Figure 13 shows a direct comparison of the time series output from the model with the observed water vapor mixing ratio (WVMR) from the DIAL. The profile plot in Fig. 14 shows the impact

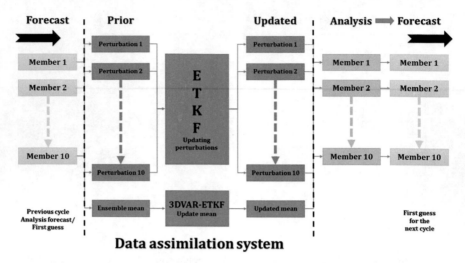

Fig. 15 Schematic of the hybrid 3DVAR-ETKF data assimilation system with 10 ensemble members

of assimilation on the vertical profile at the first assimilation and after six assimilation cycles. The 15 UTC assimilation shows good agreement with the DIAL profile although the model couldnt capture the boundary layer gradient in the first assimilation at 9 UTC.

Another data assimilation technique to incorporate flow-dependent forecast error covariances into the assimilation process is currently being tested for the same domain to alleviate the effects of the static covariance of the 3DVAR assimilation system. The ensemble technique implemented for this purpose is the hybrid 3DVAR-ensemble transform Kalman filter (ETKF) where the error covariance matrix in the assimilation system is a combination of static and flow-dependent covariances. Thus, the model has a lesser dependence on the static background error covariance matrix previously determined to minimize the cost function.

Figure 15 illustrates the data assimilation process undergone in the 3DVAR-ETKF data assimilation. The assimilation system starts with a set of currently 10 ensemble members each with the same configuration as the previous experiments with a spatial resolution of 2.5 km and vertical resolution of 100 levels. The mean of the set of ensemble members are derived along with the variance of the members as well. With the generated mean the perturbations of the individual ensemble members from the mean are calculated. These perturbations are updated by the ETKF and the mean of the ensemble members is assimilated with the combination of the analysis covariance matrix obtained from the variance of the ensemble members and the previously determined static covariance matrix.

The hybrid output generated from the assimilation can be continued for next cycles. The updated perturbations from the ETKF is added to the ensemble mean to get the next set of ensemble members which now are the analysis ensemble members.

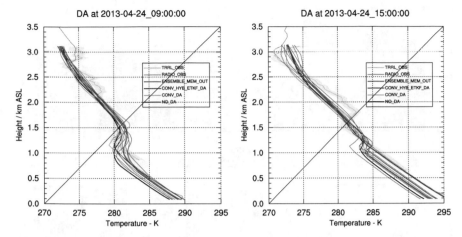

Fig. 16 Temperature profile plot for a conventional 3DVAR DA and conventional 3DVAR-ETKF DA. Left panel: Assimilation performed at 09 UTC. Right panel: Assimilation at 15 UTC after six assimilation cycles

These analysis ensemble members are forecasted along with the hybrid ensemble output to get the next first guess or the background ensemble members.

Figure 16 shows the impact of assimilation of conventional data on the temperature profiles. The hybrid 3DVAR-ETKF out which is shown in the red color has a slightly better agreement with observed data which is the radiosonde and the temperature Raman lidar (TRRL). The blue lines are the 10 ensemble member outputs, which are obtained after adding the mean to the ETKF perturbations in the data assimilation process.

2.5 LES for SABLE and the AMMER Catchment (LES-PBL)

Nested large eddy simulations (LES) with the weather research and forecast model (WRF) have been conducted. Operating LES with an NWP model has the advantage of simulating the turbulent exchange processes between the land surface and the atmosphere of the moisture, heat and momentum in the planetary boundary layer (PBL) under realistic conditions. Therefore, the impact of a specific weather situation on the development and the transport happening inside the PBL can be investigated. Simulations for two cases studies for the Ammer catchment west of Tuebingen have been made. One case was the 15th of July 2015 (AMMER2015), which was in a distinct dry period and the other the 09th of June 2016 (AMMER2016), which took place after a heavy precipitation event, with soil moistures close to saturation. Thus, the surface forcing differed for both cases, with much higher surface sensible heat flux S_h found for the 2015 case (250 W/m^2 in the domain mean) than for the 2016 case (200 W/m^2 in the domain mean). Furthermore, the 2015 case study was very

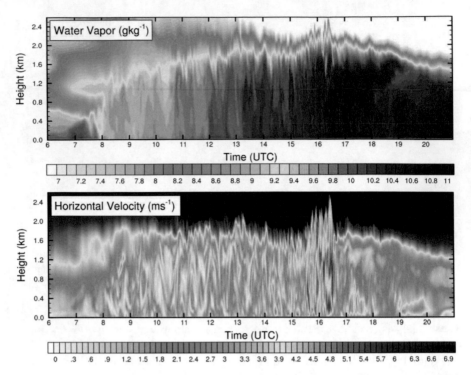

Fig. 17 Time-height cross section of the water vapor mixing ratio q_v [g/kg] (top) and the horizontal wind velocity u_h [m/s] (bottom) for the AMMER2015 case study. The temporal resolution is 10 s

calm with maximum mean horizontal wind velocities of 2 m/s compared to 9 m/s found for the 2016 case.

Figures 17 and 18 present time-height cross sections of the water vapor mixing ratio q_v and the horizontal wind velocity u_h for both the 2015 and the 2016 case study at the town of Entringen, located in the Ammer catchment. For the AMMER2015 case, an accumulation of moisture in the PBL (Fig. 17) is found. Relatively high values of around 2.0 km for the depth of the turbulent PBL (z_i) are found. The height of the PBL z_i is visible at the level with the largest decrease of q_v with height. The exchange of the air from the atmosphere above z_i and the PBL (entrainment) is limited to the upper part of the PBL. The horizontal wind velocity u_h in the PBL shows for most of the case relatively low values up to 2 m/s in the PBL.

For the AMMER2016 case, a decrease in the moisture inside the PBL during daytime is found (Fig. 18). The time-height cross section shows a growth of the PBL into an already relatively moist residual layer. Entrainment of dry air into the PBL can be identified as one factor for decreasing q_v in the PBL during daytime. After 1200 UTC, a series of dry tongues reaching down to the surface are visible in the moisture field of the PBL. Over the day, the horizontal wind velocity stays constant on relatively large values of around 7 m/s, with larger variations in the evening. Smaller

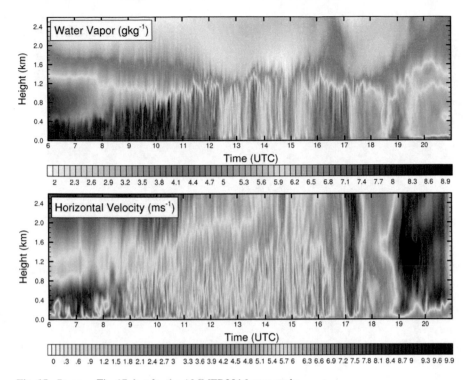

Fig. 18 Same as Fig. 17, but for the AMMER2016 case study

maximum values for z_i are found compared to the AMMER2015 case. This is in accordance with the smaller sensible heat fluxes found for the AMMER2016 case.

The differences in the horizontal wind velocity between the 2015 and the 2016 case have an influence on the organization of the convection in the PBL. Figure 19 presents the instantaneous values of the horizontal moisture flux convergence in the lower PBL for both cases.

Positive flux convergence is found in areas where updrafts are forming and negative values in areas with downdrafts. Due to the mass balance, the moist air in the lower PBL is converging in the area of updrafts, while in the area of downdrafts, dryer air from higher altitudes is transported downwards.

For the AMMER2015 case (Fig. 19a), a rather unstructured field of the moisture flux convergence is found. Clear signs for open cell convection are found, with relatively distinct areas where updrafts are forming, separated with larger areas of downdrafts. In contrast, for the 2016 case the convection pattern is stronger influenced by the larger mean wind. Here roll convection is found. The updrafts are aligned with the mean wind.

For the 2016 case study, a stronger influence of the entrained air was found in the moisture field. Thus, a connection between the organization of the convection and

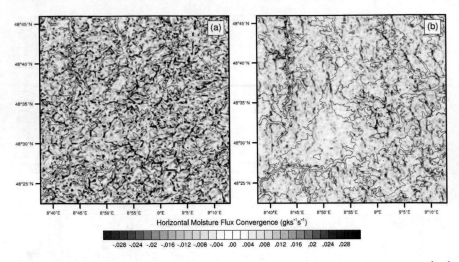

Fig. 19 Instantaneous fields of the horizontal water vapor mixing ratio convergence $[gkg^{-1}s^{-1}]$ for the 2015 case in panel (**a**) and for the 2016 case in panel (**b**). All fields where averaged vertically over the 15 lowermost model levels

the strength of the entrainment can be drawn. For roll convection, the engulfment of entrained air into the PBL seems to be stronger, than for open cell convection.

3 Computational Issues

Table 2 summarizes the applied computer resources for the whole project during the period May 2018 to April 2019, subdivided into the five sub-projects.

Table 2 Core hours used by the different sub-projects between May 2017 and April 2018

Sub-project	Used core-h May 2018 to April 2019
LES-PROC	216776
UAE-1	1004543
UAE-2	474543
VAP-DA	271594
LES-PBL	535607
Total	2503063

References

1. K. Becker, V. Wulfmeyer, T. Berger, J. Gebel, W. Münch, Carbon farming in hot, dry coastal areas: An option for climate change mitigation. Earth Syst. Dyn. **4**, 237–251 (2013)
2. O. Branch, K. Warrach-Sagi, V. Wulfmeyer, S. Cohen, Simulation of semi-arid biomass plantations and irrigation using the WRF-NOAH model—a comparison with observations from Israel. Hydrol. Earth Syst. Sci. **18**, 1761–1783 (2014)
3. V. Wulfmeyer, O. Branch, K. Warrach-Sagi, H.-S. Bauer, T. Schwitalla, K. Becker, The impact of plantations on weather and climate in coastal desert regions. J. Appl. Meteorol. Climatol. **53**, 1143–1169 (2014)
4. V. Wulfmeyer, D.D. Turner, B. Baker, A. Behrendt, T. Bonin, A. Brewer, M. Burban, A. Choukulkar, E. Dumas, R. Hardesty, T. Heus, J. Ingwersen, D. Lange, T. Lee, S. Metzendorf, S. Muppa, T. Meyers, R. Newsom, M. Osman, S. Raasch, A new research approach for obserwving and characterizing land-atmosphere feedback. Bull. Amer. Meteorol. Soc. **99**, 1639–1667 (2018)

Computer Science

Hans-Joachim Bungartz

This year, again, two reports on projects labeled as "computer science" made it into the HLRS report volume and, thus, got their own chapter.[1]

The contribution *Load Balancing and Auto-Tuning for Heterogeneous Particle Systems Using ls1 mardyn* by Steffen Seckler et al. reports on recent progress in the project GCS-MDDC using the software *ls1 mardyn*. This simulation software is the outcome of a long-term interdisciplinary collaboration of several groups to allow for highly scalable molecular dynamics simulations of multi-component and multi-phase processes in chemical engineering. *ls1 mardyn* was twice used for world-record molecular dynamics simulations (world record with respect to the number of particles involved), and it was part of the benchmark suite used by the HLRS for their procurement decision. In this paper, the authors present advances concerning software design (such as a plugin framework) and HPC (enhanced load balancing, e.g.), and they discuss preliminary results on the integration of the C++ node-level auto-tuning library *AutoPas* into *ls1 mardyn*.

The second report *Ad-hoc File Systems at Extreme Scales* by Mehmet Soysal and Achim Streit is on research done within the ForHLR-II project *ADA-FS—Advanced Data Placement via Ad-hoc File Systems at Extreme Scales*. The project's overall goal is to improve I/O performance for highly parallel applications by using distributed on-demand file systems. Such temporary file systems are created locally on the allocated compute nodes. Via integration into the scheduling system, they can be requested like any other resource. This integration, in particular in the context of an HPC system, as well as the evaluation of a possible impact of the on-demand file systems on running jobs, are the topics of the study presented here.

---

[1]Hans-Joachim Bungartz, Department of Informatics, Technische Universität München (TUM), Boltzmannstrasse 3, 85748 Garching, Germany, bungartz@in.tum.de

Load Balancing and Auto-Tuning for Heterogeneous Particle Systems Using Ls1 Mardyn

Steffen Seckler, Fabio Gratl, Nikola Tchipev, Matthias Heinen, Jadran Vrabec, Hans-Joachim Bungartz, and Philipp Neumann

Abstract *ls1 mardyn* is a molecular dynamics (MD) simulation framework that enables investigations of multicomponent and multiphase processes relevant to engineering applications, such as droplet coalescence or bubble formation. These scenarios require the simulation of ensembles containing a large number of molecules. We present recent advances in *ls1 mardyn* both from the software design and high-performance computing perspective. From the former we describe the recently introduced plugin framework, from the latter we will look at some recent load balancing improvements to *ls1 mardyn*. We further present preliminary results of the integration of AutoPas, a C++ node-level library employing auto-tuning to achieve optimal node-level performance for particle simulations, into *ls1 mardyn*.

S. Seckler (✉) · F. Gratl · N. Tchipev · H.-J. Bungartz
Department of Informatics, Technical University of Munich, Boltzmannstr. 3, 85748 Munich, Germany
e-mail: seckler@in.tum.de

F. Gratl
e-mail: gratl@in.tum.de

N. Tchipev
e-mail: tchipev@in.tum.de

H.-J. Bungartz
e-mail: bungartz@in.tum.de

M. Heinen · J. Vrabec
Technical University of Berlin, Thermodynamics and Process Engineering, Ernst-Reuter-Platz 1, 10587 Berlin, Germany
e-mail: heinen@tu-berlin.de

J. Vrabec
e-mail: vrabec@tu-berlin.de

P. Neumann
Helmut-Schmidt-Universität Hamburg, Chair for High Performance Computing, Holstenhofweg 85/Postfach 70 08 22, 22043 Hamburg, Germany
e-mail: philipp.neumann@hsu-hh.de

© Springer Nature Switzerland AG 2021
W. E. Nagel et al. (eds.), *High Performance Computing in Science and Engineering '19*,
https://doi.org/10.1007/978-3-030-66792-4_35

1 Introduction

Molecular dynamics (MD) simulations have become a valuable tool for engineering applications. They rest on molecular models that describe the molecular interactions and encode the macroscopic behavior of matter. Equilibrium MD simulations thus enable sampling of thermodynamic properties in a consistent manner. Such data can be used to develop either fully predictive equations of state (EOS) or hybrid EOS, where simulation data are combined with experimental data [11]. An important advantage is that simulations can straightforwardly be carried out under extreme conditions, i.e. high temperatures and pressures, that are hardly accessible with experiments. Beside classical equilibrium scenarios, MD simulations can also be employed to investigate systems that are not in global equilibrium so that imposed gradients drive processes like droplet coalescence [17], bubble formation [12] or interfacial flows [13]. For many phenomena concerning multi-phase systems, the interface between the phases plays a key role. The spatial extent of the interface region is often only a few molecular diameters and can therefore only be resolved on the atomistic level. Employing molecular simulation, there are no additional modeling approaches, the physical processes evolve naturally and hence can be investigated unbiasedly. For many fluids that are relevant for engineering applications, comparatively simple molecular force field models have been developed, consisting of a few interaction sites, e.g.. Lennard Jones (LJ) sites considering the dispersive interaction and point charges, dipoles or quadrupoles to model the electrostatic interaction. A typical example is the mixture of acetone (four LJ sites, one dipole and one quadrupole) and nitrogen (two LJ sites and one quadrupole) which is frequently used to model fuel injection-like scenarios in thermodynamic laboratories. However, the present simulations were conducted with a simpler molecular model, consisting of a single LJ site. This model can be parametrized such that it mimics the thermodynamic behavior of noble gases like argon, krypton or xenon as well as methane [22]. This model is well suited for investigations focusing on the basic understanding of processes like the droplet coalescence so that it was considered in the present work.

In a long-term interdisciplinary effort of computer scientists and mechanical engineers, the MD framework *ls1 mardyn* has evolved over the last decade to investigate such large systems of small molecules [14]. *ls1 mardyn* has been used in various studies [21] and has been continuously extended to optimally exploit current HPC architectures [6, 18, 20]. In the following, we detail recent developments within the framework to achieve optimal performance at node and multi-node level. After introducing the actual problem setting of short-range molecular dynamics, related work and the original implementation of *ls1 mardyn* in Sect. 2, we introduce the newly developed plugin framework of *ls1 mardyn* in Sect. 3. Improvements to the MPI load balancing are shown in Sect. 4. We report preliminary results on the integration of AutoPas in *ls1 mardyn* in Sect. 5, which have been published in [8]. We close with a summary and an outlook to future work in Sect. 6.

2 Short-Range Molecular Dynamics

2.1 Theory

In short-range MD, Newton's equations of motion are solved numerically [16]. In the following, considerations are restricted to small molecules. Due to their negligible conformational changes, molecules undergo translational or rotational motion; both are included in the equations of motion and are solved simultaneously in *ls1 mardyn* using a leapfrog time integrator, without the need for iterative procedures (such as the SHAKE algorithm) to handle geometric constraints [16].

Molecules interact via force fields. In short-range MD, arising forces are only explicitly accounted for if the distance between two considered molecules is below a specified cut-off radius r_c. There are basically two variants to efficiently implement the cut-off condition: *linked cells* and *Verlet lists* [16]. Both methods turn the actual molecule-molecule interaction complexity from $O(N^2)$ to $O(N)$. In the *Verlet list* approach, a list of all molecules within a surrounding $r_c + h$ is stored per molecule and updated regularly. Computing interactions thus reduces to traversing the list. The choice of h dictates the frequency of necessary list rebuilds on the one hand and the overall size of interaction search volume on the other hand. *ls1 mardyn* makes use of the *linked cell* approach: a Cartesian grid with cell sizes $\geq r_c$ is introduced and covers the computational domain. The molecules are sorted into these cells. Molecular interactions only need to be considered for molecules that reside within the same cell or in neighboring cells.

All simulations reported in this contribution rest on the truncated and shifted form of the LJ potential [22]

$$U_{LJ}(r_{ij}) = 4\epsilon \left(\left(\frac{\sigma}{r_{ij}} \right)^{12} - \left(\frac{\sigma}{r_{ij}} \right)^6 \right), \tag{1}$$

$$U_{LJ,trunc}(r_{ij}) = \begin{cases} U_{LJ}(r_{ij}) - U_{LJ}(r_c) & \text{for } r_{ij} \leq r_c \\ 0 & \text{for } r_{ij} > r_c \end{cases} \tag{2}$$

with species-dependent parameters for size σ and energy ϵ and the distance r_{ij} between molecules i and j, as well as the cutoff radius r_c. Due to the truncation of the potential, no long range corrections have to be considered. This simplifies the treatment of multi-phase systems, where the properties of the interface can be strongly dependent on the cut-off radius [24]. The force calculation is typically by far the most expensive part of MD simulations that often contributes $\geq 90\%$ to the overall compute time and hence is the preferential target for code optimizations.

2.2   Related Work

HPC and Related MD Implementations

Various packages efficiently and flexibly implement (short-range) molecular dynamics algorithms, with the most popular ones given by Gromacs,[1] LAMMPS[2] and NAMD.[3] Gromacs leverages particularly GPUs but also supports OpenMP and large-scale MPI parallelism, and it also exploits SIMD instructions via a new particle cluster-based Verlet list method [1, 15]. A LAMMPS-based short-range MD implementation for host-accelerator systems is reported in [2] with speedups for LJ scenarios of 3–4. A pre-search process to improve neighbor list performance at SIMD level and an OpenMP slicing scheme are presented in [10, 23]. The arising domain slices, however, need to be thick enough, to actually boost performance at shared-memory level. This restricts the applicability of the method to rather large (sub-)domains per process.

ls1 mardyn

An approach to efficient vectorization built on top of the linked cell data structure within *ls1 mardyn* is presented for single [5] and multi-site[4] molecules [4]. This method, combined with a memory-efficient storage, compression and data management scheme [7], allowed for a four-trillion atom simulation in 2013 on the supercomputer SuperMUC, phase 1 [6]. A multi-dimensional, OpenMP-based coloring approach that operates on the linked cells is provided in [20]. The method has been evaluated on both Intel Xeon and Intel Xeon Phi architectures and exhibits good scalability up to the hyperthreading regime. *ls1 mardyn* further supports load balancing. It uses k-d trees for this purpose. Recently, this approach has been employed to balance computational load on heterogeneous architectures [18]. A detailed overview of the original release of *ls1 mardyn* is provided in [14]. Various applications from process and energy engineering, including several case studies that exploit *ls1 mardyn*, are discussed in [21]. Recently, *ls1 mardyn* was used to simulate twenty trillion atoms at up to 1.33 PFLOPS performance [19].

3   Plugin Framework

ls1 mardyn has many users with different backgrounds (process engineering, computer science) which have very differing levels of C++ knowledge. To implement new features, developers had to first understand considerable parts of the program before being able to contribute to the further development of *ls1 mardyn*. Additionally, most

---

[1] www.gromacs.org.

[2] www.lammps.org.

[3] http://www.ks.uiuc.edu/Research/namd/.

[4] Molecules that consist of several interaction sites, e.g.. two LJ sites.

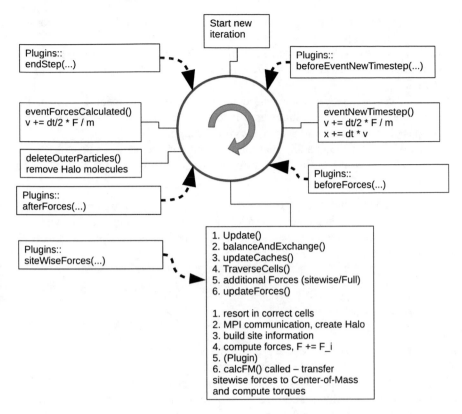

Fig. 1 Extension points of *ls1 mardyn* through plugins. The extension points are marked by dashed arrows, normal steps of the simulation are shown with solid lines

changes were done on a local copy or a private branch within the main simulation loop of the program or within some deeply coupled classes. Consequently, integrating the new code into the main source tree became a major difficulty and therefore often was rejected. If such changes were integrated anyways, they cluttered the source code and made it harder to understand. Moreover, new features often were not easily configurable and could only be disabled or enabled at compile-time.

To prevent the mentioned drawbacks, we have performed major code refactoring steps within *ls1 mardyn* to allow for both easier maintainability and extendability by introducing a plugin framework. Most user code can now be expressed as plugins that can be easily implemented, maintained, extended, integrated into *ls1 mardyn* and enabled upon startup of the simulation. Additionally, the user code is now mostly removed from the main simulation loop and main classes from *ls1 mardyn*, making maintainability more affordable and the code more readable.

ls1 mardyn provides a total of five different extension points that each prove their own purpose (c.f. Fig. 1):

beforeEventNewTimeStep   This extension point (EP) is used as legacy support for some older code parts. Mostly *endStep* can be used instead.

beforeForces   At this point the positions have been updated. Using this EP you can change positions of particles, for example to realign a droplet at the center of the domain.

siteWiseForces   This EP can be used to apply forces on specific sites of the molecules. One existing plugin uses it to implement a site-wise potential that prevents Lennard-Jones sites from moving through a wall.

afterForces   At this point additional forces to entire molecules can be added.

endStep   This step is mostly used for output. Most plugins only use this extension point.

Even though less than a year has passed since these changes were implemented (as of March 2019), we have already seen a lot of user code to actually find its way into the main source tree. Additionally, the user-base has provided very positive feedback on these changes, as their life got easier as well.

4   Load Balancing

In the previous report we presented preliminary results on the coalescence of two droplets with a diameter of $d = 50$ nm containing a number of $N = 10^6$ particles, cf. Fig. 2. These simulations were, however, only run on a fairly small amount of processes. When we tried scaling the simulation to more processes we discovered that the k-d tree-based load balancing implementation (*kdd*, see [3, 14, 18], Fig. 3) in *ls1 mardyn* at that point did not provide the performance we expected, as the load-unaware Cartesian domain decomposition (sdd, Fig. 4) outperformed the load-balancing kdd starting at around 32 nodes (see *old, sdd* in Fig. 5).

The *kdd* distributes the domain by splitting the overall domain into a grid of cells. A load c_{cell} is assigned to each cell. The grid is then split into N disjunct subdomains, such that each subdomain j contains roughly the same load

$$C_{subdomain} = \sum_{\text{cells in subdomain}} c_{cell} = C_{total}/N, \tag{3}$$

where C_{total} is the total combined cost for the entire domain

$$C_{total} = \sum_{\text{all cells}} c_{cell}.$$

To get the loads per cell a load estimation model was used, that takes the number of particles in the current cell n_{cell} and its neighbors $n_{neighbor\ cell}$ into account and uses a quadratic model:

Fig. 2 Snapshot of two argon droplets with a diameter of $d = 50$ nm containing a number of $N = 10^6$ particles in equilibrium with their vapor at a temperature of $T = 110$ K, rendered by the cross-platform visualization framework MegaMol [9]. It shows the time instance where a liquid bridge starts to grow, spanning over the initial gap of 1 nm between the droplets' interfaces. The colors red and green were selected to be able to distinguish between particles that initially constituted either the left or right droplet. To provide a clear view through the vapor, particles were rendered with a diameter of $\sigma/3$

Fig. 3 Space-partitioning using kdd. The different colors represent the different levels of the splitting hyperplanes

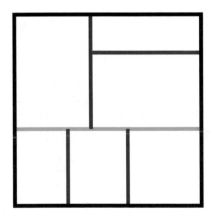

$$c_{\text{cell}} = n_{\text{cell}}^2 + \frac{1}{2} \sum_{\text{neighboring cells}} n_{\text{cell}} \cdot n_{\text{neighbor cell}} \qquad (4)$$

The investigation of the observed performance drops showed that the distribution of the loads $C_{\text{subdomain}}$ was appropriate, but the actual time spent on the calculations of a specific subdomain did not properly match the loads, indicating a poor estimation of the loads c_{cell}. We henceforth introduced three additional load estimators:

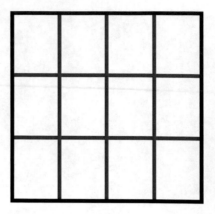

vecTuner   This load estimator evaluates the time needed for each cell by doing
a reference simulation at the beginning of the simulation. Therefore, for each
particle count n_{cell} the time needed to calculate the interactions within a cell and
the interactions across cells is measured.

measureLoadV1   This load estimator uses dynamic runtime measurements within
the actual simulation. Therefore, the time needed to calculate all interactions
within each process is measured. This time is the sum of the times needed for
each cell, similar to Eq. (3):

$$T_{subdomain} = \sum_{\text{cells in subdomain}} t_{cell} \tag{5}$$

The time for each cell t_{cell} cannot be easily measured, because these times are very
small and exhibit a high level of noise and inaccuracy. Instead of determining the
values t_{cell} we decided to introduce cell types to get better statistical properties.
One typical cell type would be characterized by the number of particles per cell,
but other characterizations are possible. Using the cell types, Eq. (5) becomes

$$T_{subdomain} = \sum_{\text{cell types}} n_{\text{cell type}} \cdot t_{\text{cell type}}. \tag{6}$$

Assuming that the processes need the same amount of time for each cell of the
same type, we can derive the matrix equation

$$\forall i : \quad T_i = \sum_j n_{i,j} \cdot t_j, \tag{7}$$

where T_i is the time needed by process i, $n_{i,j}$ is the amount of cells of type j
within rank i and t_j is the time needed to calculate the interactions of cell type
j. Hereby only t_j is an unknown and can thus be estimated by solving the matrix

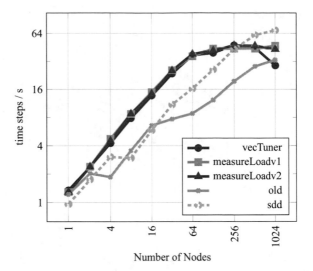

Fig. 5 The different load estimation techniques for a droplet coalescence scenario with 8 million particles using 8 OpenMP threads per rank and the full shell method

equation of the typically overdetermined system through a least squares fit. We are always using the characterization of cell type by particle number, i.e. cell type j resembles all cells with j particles.

measureLoadV2   This load estimator is based on *measureLoadV1*, but additionally assumes a quadratic dependency of t_j on the particle count j.

$$t_j = a_0 + a_1 \cdot j + a_2 \cdot j^2 \tag{8}$$

The resulting matrix equation

$$\forall i: \quad T_i = \sum_j n_{i,j} \cdot \sum_{k=0}^{2} j^k \cdot a_k \tag{9}$$

$$\forall i: \quad T_i = \sum_{k=0}^{2} (\sum_j n_{i,j} \cdot j^k) \cdot a_k \tag{10}$$

is then solved using a non-negative least squares algorithm to obtain a_k.

A comparison of the results using the different load estimation techniques is shown in Fig. 5 for a droplet coalescence scenario with 3 million particles, showing a clear improvement of all new load estimators compared to the old one. While for 64 nodes a speedup of roughly 4x over the old load estimation techniques and an improvement of 2x over the standard domain decomposition (*sdd*) is visible, the *sdd* still performs best for large process counts. This is due to better communication schemes and sub-optimal load balancing even when using the new estimators with the *kdd*.

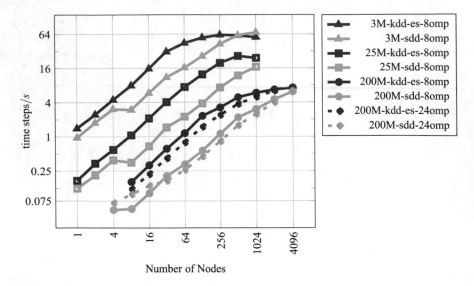

Number of Nodes

Fig. 6 Comparison of the different scenarios for the *vecTuner* load estimator. The eight-shell method has been used for the *kdd*, marked by *es* in the legend

Scaling results using *vecTuner* for scenarios with 25 million and 200 million particles are shown in Fig. 6. For these scenarios the *kdd* always outperforms the standard domain decomposition if the new load estimators are used.

Simulations over a longer time-scale have been calculated for all three scenarios. For the scenario with 25 million particles, the evolution of the droplets is show in Fig. 7. In contrast to the previous simulations, the larger simulation was able to visualize the wiggling within the droplet formation nicely.

5   Preliminary Results: AutoPas Integration

Our work further concentrated on the integration of the C++ library AutoPas [8] into *ls1 mardyn*. The library employs auto-tuning to provide close to optimal node-level performance for particle simulations, which is expected to complement the distributed-memory load balancing approach. Early studies have shown successful automatic adaptions of the employed algorithms to both varying inputs as well as dynamically changing scenarios.

Figure 8 shows how AutoPas can already be used to calculate a spinodal decomposition scenario using *ls1 mardyn*. The simulation starts with a supercritical temperature and a density close to the critical one. Then, the temperature is controlled immediately to a temperature far below the critical one by the velocity scaling thermostat, so that the state of the fluid suddenly becomes physically unstable, resulting in the decomposition of the fluid into stable vapor and liquid phases

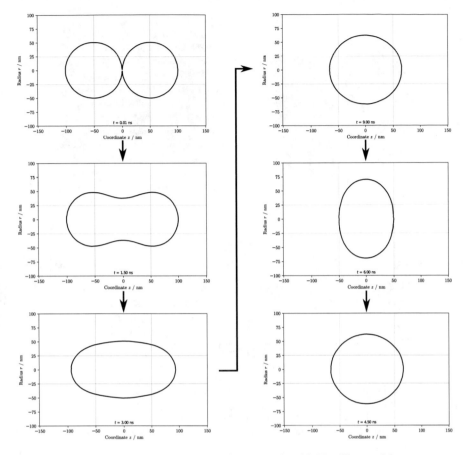

Fig. 7 Time evolution of the droplet contour of the scenario with 25 million particles

(see Fig. 8, top). During the simulation the system thus changes from a homoge-neous state to a very heterogeneous one. Looking at shared-memory parallelization strategies, in the beginning, while the system is still homogeneous, a load-unaware strategy can be used that simply splits the subdomain into even parts to be cal-culated by each thread of a cpu. Later on, when the system becomes increasingly heterogeneous a load-balancing strategy is needed. In the shown figure AutoPas is allowed to choose between two shared-memory parallelization strategies, here called traversals [19, 20]:

c08   This traversal uses coloring to split the domain into multiple groups of cells (colors), where calculation on all cells in one group can be done in parallel without any data races. The cells of each color are then distributed to the threads using OpenMP's dynamic scheduling. After one color is finished, the next color is started.

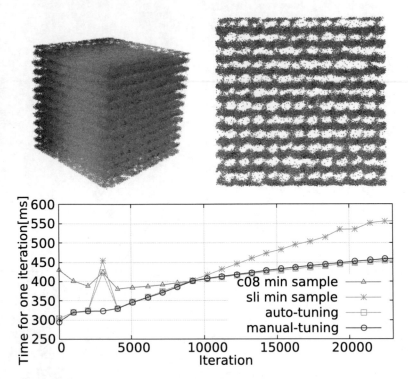

Fig. 8 Spinodal decomposition scenario with 4 million particles calculated with *ls1 mardyn* and AutoPas. The images on the top show the end configuration of the system from the side (top left) and a slice of it (top right). The bottom figure shows the time needed for each iteration for two different shared-memory parallelization strategies. AutoPas is able to automatically choose between these two strategies

sli   The sliced traversal (sli) slices the domain into multiple equally sized subdo-
       mains. Each subdomain is then calculated by one thread. Locks are employed to
       prevent data races.

Henceforth, the c08 traversal is better suited for heterogeneous scenarios, as it provides dynamic scheduling, while the sli traversal is better suited for homogeneous scenarios, as it uses less overhead. As expected, AutoPas switches the shared-memory parallelization strategy for the mentioned scenario at time step ∼9 000 from sli to c08.

6   Summary and Outlook

We have outlined recent progress in usability (plugin concept), load balancing (kdd-based decomposition and load estimation approaches) and auto-tuning (library AutoPas) to improve the molecular dynamics software *ls1 mardyn*. Load balancing improvements enabled unprecedented large-scale droplet coalescence simulations leveraging the supercomputer Hazel Hen. Yet, more work and effort is required to improve scalability of the scheme beyond O(200) nodes. The auto-tuning approach we follow by the integration of AutoPas appears promising in terms of both scenario as well as hardware-aware HPC algorithm adoption. More work in this regard is in progress, focusing amongst others on the incorporation of Verlet list options and different OpenMP parallelization schemes.

Acknowledgements The presented work was carried out in the scope of the Large-Scale Project *Extreme-Scale Molecular Dynamics Simulation of Droplet Coalescence*, acronym GCS-MDDC, of the Gauss Centre for Supercomputing. Financial support by the Federal Ministry of Education and Research, project *Task-based load balancing and auto-tuning in particle simulations (TaLPas)*, grant numbers 01IH16008A/B/E, is acknowledged.

References

1. M. Abraham, T. Murtola, R. Schulz, S. Páll, J. Smith, B. Hess, E. Lindahl, GROMACS: high performance molecular simulations through multi-level parallelism from laptops to supercomputers. SoftwareX **1–2**, 19–25 (2015)
2. W. Brown, P. Wang, S. Plimpton, A. Tharrington, Implementing molecular dynamics on hybrid high performance computers—short range forces. Comput. Phys. Commun. **182**(4), 898–911 (2011)
3. M. Buchholz, Framework zur Parallelisierung von Molekulardynamiksimulationen in verfahrenstechnischen Anwendungen. Dissertation, Institut für Informatik, Technische Universität München, 2010
4. W. Eckhardt, Efficient HPC implementations for large-scale molecular simulation in process engineering. Dissertation, Dr. Hut, Munich, 2014
5. W. Eckhar, A. Heinecke, An efficient Vectorization of Linked-Cell Particle Simulations, in *ACM International Conference on Computing Frontiers* (ACM, New York, NY, USA, 2012) pp. 241–243
6. W. Eckhardt, A. Heineck, R. Bader, M. Brehm, N. Hammer, H. Huber, H.G. Kleinhenz, J. Vrabec, H. Hasse, M. Horsch, M. Bernreuther, C. Glass, C. Niethammer, A. Bode, H.J. Bungartz, 91 TFLOPS Multi-trillion Particles Simulation on SuperMUC (Springer, Berlin, Heidelberg, 2013), pp. 1–12
7. W. Eckhardt, T. Neckel, Memory-efficient implementation of a rigid-body molecular dynamics simulation, in *Proceedings of the 11th International Symposium on Parallel and Distributed Computing (ISPDC 2012)* (IEEE, Munich 2012) , pp. 103–110
8. F.A. Gratl, S. Seckler, N. Tchipev, H.J. Bungartz, P. Neumann, Autopas: auto-tuning for particle simulations, in *2019 IEEE International Parallel and Distributed Processing Symposium (IPDPS)* (2019)
9. S. Grottel, M. Krone, C. Müller, G. Reina, T. Ertl, MegaMol—a prototyping framework for particle-based visualization. IEEE Trans. Visual Comput. Graph. **21**(2), 201–214 (2015)
10. C. Hu, X. Wang, J. Li, X. He, S. Li, Y. Feng, S. Yang, H. Bai, Kernel optimization for short-range molecular dynamics. Comput. Phys. Commun. **211**, 31–40 (2017)
11. A. Köster, T. Jiang, G. Rutkai, C. Glass, J. Vrabec, Automatized determination of fundamental equations of state based on molecular simulations in the cloud. Fluid Phase Equilib. **425**, 84–92 (2016)
12. K. Langenbach, M. Heilig, M. Horsch, H. Hasse, Study of homogeneous bubble nucleation in liquid carbon dioxide by a hybrid approach combining molecular dynamics simulation and density gradient theory. J. Chem. Phys. **148**, 124702 (2018)

13. G. Nagayama, P. Cheng, Effects of interface wettability on microscale flow by molecular dynamics simulation. Int. J. Heat Mass Transf. **47**, 501–513 (2004)
14. C. Niethammer, S. Becker, M. Bernreuther, M. Buchholz, W. Eckhardt, A. Heinecke, S. Werth, H.J. Bungartz, C. Glass, H. Hasse, J. Vrabec, M. Horsch, ls1 mardyn: the massively parallel molecular dynamics code for large systems. J. Chem. Theory Comput. **10**(10), 4455–4464 (2014)
15. S. Páll, B. Hess, A flexible algorithm for calculating pair interactions on SIMD architectures. Comput. Phys. Commun. **184**(12), 2641–2650 (2013)
16. D. Rapaport, *The Art of Molecular Dynamics Simulation* (Cambridge University Press, Cambridge, 2004)
17. L. Rekvig, D. Frenkel, Molecular simulations of droplet coalescence in oil/water/surfactant systems. J. Chem. Phys. **127**, 134701 (2007)
18. S. Seckler, N. Tchipev, H.J. Bungartz, P. Neumann, Load balancing for molecular dynamics simulations on heterogeneous architectures, in *2016 IEEE 23rd International Conference on High Performance Computing (HiPC)* (2016), pp. 101–110
19. N. Tchipev, S. Seckler, M. Heinen, J. Vrabec, F. Gratl, M. Horsch, M. Bernreuther, C.W. Glass, C. Niethammer, N. Hammer, B. Krischok, M. Resch, D. Kranzlmüller, H. Hasse, H.J. Bungartz, P. Neumann, Twetris: twenty trillion-atom simulation. Int. J. High Perform. Comput. Appl. 1094342018819,741 (0). DOI https://doi.org/10.1177/1094342018819741
20. N. Tchipev, A. Wafai, C. Glass, W. Eckhardt, A. Heinecke, H.J. Bungartz, P. Neumann, Optimized force calculation in molecular dynamics simulations for the Intel Xeon Phi (Springer International Publishing, Cham, 2015), pp. 774–785
21. J. Vrabec, M. Bernreuther, H.J. Bungartz, W.L. Chen, W. Cordes, R. Fingerhut, C. Glass, J. Gmehling, R. Hamburger, M. Heilig, M. Heinen, M. Horsch, C.M. Hsieh, M. Hülsmann, P. Jäger, P. Klein, S. Knauer, T. Köddermann, A. Köster, K. Langenbach, S.T. Lin, P. Neumann, J. Rarey, D. Reith, G. Rutkai, M. Schappals, M. Schenk, A. Schedemann, M. Schönherr, S. Seckler, S. Stephan, K. Stöbener, N. Tchipev, A. Wafai, S. Werth, H. Hasse, Skasim—scalable hpc software for molecular simulation in the chemical industry. Chem. Ing. Tech. **90**(3), 295–306 (2018)
22. J. Vrabec, G.K. Kedia, G. Fuchs, H. Hasse, Comprehensive study of the vapour-liquid coexistence of the truncated and shifted lennard-jones fluid including planar and spherical interface properties. Mol. Phys. **104**(9), 1509–1527 (2006)
23. X. Wang, J. Li, J. Wang, X. He, N. Nie, Kernel optimization on short-range potentials computations in molecular dynamics simulations. (Springer, Singapore, 2016) , pp. 269–281
24. S. Werth, G. Rutkai, J. Vrabec, M. Horsch, H. Hasse, Long-range correction for multi-site lennard-jones models and planar interfaces. Mol. Phys. **112**(17), 2227–2234 (2014)

Ad-Hoc File Systems At Extreme Scales

Mehmet Soysal and Achim Streit

Abstract This work presents the results of the project with the acronym ADA-FS (Advanced Data Placement via Ad-hoc File Systems at extreme scale). The project which has been approved for the ForHLR II aims to improve I/O performance for highly parallel applications by using distributed on-demand file systems. These temporary file systems are created on the allocated compute nodes, using the node-local disks for the on-demand file system. Through integration into the scheduling system of the supercomputer, it can be requested like any other resource. The research approach contains the design of the file system itself as well as the questions about the right planning strategy for the necessary I/O transfers. In the granted project for the ForHLR II we are investigating the methods on how to integrate the approach into a HPC system. Also, we are evaluating the impact of the on-demand created file systems to running HPC jobs and the applications.

1 Introduction

Today's HPC systems utilize parallel file systems that comply with POSIX semantics, such as Lustre [1], GPFS [2], or BeeGFS [3]. The storage subsystem within HPC systems is increasingly becoming a bottleneck. Furthermore, the performance is limited by the interface between the global file system and the compute nodes. Moreover, parallel file systems (and their I/O subsystem) are often shared by many users and their jobs. When users develop applications for HPC systems, they typically tend to optimize for computing power, sometimes disregarding the I/O behavior of the application. While the computing resources can often be allocated exclusively, the global PFS is shared by all users of a HPC system. This environment makes it difficult for the user to optimize the application concerning I/O. There are many

M. Soysal (✉) · A. Streit
Steinbuch Centre for Computing, Karlsruhe Institute of Technology,
Hermann-von-Helmholtz-Platz 1, 76344 Eggenstein-Leopoldshafen, Germany
e-mail: mehmet.soysal@kit.edu

A. Streit
e-mail: achim.streit@kit.edu

© Springer Nature Switzerland AG 2021
W. E. Nagel et al. (eds.), *High Performance Computing in Science and Engineering '19*,
https://doi.org/10.1007/978-3-030-66792-4_36

possible factors a user would have to consider. Influences from the back-end storage device, network interface, storage servers, data distribution, request size, and other applications slowing down the PFS [4].

One of the reasons why PFSs struggle with certain I/O operations is that they have to cover a wide range of applications. In addition, the PFS must be robust with high availability as the HPC system is dependent on the global storage system. But there are applications and scenarios that do not suit the general case. This includes scenarios and cases in which large amounts of data or millions of small files are generated, causing high load on the storage system. Consequently, bad behaving applications can result in poor performance affecting all users. The ADA-FS project aims to improve I/O performance for highly-parallel applications by a distributed ad-hoc overlay file systems. The results have been published [5–7] and we are showing in this paper a overview of our work. In order to achieve our goals, several challenges need to be addressed.

The first step is to find a way to deploy these on-demand file systems on production systems. It has to be minimal invasive and should not involve any changes to the operating model. This initial step also includes performance measurements with synthetic benchmarks. The results for this first step are represented in Chap. 3. Another point in our approach is the question, if the data can be pre-staged to the allocated compute nodes. For this challenge it is important to know which nodes are going to be allocated to a waiting job. To this end, we investigated how the run times of jobs can be predicted and how good they must be to allow our approach. Methods from the field of machine learning were used here, to predict run times. Also we simulated different workloads of HPC systems and evaluated the impact of improved wall time estimates. In Chap. 4 we present our evaluation regarding this part of the project. The next Chapter includes applications and use cases of our users. Here we present how much performance is achievable with our approach and what impact we have on applications and the HPC system. For this we picked three different applications from our users. We examined the application behavior with the on-demand file system and how our approach can help these use cases. We present the results of the real usage scenarios in Chap. 5. First we start with the related work in the next Chap. 2 and conclude at the end with a summary of the approved project.

2   Related Work

The project covers different scientific domains, e.g., machine learning, file systems, scheduling and a wide range of applications. In this chapter we give a brief introduction in the important parts of the related work.

2.1 I/O

In the recent past there have been many developments and innovations to improve I/O throughput and performance. We cannot cover everything and try to give a brief overview of existing solutions. Many solutions are implemented at multiple levels in the I/O stack, but four basic categories can be formed: file system features, hardware solutions, libraries and dynamic system re-configurations.

File system features

File Systems have received interesting new features to reduce I/O bottlenecks. BeeGFS offers storage pools [8] to group storage targets in different classes, e.g., one pool with very fast solid state drives. GPFS has implemented a Highly Available Write Cache (HAWC)[9]. Node-local *solid-state drives* (SSDs) are used as buffers for the global file system. As a result, random I/O patterns are processed on local storage. Lustre has the Progressive File Layouts (PFL) [10] feature, which adjusts dynamically the stripe pattern and chunk size based on I/O traffic.

However, such solution are only available when using the vendor software solution.

Hardware solutions

Today's wide spread use of SSDs in compute nodes of HPC systems has provided a new way of accelerating storage. SSDs have been considered for file system metadata [11, 12], as its meta-data performance is a major bottleneck in HPC environments.

A different kind of hardware solutions are burst buffers, which aim to reduce the load on the global file system [13].

Libraries

There is a large number of libraries available for improving I/O behavior of an application. Middleware libaries, such as MPI-IO [14], help to improve usage of parallel storage systems, e.g. collective I/O [15]. High-level libraries, such as HDF5 [16], NETCDF [17] or ADIOS [18], are trying help users to express I/O as data structures and not only as bytes and blocks. These libraries are not in contrast to our approach. The advantages of using such libraries also apply to the on-demand file systems.

System reconfiguration

Like our approach, the configuration of the system can be modified to improve I/O. There are several basic methods. A Dynamic Remote Scratch [19] implementation was developed to create an on-demand block device and use it with local SSDs as a LVM [20] device. Another software based solution is the RAMDISK Storage Accelerator [21]. It introduces a additional cache layer into HPC systems. Our approach also fits into this category.

2.2 Job Walltime Prediction

Batch schedulers are responsible for the resource planing and allocate the nodes to a job [22]. One of the factors of this resource planning is based on wall time estimates, given by the user. It is a well known problem that the user provided estimates are far from optimal. With exact information about the run time of a job, the scheduler can predict more accurately when sufficient resources are available to start queued jobs [23]. However, the user requested wall time is not close to the real used wall time. Gibbons [24, 25], and Downey [23] use historical workloads to predict the wall times of parallel applications. They predict wall times based on templates. These templates are created by analyzing previously collected metadata and grouped according to similarities. However, both approaches are restricted to simple definitions.

In the recent years, the machine learning algorithms are used to predict resource consumption in several studies [26–31].

However, all of the above mentioned studies do not try to evaluate the accuracy of the node allocation predictions. Most of the publications focus on observing the utilization of the HPC system and the reliability of the scheduler estimated job start times. In our work we focus on the node allocation prediction and how good wall time estimates have to be. This directly affects, whether a cross-node, ad-hoc, independent parallel file system can be deployed and data can be pre-staged, or not.

2.3 Machine Learning

Machine learning (ML) is about knowledge retrieval from data. It can also be understood as statistical learning and predictive analytics. In general, machine learning is a method to learn from a set of samples with a target value and use the learned data to predict target values from unknown samples. For our evaluation, we use a supervised machine learning approach [32].

In our evaluation, the AUTOML library auto-sklearn [33] (based on scikit-learn [34, 35]) is used to automate the complex work of machine learning optimization. In a classical ML process, different models and systems are explored until the best is chosen and auto-sklearn automatizes this process.

3 Deployment On-Demand File System

Usually HPC systems use a *batch system*, such as SLURM [36], MOAB [37], or LSF [38]. The batch system manages the resources of the cluster and starts the user jobs on allocated nodes. At the start of the job, a prologue script may be started on one or all allocated nodes and, if necessary, an epilogue script at the end of a job (see Fig. 1). These scripts are used to clean, prepare, or test the full functionality of the

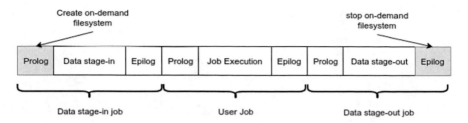

Fig. 1 Job flow for creating an on-demand file system

Table 1 BeeGFS startup and throughput

Nodes	8	16	32	64	128	256
Startup (s)	10.21	16.75	29.36	56.55	152.19	222.43
Shutdown (s)	11.90	12.13	9.40	15.96	36.13	81.06
Throughput (GiB/s)	2.79	6.74	10.83	28.37	54.06	129.95

nodes. We modified these scripts to start the on-demand file system upon request. During job submission a user can request an on-demand file system for the job. This solution has minimal impact on the HPC system operation. Users without the need for an on-demand file system are not affected.

3.1 Benchmarks

As initial benchmarks we tested the startup time of the on-demand file system and used the "iozone" benchmark for a throughput test. The Startup and shutdown times are shown in Table 1. The delivered tools in the BeeOND package have a serial part during initialization. After optimizing these regions we were able to start BeeGFS within 60 s on 512 nodes.

In Fig. 2a we show the IoZone [39] benchmark to measure the read and write throughput of the on-demand file system (solid line). The Figure show that performance increases linearly with the number of used compute nodes. The limiting factor here is the aggregate throughout of the used SSDs. A small deviation can be observed due to performance scattering of SSDs [40].

In a further test, we evaluated the storage pooling feature of BeeGFS [8]. We created a storage pool for each leaf switch (see Fig. 3). In other words, when writing to a storage pool, the data is distributed via the stripe count and chunk size, but remains physical within a switch. Only the communication with the meta data server is forwarded across the core switches. Figure 2b shows the write throughput for the scenarios. In the first experiment, with all six core switches, there is only a minimal performance loss, which indicates a small overhead when using storage pools. In the

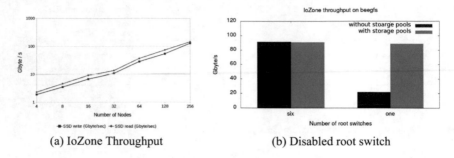

(a) IoZone Throughput   (b) Disabled root switch

Fig. 2 a IoZone Throughput **b** Disabled root switch

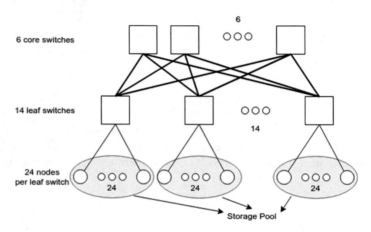

Fig. 3 Scheme of the fabric topology (small island), with a storage pool per leaf switch

second case we turned five switches off. With reduced number of core switches, the throughput drops due to the reduced network capacity. If storage pools are created accordingly to the leaf switches, it is possible to achieve the same performance.

3.2 Conclusion

Adding on-demand file system functionality to an HPC system is easy. There is no need to change the operating model. An on-demand file system is only started if it is actually requested. Startup times might be acceptable on smaller HPC systems but, they are not feasible at large scales. But what is exactly acceptable depends on several factors. For example, a few minutes start-up time may be acceptable if the jobs run for a day. But waiting an hour for the file system to start when the job itself isn't running much longer, doesn't make much sense.

Various observations show that with this approach the network is no longer the bottleneck. Since the fabric of an HPC system has a high bisection bandwidth, there is enough bandwidth left for an on-demand fs. However, if the network is designed somewhat weaker, enormous throughput can still be achieved with taking the topology into consideration.

4 Walltime Prediction

An investigation whether data can be pre-staged also belongs to the tasks of this project. One of the challenges is to know which nodes are going to be allocated to a queued job. The HPC scheduler predicts these nodes based on the user given wall times. Therefore, we have decided to evaluate whether there is an easy way to automatically predict such wall time. Our proposed approach for wall time prediction is to train an individual model for every user with methods from the machine learning domain. As historical data, we used serveral workloads from two of the HPC-systems at the Karlsruhe Institute for Technology/Steinbuch Centre for Computing [41], the ForHLR I + II [42, 43] clusters. We used Automatic machine learning (AUTOML) to pre-process the input data and selecting the correct model including the optimization of hyperparameters. In this work, the auto ML library auto-sklearn [33] is used. It is based on scikit-learn [34, 35].

Figure 4 shows the R^2 score for models of the users on ForHLR I+II with 30 min AUTOML. A concentration of the points in the upper right corner indicates a higher number of good models for the training and test data. A more descriptive illustration of the results are given in Fig. 5 for the ForHLR II. Here the median absolute error is compared between the AUTOML, the default linear regression, and the user given wall time prediction. On the ForHLR II cluster 50% of the prediction have a smaller median absolute error of around 21 min, 43 min, 186 min for the AUTOML model, the linear regression model, and the user prediction, respectively.

4.1 Conclusion

We showed that we can achieve good walltime predictions with very simple methods. We only used general meta-data and trained an individual model for each user. The results are very remarkable, considering that hardly any manual optimizations were performed on the models.

But we have shown in a futher investigation that even with almost perfect job wall time estimates the node allocation can't be predicted in a sufficient manner [7].

It has therefore been decided that further work is needed here. In this case, a modification must be made to the operational processes of the scheduler. We have achieved this by developing a plug-in (On-demand burst buffer plugin) for the SLURM scheduler. If required, this plugin starts an on-demand plugin and transfers the data on the

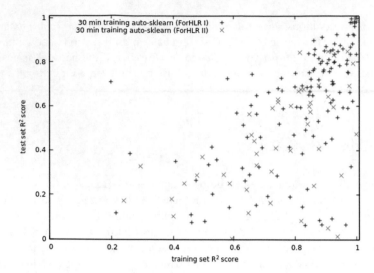

Fig. 4 X-Axis R^2 score on training samples, Y-Axis R^2 score on test samples for ForHLR I+II with 30 min AUTOML

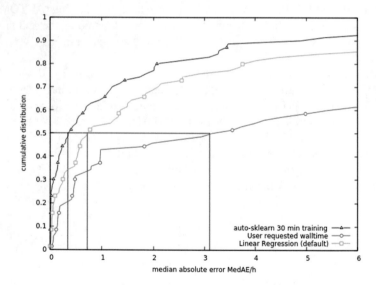

Fig. 5 Y-Axis Cumulative distribution, X-Axis Median absolute error ForHLR II

temporary fs. The challenge with the unknown node list is solved with reservations by this plugin.

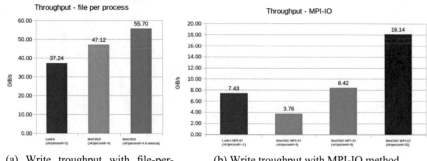

(a) Write troughput with file-per-process method

(b) Write troughput with MPI-IO method

Fig. 6 Write benchmark with super_sph

5 Scientific Applications

We have evaluated several applications regarding to on-demand file systems. We selected applications which either generate a very high load on our system or the I/O part is identified as a bottleneck. In this paper we present only a very brief overview, other results are already published [44].

5.1 Super_sph

We evaluated the application super_sph ("Simulation for Smoothed Particle Hydrodynamics") [45] which is developed at "Institut für Strömungsmaschinen" @KIT. The software scales up to 15000 Cores and 10^9 particles. The first implementation of the software created a file per process and required data-gathering as post-processing. A new implementation is now writing directly to time steps using MPI-IO which makes the data-gathering process unnecessary. From our observation—file per process method is causing heavy load on the PFS. Using MPI-IO is slower but has less impact on global PFS. Figure 6 show the results of super_sph when writing directly to the global filesystem (Lustre) and to an on-demand created filesystem (BeeOND/BeeGFS). For the benchmark we used 256 Nodes. While using the simple file-per-process method we gain a small performance increase. When using MPI-IO it is important to hoose the right parameters for the file-system. If the chunksize is not well chosen the performance loss is tremendous.

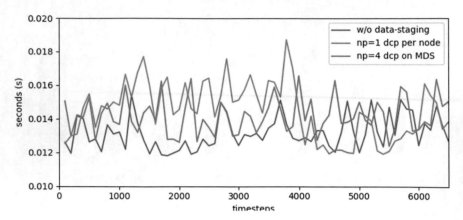

Fig. 7 Execution time per time-step. Different scenarios w/o data staging

5.2 Data Staging

We also considered the case of copying data back to the PFS while the application is running. The results are already published [44] and here we show a short summary. For this purpose, we used different NAStJA simulations on 23 nodes. The parallel copy tool dcp [46] was used to stage data. Figure 7 show the average execution time per time-step of five runs. With 16 cores for the application, from available 20 cores, the run times are similar whether the run was executed with or without data staging. If there are enough free resources on the compute nodes, the data can be staged-out without slowing down the application.

5.3 Conclusion

The results with real applications and use cases are already very good in the early phase. The use of on-demand file system immediately reduces the load on the global file system. This is of great importance for the shared HPC system and means a much more stable operation with less interference between the jobs. The impact of an on-demand file system to the application is minimal. However, we have only tested few applications and use cases to see if an on-demand file system becomes a disadvantage. Also the results for data-staging are promising, depending on whether you have much or little time to move the data, there are ways to choose the right method.

6 Summary

On-demand file systems is easy adaptable into a HPC System. It immediately reduces the load on the global file systems. Startup and shutdown times are acceptable only for long running jobs. For very short running jobs it might be senseless, but experience shows that large scale jobs usually request longer wall times. However, many more factors have to be taken into account to enable a reasonable and fast use in a wide range. There are also many factors to consider during deployment so that a user is not overwhelmed, e.g., setting the right strip-count and chunk-size parameters.

It turned out that pre-staging data to the compute nodes is not possible with the unreliable allocation prediction of the scheduler. Here a modification is needed to cope with the issue of the unknown node list. A plugin has been developed which solves this issue, by using reservations. The plugin extends the use of the built-in burst buffer concept and creates an on-demand file system and moves the required data to the temporary file system.

The trend in the HPC environment clearly shows that faster solid state disks keep coming into the compute nodes. With these, the advantages of on-demand file systems on the compute nodes should be even more significant.

Acknowledgements The project ADA-FS is funded by the DFG Priority Program "Software for exascale computing" (SPPEXA, SPP 1648), which is gratefully acknowledged. This work was supported by the Helmholtz Association of German Research Centres (HGF) and the Karlsruhe Institue of Technology. This work was performed on the computational resource ForHLR II with the acronym ADA-FS funded by the Ministry of Science, Research and the Arts Baden-Württemberg and DFG ("Deutsche Forschungsgemeinschaft"). We would like to thank the operation team of the ForHLR II cluster, which allowed us to adapt operational areas of the system to our needs.

References

1. S. Microsystems, LUSTRE™ FILE SYSTEM High-Performance Storage Architecture and Scalable Cluster File System (2007), http://www.csee.ogi.edu/zak/cs506-pslc/lustrefilesystem. pdf. Accessed 05 Sept 2016
2. F. Schmuck, R. Haskin, Gpfs: A shared-disk file system for large computing clusters, in *Proceedings of the 1st USENIX Conference on File and Storage Technologies*, ser. FAST '02 (Berkeley, CA, USA, USENIX Association, 2002)
3. J. Heichler, An introduction to BeeGFS (2014), http://www.beegfs.com/docs/Introduction_to_ BeeGFS_by_ThinkParQ.pdf. Accessed 6 Sept 2016
4. O. Yildiz, M. Dorier, S. Ibrahim, R. Ross, G. Antoniu, On the root causes of cross-application I, O interference in HPC storage systems, in *IEEE International on Parallel and Distributed Processing Symposium* (IEEE 2016), pp. 750–759
5. M. Soysal, M. Berghoff, A. Streit, Analysis of job metadata for enhanced wall time prediction, in *Job Scheduling Strategies for Parallel Processing* (2018)
6. M. Soysal, M. Berghoff, A. Streit, *Analysis of job metadata for enhanced wall time prediction, in Job Scheduling Strategies for Parallel Processing* (Springer International Publishing, Cham, 2019), pp. 1–14
7. M. Soysal, M. Berghoff, D. Klusáček, A. Streit, On the quality of wall time estimates for resource allocation prediction, in *Proceedings of the 48th International Conference on Parallel*

Processing: Workshops, ser. ICPP 2019 (ACM, New York, NY, USA, 2019) vol 23, pp. 1–23, 8. https://doi.org/10.1145/3339186.3339204

8. BeeGFS, BeeGFS Storage Pool (2018), https://www.beegfs.io/wiki/StoragePools. Accessed 18 Aug 2018

9. IBM, GPFS—highly available write cache (hawc) (2018), https://www.ibm.com/support/knowledgecenter/en/STXKQY_5.0.0/com.ibm.spectrum.scale.v5r00.doc/bl1adv_hawc.htm

10. R. Mohr, M.J. Brim, S. Oral, A. Dilger, Evaluating progressive file layouts for lustre

11. J. Xing, J. Xiong, N. Sun, J. Ma, Adaptive and scalable metadata management to support a trillion files, in *Proceedings of the Conference on High Performance Computing Networking, Storage and Analysis*, ser. SC '09 (ACM, New York, NY, USA, 2009) pp. 26:1–26:11. https://doi.org/10.1145/1654059.1654086

12. S. Lang, P. Carns, R. Latham, R. Ross, K. Harms, W. Allcock, I/O performance challenges at leadership scale, in *Proceedings of the Conference on High Performance Computing Networking, Storage and Analysis*(2009), pp. 1–12

13. N. Liu, J. Cope, P. Carns, C. Carothers, R. Ross, G. Grider, A. Crume, C. Maltzahn, On the role of burst buffers in leadership-class storage systems, in *IEEE 28th Symposium on Mass Storage Systems and Technologies (MSST)*, (IEEE, 2012), pp. 1–11

14. R. Thakur, W. Gropp, E. Lusk, On implementing MPI-IO portably and with high performance, in *Proceedings of the Sixth Workshop on I/O in Parallel and Distributed Systems* (ACM, 1999), pp. 23–32

15. R. Thakur, W. Gropp, E. Lusk, Data sieving and collective I, O in ROMIO, in *The Seventh Symposium on the Frontiers of Massively Parallel Computation, Frontiers' 99* (IEEE, 1999), pp. 182–189

16. M. Folk, G. Heber, Q. Koziol, E. Pourmal, D. Robinson, An overview of the HDF5 technology suite and its applications, in *Proceedings of the EDBT/ICDT 2011 Workshop on Array Databases* (ACM, 2011), pp. 36–47

17. R. Rew, G. Davis, Netcdf: an interface for scientific data access. IEEE Comput. Graphics Appl. **10**(4), 76–82 (1990)

18. J.F. Lofstead, S. Klasky, K. Schwan, N. Podhorszki, C. Jin, Flexible IO and integration for scientific codes through the adaptable IO system (ADIOS), in *Proceedings of the 6th International Workshop on Challenges of Large Applications in Distributed Environments* (ACM, 2008), pp. 15–24

19. M. Neuer, J. Salk, H. Berger, E. Focht, C. Mosch, K. Siegmund, V. Kushnarenko, S. Kombrink, S. Wesner, Motivation and implementation of a dynamic remote storage system for I/O demanding HPC applications, in *International Conference on High Performance Computing* (Springer, 2016), pp. 616–626

20. D. Teigland, H. Mauelshagen, Volume managers in linux, in USENIX Annual Technical Conference. FREENIX Track 185–197 (2001)

21. T. Wickberg, C. Carothers, The RAMDISK storage accelerator: a method of accelerating I/O performance on HPC systems using RAMDISKs, in *Proceedings of the 2nd International Workshop on Runtime and Operating Systems for Supercomputers*, ser. ROSS '12 (ACM, New York, NY, USA, 2012), pp. 5:1–5:8. https://doi.org/10.1145/2318916.2318922

22. M. Hovestadt, O. Kao, A. Keller, A. Streit, Scheduling in hpc resource management systems: queuing vs planning, in *Job Scheduling Strategies for Parallel Processing*, ed. by D. Feitelson, L. Rudolph, U. Schwiegelshohn (Springer, Berlin, 2003), pp. 1–20

23. A.B. Downey, Predicting queue times on space-sharing parallel computers, in *Proceedings of the 11th International Parallel Processing Symposium* (IEEE, 1997), pp. 209–218

24. R. Gibbons, A historical profiler for use by parallel schedulers. Master's thesis, University of Toronto, 1997

25. R. Gibbons, A historical application profiler for use by parallel schedulers, in *Job Scheduling Strategies for Parallel Processing* (Springer, 1997), pp. 58–77

26. A. Matsunaga, J.A. Fortes, On the use of machine learning to predict the time and resources consumed by applications, in *Proceedings of the 2010 10th IEEE/ACM International Conference on Cluster, Cloud and Grid Computing* (IEEE Computer Society, 2010), pp. 495–504

27. N.H. Kapadia, J.A. Fortes, On the design of a demand-based network-computing system: the purdue university network-computing hubs, in *Proceedings of the Seventh International Symposium on High Performance Distributed Computing* (IEEE, 1998), pp. 71–80
28. A.W. Mu'alem, D.G. Feitelson, Utilization, predictability, workloads, and user runtime estimates in scheduling the IBM SP2 with backfilling. IEEE Trans. Parallel Distrib. Syst. **12**(6), 529–543 (2001)
29. F. Nadeem, T. Fahringer, Using templates to predict execution time of scientific workflow applications in the grid, in *Proceedings of the 2009 9th IEEE/ACM International Symposium on Cluster Computing and the Grid* (IEEE Computer Society, 2009), pp. 316–323
30. W. Smith, Prediction services for distributed computing, in *IEEE International on Parallel and Distributed Processing Symposium, (IPDPS 2007)* (IEEE, 2007), pp. 1–10
31. D. Tsafrir, Y. Etsion, D.G. Feitelson, Backfilling using system-generated predictions rather than user runtime estimates. IEEE Trans. Parallel Distribut. Syst. **18**(6), (2007)
32. M. Mohri, A. Rostamizadeh, A. Talwalkar, *Foundations of Machine Learning*, (MIT press, 2012)
33. M. Feurer, A. Klein, K. Eggensperger, J. Springenberg, M. Blum, F. Hutter, Efficient and robust automated machine learning, in *Advances in Neural Information Processing Systems*, ed. by C. Cortes, N.D. Lawrence, D.D. Lee, M. Sugiyama, R. Garnett (Curran Associates, Inc., 2015), pp. 2962–2970. http://papers.nips.cc/paper/5872-efficient-and-robust-automated-machine-learning.pdf
34. F. Pedregosa, G. Varoquaux, A. Gramfort, V. Michel, B. Thirion, O. Grisel, M. Blondel, P. Prettenhofer, R. Weiss, V. Dubourg, J. Vanderplas, A. Passos, D. Cournapeau, M. Brucher, M. Perrot, E. Duchesnay, Scikit-learn: machine learning in Python. J. Mach. Learn. Res. **12**, 2825–2830 (2011)
35. L. Buitinck, G. Louppe, M. Blondel, F. Pedregosa, A. Mueller, O. Grisel, V. Niculae, P. Prettenhofer, A. Gramfort, J. Grobler, R. Layton, J. VanderPlas, A. Joly, B. Holt, G. Varoquaux, API design for machine learning software: experiences from the scikit-learn project, in *ECML PKDD Workshop: Languages for Data Mining and Machine Learning* (2013), pp. 108–122
36. Slurm-schedmd, http://www.schedmd.com
37. Adaptive Computing, http://www.adaptivecomputing.com
38. IBM—platform computing, http://www.ibm.com/systems/platformcomputing/products/lsf/
39. D. Capps, W. Norcott, Iozone filesystem benchmark (2008), http://iozone.org/
40. E. Kim, SSD performance-a primer: an introduction to solid state drive performance, evaluation and test, Tech. rep. (Storage Networking Industry Association, 2013)
41. Steinbuch Center for Computing, Scc (2016), http://www.scc.kit.edu. Accessed 16 Aug 2016
42. Forschungshochleistungsrechner ForHLR 1 (2018), www.scc.kit.edu/dienste/forhlr1.php
43. Forschungshochleistungsrechner ForHLR 2 (2018), www.scc.kit.edu/dienste/forhlr2.php
44. M. Soysal, M. Berghoff, T. Zirwes, M.A. Vef, S. Oeste, A. Brinkman, W. E. Nagel, A. Streit, Using On-demand File Systems in HPC Environments, *Accepted @ The 2019 International Conference on High Performance Computing and Simulation (HPBench@HPCS)*
45. S. Braun, R. Koch, H.J. Bauer, Smoothed particle hydrodynamics for numerical predictions of primary atomization **15**(1), 56–60 (2017)
46. D. Sikich, G. Di Natale, M. LeGendre, A. Moody, mpifileutils: a parallel and distributed toolset for managing large datasets, Lawrence Livermore National Lab (LLNL) (Livermore, CA, United States, Tech. Rep., 2017)

Miscellaneous Topics

Univ.-Prof. Dr.-Ing. Wolfgang Schröder

In this chapter, contributions on materials science, molecular thermodynamics, neuroscience, and geophysics complement the research fields which have been tackled in the previous chapters. The articles widen the field where numerical analyses are useful to gain novel results and to improve the scientific knowledge. The findings evidence the close link between natural science in general and computer science in particular. Furthermore, these results provide a sound basis to develop new scientific models. Compact mathematical descriptions will be solved by highly sophisticated and efficient algorithms using high-performance computers. This interdisciplinary collaboration between several scientific fields defines the extremely intricate numerical challenges and as such drives the progress in fundamental and applied research. The subsequent articles represent an excerpt of the projects being linked to HLRS and SSC. The computations are used to obtain some quantitative results and to corroborate physical models and to even derive new theoretical approaches. However, whenever possible experimental investigations and analytical solutions are to be taken into account to complement the physical knowledge.[1]

The first article is from the Institute of Applied Materials and the Institute of Applied and Numerical Mathematics of the Karlsruhe Institute of Technology in cooperation with the Institute of Digital Materials of the Karlsruhe University of Applied Sciences. In materials science, simulations are used to improve the properties of materials. These properties depend on the chemical composition and on the evolving microstructure during the manufacturing process. To investigate the microstructure evolution, multi-physics phase-field simulations are used. For the directional solidification of alloys, large scale and massive parallel multi-physics phase-field simulations are required. During directional solidification, a liquid melt solidifies into one or more phases and thereby forms various microstructural patterns. This solidification process can be described by a coupled system of phase-field equations based on an Allen-Cahn approach and a concentration equation. Both are solved using a finite difference scheme for the spatial discretization and an explicit Euler scheme for the time discretization. Usually, the same time step width and grid res-

---

[1]Univ.-Prof. Dr.-Ing. Wolfgang Schröder, Institute of Aerodynamics, RWTH Aachen University, Wüllnerstr. 5a, 52062 Aachen, Germany, office@aia.rwth-aachen.de

olution is employed on all coupled equations. In this contribution, an explicit local time-stepping approach is applied to calculate the larger time steps in the region without diffuse interface. Furthermore, the domain decomposition for the parallelization and a moving window approach are utilized to ensure that the diffuse interface is limited to a defined region.

Another contribution is from the Chair of Thermodynamics and Thermal Process Engineering of TU Berlin, the Institute of Aerodynamics and Gas Dynamics of the University of Stuttgart, and the Chair Thermodynamics and Energy Systems of the University of Paderborn. High-performance computing has instigated progress in thermodynamic research for over half a century. Atomistic simulations are now well established, providing accurate data for thermodynamic states that are unattainable by current experiments. In fact, the development of modern equations of state is based on hybrid datasets, in which experimental results are complemented by extensive molecular dynamics (MD) simulations. Due to a sound physical basis, atomistic simulations additionally enable insight into phenomena that are still insufficiently explained. Transport diffusion in the vicinity of the mixture's critical point with one diluted component as well as the emergence of hydrogen bonds in aqueous mixtures constitute prominent examples. The successive increase in computational resources has thereby facilitated the investigation of phenomena at hydrodynamic length and time scales, even for ensemble sizes containing $N = 2 \cdot 10^{13}$ individual particles and has most recently lead to quantitative agreement between results obtained by large-scale MD and computational fluid dynamics (CFD).

Next, the first step toward a scaling analysis of an electrical neuron model from the Center of Super Computing of the Goethe University Frankfurt are presented. In computational neuroscience, the transmission of electrical signals of neurons is normally simulated by means of point process neurons, which mainly reflect the scale in time, or the classical cable equation which additionally introduces one space dimension. A fully resolved electrical model based on Gauss' law and the conservation of charges is presented which considers all space dimensions and is capable to simulate the extracellular and intracellular potential. For these simulations, three-dimensional volume meshes are required and due to the inherent complexity of the neuronal structure, these 3D-reconstructions yield large data-sets and need efficient solving strategies. The UG4-simulation framework is a powerful software for the computation of partial differential equations on unstructured grids in one, two, and three space dimensions and with its efficient, parallel solvers is well suited for this task. Computations of the 3D-cable equation on a simple geometry and on a three-dimensionally reconstructed neuron were performed.

Finally, a study on adaptive implicit space–time discretization from the Institute for Applied and Numerical Mathematics of the Karlsruhe Institute of Technology is presented. Classically, wave equations are considered as evolution equations where the derivative with respect to time is treated in a stronger way than the spatial differential operators. This results in an ordinary differential equation (ODE) with values in a function space with respect to the spatial variable. In order to analyze the ODE, space and time are treated separately and hence tools for partial differential equations are used in space and tools for ODEs are used in time. Typically, this separation carries over to the analysis of numerical schemes to approximate solutions of the equation. In the contribution of the Institute of Applied and Numerical Mathematics, the space–time operator is considered as a whole treating time and space dependence simultaneously in a variational manner. Using this approach, a space–time Hilbert space setting is constructed that allows for irregular solutions. A variant of the discontinuous Petrov-Galerkin method (DPG) for acoustic waves is considered. The construction of the method heavily relies on the generalized space–time traces. Expecting that an ansatz space containing functions of low regularity can lead to improved approximation quality for irregular solutions, a non-conforming variant of the DPG method is used. This method allows for face-wise defined traces on the space–time skeleton.

Accelerating Phase-Field Simulations for HPC-Systems

M. Seiz, J. Hötzer, H. Hierl, A. Reiter, K. Schratz, and B. Nestler

Abstract In the last decades simulations have become a powerful tool to understand physical processes and accelerate the development of new products. A high computational effort is necessary to resolve these processes accurately with quantitative models. Especially for large-scale and massive parallel simulations a reduction of the total computational time results in a large reduction of the computational effort. In this work we present an explicit local time stepping approach to accelerate phase-field simulations of directional solidification. To efficiently store the simulation data, we also investigate the MPI-I/O write performance of voxel data on Hazel Hen's ws8 Lustre filesystem.

1 Introduction

In materials science, simulations are used to improve the properties of materials. These properties depend on the one hand on the chemical composition and on the other hand on the evolving microstructure during the manufacturing process. To investigate the microstructure evolution, multi-physics phase-field simulations are used [13]. For the directional solidification of alloys, large scale and massive parallel multi-physics phase-field simulations are required. These are typically solved with tens of thousands of compute cores due to the computationally expensive evolution

M. Seiz (✉) · J. Hötzer · B. Nestler
Institute of Applied Materials (IAM), Karlsruhe Institute of Technology (KIT),
Straße am Forum 7, 76131 Karlsruhe, Germany
e-mail: marco.seiz@kit.edu

J. Hötzer
e-mail: johannes.hoetzer@kit.edu

J. Hötzer · H. Hierl · A. Reiter · B. Nestler
Institute of Digital Materials, Hochschule Karlsruhe Technik und Wirtschaft,
Moltkestr. 30, 76131 Karlsruhe, Germany

K. Schratz
Faculty of Mathematics, Karlsruhe Institute of Technology, Englerstraße 2, 76131 Karlsruhe,
Germany

© Springer Nature Switzerland AG 2021
W. E. Nagel et al. (eds.), *High Performance Computing in Science and Engineering '19*,
https://doi.org/10.1007/978-3-030-66792-4_37

equations and large domain sizes required for representative volume elements [2, 12, 14]. During directional solidification, a liquid melt solidifies into one or more phases and thereby forms various microstructural patterns [7, 9]. This solidification process can be described by a coupled system of phase-field equations based on an Allen-Cahn approach and a concentration equation. Both are solved using a finite difference scheme for the spatial discretization and an explicit Euler scheme for the time discretization [15]. Usually the same time step width and grid resolution is employed on all coupled equations. The biggest stable time step per evolution equation can be approximated with a von Neumann stability analysis. Depending on the material system, the stable time step widths can differ strongly between evolution equations. Phase-field models with an obstacle potential term have a limited diffuse interface width between the phases, which allows the skipping of calculation in those grid cells where there is no diffuse interface, i.e. the phase-field variables attain the values 0 and 1. Hence a different time step width could be used in domains without any diffuse interface.

In order to exploit this, an explicit local time stepping approach [1, 10] is applied to calculate the larger time steps in the region without diffuse interface. Furthermore the domain decomposition for the parallelization and a moving window approach are utilized to ensure that the diffuse interface is limited to a defined region.

Another point for large-scale phase-field simulations is the time taken for I/O. Without exploiting the parallelism provided by a parallel filesystem such as Lustre, large-scale simulation runs would spend significant computational time in writing simulation data to disk. Hence the I/O characteristics of ws8 were investigated in order to improve the I/O of PACE3D and develop heuristics.

2   Phase-Field Model for Solidification

To study the directional solidification of alloys, the phase-field model of [4, 12, 18] based on the grand potential approach is used. The N different phases are described by the order-parameters ϕ_α, with $\alpha \in \{1, \ldots, N\}$ [15]. Each ϕ_α can be interpreted as the local volume fraction of a phase α, each of which may differ in its state of matter, crystallographic arrangement and orientation. These order-parameters are collected in the phase-field vector ϕ. The K concentrations c_i and their corresponding chemical potential μ_i of the components $i \in \{1, \ldots, K\}$ are described by the concentration vector $\mathbf{c}$ and chemical potential vector μ, respectively [6, 18].

The evolution equations following [12] are:

$$\tau\varepsilon\frac{\partial\phi_\alpha}{\partial t} = -\varepsilon\left(\frac{\partial a(\phi,\nabla\phi)}{\partial\phi_\alpha} - \nabla\cdot\frac{\partial a(\phi,\nabla\phi)}{\partial\nabla\phi_\alpha}\right) - \frac{1}{\varepsilon}\frac{\partial\omega(\phi)}{\partial\phi_\alpha}$$
$$- \sum_{\beta=1}^{N}\psi_\beta(\mu,T)\frac{\partial h_\beta(\phi)}{\partial\phi_\alpha} - \lambda, \tag{1}$$

$$\frac{\partial\mu}{\partial t} = \left[\sum_{\alpha=1}^{N}h_\alpha(\phi)\left(\frac{\partial\mathbf{c}_\alpha(\mu,T)}{\partial\mu}\right)\right]^{-1}\left(\nabla\cdot\left(\mathbf{M}(\phi,\mu,T)\nabla\mu - \mathbf{J}_{at}(\phi,\mu,T)\right)\right.$$
$$\left. - \sum_{\alpha=1}^{N}\mathbf{c}_\alpha(\mu,T)\frac{\partial h_\alpha(\phi)}{\partial t} - \sum_{\alpha=1}^{N}h_\alpha(\phi)\left(\frac{\partial\mathbf{c}_\alpha(\mu,T)}{\partial T}\right)\frac{\partial T}{\partial t}\right), \tag{2}$$

$$\frac{\partial T}{\partial t} = \frac{\partial}{\partial t}(T_0 + G(z - vt)) = -Gv. \tag{3}$$

The kinetics of the interfaces in the phase-field evolution equation Eq. (1) are described by

$$\tau = \left(\sum_{\substack{\alpha,\beta=1 \\ (\alpha<\beta)}}^{N,N}\tau_{\alpha\beta}\phi_\alpha\phi_\beta\right)\left(\sum_{\substack{\alpha,\beta=1 \\ (\alpha<\beta)}}^{N,N}\phi_\alpha\phi_\beta\right)^{-1} \tag{4}$$

with $\tau_{\alpha\beta}$ as the reciprocal mobility, which can be derived following [5] based on the Gibbs energies and the diffusion coefficient.

The diffuse interface is controlled by the gradient energy density a and the potential energy ω, which are dependent on the interface energies $\gamma_{\alpha\beta}$ [12]. The interface thickness is related to the parameter ε. The phase transition between the different phases is described by the phase dependent grand potentials ψ_β and the interpolation h_β of [17]. The Lagrange multiplier $\lambda = \frac{1}{N}\sum_{\beta=1}^{N}rhs_\beta$ is accounting for the constraint $\sum_{\alpha=1}^{N}\partial\phi_\alpha/\partial t = 0$.

The evolution equations for the chemical potentials in Eq. (2) consist of a concentration flux corrected with an anti-trapping current $\mathbf{J}_{at}$ to balance effects of the artificially enlarged interface [4, 8, 16]. This flux-divergence is then converted into a chemical potential change by the susceptibility $\partial\mathbf{c}_\alpha(\mu,T)/\partial\mu$. The source terms in Eq. (2) can be derived from the Gibbs energies and describe the change of the chemical potential due to the order parameter as well as the temperature. The mobility function is defined as $\mathbf{M}(\phi,\mu,T) = \sum_{\alpha=1}^{N}h_\alpha(\phi)\mathbf{D}_\alpha\partial\mathbf{c}_\alpha(\mu,T)/\partial\mu$ with $\mathbf{D}_\alpha$ as the diffusion matrix of the phase α with the dimensions $K \times K$. A frozen temperature approximation [3] is used for the temperature evolution Eq. (3), employing the velocity v and the temperature gradient G.

The model is discretized with a finite differences scheme in space and an explicit Euler scheme in time as described in [2, 15]. This results in typical stencil schemes which are denoted in the following as DxCy, where x describes the number of Dimensions and y the number of adjacent Cells as introduced in [2]. The stencils

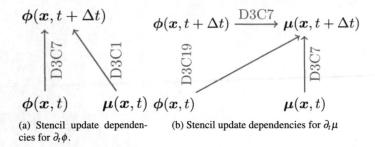

(a) Stencil update dependencies for $\partial_t \phi$.

(b) Stencil update dependencies for $\partial_t \mu$

Fig. 1 Stencil dependencies to calculate the next time step

and the dependencies to calculate the next time step $t + \Delta t$ for the phase-fields ϕ and the chemical potential μ are depicted in Fig. 1.

Based on a von Neumann stability analysis the maximal time step width for the phase-field Δt_ϕ and the chemical potential Δt_μ are derived as

$$\Delta t_\phi = \frac{\min(\tau_{\alpha\beta})\Delta x^2}{2 \cdot 2 \cdot d \cdot \max(\gamma_{\alpha\beta})} \tag{5}$$

$$\Delta t_\mu = \frac{\Delta x^2}{2 \cdot d \cdot \max(D)} \tag{6}$$

with d as the number of dimensions [19, 20]. Since both equations are solved simultaneously, the largest stable time step is $\Delta t = \min(\Delta t_\phi, \Delta t_\mu)$. Depending on the material parameters, the maximal time step between the evolution equations can differ. To efficiently solve both evolution equations we employ an explicit local time stepping scheme in the existing PACE3D framework [11].

3   The PACE3D Framework

The described phase-field model of Sect. 2 is implemented in the massive parallel framework PACE3D [11]. The solver consists of a wide range of multi-physics models to study the microstructure evolution for many processes. The framework uses the message passing interface (MPI) and domain decomposition to calculate large domains in parallel. Each rectangular subdomain is assigned to a rank and extended by ghost layers. The communication of the ghost layers is realized by blocking send and receive functions which are executed separately for each dimension to avoid diagonal exchanges between the subdomains. The solver scales almost ideally on up to 172 032 cores on different German supercomputers like the Hazel Hen, the SuperMUC as well as the ForHLR I and II. To store the simulation data MPI-I/O is used for the voxel fields.

4 Implementation of the Explicit Local Time Stepping

The explicit local time stepping scheme is implemented by using different time step widths per MPI process and per evolution equation. Each MPI process can be assigned an individual time step width per evolution equation by the user. An example for two processes with different time step widths is shown in Fig. 2. In the upper part, the directionally growing solid phases α, β, γ (red, green and blue) and the liquid l (yellow) are shown. Below, the profiles of the phase-fields ϕ_α and ϕ_ℓ (dashed lines) as well as the concentration c (solid line) are shown along the red line of the upper setting. Only in the interface region between the solids and melt a strong change of the profiles can be seen. In the lower part, the two subdomains and the ghost layers for multiple time steps are shown. In green, the time step widths for the phase-field evolution equation Δt_ϕ and in blue for the concentration evolution equation Δt_c respectively are shown. For both MPI processes the time step width Δt_c is $2\Delta t_\phi$. Furthermore, the Δt_ϕ of rank M (Δt_ϕ^M) is half that of rank $M + 1$ (Δt_ϕ^{M+1}). Hence, for this setting, after four time steps of the smallest Δt the ghost layers need to be synchronized as indicated by the orange cells. However, due to the different time step widths, a different number of ghost layers is needed per side. The evolution equations are also solved in the ghost layers to avoid synchronization until both workers are at the same simulation time again. The number of calculated ghost layers is reduced every time step as their update will not influence the inner subdomain. Hence, rank $M + 1$ updates the concentration once and the phase-field twice. In contrast, rank M updates the concentration twice and the phase-field four times. This results in a different number of ghost layers depending on the used time step widths and the time step widths of neighboring workers. In order to overcome the different loads resulting from the different time steps, the size of the subdomains can be statically adjusted.

5 Results of the Explicit Local Time Stepping

In order to calibrate the subdomain sizes to ensure similar loads per MPI process, different settings are analyzed based on the directional solidification of Ni-Al-Mo. For the considered material system the time step width for Δt_c can be chosen three times larger than Δt_ϕ. The time-to-solution behavior for three time step widths and different subdomain ratios is analyzed. Additionally the influence of the core count (2 cores and 24 cores) and hence the influence of different domain decompositions is investigated. For all measurements a total domain size of $400 \times 160 \times (1 \, or \, 12)$ voxel cells is used to ensure the same amount of voxel cells for the two core (decomposed in $2 \times 1 \times 1$ cores) and 24 core (decomposed in $2 \times 1 \times 12$ cores) case. A target simulation time of $2\,000 \times min(\Delta t_\phi)$ was established and hence $2\,000$ time steps were calculated. The domain is decomposed into two workers along the growth direction and up to 12 workers in the transverse direction. The size of the subdomains

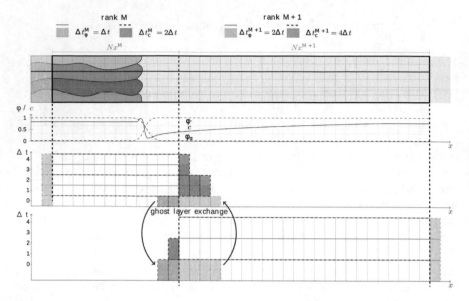

Fig. 2 Schematic illustration of the explicit local time stepping for directional solidification for two subdomains with different time step widths for the two evolution equations

along the growth direction is varied from an initial edge length of 200 voxel cells to 20 voxel cells. The ratio is calculated by the edge length of the small subdomain (Nx^M) divided by the total domain size $(Nx^M + Nx^{M+1})$.

All measurements are conducted on the Hazel Hen system at the HLRS. In Fig. 3 the results are compiled for the two and 24 cores setting. The domain size ratio shows a direct influence on the time-to-solution. Furthermore, the time-to-solution shows a minimum whose domain size ratio is dependent on the time step width ratio. Due to the larger time step width in one region, the computational load per process decreases. Hence the subdomain sizes need to be changed to reach a minimal runtime. Larger time step ratios exhibit the minimal runtime at smaller domain size ratios. For a time step ratio of two, the minimal runtime is found for a ratio of ~ 0.3 and for a time step ratio of three the minimum is observed at a ratio of 0.225 between the small and large subdomains, which holds for both the two and the 24 core case. For the two core case, the minimal runtime decreases from 80.9 s to 56.38 s and 44.8 s for a time step width of 1, 2 and 3 respectively. For the 24 cores case, the minimal runtime decreases from 66.44 s, 47.98 s to 35.49 s with larger time step widths. In the best case of three times larger time step width, the runtime is reduced in both settings by over 87 %.

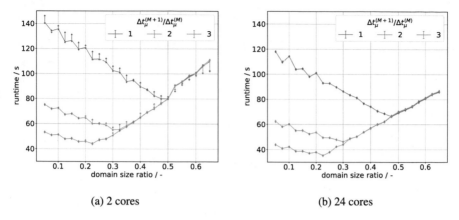

(a) 2 cores   (b) 24 cores

Fig. 3 Schematic illustration of the explicit local time stepping for directional solidification for two subdomains with different time step widths for the two evolution equations

6 ws8 is dead, long live ws8!

In order to optimize the I/O for PACE3D, a small MPI-I/O test program called mpiiotester was developed. It allows for different reading and writing styles (separate files per frame, all frames in one file, with and without headers, …) and outputs process rank based measurement data in an SQL format. The separate file per frame approach mimics file formats like VTK, whereas the single file approach for all frames mimics that of the proprietary format PACE3D employs. Besides the timers within mpiiotester, the CRAY timers and statistics were gathered as well (MPICH_MPIIO_STATS=1, MPICH_MPIIO_TIMERS=1). In this report, we only consider writing all frames into a single file without headers. Furthermore, the tests performed here only use a 1D domain decomposition, which generally shows better write rates than equivalent 3D domain decompositions (3DDD), but the principal results are transferable to 3DDD as PACE3D showed comparable speedups when a 3DDD was used. The data written is generated randomly on-the-fly and changed between frames as to prevent file system caching. All the shown write rates are the median values of at least 50 written frames with the error bars indicating the 95 % confidence interval of the median calculated by bootstrapping.

Simply applying this program on ws8 without any optimization hints leads to Fig. 4. Herein the block size refers to the amount of data written per MPI process. The employed block sizes are based on the block sizes typically present in real simulations, e.g.. a block size of 512 KB corresponds to a simulation domain of 40^3 voxel cells per worker in double precision, whereas 64 MB corresponds to a simulation domain of 200^3 voxel cells per worker. A maximum write performance of 2 GB/s is observed once 96 processes (4 nodes) are reached. Furthermore, increasing the block size beyond 512 KB seems to show little improvement. In order to overcome this barrier, the striping of the file has to be adjusted such that the parallelism of the

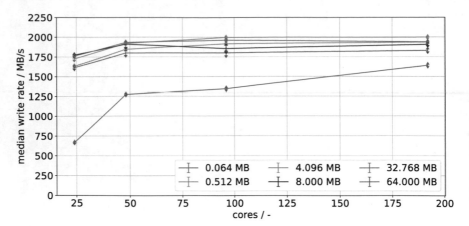

Fig. 4 Write performance of `mpiiotester` for standard striping, from 1 node to 8 nodes for different block sizes

Lustre file system is fully exploited. The striping consists of two main parameters, the stripe count and the stripe size. The stripe count is the number of object storage targets (OST) being written to in parallel, whereas the stripe size is the size of an individual stripe on an OST. Both can be adjusted on a per-file and per-directory basis, which is achieved with the `lfs setstripe` command, but only at file creation.

To gather data for a striping heuristic, different stripe counts on a single node were considered, leading to Fig. 5. There is an obvious peak at 8 OSTs for block sizes at or above 512 KB, implying a heuristic of 8 OSTs per node. Multi-node runs up to 8 nodes were conducted for validation. These results are shown in Fig. 6. The vertical lines show the predicted optimal OST count, which tends to overestimate the optimal count, especially for larger node counts. Hence we require more data especially at higher node counts to get a better heuristic. However, before doing this, it is necessary to consider the influence of stripe size.

The stripe size was varied together with the block size on a single node, leading to Fig. 7. It can be seen that for stripe sizes slightly smaller or comparable to the block size, optimal performance is achieved. This can be understood when considering both the number of system writes and stripe sized writes: The number of system writes should be kept minimal (less latency) while keeping most of the writes stripe-aligned (higher throughput). Figure 8 illustrates the data for 4.096 MB from Fig. 7 together with the stripe sized write to system write ratio. As shown by the green line, we can easily see that the optimal ratio of stripe sized writes to system writes is 1.0, which can be achieved by setting the stripe size to some divisor of the block size. Note that the stripe size has to be an integer multiple of $2^{16} = 65536$; for actual applications, the write size is seldom an integer multiple of 2^{16}. However, already getting close to only stripe-sized writes closes the gap to optimal performance significantly. A ratio of 4 of block size to stripe size was chosen for the rest of the investigation.

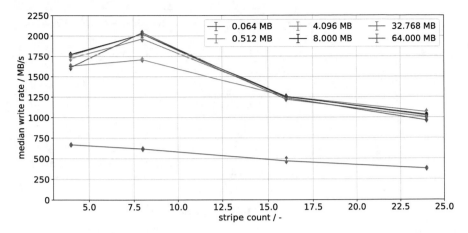

Fig. 5 Write performance of mpiiotester for varying stripe counts on a single node (24 processes)

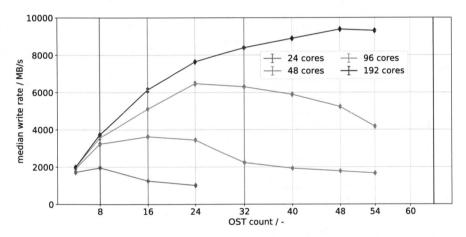

Fig. 6 Write performance of mpiiotester for varying stripe counts for a block size of 4.096 MB

With the influence of stripe size figured out, we shall now investigate the write rate up to 256 nodes (6144 processes) while varying the stripe count. Figure 9 displays the results: Contrary to the earlier investigation, there is only a clear maximum for 384 processes; the line for 768 processes exhibits a very stretc.hed out maximum between a stripe count of 112 and 160. This implies that the optimal stripe count as a function of processes reaches a saturation around 768 processes, i.e. it is optimal to use almost all available OSTs when using more than the corresponding number of nodes (32). Considering this, we can adjust our earlier heuristic and use an affine linear function $mx + b$ and input the two boundary values, i.e. 8 OSTs at 1 node and 160 OSTs at 32 nodes. This results in the function

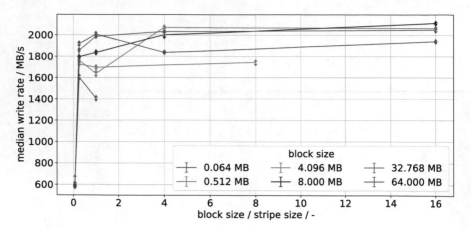

Fig. 7 Write performance of `mpiiotester` for varying stripe sizes and block sizes on a single node with a stripe count of 8

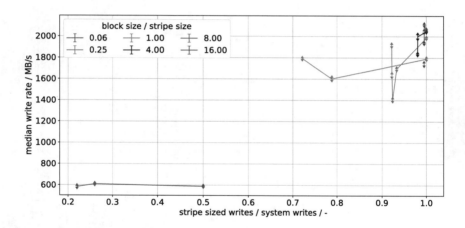

Fig. 8 Write rates for a block size of 4.096 MB from Fig. 7 over the the block size to stripe size ratio. Each block size to stripe size ratio leads to a different number of stripe sized writes to system writes

$$\#OST_{opt}(n) = \frac{152}{31} OST/node \times n + \frac{96}{31} OST, \tag{7}$$

which underestimates the optimal point for lower nodes count n. A simpler (and more accurate at lower node counts) heuristic is achieved by using integer coefficients, i.e.

$$\#OST_{opt}(n) = 5\,OST/node \times n + 3\,OST. \tag{8}$$

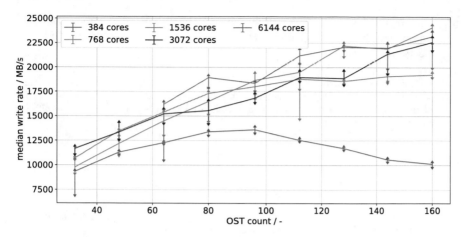

Fig. 9 Write performance of `mpiiotester` for varying stripe counts up to 256 nodes (6144 processes), with a block size of 4.096 MB and a stripe size of 1.024 MB corresponding to a block size to stripe size ratio of 4

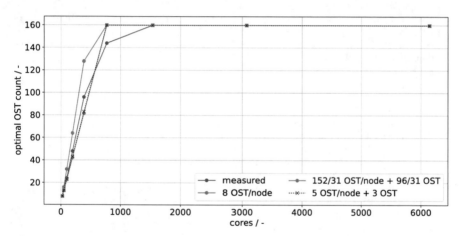

Fig. 10 Prediction of optimal OST count and actual data

The measured optimal OST count as well as the predictions are shown in Fig. 10. Note that it is not necessary to enforce an upper limit as trying to stripe a file with more OSTs than available reverts to the maximum OST count.

7 Conclusion

In order to accelerate phase-field simulations on current HPC systems, an explicit local time stepping approach and MPI-I/O optimizations were investigated. By applying the explicit local time stepping approach, the simulations could be accelerated by almost a factor of two. The MPI-I/O performance on ws 8 was investigated with a test program and improved by a factor of more than 10 by tuning the striping. A heuristic for striping was developed and applied to the simulation framework PACE3D which showed similar speedups.

Acknowledgements We thank Kaveh Dargahi Noubary for helping to validate the simulation results. We thank the High Performance Computing Center Stuttgart and the SCC at the KIT for computational resources. Funding of the research work through the BMBF project "SKAMPY" is gratefully acknowledged.

References

1. M. Almquist, M. Mehlin, Multilevel local time-stepping methods of runge-kutta-type for wave equations. SIAM J. Sci. Comput. **39**(5), A2020–A2048 (2017)
2. M. Bauer, J. Hötzer, M. Jainta, P. Steinmetz, M. Berghoff, F. Schornbaum, C. Godenschwager, H. Köstler, B. Nestler, U. Rüde, Massively parallel phase-field simulations for ternary eutectic directional solidification, in *Proceedings of the International Conference for High Performance Computing, Networking, Storage and Analysis* (ACM, 2015), 8 p
3. Z. Bi, R.F. Sekerka, Phase field modeling of shallow cells during directional solidification of a binary alloy. J. Cryst. Growth **237**, 138–143 (2002)
4. A. Choudhury, B. Nestler, Grand-potential formulation for multicomponent phase transformations combined with thin-interface asymptotics of the double-obstacle potential. Phys. Rev. E **85**(2), 021602 (2012)
5. A. Choudhury, B. Nestler, Grand-potential formulation for multicomponent phase transformations combined with thin-interface asymptotics of the double-obstacle potential. Phys. Rev. E **85**, 021602 (2012)
6. J.A. Dantzig, M. Rappaz, *Solidification* (EPFL Press, 2009)
7. A. Dennstedt, L. Ratke, Microstructures of directionally solidified Al-Ag-Cu ternary eutectics. Trans. Indian Inst. Met. **65**(6), 777–782 (2012)
8. B. Echebarria, R. Folch, A. Karma, M. Plapp, Quantitative phase-field model of alloy solidification. Phys. Rev. E **70**(6), 061604 (2004)
9. A. Genau, L. Ratke, Morphological characterization of the al-ag-cu ternary eutectic. Int. J. Mater. Res. **103**(4), 469–475 (2012)
10. M. Grote, T. Mitkova, Explicit local time-stepping methods for time-dependent wave propagation. *arXiv preprint* arXiv:1205.0654 (2012)
11. J. Hötzer, A. Reiter, H. Hierl, P. Steinmetz, M. Selzer, B. Nestler, The parallel multi-physics phase-field framework pace3d. J. Comput. Sci. **26**, 1–12 (2018)
12. J. Hötzer, M. Jainta, P. Steinmetz, B. Nestler, A. Dennstedt, A. Genau, M. Bauer, H. Köstler, U. Rüde, Large scale phase-field simulations of directional ternary eutectic solidification **93**(0), 194–204 (2015)
13. J. Hötzer, M. Kellner, P. Steinmetz, B. Nestler, Applications of the phase-field method for the solidification of microstructures in multi-component systems **96**(3), 235–256 (2016)
14. J. Hötzer, P. Steinmetz, A. Dennstedt, A. Genau, M. Kellner, I. Sargin, B. Nestler, Influence of growth velocity variations on the pattern formation during the directional solidification of ternary eutectic Al-Ag-Cu **136**, 335–346 (2017)
15. J. Hötzer, O. Tschukin, M. Ben Said, M. Berghoff, M. Jainta, G. Barthelemy, N. Smorchkov, D. Schneider, M. Selzer, B. Nestler, Calibration of a multi-phase field model with quantitative angle measurement **51**(4), 1788–1797 (2015)
16. A. Karma, Phase-field formulation for quantitative modeling of alloy solidification. Phys. Rev. Lett. **87**(11), 115701 (2001)
17. N. Moelans, A quantitative and thermodynamically consistent phase-field interpolation function for multi-phase systems. Acta Mater. **59**(3), 1077–1086 (2011)

18. Mathis Plapp, Unified derivation of phase-field models for alloy solidification from a grand-potential functional. Phys. Rev. E **84**, 031601 (2011)
19. M. Selzer, Mechanische und Strömungsmechanische Topologieoptimierung mit der Phasen-feldmethode. PhD thesis (2014)
20. R. Siquieri, H. Emmerich, Phase-field investigation of microstructure evolution under the influence of convection. Phil. Mag. **91**(1), 45–73 (2011)

Atomistic Simulations: The Driving Force Behind Modern Thermodynamic Research

René Spencer Chatwell, Robin Fingerhut, Gabriela Guevara-Carrion,
Matthias Heinen, Timon Hitz, Y. Mauricio Muñoz-Muñoz,
Claus-Dieter Munz, and Jadran Vrabec

Abstract We are outlining our most recent findings, covering Kirkwood-Buff integration in ternary mixtures, transport diffusion of a diluted component in supercritical carbon dioxide, molecular dynamics simulations of the classical Riemann problem and the development of highly accurate force fields for aqueous mixtures with alcohols, that were made possible by computationally demanding atomistic simulations.

1 Introduction

High performance computing has instigated progress in thermodynamic research for over half a century [1]. Atomistic simulations are now well established, providing accurate data for thermodynamic states that are unattainable by current experiments. In fact, the development of modern equations of state is based on hybrid data sets, in which experimental results are complemented by extensive molecular dynamics (MD) simulations. Consequently, accurate force fields have been developed for a considerable range of substances of varying complexity [2].

Due to a sound physical basis, atomistic simulations additionally enable insight into phenomena that are still insufficiently explained. Transport diffusion in the vicinity of the mixture's critical point with one diluted component, as well as the emergence of hydrogen bonds in aqueous mixtures constitute prominent examples. The successive increase in computational resources has thereby facilitated the investigation of phenomena at hydrodynamic length and time scales [3], even for ensemble sizes containing $N = 2 \cdot 10^{13}$ individual particles [4] and has most recently lead to

R. S. Chatwell · R. Fingerhut · G. Guevara-Carrion · M. Heinen · J. Vrabec (✉)
Thermodynamik und Thermische Verfahrenstechnik, Technische Universität Berlin, Ernst-Reuter Platz 1, 10587 Berlin, Germany
e-mail: vrabec@tu-berlin.de

T. Hitz · C.-D. Munz
Institut für Aerodynamik und Gasdynamik (IAG), Universität Stuttgart, Pfaffenwaldring 21, 70569 Stuttgart, Germany

Y. M. Muñoz-Muñoz
Universität Paderborn, Warburger Str. 100, 33098 Paderborn, Germany

© Springer Nature Switzerland AG 2021
W. E. Nagel et al. (eds.), *High Performance Computing in Science and Engineering '19*,
https://doi.org/10.1007/978-3-030-66792-4_38

quantitative agreement between results obtained by large scale MD and computational fluid dynamics (CFD) simulations for propagating shock waves in an argon like model fluid.

2 Thermodynamic Factor via Kirkwood-Buff Integration

Transport diffusion is a challenging phenomenon in general. Instead of very time consuming experimental work, molecular modelling and simulation provides a valuable alternative. The diffusion coefficients that are sampled by MD simulations, i.e. those according to Maxwell-Stefan D_{ij}, are distinct from those that are determined experimentally, i.e. those according to Fick D_{ij}. However, both coefficients are related by the thermodynamic factor matrix Γ

$$D = D \cdot \Gamma . \tag{1}$$

In the case of a binary mixture, this thermodynamic factor reduces to a scalar and is given by [5]

$$\Gamma = 1 + x_1 \left(\frac{\partial \ln \gamma_1}{\partial x_1} \right) \Bigg|_{T,p,\Sigma} = 1 + x_2 \left(\frac{\partial \ln \gamma_2}{\partial x_2} \right) \Bigg|_{T,p,\Sigma} , \tag{2}$$

where γ_i constitutes the activity coefficient of component i and x_i its mole fraction. The thermodynamic factor is not directly accessible via experiment, however, MD simulations offer two possibilities for its computation. On the one hand, the fluid mixture's activity coefficient can be determined first by dedicated MD simulations and subsequently the derivative in Eq. (2) is computed numerically. Alternatively, Γ can be sampled directly via Kirkwood-Buff integration (KBI) [6]. KBI is inherently linked to the fluid's microscopic structure, represented by the radial distribution function (RDF)

$$g_{ij}(r) = \frac{\rho_{L,i}(r)}{\rho_j} , \tag{3}$$

where the local density of component i in a spherical shell centered around species j is given as

$$\rho_{L,i}(r) = \frac{dN_i}{4\pi r^2 dr} , \tag{4}$$

with dN_i denoting the i-th component's particle number and $\rho_j = x_j \rho$ the overall partial density of component j. Strictly speaking, KBI constitutes an improper integral

$$G_{ij} = 4\pi \lim_{r \to \infty} \int_0^r \left[g_{ij}(r) - 1 \right] r^2 dr , \tag{5}$$

that is formally defined only in the grand canonical (μVT) ensemble [6], in which MD simulations of dense liquids can hardly be performed. Alternatively, the integral was implemented in the canonical (NVT) ensemble and truncated to finite radii. This truncation was considered by the mathematical procedure developed by Krüger et al. [7]

$$G_{ij}(R) = 4\pi \int_0^{2R} \left(1 - 3x/2 + x^3/2\right) \cdot \left[g_{ij}(r) - 1\right] r^2 dr, \tag{6}$$

where $x = r/2R$ constitutes the reduced radius. However, extrapolating to the thermodynamic limit, i.e to infinite system size, turns out to be vital which was laid out conclusively in a recent study [8], yielding the well suited approximation

$$G_{ij}^{\infty}(R) = 4\pi \int_0^{2R} \left(1 - 23x^3/8 + 3x^4/4 + 9x^5/8\right) \cdot \left[g_{ij}(r) - 1\right] r^2 dr. \tag{7}$$

On that basis, it is conceptually straightforward to obtain liquid solution properties, like the thermodynamic factor Γ or the partial molar volumes v_i. Ben-Naim [9] provided an explicit expression for Γ via KBI for a binary mixture

$$\Gamma = 1 - \frac{x_1 x_2 \rho \left(G_{11} + G_{22} - 2G_{12}\right)}{1 + x_1 x_2 \rho \left(G_{11} + G_{22} - 2G_{12}\right)}. \tag{8}$$

Liu et al. [10] utilized Ben-Naim's general mathematical formalism to calculate the thermodynamic factor for a ternary mixture, which for this case constitutes a 2×2 matrix.

The KBI formalism to determine the thermodynamic factor of binary and ternary mixtures was successfully implemented into the molecular simulation tool *ms2* [11]. Users can invoke KBI in the canonical ensemble both for Monte Carlo (MC) and MD simulations. Details on the implementation and parallelization are given in our recent study [12], confirming previous findings [13] that RDF corrections are crucial and that RDF corrected by the method outlined by Ganguly and van der Vegt [14] are superior to other schemes, given the finite size correction is taken into account [15]. The correction additionally considers that molecules within a distance r around a central molecule may experience excess or depletion phenomena.

Ternary mixture of water + methanol + ethanol

In order to validate the implementation in *ms2*, thermodynamic factor results Γ_{ij} for the ternary mixture water + methanol + ethanol were assessed, cf. Fig. 1. Values were either calculated on the basis of precise chemical potential data [16] in combination with the Wilson model [17] or sampled directly via KBI for different mole fractions,

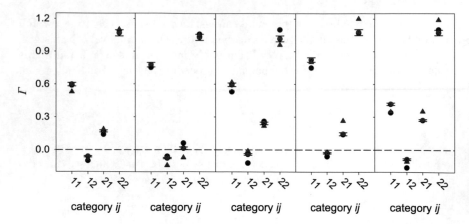

Fig. 1 Comparison of the thermodynamic factor matrix elements Γ_{11}, Γ_{12}, Γ_{21} and Γ_{22} for the considered ternary mixture at five different state points (from left to right: cf. Table 1); black symbols: Γ calculated by the Wilson model [17] fitted to chemical potential data [16]; red symbols: Γ sampled by KBI in the NVT ensemble with $ms2$ [11]; circles/triangles: non-extrapolated/extrapolated data to macroscopic system size by Eqs. (6) and (7), where the triangles have the same error bars as the circles

Table 1 Studied compositions of the considered mixture, sampling the density with dedicated NpT simulations [16]

x_1 / mol mol^{-1}	x_2 / mol mol^{-1}	x_3 / mol mol^{-1}	ρ / mol dm^{-3}
0.33	0.33	0.34	26.09
0.2	0.4	0.4	23.39
0.4	0.4	0.2	29.05
0.2	0.6	0.2	25.48
0.6	0.2	0.2	33.53

cf. Table 1. The MD simulations at these five state points were performed in the canonical ensemble containing $N = 4000$ particles. The mixtures' density was sampled with preceeding isothermal-isobaric (NpT) ensemble simulations at constant pressure $p = 0.1$ MPa and temperature $T = 298.15$ K [16]. All NVT simulations were carried out with $8 \cdot 10^5$ equilibration and $5 \cdot 10^7$ production steps. The equations of motion were solved numerically by applying the Gear predictor-corrector method [18] with an integrator time step of $t = 0.877$ fs. The cutoff radius was set to $r_c = 17.5$ Å and the intermolecular dispersion and repulsion interactions were described by the Lennard-Jones (LJ) 12–6 potential. Unlike interactions were specified by the Lorentz-Berthelot combination rules [19, 20]. Beyond the cutoff radius, the LJ interactions were corrected by applying angle averaging [21] and the long-range electrostatic interactions were considered via the reaction field method [18]. RDF were calculated in each time step and the maximum radius was set to half the edge length of the cubic simulation volume, which was subdivided into 300 shells.

The present results emphasize how well the KBI is suited to determine the thermodynamic factor Γ_{ij} for a ternary mixture. Due to its parallelization and implementation in *ms2*, the sampling of KBI data is computationally efficient. It was shown that the additional computational effort is negligibly small when Γ_{ij} are sampled by KBI, compared to calculations via the chemical potentials route [12].

3   Transport Diffusion in the Dilute Limit

The thermodynamic factor of a binary mixture at infinite dilution is asymptotically constrained

$$\lim_{x_1 \to 0} \Gamma = 1 , \tag{9}$$

irrespective of temperature and pressure. It is common practice, however, to utilize Eq. (9) even for finitely diluted mixtures. As mentioned above, there are two different methods to compute the thermodynamic factor and in this study on the dilute limit Γ was determined by numerical differentiation. The molar derivative of the activity coefficient i can alternatively be written in terms of its chemical potential μ_i

$$\left. \frac{\partial \ln \gamma_i}{\partial x_j} \right|_{T,p,\Sigma} = \left. \frac{\partial (\tilde{\mu}_i/(kT) - \ln x_i)}{\partial x_j} \right|_{T,p,\Sigma} , \tag{10}$$

with

$$\tilde{\mu}_i(T, p, \mathbf{x}) \equiv \mu_i(T, p, \mathbf{x}) - \mu_i^{id}(T) , \tag{11}$$

and $\tilde{\mu}_i(T, p, \mathbf{x})$ can be sampled directly via MD simulations applying the Widom test particle method [22]. The numerical differentiation was performed by calculating the activity coefficient at two different molar fractions near the target one, i.e. $x_{a+h} = x_i + \Delta x$ and $x_{a-h} = x_i - \Delta x$, with $\Delta x = 0.003 \; \text{mol} \, \text{mol}^{-1}$.

Benzene diluted in supercritical carbon dioxide

Diluted benzene (C_6H_6) in supercritical carbon dioxide (CO_2) constitutes a relevant test case for $\Gamma \neq 1$ even for very low benzene concentrations. The considered force fields were chosen to be rigid, non-polarizable, LJ based and the corresponding parameters can be found in publications [23, 24]. The thermodynamic factor was assessed around the mixture's critical point, i.e. at two different mole fractions x_{C6H6} $= 0.01$ and $0.03 \, \text{mol} \, \text{mol}^{-1}$ in the temperature range between $T = 293.15$ and 335 K along the $p = 9$ MPa isobar. It is evident, cf. Fig. 2, that the thermodynamic factor differs significantly from unity even for benzene mole fractions as low as x_{C6H6}

Fig. 2 Temperature variation of the thermodynamic factor for diluted benzene in supercritical carbon dioxide at $p = 9$ MPa for $x_{C6H6} = 0.03$ (blue symbols) and 0.01 mol mol^{-1} (red symbols)

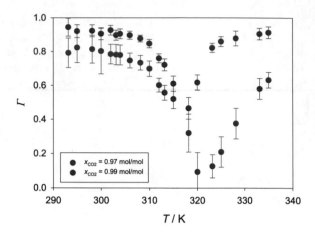

$= 0.01$ mol mol^{-1}. It can further be seen that Γ tends to unity for lower benzene concentrations.

Relation (1) allows to validate these findings with experimentally determined Fick diffusion data. For that reason, Maxwell-Stefan diffusion coefficients were sampled by MD simulations employing the Green-Kubo formalism [25]. The temperature dependence of the predicted Fick diffusion coefficients is shown in Fig. 3 for varying benzene mole fractions. Generally speaking, the diffusion coefficients increase with rising temperature, however, between temperatures $T = 310$ and 320 K an anomaly was observed at $x_{C6H6} = 0.02$ mol mol^{-1}. This decrease correlates well with the already observed decrease in the thermodynamic factor, cf. Fig. (2).

In general, a good agreement between literature data [26–28] and simulation results is found, cf. Fig. 4, although experimental data scatter quite significantly near the critical point. In contrast to MD simulations, the mixture's molar composition is not precisely known for most of the reported experimental data, yet the results can be explained in light of Fig. 3, where strong changes of the Fick diffusion coefficients are caused by small changes in benzene mole fraction.

4   Shock Wave Propagation in Supercritical Fluids

The shock tube scenario is a manifestation of the classical Riemann problem that has already been solved conclusively and now acts as a test case to validate solvers in modern CFD codes. Recent progress in our massively parallelized MD program *lsl mardyn* [29] allows to simulate hydrodynamic length and time scales atomistically [3, 4]. In our most recent study [30], MD predictions were successfully compared to CFD results that were obtained with *FLEXI* [31], in which a discontinuous Galerkin spectral element method (DGSEM) solves the hydrodynamic equations numerically [32]. All MD simulations were based on the LJ truncated and shifted (LJTS) model

Fig. 3 Temperature dependence of the Fick diffusion coefficient of the mixture $CO_2 + C_6H_6$ along the $p = 9$ MPa isobar for $x_{C6H6} = 0.005, 0.015$ and 0.02 mol mol^{-1}

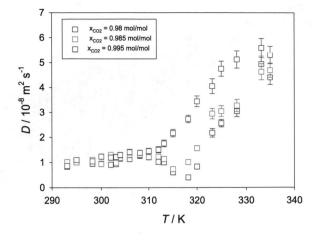

Fig. 4 Fick diffusion coefficient of benzene in supercritical carbon dioxide as a function of density. Comparison between simulation results and experimental literature data [26–28]

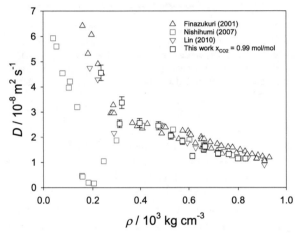

that is well suited to describe the thermodynamic properties of simple fluids, such as noble gases and methane [33, 34].

As outlined in Fig. 5, a symmetric setup was used that contains $N = 3.2 \cdot 10^7$ particles distributed among a liquid-like phase of higher density $\rho_1 = 0.6$ and a gas-like phase of lower density $\rho_2 = 0.2$. Prior to bringing the phases in physical contact with each other, both were equilibrated separately at the same supercritical temperature $T = 3.0$ (the critical temperature and density for the LJTS fluid are $T_c = 1.0779$ and $\rho_c = 0.319$, respectively [33]). Due to the emerging dynamics under these extreme conditions a comparatively large system size in the relevant z direction was necessary, allowing the shock waves that originate at the phase boundaries, i.e. at $z = 2000$ and $z = -2000$, to properly develop before reaching the periodic boundary condition at the system's outer limits. Obtaining good statistics in such a highly dynamic scenario proves to be a challenging task and consequently a large cross-sectional

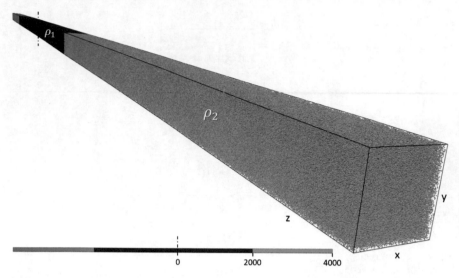

Fig. 5 *Top*: Snapshot of the atomistic representation of the shock tube scenario. *Bottom*: plan view clarifying the system dimensions. The system was set up symmetrically around its origin, i.e at $x = 0$. Particles constituting the high density state $\rho_1 = 0.6$ are colored red, while those constitutung the lower density state $\rho_2 = 0.2$ are colored green

area of $A_{xy} = L_x \cdot L_y = 100^2$ particle diameters was necessary, while additionally exploiting the system's symmetry to increase all sampling results.

The integrator time step was set to $\Delta t = 2$ fs (for argon), while the overall simulated time was limited to $t = 0.4$ ns, right after both shock waves had reached the system's boundary. During simulation, the temperature T, density ρ and hydrodynamic velocity v profiles were sampled by a classical binning approach, dividing the system in z direction into $3.2 \cdot 10^4$ bins of width $\delta = 0.25$ particle diameters. All profiles were averaged over a period of $t = 10$ ps to yield reasonably good statistics on the one hand, while simultaneously minimizing the effect of blurring out rapidly changing quantities. The LJTS force field is not only computationally cheap, since it allows for a small cutoff radius of $r_c = 2.5$ particle diameters, there are also highly accurate equations of state (EOS) available that were necessary to obtain the CFD solutions, i.e. the LJTS EOS [35] and the perturbed truncated and shifted (PeTS) EOS [36].

The results of the MD and CFD simulations are in excellent agreement with each other, cf. Fig. 6. The shock and rarefaction waves, as well as the contact discontinuity can clearly be identified in both simulation approaches. The largest discrepancy between MD and CFD results was observed in the velocity and temperature profiles, when the PeTS EoS was used in the CFD simulations, showing a small spatial shift of the rarefaction wave.

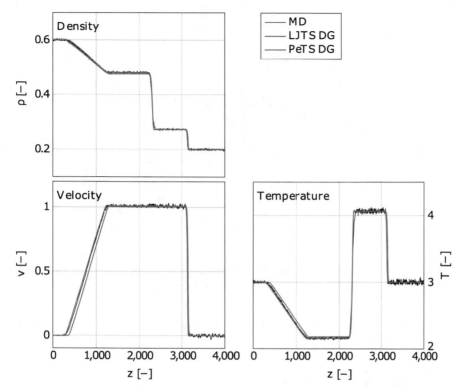

Fig. 6 Results for the one-dimensional Riemann problem obtained by MD simulations as well as by the numerical DGSEM scheme using the Suliciu relaxation solver, while both equations of state, i.e LJTS EOS and PeTS EOS, were used

5 Parametrization Scheme to Develop Force Fields of Alcohols: Isopropanol as Case Study

Atomistic simulations allow to determine thermodynamic properties precisely and unconstrainedly, yet they also help to understand phenomena such as clustering, hydrogen bonding in aqueous mixtures and molecular segregation [25]. However, in order to observe such association patterns in the RDF, highly accurate force fields have to be employed. A suitable parametrization scheme to develop such accurate force fields is shown in Fig. 7. Each parametrization procedure starts with an *ab initio* determination of the atom's spatial coordinates within the molecule as well as its electronic structure by solving the approximated Schrödinger equation numerically. The internal coordinates are most suitably given in z Matrix representation [37], allowing update the internal angles and/or bond lengths straightforwardly within the optimization procedure [38]. A molecule's electrostatic potential is represented by placing point charges at distinct intramolecular positions. In the case of isopropanol, a point charge was placed at the center of the hydrogen, oxygen and methanetriyl

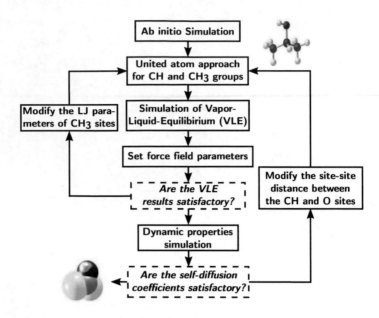

Fig. 7 Parametrization scheme to determine highly accurate for fields alcohols in general and isopropanol in particular

group, respectively, cf. Fig. 7 in which the CH site is depicted in orange, oxygen in red, hydrogen in white and the CH_3 sites in yellow. The configurational internal energy of an ensemble containing N particles is given as a superposition of its dispersive and electrostatic contributions

$$U = U_{disp.} + U_{elec.}$$

$$= \sum_{i=1}^{N-1} \sum_{j=i+1}^{N} \left[\sum_{a=1}^{s_i^{LJ}} \sum_{b=1}^{s_j^{LJ}} 4\varepsilon_{ijab} \left[\left(\frac{\sigma_{ijab}}{r_{ijab}} \right)^{12} - \left(\frac{\sigma_{ijab}}{r_{ijab}} \right)^6 \right] + \sum_{c=1}^{s_i^e} \sum_{d=1}^{s_j^e} \frac{1}{4\pi\varepsilon_0} \frac{q_{ic}q_{jd}}{r_{ijcd}} \right],$$

where r_{ijab}, ε_{ijab}, σ_{ijab} constitute the distance, the LJ energy and size parameters respectively, for the pair-wise interaction between the i-th molecule's LJ site a and the j-th molecule's LJ site b. The vacuum permittivity is denoted by ε_0 and q_{ic} represents the point charge of the i-th molecule interaction site c and so forth. The parameters of the isopropanol force field were recently reported [25] and it was found that the LJ parameters σ and ε of the CH_3 sites, as well as the site-site distance between oxygen and the CH group were the most crucial to be optimized. Optimization was achieved by minimizing the deviations between experimental and simulation data for vapor-liquid equilibrium and self-diffusion coefficient.

It was extensively shown in Ref. [25] that MD simulation results are in excellent agreement with experimental data, i.e. the average relative deviation for the saturated liquid density, vapor pressure and enthalpy of vaporization are 0.4%, 7.0% and

4.8%, respectively. Additionally, the average relative deviation of the self-diffusion coefficient does not exceed 2.6% in temperature range between $T = 270$ and 340 K. Once the parameters have been optimized, further MD simulations with $N = 1000$ particles were carried out in the NpT ensemble to predict the microscopic structure of isopropanol at temperatures between $T = 278$ and 338 K at $p = 1$ atm. The optimized force field accurately reproduced published X-ray scattering data [39], revealing hydrogen bonds by peaks in the RDF $g_{OH}(r)$ at $r \sim 1.9$ Å. The MD simulations revealed further that the average number of hydrogen bonds per isopropanol molecule is $\langle n_{OH} \rangle \approx 2$ at $T = 278$ K.

Acknowledgements We gratefully acknowledge support by the Deutsche Forschungsgemeinschaft (DFG). This work was carried out under the auspices of the Boltzmann-Zuse Society (BZS), while all simulations were performed on the CRAY XC40 (Hazel Hen) at the High Performance Computing Center Stuttgart (HLRS) within the project *molecular models for hydrogen bonding fluids* (MMHBF2).

References

1. E. Forte, F. Jirasek, M. Bortz, J. Burger, J. Vrabec, H. Hasse, Digitalization in thermodynamics. Chem. Ing. Tech. **91**, 201 (2019)
2. S. Stephan, M. Horsch, J. Vrabec, H. Hasse, MolMod: an open access database of force fields for molecular simulations of fluids. Mol. Simul. **4**, 1 (2019)
3. R.S. Chatwell, M. Heinen, J. Vrabec, Diffusion limited evaporation of a binary liquid film. Int. J. Heat Mass Transf. **132**, 1296 (2019)
4. N. Tchipev, S. Seckler, M. Heinen, J. Vrabec, F. Gratl, M. Horsch, M. Bernreuther, C.W. Glass, C. Niethammer, N. Hammer, B. Krischok, M. Resch, D. Kranzlmüller, H. Hasse, H.J. Bungartz, P. Neumann, TweTriS: twenty trillion-atom simulation. Int. J. High Perform. Comput. Appl. **33**, 838 (2019)
5. R. Taylor, H.A. Kooijman, Composition derivatives of activity coefficient models for the estimation of thermodynamic factors in diffusion. Chem. Eng. Comm. **102**, 87 (1991)
6. J.G. Kirkwood, F.P. Buff, The statistical mechanical theory of solutions. Int. J. Chem. Phys. **19**, 774–778 (1951)
7. P. Krüger, S.K. Schnell, D. Bedeaux, S. Kjelstrup, T.J.H. Vlugt, J.M. Simon, Kirkwood-Buff integrals for finite volumes. J. Phys. Chem. Lett. **4**, 235 (2013)
8. P. Krüger, T.J.H. Vlugt, Size and shape dependence of finite volume Kirkwood- Buff integrals. Phys. Rev. E. **97**, 051301 (2018)
9. A. Ben-Naim, *Molecular Theory of Solutions* (Oxford University Press, Oxford, 2006)
10. X. Liu, A. Martin-Calvo, E. McGarrity, S.K. Schnell, S. Calero, J.M. Simon, D. Bedeaux, S. Kjelstrup, A. Bardow, T.J.H. Vlugt, Fick diffusion coefficients in ternary liquid systems from equilibrium molecular dynamics simulations. Ind. Eng. Chem. Res. **51**, 10247 (2012)
11. G. Rutkai, A. Köster, G. Guevara-Carrion, T. Janzen, M. Schappals, C.W. Glass, M. Bernreuther, A. Wafai, S. Stephan, M. Kohns, S. Reiser, S. Deublein, M. Horsch, H. Hasse, J. Vrabec, ms2: a molecular simulation tool for thermodynamic properties, release 3.0. Comp. Phys. Commun. **221**, 343 (2017)
12. R. Fingerhut, J. Vrabec, Kirkwood-Buff integration: a promising route to entropic properties. Fluid Phase Equilib. **485**, 270 (2019)
13. J. Milzetti, D. Nayar, N.F.A. van der Vegt, Convergence of Kirkwood-Buff integrals of ideal and nonideal aqueous solutions using molecular dynamics simulations. J. Phys. Chem. B **122**, 5515 (2018)

14. P. Ganguly, N.F.A. Van Der Vegt, Convergence of sampling Kirkwood-Buff integrals of aqueous solutions with molecular dynamics simulations. J. Chem. Theor. Comput. **9**, 1347 (2013)

15. S.K. Schnell, X. Liu, J.-M. Simon, A. Bardow, D. Bedeaux, T.J.H. Vlugt, S. Kjelstrup, Calculating thermodynamic properties from fluctuations at small scales. J. Phys. Chem. B **115**, 10911 (2011)

16. S. Pařez, G. Guevara-Carrion, H. Hasse, J. Vrabec, Mutual diffusion in the ternary mixture of water + methanol + ethanol and its binary subsystems. Phys. Chem. Chem. Phys. **15**, 3985 (2013)

17. G.M. Wilson, Vapor-liquid equilibrium XI: a new expression for the excess free energy of mixing. J. Am. Chem. Soc. **86**, 127 (1964)

18. M.P. Allen, D.J. Tildesley, *Computer Simulation of Liquids* (Clarendon Press, Oxford, 1987)

19. H.A. Lorentz, Über die Anwendung des Satzes vom Virial in der kinetischen Theorie der Gase. Ann. Phys. **248**, 127 (1881)

20. D. Berthelot, Sur le melange des gaz C. r. hebd. Seances Acad. Sci. **126**, 1703 (1898)

21. R. Lustig, Angle-average for the powers of the distance between two separated vectors. Mol. Phys. **65**, 175 (1988)

22. B. Widom, Some topics in the theory of fluids. J. Chem. Phys. **39**, 2808 (1963)

23. T. Merker, C. Engin, J. Vrabec, Molecular model for carbon dioxide optimized to vapor-liquid equilibria. J. Chem. Phys. **132**, 34512 (2010)

24. G. Guevara-Carrion, T. Janzen, Y.M. Munoz-Munoz, J. Vrabec, Mutual diffusion of binary liquid mixtures containing methanol, ethanol, acetone, benzene, cyclohexane, toluene, and carbon tetrachloride. J. Chem. Phys. **144**, 124501 (2016)

25. Y.M. Muñoz-Muñoz, G. Guevara-Carrion, J. Vrabec, Molecular insight into the liquid 2-propanol + water mixture. J. Phys. Chem. B **122**, 8718 (2018)

26. T. Funazukuri, C.Y. Kong, S. Kagei, Infinite dilution binary diffusion coefficients of benzene in carbon dioxide by the taylor dispersion technique at temperatures from 308.15 to 328.15 k and pressures from 6 to 30 MPa. Int. J. Thermophys. **22**, 1643 (2001)

27. H. Nishiumi, T. Kubota, Fundamental behavior of benzene-CO2 mutual diffusion coefficients in the critical region of CO2. Fluid Phase Equilib. **261**, 146 (2007)

28. R. Lin, L. Tavlarides, Diffusion coefficients of diesel fuel and surrogate compounds in supercritical carbon dioxide. J. Supercrit. Fluids **52**, 47 (2010)

29. C. Niethammer, S. Becker, M. Bernreuther, M. Buchholz, W. Eckhardt, A. Heinecke, S. Werth, H.J. Bungartz, C.W. Glass, H. Hasse, J. Vrabec, M. Horsch, s1 mardyn: the massively parallel molecular dynamics code for large systems. J. Chem. Theor. Comput. **10**, 4455 (2014)

30. T. Hitz, M. Heinen, J. Vrabec, C.D. Munz, Comparison of macro and microscopic solutions of the riemann problem i. supercritical shock tube and expansion into vacuum. J. Comput. Phys., 109077 (2019)

31. FLEXI—Description and source code, https://www.flexi-project.orgAccessed 31 Jan 2019

32. F. Hindenlang, G.J. Gassner, C. Altmann, A. Beck, M. Staudenmaier, C.-D. Munz, Explicit discontinuous Galerkin methods for unsteady problems. Comput. Fluids **61**, 86 (2012)

33. J. Vrabec, G.K. Kedia, G. Fuchs, H. Hasse, Comprehensive study of the vapour-liquid coexistence of the truncated and shifted Lennard-Jones fluid including planar and spherical interface properties. Mol Phys. **104**, 1509 (2006)

34. G. Rutkai, M. Thol, R. Span, J. Vrabec, How well does the Lennard-Jones potential represent the thermodynamic properties of noble gases? Mol. Phys. **115**, 1104 (2017)

35. M. Thol, G. Rutkai, R. Span, J. Vrabec, R. Lustig, Equation of state for the lennard-jones truncated and shifted model fluid. Int. J. Thermophys. **36**, 25 (2015)

36. M. Heier, S. Stephan, J. Liu, W.G. Chapman, H. Hasse, K. Langenbach, Equation of state for the Lennard-Jones truncated and shifted fluid with a cut-off radius of 2.5 sigma based on perturbation theory and its applications to interfacial thermodynamics. Mol. Phys. **116**, 2083 (2018)

37. H. Essén, M. Svensson, Calculation of coordinates from molecular geometric parameters and the concept of a geometric calculator. Comput. Chem. **20**, 389 (1996)

38. Y.M. Muñoz-Muñoz, G. Guevara-Carrion, M. Llano-Restrepo, J. Vrabec, Lennard-Jones force field parameters for cyclic alkanes from cyclopropane to cyclohexane. Fluid Phase Equilib. **404**, 150 (2015)
39. T. Takamuku, K. Saisho, S. Aoki, T. Yamaguchi, Large-Angle X-Ray Scattering Investigation of the Structure of 2-Propanol-Water Mixtures Zeitschrift für Naturforsch. A **57**, 982 (2002)

First Steps Towards a Scaling Analysis of a Fully Resolved Electrical Neuron Model

Myra Huymayer, Michael Lampe, Arne Nägel, and Gabriel Wittum

Abstract In computational neuroscience the transmission of electrical signals of neurons is normally simulated by means of point process neurons, which mainly reflect the scale in time, or the classical cable equation which additionally introduces one space dimension. Here we present a fully resolved electrical model based on Gauss' law and the conservation of charges which considers all space dimensions and is capable to simulate the extracellular and intracellular potential. For these simulations three dimensional volume meshes are required and due to the inherent complexity of the neuronal structure, these 3D-reconstructions yield large data-sets and need efficient solving strategies. The UG4-simulation framework is a powerful software for the solution of partial differential equations on unstructured grids in one, two and three space dimensions and with its efficient, parallel solvers is well suited for this task. Computations of the 3D-cable equation on a simple geometry and on a three-dimensionally reconstructed neuron were performed on the *Hazel Hen* supercomputer, testing for weak scalability.

1 Introduction

The human brain consists of $\approx 8.6 \cdot 10^{10}$ neuronal cells and modeling neuronal signal transmission in large networks of neurons has become extremely important recently. The electrical signal is often described by means of point process simulations, which describes neuronal activity by ordinary partial differential equations, i.e. only the

M. Huymayer · M. Lampe (✉) · A. Nägel · G. Wittum
G-CSC, Goethe-Universität Frankfurt, Kettenhofweg 139, 60325 Frankfurt (M.), Germany
e-mail: lampe@gcsc.uni-frankfurt.de

M. Huymayer
e-mail: mhuymayer@gcsc.uni-frankfurt.de

A. Nägel
e-mail: naegel@gcsc.uni-frankfurt.de

G. Wittum
e-mail: wittum@gcsc.uni-frankfurt.de

© Springer Nature Switzerland AG 2021   583
W. E. Nagel et al. (eds.), *High Performance Computing in Science and Engineering '19*,
https://doi.org/10.1007/978-3-030-66792-4_39

change of the membrane potential in time is considered. The traditional cable equation describes the electrical signal by means of a partial differential equation, which considers the changes in the membrane potential in time and in one space dimension. The underlying assumption states that current of the core conductor flows parallel to the cylinder axis along the x-axis, neglecting the other space dimensions.

An electrical model considering all three space dimensions was first introduced by [1]. The three dimensional cable equation can be derived from Gauss' law, which states: $\nabla \cdot \vec{E} = \frac{\rho}{\epsilon_0}$ and conservation of charge: $\nabla \cdot \vec{J} = -\frac{\partial \rho}{\partial t}$. Since we assume static magnetic fields, the electric field can be expressed as the gradient of the potential: $\vec{E} = -\nabla \Phi$. The membrane potential is defined as the potential difference between the intra- and extracellular space: $V_m = \Phi_{in} - \Phi_{out}$. Thus a three dimensional model of the cable equation can be described by the following system of PDEs:

$$-\nabla \cdot (\sigma_{in} \nabla \Phi_{in}) = 0 \qquad\qquad \text{in } \Omega_{in}$$

$$-\sigma_{in} \nabla \Phi_{in} \cdot \vec{n}_{in \to out} = c_m \frac{dV_m}{dt} + j_{mem} \qquad\qquad \text{on } \Gamma \qquad (1)$$

$$-\nabla \cdot (\sigma_{out} \nabla \Phi_{out}) = 0 \qquad\qquad \text{in } \Omega_{out}$$

$$-\sigma_{out} \nabla \Phi_{out} \cdot \vec{n}_{out \to in} = -c_m \frac{dV_m}{dt} - j_{mem} \qquad\qquad \text{on } \Gamma \qquad (2)$$

σ_{in} and σ_{out} describe the conductivity in the intracellular and extracellular medium. The membrane currents are composed of an input current and Hodgkin Huxley currents [2]:

$$j_{mem} = g_K \cdot n^4 \cdot (V_m - E_K) + g_{Na} \cdot m^3 \cdot h \cdot (V_m - E_{Na}) + g_L \cdot (V - E_L)$$

where

$$\frac{di}{dt} = \frac{i_\infty - i(t)}{\tau_i}$$

$$\tau_i = \frac{1}{\alpha_i(V_m) + \beta_i(V_m)}$$

$$i_\infty = \frac{\alpha_i(V_m)}{\alpha_i(V_m) + \beta_i(V_m)}$$

where $i = \{n, m, h\}$. The result of $\frac{di}{dt} = \frac{i_\infty - i(t)}{\tau_i}$ is computed in such a way that it is used as right hand side in $\frac{di}{dt} = f$, which will make the solution of that ODE with the explicit Euler method the analytical solution for one time step with constant V_m:

$$\frac{di}{dt} = \frac{i_\infty - i(t)}{\tau_i}$$

$$\Leftrightarrow \frac{-\frac{di}{dt}}{i_\infty(V_m) - i} = -\frac{1}{\tau_i(V_m)}$$

$$\Leftrightarrow \int_t^{t+\Delta t} \frac{-\frac{di}{dt}}{i_\infty(V_m) - i} = -\int_t^{t+\Delta t} \frac{1}{\tau_i(V_m)}$$

$$\Leftrightarrow \log\left(\frac{i_\infty(V_m) - i_{k+1}}{i_\infty(V_m) - i_k}\right) = \frac{-\Delta t}{\tau_i(V_m)}$$

$$\Leftrightarrow \frac{i_\infty(V_m) - i_{k+1}}{i_\infty(V_m) - i_k} = \exp\left(\frac{-\Delta t}{\tau_i(V_m)}\right)$$

$$\Leftrightarrow i_{k+1} = i_\infty(V_m) - (i_\infty(V_m) - i_k)\exp\left(\frac{-\Delta t}{\tau_i(V_m)}\right)$$

Since neurons have a complex three dimensional geometry resulting in a huge number of degrees of freedom (DoF) efficient solving strategies are necessary. Weak scaling was tested for an AMG-GMG hybrid solver setup on a relatively simple geometry.

2 Neuronal Reconstructions for Simulations and Geometries

The database neuromorpho.org gives free access to a wide range of experimentally reconstructed neurons in terms of swc files [3], which contain coordinate and diameter information. Since neuronal reconstructions often display errors, such as neurite-neurite intersections or sharp curves, with regard to 3D-reconstructions, corrections were performed interactively in ProMesh [4]. To solve the partial differential equations on a suitable domain a water-tight surface mesh was created from a layer 3 pyramidal cell from primary rat motor cortex (M1) [5] downloaded from Neuromorpho.org using AnaMorph [6] in a first step. This tool ensures that the aspect ratio is close to 1 which is necessary in Finite Volume discretization. Subsequently, a subset for the outer space was created in and tetrahedrization was realized with ProMesh [4, 7] (see Fig. 1). As this results in quite complex geometries with a total of 3.915.236 DOFs (increasing with an increase in refinements) for the described problem, a minimal test geometry, replacing the neuronal geometry with a simple cube (see Fig. 1 right), was built with ProMesh to investigate weak scaling for this model.

 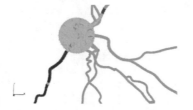 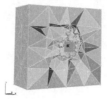

Fig. 1 3D reconstruction of a pyramidal cell from rat motor cortex in extracellular medium (left). Close up of section of reconstructed and tetrahedralized cell (middle). Simplified geometry for test computations (right). color map: lime green: dendrite, red: myelin, purple: conducting membrane, and pink: injection site, lavender: extracellular medium, sky blue: intracellular medium

3   Simulation Setup

UG4 is a simulation environment designed to efficiently solve partial differential equations on unstructured grids, with a strong focus on high performance computing. It provides a great variety of numerical solvers. For computation of the system of equations (1–2), the domain of the cube was distributed to 6–3072 cores on *Hazel Hen*. A well-balanced distribution of the neuronal surface components (Γ) was ensured by the ParMetis library, which becomes especially important when calculating on the complex structure of neurons. The change of the membrane potential in time was investigated for 10 ms, with a 5–6 ms long stimulus of $900 \ \frac{C}{mV \, ms \, \mu m^2}$.

The system of PDEs was solved with a geometric multigrid (GMG) with a Gauss Seidel preconditioner. GMG convergence was strongly increased with an increasing number of pre- and post-smoothing steps (optimal = 80). An algebraic multigrid (AMG), with Gauss Seidel smoother, was chosen as base solver for the GMG. As base solver for the AMG SuperLU [8] was selected.

4   Results

As shown in Fig. 2 an action potential can be observed on the cube membrane (Φ_{in}). The extracellular signal (Φ_{out}) displays the same behavior as in neuron morphologies, but is much smaller (minimal amplitude in neuron: $\approx$-0.03 mV; minimal amplitude in cube: $-4.8 \cdot 10^{-5}$ mV). Thus it can be assumed that the morphology has an influence on the characteristics and size of the extracellular potential.

It has been shown in the past, that the GMG implemented in UG4 shows good weak scalability [4] and as we want to calculate large realistic networks in the future, we tested the GMG-AMG hybrid for our coupled system of partial differential equations on the cube grid. Table 1 shows the time taken for certain solving steps in seconds. As this is a time dependent problem, the mean time of time step preparation, right hand side assembly and solving is given. Overall 500 time steps were performed.

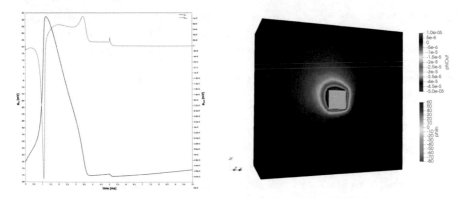

Fig. 2 Intracellular and extracellular potential changes in time (left). Cross section of intracellular and extracellular potential after 1.14 ms (right)

Table 1 Study of weak scalability

DOFs on Domain	7145		50628		379714		2937982
#Processors	6	×	48	×	384	×	3072
Loading domain	0.02888	1.69	0.04880	1.06	0.05166	1.00	0.05168
domain distribution	0.06721	2.24	0.15070	1.11	0.16737	1.06	0.17738
domain refinement	0.03010	1.90	0.05721	5.73	0.32765	7.73	2.53434
Preparing time step	0.00005	3.12	0.00017	1.08	0.00018	1.02	0.00018
Assembling system	0.08562	1.99	0.17075	2.87	0.49044	5.50	2.69781
Assembling rhs	0.00752	1.97	0.01477	5.92	0.08746	7.63	0.66731
Applying solver	0.14099	2.57	0.36237	5.06	1.83253	5.74	10.52510
Total time	74.63698	2.54	189.33351	5.08	961.53549	5.83	5603.41139

Domain loading, distribution and refinement and system assembly is only performed once.

It can be assumed from Table 1 that solving and assembly, which are of major importance, are not yet scaling well.

For the given geometry the convergence rate of the AMG is ≈ 0.15, the convergence rate of the GMG is ≈ 0.003 − 0.008. AMG convergence becomes even worse for the more complex geometries of neuronal reconstructions and thus improving AMG convergence is necessary.

5 Discussion

This was the first investigation of weak scalability of the system of partial differential equations describing the electric properties of neurons in three space dimensions. Currently the solver setup is still optimized. The investigation of the given problem shows no weak scalability, which requires a closer investigation of the properties of the assembled matrix. Furthermore, AMG convergence has to be improved in order to compute digitally reconstructed neurons in three space dimensions.

Acknowledgements We thank the HLRS for the opportunity to use *Hazel Hen* and their kind support.

References

1. K. Xylouris, G. Wittum, A three-dimensional mathematical model for the signal propagation on a neuron's membrane. Front. Comput. Neurosci. **9**, 1–9 (2015)
2. A.L. Hodgkin, A.F. Huxley, A quantitative description of membrane current and its application to conduction and excitation in nerve. J. Physiol. **117**, 500–544 (1952)
3. G. A. Ascoli, Mobilizing the base of neuroscience data: the case of neuronal morphologies. Nat. Rev. Neurosci. **7**(4), 318–324 (2006)
4. Sebastian Reiter, Andreas Vogel, Ingo Heppner, Martin Rupp, Gabriel Wittum, A massively parallel geometric multigrid solver on hierarchically distributed grids. Comp. Vis. Sci. **16**(4), 151–164 (2013)
5. T. Radman, R.L. Ramos, J.C. Brumberg, M. Bikson, Role of Cortical Cell Type and Morphology in Sub- and Suprathreshold Uniform Electric Field Stimulation. Brain Stimul. **2**(4), 215–228 (2009)
6. K. Mörschel, M. Breit, G. Queisser, Generating neuron geometries for detailed three-dimensional simulations using anamorph. Neuroinformatics (2017)
7. A. Vogel, S. Reiter, M. Rupp, A. Nägel, G. Wittum, UG 4: A novel flexible software system for simulating PDE based models on high performance computers. Comp. Vis. Sci. **16**(4), 165–179 (2013)
8. X.S. Li, An overview of superlu: Algorithms, implementation, and user interface. Toms (2005)

Parallel Space-Time Solutions for the Linear Visco-Acoustic and Visco-Elastic Wave Equation

W. Dörfler, C. Wieners, and D. Ziegler

Abstract We present parallel adaptive results for a discontinuous Galerkin space-time discretization for acoustic and elastic waves with attenuation. The method is based on p-adaptive polynomial discontinuous ansatz and test spaces and a first-order formulation with full upwind fluxes. Adaptivity is controlled by dual-primal error estimation, and the full linear system is solved by a Krylov method with space-time multilevel preconditioning. The discretization and solution method is introduced in Dörfler-Findeisen-Wieners (Comput. Meth. Appl. Math. 2016) for general linear hyperbolic systems and applied to acoustic and elastic waves in Dörfler-Findeisen-Wieners-Ziegler (Radon Series Comp. Appl. Math. 2019); attenuation effects were included in Ziegler (PhD thesis 2019, Karlsruhe Institute of Technology). Here, we consider the evaluation of this method for a benchmark configuration in geophysics, where the convergence is tested with respect to seismograms. We consider the scaling on parallel machines and we show that the adaptive method based on goal-oriented error estimation is able to reduce the computational effort substantially.

1 Introduction

Classically, wave equations are considered as evolution equations where the derivative with respect to time is treated in a stronger way than the spatial differential operators. This results in an ordinary differential equation (ODE) with values in a function space with respect to the spatial variable. For example, acoustic waves in a spatial domain $\Omega \subset \mathbb{R}^d$ for a given right-hand side $\mathbf{b}$ can be considered in terms of the following ODE

$$\partial_t \mathbf{u} = A\mathbf{u} + \mathbf{b} \quad \text{in } [0, T], \quad \mathbf{u}(0) = \mathbf{0}, \quad A = \begin{pmatrix} 0 & \operatorname{div} \\ \nabla & 0 \end{pmatrix},$$

W. Dörfler · C. Wieners · D. Ziegler (✉)
KIT, Fakultät für Mathematik, Englerstraße 2, 76131 Karlsruhe, Germany
e-mail: daniel.ziegler@kit.edu

C. Wieners
e-mail: christian.wieners@kit.edu

© Springer Nature Switzerland AG 2021   589
W. E. Nagel et al. (eds.), *High Performance Computing in Science and Engineering '19*,
https://doi.org/10.1007/978-3-030-66792-4_40

where the solution $\mathbf{u}$ is an element of the space $C^0(0, T; \mathcal{D}(A)) \cap C^1(0, T; L_2(\Omega)^{1+d})$ with $\mathcal{D}(A) \subset H^1(\Omega) \times H_0(\text{div}, \Omega)$. In order to analyze this ODE, space and time are treated separately and hence tools for partial differential equations are used in space and tools for ODEs are used in time.

Since time integration is a sequential process, we consider here for the parallelization the space-time operator

$$L(p, \mathbf{v}) = \begin{pmatrix} \partial_t p + \text{div } \mathbf{v} \\ \partial_t \mathbf{v} + \nabla p \end{pmatrix},$$

in $Q = (0, T) \times \Omega$ as a whole treating time and space dependence simultaneously in a variational manner. Using this approach, we construct the space-time Hilbert space $H(L, Q)$ that allows for irregular solutions, e.g.., with space-time discontinuities. We select a space $V \subset H(L, Q)$ including homogeneous initial and boundary conditions such that the full space-time operator $L: V \longrightarrow L_2(\Omega)^{d+1}$ defines an isomorphism. As a result, for every given right-hand side $\mathbf{b} \in L_2(\Omega)^{d+1}$, the problem of finding $\mathbf{u} \in V$ such that

$$L\mathbf{u} = \mathbf{b}$$

is well-posed in our framework, see [1].

Many applications rely on accurate numerical simulations of waves through complex material structures. For instance, geophysical structures like the earth's crust below the sea bed feature complex varying material properties. A typical example is the problem of full waveform inversion (FWI), where the material distribution is reconstructed from measurements of the wave field close to the surface. This is achieved by minimizing a misfit functional. The evaluation of this functional and its gradient require wave solutions forward and backward, where the adjoint problem relies on the full information in space and time.

The forward problem for this application is considered in this work for acoustic, visco-acoustic, and visco-elastic waves. A discretization in space and time is provided by a Discontinuous Galerkin approach presented in [2]. An alternative method is the Discontinuous Petrov-Galerkin (DPG) method, which was introduced by Demkowicz et al. and provides a framework for the discretization of general linear first-order systems, see [3]. The application to acoustic waves is presented in [4, 5] and extended to heterogeneous media in [1].

2 Solving the Visco-Elastic Wave Equation in Space and Time

For the parallel approximation we use a space-time discretization with discontinuous ansatz functions in space and continuous ansatz functions in time together with an adaptive algorithm using dual weighted residual estimators and a multilevel preconditioner. We call this discretization the dG-cPG method, since it is discontinuous in space and continuous in time, but combined with discontinuous test functions in space and time, resulting in an Petrov–Galerkin method. A more detailed version can be found in [6]. This discretization is applied to the acoustic and elastic wave equation in [7]. A discretization using discontinuous ansatz functions in space and time is presented in [8] and applied to the visco-acoustic and visco-elastic wave equation. We call this discretization the dG-dG method. We want to remark that the dG-cPG(q) and dG-dG($q-1$) method have the same amount of degrees of freedom, where q denotes the polynomial degree in time.

The wave equations including attenuation effects can be written in the compact operator formulation

$$L\mathbf{u} = \mathbf{b} \qquad \text{a.e. in } (0, T)$$

with $L = M\partial_t + A + D$. In the case of visco-elasticity we get

$$M(\mathbf{v}, \boldsymbol{\sigma}_0, \ldots, \boldsymbol{\sigma}_G) = \left(\rho\mathbf{v}, \mathbf{C}_0^{-1}\boldsymbol{\sigma}_0, \ldots, \mathbf{C}_G^{-1}\boldsymbol{\sigma}_G\right),$$
$$A(\mathbf{v}, \boldsymbol{\sigma}_0, \ldots, \boldsymbol{\sigma}_G) = -(\nabla \cdot (\boldsymbol{\sigma}_0 + \cdots + \boldsymbol{\sigma}_G), \boldsymbol{\varepsilon}(\mathbf{v}), \ldots, \boldsymbol{\varepsilon}(\mathbf{v})),$$
$$D(\mathbf{v}, \boldsymbol{\sigma}_0, \ldots, \boldsymbol{\sigma}_G) = \left(\mathbf{0}, \mathbf{0}, \tau_1^{-1}\mathbf{C}_1^{-1}\boldsymbol{\sigma}_1, \ldots, \tau_G^{-1}\mathbf{C}_G^{-1}\boldsymbol{\sigma}_G\right).$$

In the special case of isotropic materials the elasticity tensors $\mathbf{C}_g = \mathbf{C}(\mu_g, \kappa_g)$ for $g = 0, \ldots, G$, with

$$\mathbf{C}(\mu, \kappa)\boldsymbol{\varepsilon} = 2\mu \, \mathrm{dev}(\boldsymbol{\varepsilon}) + \kappa \, \mathrm{trace}(\boldsymbol{\varepsilon})\mathbf{I}, \qquad \mathrm{dev}(\boldsymbol{\varepsilon}) = \boldsymbol{\varepsilon} - \frac{1}{3} \mathrm{trace}(\boldsymbol{\varepsilon})\mathbf{I},$$

only depend on the shear moduli μ_g and the compression moduli $\kappa_g = \lambda_g + \frac{2}{3}\mu_g$. In the limit of vanishing shear forces one obtains for the *hydrostatic pressure* $p = \frac{1}{3} \mathrm{trace}(\boldsymbol{\sigma}) = p_0 + \cdots + p_G$ the *visco-acoustic system*

$$\rho \, \partial_t \mathbf{v} = \nabla p_0 + \cdots + \nabla p_G + \mathbf{b},$$
$$\partial_t p_0 = \kappa_0 \nabla \cdot \mathbf{v},$$
$$\partial_t p_g = \kappa_g \nabla \cdot \mathbf{v} - \frac{1}{\tau_g} p_g, \qquad g = 1, \ldots, G.$$

For the discretization, we assume that Ω is a bounded polyhedral Lipschitz domain decomposed into a finite number of open elements $K \subset \Omega$ such that $\overline{\Omega} = \bigcup_{K \in \mathcal{K}} \overline{K}$, where $\mathcal{K}$ is the set of elements in space. Let $\overline{Q} = \bigcup_{R \in \mathcal{R}} \overline{R}$ be a decomposition of the space-time cylinder into space-time cells $R = K \times I$ with $K \in \mathcal{K}$ and $I \subset [0, T]$ an interval; $\mathcal{R}$ denotes the set of space-time cells. For the fixed mesh $\mathcal{K}$ in space and a time series $0 = t_0 < t_1 < \cdots < t_N = T$, the space-time mesh is defined by

$$\mathcal{R} = \bigcup_{n=1,\ldots,N} \mathcal{R}_n, \qquad \mathcal{R}_n = \left\{K \times I_n : I_n := (t_{n-1}, t_n], \ K \in \mathcal{K}\right\}.$$

Algorithm 1

RCB_st(cells $\mathcal{R}$, weights $\mathcal{W}$, factor m, bisections b, sort c)

Require: $m, b \in \mathbb{N}, \ c \in \{t, x, y, z\}$
1: **if** $b == 0$ **then**
2:    send cells in $\mathcal{R}$ to process m *distribute cells*
3:    **return**
4: **end if**
5:   *sort and bisect set of cells*
6: sort $\mathcal{R}$ by coordinate c
7: split $\mathcal{R}$ into $\mathcal{R}_1$ and $\mathcal{R}_2$ such that
8:    $\sum_{R_1 \in \mathcal{R}_1} \mathcal{W}_{R_1} \approx \sum_{R_2 \in \mathcal{R}_2} \mathcal{W}_{R_2}$
9:                                  *define coordinate for next bisection*
10: **if** $c == z$ **then**
11:    $c := t$
12: **else if** $c == y$ **then**
13:    **if** dim $== 3$ **then**
14:      $c := z$
15:    **else**
16:      $c := t$
17:    **end if**
18: **else if** $c == x$ **then**
19:    **if** dim > 1 **then**
20:      $c := y$
21:    **else**
22:      $c := t$
23:    **end if**
24: **else**
25:    $c := x$
26: **end if**
27:   *recursive call*
28: RCB_st($\mathcal{R}_1, \mathcal{W}, m, b - 1, c$)
29: RCB_st($\mathcal{R}_2, \mathcal{W}, m + 2^{b-1}, b - 1, c$)

On each cell R we define the local space

$$V_{h,R} = \mathbb{P}_{p_R}(K; \mathbb{R}^J) \otimes \mathbb{P}_{q_R}(I_n; \mathbb{R}^J) \subset L_2(R; \mathbb{R}^J)$$

and the global space

$$V_h = \left\{ \mathbf{v}_h \in L_2((0, T); L_2(\Omega; \mathbb{R}^J)) : \mathbf{v}_{h,R} = \mathbf{v}_h|_R \in V_{h,R} \right\}.$$

The polynomial degree in space and time (p_R, q_R) in each cell can be arbitrary and is chosen by an (p, q)-adaptive algorithm. The space-time cells are distributed on 2^b processes using the recursive bisection algorithm presented in Alg. 1. The algorithm combined with the appropriate choice of weights $\mathcal{W}_R$, based on the polynomial degrees (p_R, q_R), leads to a distribution, where every process has to handle the same computational effort. The space-time system is solved with a multilevel preconditioner in space and time, see [6, Chap. 6]. The parallel direct solver [9–11] is used on the coarse level.

Fig. 1 Density distribution for the Marmousi II benchmark: The graphic shows the full Marmousi II benchmark with a domain size of 17 km × 3.5 km. The red subdomain 10 km × 3 km is used in the adaptive numerical experiments and the smaller yellow subdomain 3 km × 3 km for the convergence tests in space and time on uniform discretizations

2.1 A Geophysical Benchmark in Heterogeneous Media

The benchmark problem Marmousi II [12] for geophysical applications provides realistic structures in two space dimensions with heterogeneous media, see Fig. 1 for the density distribution in this benchmark configuration.

For the numerical experiments, we simulate maritime measurements in seismic exploration with a local source initiating a wave by a smooth pulse in space of width $w_s = 100$ [m] located at $\mathbf{x}_s \in \Omega$

$$\phi(\mathbf{x}) = \begin{cases} \cos^6\left(\dfrac{\pi|\mathbf{x}_s - \mathbf{x}|}{2w_s}\right) & |\mathbf{x}_s - \mathbf{x}| < w_s, \\ 0 & \text{else} \end{cases} \tag{1}$$

and a Ricker wavelet in time

$$\psi(t) = \left(1 - 2\pi^2(t - t_s)^2 f^2\right)\exp\left(-\pi^2(t - t_s)^2 f^2\right)$$

with frequency f and time delay $t_s = 0.15$ [s]. This results in the right-hand side $\mathbf{b}(t, \mathbf{x}) = \psi(t)\,\phi(\mathbf{x})\,\mathbf{e}$ with $\mathbf{e} = (0, 1, 0, \ldots, 0) \in \mathbb{R}^{\dim +1+G}$ in the acoustic case, and $\mathbf{e} = (0, \mathbf{I}_3, 0, \ldots, 0) \in \mathbb{R}^{\dim} \times \mathbb{R}^{\dim \times \dim}_{\mathrm{sym}} \times \cdots \times \mathbb{R}^{\dim \times \dim}_{\mathrm{sym}}$ for elasticity.

In our tests, the solution is compared for different discretizations by the resulting pressure evaluated at the receivers positions $\mathbf{x}_{r,i} \in \Omega$, $i = 0, \ldots, N_r$. This defines a seismogram $\mathbf{s} \in L_2(0, T; \mathbb{R}^{N_r})$, i.e., $s_i(t) = p(t, \mathbf{x}_{r,i})$.

The Marmousi model prescribes a density distribution $\rho \in (1010, 2627)$ [kg/m³] and reference values for the velocities of shear waves $v_S \in (0, 2802)$ [m/s] and compressional waves $v_P \in (1028, 4700)$ [m/s]. This defines the parameters $\mu = \rho v_S^2$ and $\kappa = \rho v_P^2 - \frac{4}{3}\mu$ for isotropic elasticity. We fix this material parameters cellwise constant on a spatial mesh with mesh size 125 [m].

We set $\kappa_0 = \frac{\kappa}{1+G\tau_P}$ and $\kappa_1 = \cdots = \kappa_G = \kappa_0 \tau_P$ with $\tau_P = 0.1$, and we set $\mu_0 = \frac{\mu}{1+G\tau_S}$ and $\mu_1 = \cdots = \mu_G = \mu_0 \tau_S$ with $\tau_S = 0.1$, Furthermore, we use the relaxation time $\tau_g = \frac{1}{2\pi f_g}$ with reference frequencies $f_1 = 0.151$ [Hz], $f_2 = 1.93$ [Hz] and $f_3 = 18.9$ [Hz] for $G = 3$ and $f_1 = 10$ [Hz] for $G = 1$ (Fig. 2).

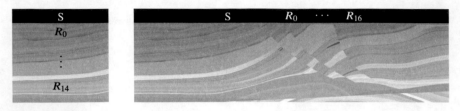

Fig. 2 Marmousi II: Sketc.h of location of source and receivers for the uniform computations used in the first numerical experiment on the left and for the adaptive computations used in the second experiment on the right

2.2 Visco-Acoustic Equation with Three Damping Mechanisms and Uniform p-refinement

We compare the dG-dG method with the dG-cPG method on uniform discretizations with polynomial degrees p and q in space and time for the visco-acoustic model with three damping mechanisms ($G = 3$) in this numerical test. Here, we use from the full Marmousi II benchmark configuration the subdomain $\Omega = (4000, 7000) \times (-3000, 0) \subset (0, 17000) \times (-3500, 0)$ [m^2] (see the yellow dashed box in Fig. 1) and the time interval $(0, T)$ with $T = 1.5$ [s]. We use a coarse mesh in space and time with $h_0 = 1000$ [m] and $\Delta t_0 = 0.25$ [s]. The initial pulse is located at $\mathbf{x}_s = (5500, -250)$. The seismograms are measured at the receivers with the positions $\mathbf{x}_{r,i} = (5500, -750 - 125i)$ for $i = 0, \ldots, 14$.

Since we have no analytical solution for the problem, we compute the reference seismogram by extrapolation. The order of convergence on the space-time mesh of level l can be estimated from the factor

$$f_l = \frac{\|\mathbf{s}_{l-1} - \mathbf{s}_{l-2}\|_{(0,T)}}{\|\mathbf{s}_l - \mathbf{s}_{l-1}\|_{(0,T)}},$$

where $\mathbf{s}_l$ denotes the seismogram on level l combined with the L$_2$-norm. With this factor a better approximation can be constructed by extrapolation as

$$\mathbf{s}_{ex} = \frac{f_l}{f_l - 1}\mathbf{s}_l - \frac{1}{f_l - 1}\mathbf{s}_{l-1}.$$

Here we choose the fixed polynomial degrees $(p, q) = (3, 2)$ and the space-time levels $l = 3, \ldots, 5$ obtained by uniform refinement in space-time. All quantities in this test are normalized with respect to the reference value $\|\mathbf{s}_{ex}\|_{(0,T)}$.

A selection of the results of this numerical experiment are shown in Table 1. The results indicate that the cPG version gives more accurate results than the dG method with one order lower in time, although both methods have the same amount of degrees of freedom. The advantage of the dG-dG method over the dG-cPG method is that the system matrix is less dense. As a result, the total time to solve the system is lower. In

Table 1 Marmousi II dG vs. cPG: comparison of the two methods on uniform discretizations. The error $e = \|s - s_{ex}\|_{(0,T)}/\|s_{ex}\|_{(0,T)}$ is given in percent and in case of using the dG-dG method also the error of the seismogram obtained by evaluation of the conforming reconstruction $\hat{e} = \|\hat{s} - s_{ex}\|_{(0,T)}/\|s_{ex}\|_{(0,T)}$. ML denotes the GMRES steps with the multilevel preconditioner. We use 10 smoothing steps if coarsened in time and 20 if coarsened in space. The time to solve the space-time system on 256 parallel cores is given in [hh:mm:ss]

dG-cPG on space-time mesh level 4

(p, q)	e	RAM	DoF	ML	Time
(2, 2)	26.9%	387 GB	23 887 872	10	0:15:04
(2, 3)	28.7%	753 GB	35 831 808	9	0:27:35
(3, 2)	4.6%	1.0 TB	42 467 328	15	1:06:22
(3, 3)	4.8%	2.2 TB	63 700 992	15	2:21:11

dG-dG on space-time mesh level 4

(p, q)	e	$\hat{e}$	RAM	DoF	ML	Time
(2, 1)	39.4%	39.1%	248 GB	23 887 872	10	0:06:48
(2, 2)	28.8%	28.8%	473 GB	35 831 808	10	0:13:38
(2, 3)	28.7%	28.7%	768 GB	47 775 744	10	0:22:55
(3, 1)	31.1%	30.9%	636 GB	42 467 328	16	0:29:54
(3, 2)	5.1%	5.1%	1.3 TB	63 700 992	15	1:02:34

addition, less total system memory is required in particular with higher polynomials in time, since the dG-dG scheme has fewer coupling between the space-time cells and thus the matrix graph is more sparse.

In Fig. 3 we illustrate the conforming reconstruction working on linear ansatz functions in time and resulting in conforming quadratic functions. The conforming reconstruction operator interpolates the ansatz functions, which are discontinuous in time, resulting in functions, which are continuous in time of one order higher as the ansatz functions. This serves the approximation order of the solution.

2.3 A Parallel Adaptive Visco-Elastic Computation

This numerical test demonstrates the parallel efficiency of the method. The visco-elastic system with one damping mechanism ($G = 1$) is solved using one adaptive step and the dG-cPG method.

Here we choose the domain $\Omega = (4000, 13000) \times (-3000, 0) \subset (0, 17000) \times (-3500, 0)$ [m^2] (marked red in Fig. 1) and the time interval $(0, T)$ with $T = 3$ [s]. The source is located at $x_s = (7000, -250)$, and the receivers positions are $x_{r,j} = (9000 + 125j, -250)$ for $j = 0, \ldots, 16$. For the adaptive simulations we use the goal functional

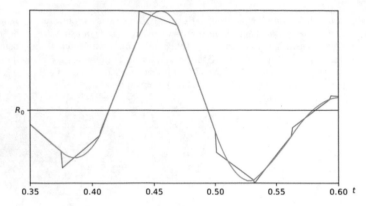

Fig. 3 Sketch of the feature using conforming reconstruction: the solution discontinuous in time obtained by the dG(p)-dG(q) method with $(p, q) = (3, 1)$ (blue) is reconstructed with Radau IIA integration points (orange)

$$\mathcal{J}_{\text{elastic}}(\mathbf{v}, \boldsymbol{\sigma}) = \frac{1}{|\Omega_{\text{RoI}}|} \int_{\Omega_{\text{RoI}} \times \{T\}} \text{trace}\, \boldsymbol{\sigma}\, d\mathbf{x} \qquad \text{with } \boldsymbol{\sigma} = \boldsymbol{\sigma}_0 + \boldsymbol{\sigma}_1,$$

together with the region of interest $\Omega_{\text{RoI}} = (4750, 100) \times (7250, 400)$. We start with piecewise linear functions in space and time and solve the primal and dual problem. In all space-time cells where the error indicator η_R is larger than $\theta = 1 \cdot 10^{-9}$ times the largest error indicator $\eta_{\max} = \max_{R \in \mathcal{R}} \eta_R$, i.e., $\eta_R > \eta_{\text{crit}} = \theta \eta_{\max}$, the polynomial degree is increased in space and time. In contrast the polynomial degree is decreased if $\eta < 0.01 \cdot \eta_{\text{crit}}$.

The visco-elastic adaptive space-time dG-cPG simulation tracks the propagation of the wave from the source to the receivers. The first stress component (column 1) and the distribution of the polynomial degrees (p, q) (column 2) are visualized in Fig. 4. In the blue area we have $(p, q) = (0, 1)$, gray $(p, q) = (1, 1)$ and red $(p, q) = (2, 2)$.

This results into 364 Mio. degrees of freedom and the full linear space-time system is solved with 14 GMRES steps using the multilevel preconditioner (50 Gauss–Seidel smoothing steps in space and 25 Jacobi smoothing steps in time). The p-adaptive method reduces the degrees of freedom by approximately 78% compared to a uniform computation (1 968 Mio. degrees of freedom). On 4096 parallel processes the system was solved in 30 min and 53 s whereas on 8192 parallel processes the time was 15 min and 47 s. The solving time was cut nearly in half by doubling the number of processes demonstrating very good strong scaling behavior.

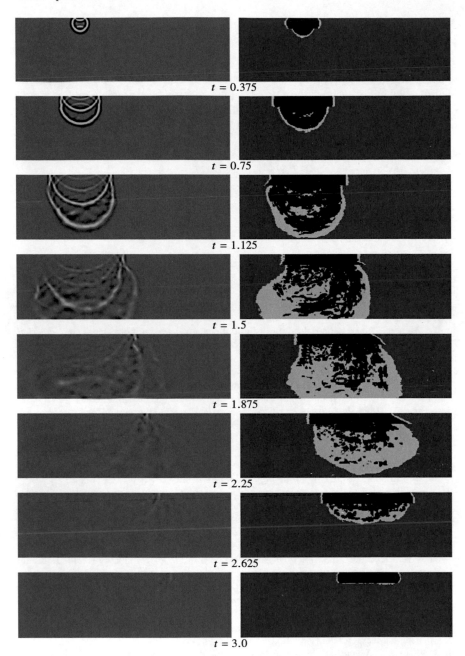

t = 0.375

t = 0.75

t = 1.125

t = 1.5

t = 1.875

t = 2.25

t = 2.625

t = 3.0

Fig. 4 Slices through the space-time solution for the visco-elastic adaptive computation at different times. On the left is the first stress component and on the right the corresponding polynomial order in space and time

3 Conclusion

We demonstrated a nearly optimal scaling behavior for an adaptive space-time method for first-order linear hyperbolic systems. The method is realized in the parallel finite element software system M++ [13], which provides a framework for various numerical challenges such as elasticity, plasticity and electromagnetic waves. The parallel scaling for these applications will be considered in future work.

Acknowledgements This work is funded by the Deutsche Forschungsgemeinschaft (DFG, German Research Foundation)—Project-ID 258734477—SFB 1173.

References

1. J. Ernesti, C. Wieners, Space-time discontinuous Petrov-Galerkin methods for linear wave equations in heterogeneous media. Comput. Method Appl. Math. **19**(3), 465–481 (2019)
2. W. Dörfler, S. Findeisen, C. Wieners, Space-time discontinuous Galerkin discretizations for linear first-order hyperbolic evolution systems. Comput. Method Appl. Math. **16**, 409–428 (2016)
3. C. Wieners, The skeleton reduction for finite element substructuring methods, in *Numerical Mathematics and Advanced Applications ENUMATH 2015*, ed. by B. Karasözen, M. Manguoğlu, M. Tezer-Sezgin, S. Göktepe, Ö. Uğur (Springer International Publishing, 2016), pp. 133–141
4. J. Ernesti, Space-time methods for acoustic waves with applications to full waveform inversion. PhD thesis, Karlsruhe Institute of Technology (KIT), Karlsruhe (2018)
5. J. Ernesti, C. Wieners, A space-time discontinuous Petrov–Galerkin method for acoustic waves, in *Space-Time Methods*, ed. by U. Langer, O. Steinbach. Applications to Partial Differential Equations, vol. 25 (Walter de Gruyter, Radon Series on Computational and Applied Mathematics, 2019), pp. 89–116
6. S. Findeisen, A Parallel and Adaptive Space-Time Method for Maxwell's Equations. PhD thesis, Karlsruhe Institute of Technology (KIT), Karlsruhe (2016)
7. W. Dörfler, S. Findeisen, C. Wieners, D. Ziegler, Parallel adaptive discontinuous Galerkin discretizations in space and time for linear elastic and acoustic waves, in *Space-Time Methods*, ed. by U. Langer, O. Steinbach. Applications to Partial Differential Equations, vol. 25 (Walter de Gruyter, Radon Series on Computational and Applied Mathematics, 2019), pp. 61–88
8. D. Ziegler, A parallel and adaptive space-time discontinuous galerkin method for visco-elastic and visco-acoustic waves. PhD thesis, Karlsruhe Institute of Technology (KIT), Karlsruhe (2019)
9. D. Maurer, C. Wieners, Parallel Multigrid Methods and Coarse Grid LDL^T solver for maxwell's eigenvalue problem, in *Competence in High Performance Computing 2010*, ed. by C. Bischof, H. Hegering, W. Nagel, G. Wittum (Springer, 2010), pp. 205–213
10. D. Maurer, C. Wieners, A parallel block LU decomposition method for distributed finite element matrices. Parallel Comput. **37**, 742–758 (2011)
11. D. Maurer, C. Wieners, A highly scalable multigrid method with parallel direct coarse grid solver for maxwell's equations, in *High Performance Computing in Science and Engineering '13* ed. by W. Nagel, D. Kröner, M. Resch (Transactions of the High Performance Computing Center, Stuttgart (HLRS), Springer, 2013) pp. 671–677

12. G. Martin, R. Wiley, K. Marfurt, Marmousi2: An elastic upgrade for Marmousi. Lead. Edge **25**, 156–166 (2006)
13. C. Wieners, A geometric data structure for parallel finite elements and the application to multigrid methods with block smoothing, in *Computing and Visualization in Science* vol. 13 (Springer, 2010), pp. 161–175

Printed in the United States
by Baker & Taylor Publisher Services